Advanced Structured Materials

Volume 167

Common engineering materials reach in many applications their limits and new developments are required to fulfil increasing demands on engineering materials. The performance of materials can be increased by combining different materials to achieve better properties than a single constituent or by shaping the material or constituents in a specific structure. The interaction between material and structure may arise on different length scales, such as micro-, meso- or macroscale, and offers possible applications in quite diverse fields.

This book series addresses the fundamental relationship between materials and their structure on the overall properties (e.g. mechanical, thermal, chemical or magnetic etc.) and applications.

The topics of *Advanced Structured Materials* include but are not limited to

- classical fibre-reinforced composites (e.g. glass, carbon or Aramid reinforced plastics)
- metal matrix composites (MMCs)
- micro porous composites
- micro channel materials
- multilayered materials
- cellular materials (e.g., metallic or polymer foams, sponges, hollow sphere structures)
- porous materials
- truss structures
- nanocomposite materials
- biomaterials
- nanoporous metals
- concrete
- coated materials
- smart materials

Advanced Structured Materials is indexed in Google Scholar and Scopus.

More information about this series at https://link.springer.com/bookseries/8611

Azman Ismail · Wardiah Mohd Dahalan ·
Andreas Öchsner
Editors

Design in Maritime Engineering

Contributions from the ICMAT 2021

Springer

Editors
Azman Ismail
Malaysian Institute of Marine Engineering Technology
Universiti Kuala Lumpur
Lumut, Perak, Malaysia

Wardiah Mohd Dahalan
Malaysian Institute of Marine Engineering Technology
Universiti Kuala Lumpur
Lumut, Perak, Malaysia

Andreas Öchsner
Faculty of Mechanical Engineering
Esslingen University of Applied Sciences
Esslingen am Neckar, Baden-Württemberg
Germany

ISSN 1869-8433 ISSN 1869-8441 (electronic)
Advanced Structured Materials
ISBN 978-3-030-89987-5 ISBN 978-3-030-89988-2 (eBook)
https://doi.org/10.1007/978-3-030-89988-2

This Springer imprint is published by the registered company Springer Nature Switzerland AG
The registered company address is: Gewerbestrasse 11, 6330 Cham, Switzerland

Preface

The *Design in Maritime Engineering* book covers several research outcomes of various fields and school of thoughts particularly related to maritime operation, applications and materials science. Thirty-four research papers have been compiled from the 2nd International Conference on Marine and Advanced Technologies 2021 (ICMAT 2021) which was organized by the Research and Innovation Section of the Universiti Kuala Lumpur—MIMET. The authors were experienced lecturers from various universities in Malaysia discussing various topics and sub-topics related to maritime engineering and materials science. These chapters portray the actual knowledge on the latest developments and trends of technologies in maritime industries. This new development of marine-related matters will inculcate greater interest and ideas.

Lumut, Malaysia	Azman Ismail
Lumut, Malaysia	Wardiah Mohd Dahalan
Esslingen am Neckar, Germany	Andreas Öchsner

Contents

About the Editors

Azman Ismail is Senior Lecturer at Malaysian Institute of Marine Engineering Technology, Universiti Kuala Lumpur, Malaysia. He received his PhD in Mechanical Engineering from Universiti Teknologi PETRONAS and Master of Engineering in Mechanical—Marine Technology from Universiti Teknologi Malaysia. Prior to that, he was awarded a Bachelor of Engineering (Hons) in Electrical, Electronics and System from Universiti Kebangsaan Malaysia and Graduate Diploma in Industrial Education and Training from the Royal Melbourne Institute of Technology, Australia. He grooms his technical skill at Victoria University of Technology, Australia, for Advanced Diploma in Construction and Repair Technology (Marine Vessels). He is also active in research and development for welding and joining technologies especially for friction stir welding on tubular sections and flat panels. This also includes green technologies for sustainable marine and coastal development. He is currently leading a research cluster of Advanced Maritime Industries Sustainability at his university. He has published his research findings in indexed journals and chapters and actively competes at international- and national-level innovation competitions. In addition to his achievements, he has been a reviewer and editor for some international journals including Springer. Besides that, he is also active in conservation works as committee member for the National Eco-Campus Program with the World Wide Fund for Nature of Malaysia (WWF-Malaysia).

Wardiah Mohd Dahalan had received her Ph.D. in Power System from the University of Malaya, Kuala Lumpur, Malaysia, in 2013 and Bachelor's Degree (Hons) in Electrical and Electronics Engineering from the University of Dundee, Scotland, UK, in 1996. She is currently appointed as Senior Lecturer in the Department of Marine Electrical and Electronics Engineering of University Kuala Lumpur (UniKL-MIMET). As Head of Research & Innovation Department since 2015, she actively administers all research activities in UniKL-MIMET. Supervising of internal and external research grants and organization of all activities related to innovation such as exhibitions, competition, and conferences are her core activities at UniKL-MIMET. She actively participates in research as a principle or co-principle in many research grants. At the same time, she also shares her expertise by becoming either the author or co-author of the publications of local as well as international journals, books and proceedings especially in the area of power system and energy. Her deep research interest includes network reconfiguration, optimization techniques, and renewable energy. She is also Member of IEEE, Rina-IMARest, Malaysian Society for Engineering and Technology (MySET), and Malaysia Board of Technologist (MBOT).

Andreas Öchsner is Full Professor for Lightweight Design and Structural Simulation at the Esslingen University of Applied Sciences, Germany. Having obtained a Dipl.-Ing. degree in Aeronautical Engineering at the University of Stuttgart (1997), Germany, he served as a research and teaching assistant at the University of Erlangen-Nuremberg from 1997 to 2003 while working to complete his Doctor of Engineering Sciences (Dr.-Ing.) Degree. From 2003 to 2006, he was Assistant Professor in the Department of Mechanical Engineering and Head of the Cellular Metals Group affiliated with the University of Aveiro, Portugal. He spent seven years (2007–2013) as Full Professor in the Department of Applied Mechanics, Technical University of Malaysia, where he was also Head of the Advanced Materials and Structure Lab. From 2014 to

2017, he was Full Professor at the School of Engineering, Griffith University, Australia, and Leader of the Mechanical Engineering Program (Head of Discipline and Program Director).

Chapter 1
Review of the Current Trends on Autonomous Vehicles in Developing Countries

Shamini Janasekaran, Jean-Philippe Eulentin, Brian Tapiwa Dumba, Prasath Reuben Mathew, Nurwahdina Tahir, and Lenorr Emric Cordel Gabbidon

Abstract Various researches need to be done to create fully reliable autonomous vehicles. Autonomous vehicles (AVs) focus on upgrading cars into automatic functions such as driving without human intervention. During the conventional days, cars were powered by steam, but in present times, vehicles are powered by either gasoline and electricity or by both. Car industries and engineers are now finding ways how to create a fully reliable AVs without human intervention. Even though they already invented one, the research and experiments are still ongoing so that society can put their trust on it. This paper focuses on the consumers' acceptance and perception of AVs in the developing countries.

Keywords Autonomous vehicle · Autonomous technology · Consumer perception

1.1 Consumer Acceptances of Autonomous Vehicles

In the broader acceptance of autonomous vehicles (AVs), public perceptions play a key role. With little or no human interaction, autonomous vehicles are capable of detecting their surroundings and handling various traffic conditions. In general, the human driver is still responsible for controlling driving and the atmosphere between levels 0 and 2. The primary change happens within levels 3 and 4, during which the driver relieves his or her obligation to oversee the system's driving environment [1]. With regard to the dynamic driving task (DDT), each level reflects various types of automation. Via the input of a human operator, autonomous vehicles have the ability to minimize road congestion while reducing air emissions and improving fuel efficiency combating climate change. Study results suggest that more than 90% of all car accidents are due to human causes [2]. Most industry analysts agree that road safety will be greatly enhanced by the transition to autonomous by increased well-being across different realms. First of all, the wide-scale use of automated vehicles

S. Janasekaran (✉) · J.-P. Eulentin · B. T. Dumba · P. R. Mathew · N. Tahir · L. E. C. Gabbidon
Faculty of Engineering, Centre for Advanced Materials and Intelligent Manufacturing, Built Environment & IT, SEGi University Sdn Bhd, 47810 Petaling Jaya, Selangor, Malaysia
e-mail: shaminijanasekaran@segi.edu.my

A. Ismail et al. (eds.), *Design in Maritime Engineering*, Advanced Structured Materials 167, https://doi.org/10.1007/978-3-030-89988-2_1

is supposed to drastically decrease the number of road collisions and to prevent much of the fatalities and injuries usually caused by accidents [3]. Secondly, a rapid change was noticeable in the flow of traffic, and the usage of other sources of energy such as battery to power cars, especially where autonomous vehicles are used as public transportation, is expected to contribute to significant pollution reductions [4]. Furthermore, increased movement for the aged and the disabled would make it easier for any of them to have greater access to medical care and higher levels of social inclusion. Finally, because of improved protection for pedestrians and cyclists, autonomous cars have expanded options for active travel.

The knowledge about the benefits and fast growth of AVs technology has led to an evolution in the transportation industries. Posing many positive impact to the society, economy, environment and energy usage, several issues arise such as pricing, privacy, availability and insurance, maintenance, regulation and safety that have a crucial impact in discouraging user to explore the AVs. These issues put users in a crucial position to relate into purchasing an AV with their integrated technology, policies and will to share roads with other vehicles. Numerous researches have been done in different aspects to gain public opinion, analysis, concern and feeling about AVs.

1.2 Effects of Autonomous Vehicles

In order to measure its progress and shape its potential, the acceptance rate by locals and visitors remains a key concern with the on-demand self-driving taxi service going into effect. Given the advantages, autonomous driving issues among the general population continue to escalate, creating reluctance to autonomous vehicles. Such concerns are also embedded in the apprehension of technology taking its own path, regardless of human guidance and the inability to grant a computer autonomy and control of an essential part of human life. The use of autonomous vehicles often involves a reduction in demand for experienced drivers, which results in the understanding of technology as dehumanizing. This study examines consumers' general views of technology and how it influences the desire to use self-driving taxis at home (as residents) and while commuting (as tourists) to assess previous acceptance of self-driving taxi services [5]. The findings pave the way for a deeper understanding of the actions of customers with regard to the use of autonomous personal travel vehicles, which will benefit policymakers, including tourist destinations.

1.3 Interaction of AVs and Driver

AVs exist at different levels regarding its automation. Vehicle owners (drivers) are having direct contact to the AVs' technology in various aspects such as adaptive cruise control, parking assistance, blind spot and collision warning system. Within the ability of a driver to be aware of and use these technological system functions, the benefits and usage of these functions should be taught to the drivers in order to accept and use the opportunities as a trial. The inability to bypass the autonomous vehicle system under some cases is one of the crucial and substantial concerns posed by customers who normally encounter this while driving. Response of the driver to errors and delay in judgment after being alerted by the vehicle warning are still under consideration in relation to the aviation industry [6]. The idea of the AV system needs less human contact and is particularly likely to be diverted from practices such as social activities, watching movies and sleep due to driver factors, thereby becoming unaware and ignorant of autonomous vehicle system errors.

The very essence of autonomous systems poses financial, legal and social problems. AVs exist at different levels in regard to their automation. In recent researches, the exchange from AV systems to the drivers under unfortunate circumstances shows that the participation of the driver to fully take control of the vehicle and taking the required steps could take place within a short of period time after being alerted by the vehicle. A driving simulation approach was performed to test the capabilities of a driver to operate autonomous cars after transitioning to manual control of the automation system in order to assist it. Autonomous vehicles were used at level 3, and ten drivers were tested. As a result, several questions emerged about the amount of time the driver partly gained control of the car and the concentration and focus on their environment while manual control was in command and their visual attention. The partial control of the vehicle was detected within 10 s, and the effective control, including stabilization, was roughly 35–40 s, thereby distracting the driver from the vehicle mechanism rather than concentrating on the nearby road activities [7].

As mentioned earlier, autonomous vehicles enable a driver to be diverted from road activities, and a troubling factor in alerting them to take charge of the vehicle in crucial cases was assessed by numerous recommendations and the use of airwaves dialect notification, facial alert (eye and face recognition) and sensory motion. In addition, specialists in the field of human science concluded that drivers are qualified to understand the operation and to sustain control of an AV in manual transition from the initial autonomous state [8].

Different degrees of reaction and diversion depending on age and further autonomous vehicles indicate that younger drivers are more distracted than older drivers, so eye and vision recognition technologies are proposed as primary to catch the attention of the driver [6].

1.4 Privacy and Safety

The AV cars gather vast quantities of data, which is important for assessing their performance and optimizing their safety. Certain details can be very private, monitoring the positions of the passengers and the driver's actions. This information could also be used for applications not intended for [9]. Consequently, starting an AV engine is likely to mean that all drivers and passengers' consent to the terms and conditions of the vehicle manufacturer requesting access to those details. It should also be remembered that in various regions across the globe, this issue is viewed from different viewpoints, especially in the USA or China, where data collection is seen as less troublesome than, for example, in Germany or other European countries [10]. This data may include observations of the individuals inside the car. Moreover, observations can be used to evaluate whether or not a person should be allowed to drive the car [11]. In order to discourage people with expired licenses or alcohol disorders from driving, technologies may be used. The car itself will determine the behavior of the driver as triggers intoxication status for an autonomous vehicle and decide whether or not to allow a driver to drive. Presumably, to identify the problem, an autonomous vehicle requires knowledge about the behavior, emotional state or attitude of the person to determine whether to restrict their driving rights.

Hackers compromising the safety of the vehicle is a huge concern because it can do significant harm to the safety mechanism of the vehicle. Cars, also on a large scale, could be reprogrammed to purposely crash. Therefore, to maintain the AVs' security protocol, security agencies were involved to deal with this matter. The knowledge about the benefits and fast growth of AVs' technology has led to an evolution in the transportation industries. Posing many positive impacts to the society, economy, environment and energy usage, several issues arise such as pricing, privacy, availability and insurance, maintenance, regulation and safety that have a crucial impact in discouraging users to explore the AVs [11].

These issues put users in a crucial position to relate into purchasing an AV with their integrated technology, policies and will to share roads with other vehicles. Numerous researches have been done in different aspects to gain public opinion, analysis, concern and feeling about AVs. Frequent studies used demographic data to analyze the adoption of AVs where age, gender, nationality, education level and employment status are the factors accepting the acceptance of AV adoption. It is found that female gender has more interest than males, and females appreciate the technological benefit of AVs where man is concerned about speed [11]. Also through another form of assessment, public surveys are used to analyze points such as public feelings, beliefs, expectation, prediction, efficiency and environmental impact which are the main focus points to rate the AV system toward the public.

1.5 Conclusion

Instead of directly transitioning to fully autonomous vehicles, users are exposed within their current vehicles to autonomous driving technology according to different functions such as automatic cruise control, lane-keeping systems, automated skidding brakes, park assistance, blind spot and crash alert systems depending on their vehicle autonomous level [12]. More caution is needed before introducing the AV system to developing society.

Acknowledgements The authors would like to thank anonymous reviewers for their input and SEGi University for giving platform to perform this study.

References

1. Fagnant DJ, Kockelman K (2015) Preparing a nation for autonomous vehicles: opportunities, barriers and policy recommendations. Transp Res A-Pol Pract 77:167–181
2. Pettigrew S, Worrall C, Talati Z, Fritschi L, Norman R (2019) Dimensions of attitudes to autonomous vehicles. Urb Plng Transp Res 7(1):19–33
3. Hulse LM, Xie H, Galea ER (2018) Perceptions of autonomous vehicles: relationships with road users, risk, gender and age. Saf Sci 102:1–13
4. Igliński H, Babiak M (2017) Analysis of the potential of autonomous vehicles in reducing the emissions of greenhouse gases in road transport. Procedia Eng 192:353–358
5. Tussyadiah IP, Zach FJ, Wang J (2017) Attitudes toward autonomous on demand mobility system: the case of self-driving taxi. In: Information and communication technologies in tourism 2017: Springer Int'l Pub: 755–766
6. Jafary B, Rabiei E, Diaconeasa MA, Masoomi H, Fiondella L, Mosleh (2018) A survey on autonomous vehicles interactions with human and other vehicles. In: PSAM international conference on probabilistic safety assessment and management
7. Merat N, Jamson AH, Lai F C H, Daly M, Carsten, OMJ (2014) Transition to manual: driver behaviour when resuming control from a highly automated vehicle. Transp Res Part F, Tfc Psy Beh 27:274–282
8. Kyriakidis M, de Winter JC, Stanton N et al (2019) A human factors perspective on automated driving. Theor Issues Ergon Sci 20(3):223–249
9. Šinko S, Knez M, Obrecht M (2017) Analysis of public opinion on autonomous vehicles. In: Challenges of Europe: international conference proceedings. Sveuciliste u Splitu., pp 219–230
10. Bartneck C, Lütge C, Wagner A, Welsh S (2021) An introduction to ethics in robotics and AI Springer Nature: 117
11. Nordhoff S, Stapel J, van Arem B, Happee R (2020) Passenger opinions of the perceived safety and interaction with automated shuttles: a test ride study with 'hidden' safety steward. Transp Res A-Pol Pract 13:508–524
12. Matthaei R, Reschka A, Rieken J, Dierkes F, Ulbrich S, Winkle T, Maurer M (2015) Autonomous driving. In: Handbook of driver assistance systems. Springer Int'l Pub.: 1–31

Chapter 2
Review of Current Trends in Marine Energy: Large Tidal Current Turbines

Shamini Janasekaran, Jagadishraj Selvaraj, Saleh Alyazidi, and Salem Naeem

Abstract The world is making its transition to sustainable renewable energy. Researchers have studied the behavior of tidal current turbines (TCT) under various conditions as well as numerous approaches for application. Using Betz limit theory, many studies were conducted to analyze the usage of TCT and modification of TCT for all the living things without affecting the environment. With regards to the design development of TCT, the various technological trends such as yaw drive mechanism and horizontal axis turbine have been discussed. The paper has reviewed the latest update on various TCT energy generation capacities by different companies at differing locations.

Keywords Marine energy · Tidal current turbine · Renewable · Sustainability

2.1 Introduction

The world is making a shift from energy sources that release greenhouse gases to renewable energy sources. Many countries via their respective organizations have already set roadmaps to slowly transition to higher reliability on renewable energy sources such as the European Union [1], ASEAN [2], G20 [3] and the UAE [4]. According to the Institute of Renewable Energy, IEO in year 2020 renewables are projected to be the dominant energy supply by 2050 at the current trajectory of increased global energy consumption of 50% by 2050, a 3% global GDP growth per year and a 0.7% global population growth per year. There is a plethora of renewable energy sources available, where the more commonly known ones are photovoltaic (PV) cells, wind turbines, hydroelectric or nuclear plants, etc. The less commonly known which has the potential for scalability is based on tidal current turbines. A tidal current turbine is a subset of marine energy generation.

When a certain individual makes a recreational visit to the beach, the idea that crosses one's mind is the serenity of the sound produced by the crashing waves.

S. Janasekaran (✉) · J. Selvaraj · S. Alyazidi · S. Naeem
Centre for Advanced Materials and Intelligent Manufacturing, Faculty of Engineering, Built Environment & IT, SEGi University Sdn Bhd, 47810 Petaling Jaya, Selangor, Malaysia
e-mail: shaminijanasekaran@segi.edu.my

A. Ismail et al. (eds.), *Design in Maritime Engineering*, Advanced Structured Materials 167, https://doi.org/10.1007/978-3-030-89988-2_2

However, to the engineer, this sound of crashing waves is the sound of wasted unharvested energy produced by ocean. The tidal current turbines (TCT), also known as underwater turbine, marine current turbine and tidal stream generator are similar to concepts of wind turbines; however, instead of air moving the blades of the turbine, the seawater is replaced as a medium.

Alternative versions of TCT such as wave generators are also able to generate electricity. However, the eye sore produced by large civil and mechanical structures in the middle of the sea can be a predicament. The tidal current turbine is able to overcome this predicament as the turbines are installed underwater and at the seabed. Energy is generated through movement of the TCT blades by moving seawater attributing to changing tides. This movement is highly predictable; which is a result of the earth-Sun astronomical tide (known as solar tide) and earth-Moon astronomical tide (known as lunar tide). This predictability makes the tidal current a more reliable source as compared to solar and wind energy [5]. Reliability and predictability are an important factor in power generation so that the existing power grid running on non-renewable energy sources is able to adequately supplement the power generated to prevent an electricity blackout. When wind speeds are low or during a cloudy day, the electricity generation from these renewable sources is low. This will cause a power consumption spike in the electricity grid which can be harmful. On top of that, wind fluctuation affects voltage and frequency quality with their large deviations [6].

2.2 Betz Limit for Turbine Efficiency

The German physicist Albert Betz produced a law that applies to all Newtonian fluids, which states that the theoretical maximum efficiency of a turbine is 16/27 or 59.3% [7]. Recent theoretical framework has proposed that unsteady flow is able to surpass the Betz limit [8]. On the other hand, Young [9] states that in practice, a single foil system will not exceed the Betz limit even in unsteady flow effects. The Betz limit is commonly applied in the design of wind turbines. However, because air and water are both Newtonian fluids, therefore, for TCT, the Betz limit applies as well. According to the study of the Betz limit in wind turbines, the velocity of air drops as it passes through the turbine. Thus, 100% of the energy cannot be absorbed by the wind turbine allowing the wind flow through the exit of the turbine. When wind stops moving at the exit of the turbine, no fresher wind could get in [10].

Betz's law also states that as wind velocity decreases due to the loss of energy by the extraction from a turbine, the airflow must distribute to a wider area. Betz's law also states that as the rotational speed of the turbine blade increases, the blockage of the flow increases. This means that faster rotating turbine blades do not necessarily indicate more power generation [11]. For each speed of medium, there is an optimal rotational speed that maximizes the generation of power. The tip speed ratio as shown in Eq. (2.1), which is the velocity at the tip of the blade, must be constant.

$$\text{tip speed ratio} = \frac{\text{velocity of the blade}}{\text{velocity of the water}} = \frac{\omega r}{u} \tag{2.1}$$

where w is the rotational speed of the blade (rad/s), r is the radius of the blade (m), and u is the initial velocity of the turbine (m/s).

At low velocities of water of 0.4–0.8 m/s, the tip speed ratio should be increased to increase the coefficient of performance, Cp for energy generation [12]. The Betz limit, however, can be overcome by placing a turbine in a duct. However, this is still under discussion [13]. On top of that, studies have shown that an optimally blocked channel with turbines can increase the Cp of the Betz law by eight times [14].

2.3 Tidal Current Turbine (TCT) Conditions

The arrangement of the turbine depends on four important factors which are the blockage ratio, the tidal current turbine (TCT) horizontal distance, the tidal channel cross-section and the power fluctuations.

2.3.1 Turbine Arrangements for Tidal Current Turbine

A few studies have proposed the arrangement of multiple turbines arranged to increase the efficiency of the tidal current turbine. This is also known as farm arrangement. Arrangement patterns include double row centered (also known as double row parallel or double row fence) and double row staggered [15]. The turbine arrangement depends on the features of the tidal channel. If the channel is long but narrow, multiple staggered layouts are suitable [16]. The best arrangement for TCT in a tidal channel depends on the blockage ratio of the TCT. No two tidal channels are the same across the width and length as most seabeds have uneven floors.

2.3.2 Horizontal Distance Between Turbine Rows

Ouro et al. [17] discovered that when the horizontal distance between two rows of turbines is too near, at 4D (4 times the diameter of the rotor), the wake of the flow causes the velocity of the water to be reduced as it enters the turbines of the second row. Therefore, the adequate horizontal distance must be prioritized to ensure that the flow of the fluid that is entering the second row of turbines is fully formed and free of vortices. This is because, the efficiency of the turbine decreases when the turbine rows are placed too near to each other. It is imperative to ensure that the turbines have the adequate spacing between the adjacent rows so that the downstream turbines can benefit from the great tidal flow velocity recovery [18]. On top of reduced efficiency, lack of horizontal spacing between turbine rows can increase the stress acting on the blades of the second row of turbines at 4D. Vortices produced due to insufficient horizontal distance can cause unsteady loading on the blades, thus leading to premature failure [19].

2.3.3 Tidal Channel Cross-section

The peak flow through the channel is reduced by 71% of the initial value. When considering the flow from one end of a sea channel to another, the maximum average power produced is between 20 and 24% for a sinusoidal tidal head at its peak. The maximum average power does not depend on the location of the turbines along the channel [20]. On top of that, the model requires that the entire channel is blocked by the turbines, which is improbably; thus, the power produced would be further reduced in this case. Few turbines are only needed in a small cross-sectional area where the currents are the strongest because the increase in the number of turbines in a small cross-section can choke the flow [21].

2.3.4 Power Fluctuations and Design

Power fluctuations can occur that will affect the efficiency of the TCT. This is due to stream shear acting on the turbine blade. This problem was theorized to be magnified due to the large rotor diameter and two-blade design of the turbine system implemented [22]. Therefore, the turbine may experience severe periodic fatigue loads. Morandi et al. [23] showed that a high number of inflow kinetic energy (approximating 13–15%) tends to be displaced outside the rotor disk through experimentation. These are inflow power lost in the radial and tangential components. Ergo proposed a second counter rotating rotor of slightly larger diameter could be useful to recover the lost inflow kinetic energy [23].

2.4 Mechanical Properties of Composite Tidal Current Turbine

Growing priority is assigned for developing and promoting renewable energy technologies. The capacity for fine energy within tidal currents is a renewable source of energy. Wherever an effective way can be enhanced to obtain this capacity [24], tidal streams can be used to better meet the increasing energy requirements of the planet. Several studies have verified the great potential of marine current as a predictable, renewable resource for the development of energy on an industrial scale. The use of marine urgent turbines for electrical electricity generation has grown, and the horizontal axes of marine turbines are one good solution for this reason. Many types were considered and evaluated for hydrokinetic turbines, metal materials or composites. Composites have been produced in marine structures, particularly for offshore use. Composite technologies provide new possibilities for maritime and submarine clean energy. In addition, flood effects and seawater sprinkling can be experienced in the aquatic environment [24]. Composite materials play a vital role in the production of marine renewable energy converting systems under extreme environmental conditions. For the hydrokinetic nozzle turbine, these materials are employed. Its attractiveness, including lightweight, high strength and excellent corrosion resistance, makes it the perfect alternative for hydrokinetic designers compared with metallic materials. These criteria dictate the use of composite laminate materials like glass or polymer-reinforced carbon fibers. Efforts to monitor the actions of turbine-based composites have been produced in recent years. The study of impact on induced damaged composites for turbine pads. The components are composed with composites of glass fibers. The force interaction, distortion and damage to the delamination following dynamic loading were studied by the development in Abaqus/Explicit software of three-dimensional finite element models. To measure the initiation of a progression of injury, stress parameters and mechanical fracture techniques were employed. The technique of construction focused on the design of harm mechanics for the composite tidal turbine blades was explored by Fagan et al. [25]. The numerical model is based on Puck's phenomenological fiber failure parameters used in the user-defined 3D shell subroutine. Due to the difficulty of morphology in matrixes, on the other hand, some research studies have included several modes for matrix failure criteria. Different models of failure parameters were modified and used for the prediction of damage to laminate composites [26]. In the work of Mahfouz et al. [27], a 3-m long sea turbine blade made up of composites with polymeric foam as the center's foundation and carbon-epoxy as the face layer is tired and damaged. The technique for fatigue and damage management of the ocean turbine consisting of glass fiber enforced polymer composites is proposed by Kennedy et al. [28]. It is necessary, therefore, to research the damage to laminate composites and delamination, particularly when the turbines are mounted in the marine environment.

2.4.1 Structural Analysis and Development

The sophistication of structures such as marine structures for renewable energy makes the study of dynamic reaction and kinetics of damage of composite structures under impact not easy and sometimes even impossible. Present solutions are of course best suited to such modeling, including numerical codes and explicit approaches. It is simulated that lightweight composites can be incorporated into the ocean program. It is important to present data explaining the behavior of the system to accurately model our structure. They may derive from various sources: materials, geometry, boundaries, loads and interaction with the affected organisms.

2.4.2 Structure and Materials

A 3D structure of the turbine is then created using the Helical software and the FE code Abaqus. The nozzle has an overall diameter of 20 m in the present case. The effect experiments have been performed to provide a superior configuration and the behavior of the hydrokinetic turbine under dynamic pressure. A different impact load was placed on the composite marine turbine structure. Two types of impactor structures, including hemispheric and flat, have been taken into consideration. The effect of impact velocity and impactor structure is studied in a parametric analysis. In modeling the impactor, it is presumed that the module of Young is static and endless [26].

2.4.3 Boundary Conditions

The high density of seawater and unintended impacts on how the marine turbines work are exposed to critical loads. Composite materials are a tremendous advantage because of their outstanding mass/resistance and mass/rigidity connections to satisfy the needs of producers of tidal current turbines, which are usually related to mass gain problems. A hydrodynamic study of the turbine nozzle was used to develop a panel method software in conjunction with the blade factor momentum theory (BEM). The resulting hydrodynamic pressure was then used to predict the initiation of harm under border conditions of the FE code and the Hashin criteria. The loads arising from running water toward and around our punch mechanism are hydrodynamic loads.

2.4.4 Turbine Maintenance

Maintenance is a crucial component of TCT, and thus, certain expectations are to be taken into account. Firstly, ease of replacing the TCT structure. The ability to replace a heavy turbine in an efficient way is always preferred as the cost of installation is high. Secondly, cavitation corrosion also plays a big part in the lifespan of the turbine. This is due to the inability of metal to withstand corrosion. TCT has to deal with certain challenges associated with maintenance, loading conditions, installation, electricity transmission, as well as environmental impacts in order to achieve economic requirements [29]. Operation of TCT should be monitored closely. This is due to the high maintenance cost if an error is not detected during the early stage

[30]. The operation and maintenance cost per unit is calculated by Eq. (2.2).

$$\text{operation and maintenance cost} = \frac{\Sigma(\text{omci})}{\text{energy}} \tag{2.2}$$

where omci is the operation and maintenance cost per year and energy is the lifetime energy output of the whole farm.

2.4.5 Rotor Blade Maintenance

Tidal turbines are able to generate clean and renewable energy. The anticipated service life of the turbines is twenty years [31]. The most important piece of the entire structure is the turbine blade. Reinforcing critically loaded sections of the blade in order to improve longevity and upgrading the blade to achieve higher efficiency are both critical matters. The rotor blades' function is to reap the kinetic energy from the water flow and channel it toward the generator. The blade is able to rotate due to the water current hitting the surface of the blade. The latest designs of tidal turbine have variable pitch in order to adjust according to the water current. Apart from that, the turbines are able to rotate in both directions via bidirectional blades. This feature enables the turbines to harvest the maximum amount of energy, thus increasing the overall efficiency [32]. The material that makes up the blade is also crucial because not all materials are able to withstand the stress that act on the TCT during abnormal loading conditions. It is also found that glass fiber reinforced polymers (GFRP) and carbon fiber reinforced polymers (CFRP) are materials used to fabricate rotor blades for tidal turbines. On the other hand, Chen and Lam [33] found that GFRP and CFRP are less suitable due to their inability to endure harsh weathers, therefore resulting in a low service life. More research is required in the field of material as well as dynamics of the tidal turbine in order to enhance the service life of the rotor blade. As for now, research has proven that heat and water will contribute to the performance of carbon fiber reinforced polymer (CFRP). Firstly, based on the total water that is absorbed by the composite. Secondly, the overall temperature and exposure time of the material. Finally, the sort of polymer utilized as a matrix material [31].

2.4.6 Ability to Withstand Corrosive Saltwater

The corrosion of metal contributes greatly to the economic loss as well as the social harm in daily life and production. Approximately, one-third of the metal scrap in the world accredited to the corrosion of metal. Around 2–4% of a country's GDP is also affected due to this issue. This loss is six times more than the losses caused by typhoons, floods, earthquake and other natural disasters [34]. Saline water, also known as saltwater, is from the sea. Seawater has a salinity of 3.5%, which means for every kg of seawater, approximately 35 g of dissolved salt [35]. When comparing seawater with fresh water, saltwater is able to corrode metal five times quicker than fresh water [36]. Apart from corrosion, saltwater also consists of bacteria that are able to consume iron and turn their excretion into rust. The conductivity and oxygen content of water are directly affected by the salt content [37]. Most materials are weak against saltwater as they corrode within a short period of time. In which, it is not suitable for constructing turbine blades. With regards to the ability of the turbine blade to resist the corrosive saltwater, black anticorrosive coating can wear out after six months of deployment, and the internal composite material becomes uncoated [38].

2.4.7 Corrosion Protection Against Seawater

Material selection should be the most important aspect to ensure long-term safe performance of the turbine. The changes of mechanical, chemical and physical properties of the material should be properly analyzed. Apart from that, new technologies have developed to pay attention and improve on effects caused by contact of metals and corrosive medium. For example, surface coating and treatment technologies can accomplish corrosion protection by coating metal with a non-metallic coat [39]. Other methods such as cathodic protection are also used in order to protect the metal from corrosion. This method utilizes an electrochemical process to protect metal from corrosion. By enabling cathodic polarization via providing current, the potential of the metal turns to negative. The cathodic protection method is also known as the sacrificial anode cathodic protection [40]. Adding small chemicals to the corrosive media is known as the corrosion inhibitor method. This method relies on the physical or chemical reaction in order to slow down the rate of corrosion. All these are accomplished while maintaining the physical, chemical as well as the mechanical properties of the metal. This method is one of the most convenient and low in cost methods used to prevent corrosion [41].

2.5 Conclusion

Current trends in marine energy harvesting are still in its developmental stages. Apart from carbon fiber reinforced polymer (CFRP) and glass fiber reinforced polymers (GFRP), new materials are being researched in order to improve the service life of turbine blades exposed to saltwater. The location of the turbine is also a crucial factor as not all sites have a continuous tidal flow with suitable flow velocity. In order to reduce the maintenance and operation cost, placement of the tidal turbine farm should be near the shore.

The layout (positioning) and planning of the turbine farm play a big role in the efficiency of the farm as this may affect the future development if not planed accurately. The layout design for a turbine farm for maximum energy harvesting depends on a plethora of factors which include blockage ratio, horizontal distance between turbine rows, tidal channel cross-section; all of which have been discussed.

Acknowledgements The authors would like to thank anonymous reviewers for their input and thank SEGi University for giving platform to perform this study.

References

1. Cucchiella F, D'adamo I, Gastaldi M (2018) Future trajectories of renewable energy consumption in the European Union. Res. https://doi.org/10.3390/resources7010010
2. Erdiwansyah MR, Sani MSM, Sudhakar K (2019) Renewable energy in Southeast Asia: policies and recommendations. Sci Total Environ 670:1095–1102
3. Saygin D, Rigter J, Caldecott B, Wagner N, Gielen D (2019) Power sector asset stranding effects of climate policies. Ene Src Part B: Econ Plng and Plcy 14(4):99–124
4. Sgouridis S, Abdullah A, Griffiths S, Saygin D, Wagner N, Gielen D, Reinisch H, McQueen D (2016) RE-mapping the UAE's energy transition: an economy-wide assessment of renewable energy options and their policy implications. Renew Sustain Energy Rev 55:1166–1180
5. Marta-Almeida M, Cirano M, Guedes Soares C, Lessa GC (2017) A numerical tidal stream energy assessment study for Baía de Todos os Santos. Brazil. Renew Energy 107:271–287
6. Schmietendorf K, Peinke J, Kamps O (2017) The impact of turbulent renewable energy production on power grid stability and quality. Eur Phys J B 90(11):1–6
7. Ranjbar MH, Nasrazadani SE, Kia HZ, Gharali K (2019) Reaching the betz limit experimentally and numerically. Ene Equip Sys 7(3):271–278
8. Dabiri JO (2020) Theoretical framework to surpass the Betz limit using unsteady fluid mechanics. Phys Rev Fluids 5(2):022501
9. Young J, Tian FB, Liu Z, Lai JC, Nadim N, Lucey AD (2020) Analysis of unsteady flow effects on the Betz limit for flapping foil power generation. J Fluid Mech 902
10. Zhou Z, Benbouzid M, Charpentier JF, Scuiller F, Tang T (2017) Developments in large marine current turbine technologies—a review. Renew Sustain Energy Rev 71(December):852–858
11. Rupp Carriveau (2011) Fund. Adv. Topic.Wnd.Pwr. Winchester, UK
12. Encarnacion JI, Johnstone C (2018) Preliminary design of a horizontal axis tidal turbine for low-speed tidal flow. In: Proceedings of the 4th AWEC: 1–8
13. Venters R, Helenbrook BT, Visser KD (2018) Ducted wind turbine optimization. J Sol Energy Eng Trans ASME. https://doi.org/10.1115/1.4037741
14. Vennell R (2013) Exceeding the Betz limit with tidal turbines. Renew Energy 55:277–285
15. An Q, Wu S, Liu Y (2020). Numerical investigations on influence of spacing of tidal current turbine array. J Phys Conf Ser 1600 012010. https://doi.org/10.1088/1742-6596/1600/1/012010
16. Almoghayer MA, Woolf DK (2019). An assessment of efficient tidal stream energy extraction using 3D numerical modelling techniques. In: Proceedings of the 13th EWTEC: 1–10
17. Ouro P, Ramírez L, Harrold M (2019) Analysis of array spacing on tidal stream turbine farm performance using Large-Eddy simulation. J Fluids Struct 91:102732
18. Zhang C, Zhang J, Tong L, Guo Y, & Zhang P (2020) Investigation of array layout of tidal stream turbines on energy extraction efficiency. Ocean Eng 196(November):106775
19. Thomas Scarlett G, Viola IM (2020) Unsteady hydrodynamics of tidal turbine blades. Renew Energy 146:843–855

20. Bonar PAJ (2017) Toward best practice in the design of tidal turbine arrays. PhD Thesis, University of Edinburgh, Edinburgh
21. Garrett C, Cummins P (2005) The power potential of tidal currents in channels. Proc R Soc A: Math Phys Eng Sci 461(2060):2563–2572
22. Li Y, Liu H, Lin Y, Li W, Gu Y (2019) Design and test of a 600-kW horizontal-axis tidal current turbine. Ene 182:177–186
23. Morandi B, Di Felice F, Costanzo M, Romano GP, Dhomé D, Allo JC (2016) Experimental investigation of the near wake of a horizontal axis tidal current turbine. Int J Mar Energy 14:229–247
24. Aranake A, Duraisamy K (2017) Aerodynamic optimization of shrouded wind turbines. Wind Energy 20(5):877–889
25. Fagan EM, Kennedy CR, Leen SB, Goggins J (2016) Damage mechanics based design methodology for tidal current turbine composite blades. Renew Energy 97:358–372
26. Nachtane M, Tarfaoui M, Saifaoui D, El Moumen A, Hassoon OH, Benyahia H (2018) Evaluation of durability of composite materials applied to renewable marine energy: Case of ducted tidal turbine. Energy Rep 4:31–40
27. Mahfouz MMA, Amin AM, Youssef EB (2011) Improvement the integration of Zafarana wind farm connected to egyptian unified power grid. In: 2011 46th (UPEC), 46:1–6
28. Kennedy CR, Leen SB, Brádaigh CM (2012) A preliminary design methodology for fatigue life prediction of polymer composites for tidal turbine blades. Proc Inst Mech Eng, Part L: J Mater: Des Appl 226(3):203–218
29. Mérigaud A, Ringwood J (2016) Condition-based maintenance methods for marine renewable energy. Renew Sustain Energy Rev 66:53–78
30. De Nie R, Leontaris G, Hoogendoorn D, Wolfert A (2019) Offshore infrastructure planning using a vine copula approach for environmental conditions: an application for replacement maintenance of tidal energy infrastructure. Struct Infrastruct Eng, pp 1–18
31. Alam P, Robert C, Brádaigh CMÓ (2018) Tidal turbine blade composites - A review on the effects of hygrothermal aging on the properties of CFRP. Comp Part B: Engr 149:248–259
32. Roshanmanesh S, Hayati F, Papaelias M (2020) Utilisation of ensemble empirical mode decomposition in conjunction with cyclostationary technique for wind turbine gearbox fault detection. Appl Sci 10(9):3334
33. Chen L, Lam W (2015) A review of survivability and remedial actions of tidal current turbines. Renew Sustain Energy Rev 43:891–900
34. Milinković O, Brzaković P, Milošević O, Brzaković M (2020) Property Insurance and innovative building techniques reducing the consequences of climate change. Економика пољопривреде. https://doi.org/10.5937/ekoPolj2001269M
35. Ummah H, Suriamihardja D, Selintung M, Wahab A (2015) Analysis of chemical composition of rice husk used as absorber plates sea water into clean water. ARPN J Engr Appl Sci 10(14):6046–6050
36. Venâncio C, Castro B, Ribeiro R, Antunes S, Abrantes N, Soares A, Lopes I (2019) Sensitivity of freshwater species under single and multigenerational exposure to seawater intrusion. Philos Trans R Soc B 374(1764):20180252
37. Mohammadkhani R, Ramezanzadeh M, Saadatmandi S, Ramezanzadeh B (2020) Designing a dual-functional epoxy composite system with self-healing/barrier anti-corrosion performance using graphene oxide nano-scale platforms decorated with zinc doped-conductive polypyrrole nanoparticles with great environmental stability and non-tox. Chem. Eng J 382:122819
38. Li W, Zhou H, Liu H, Lin Y, Xu Q (2016) Review on the blade design technologies of tidal current turbine. Renew Sustain Energy Rev 63:414–422
39. Bag PP, Roymahapatra G (2019) Surface engineering for coating: a smart technique. Adv Surf Coat Tech Mod Ind Appl. https://doi.org/10.4018/978-1-7998-4870-7.ch012
40. Refait P, Jeannin M, Sabot R, Antony H, Pineau S (2015) Corrosion and cathodic protection of carbon steel in the tidal zone: products, mechanisms and kinetics. Corros Sci 90:375–382
41. MacKenzie C, Magdalenic V, Moussavi A, Joosten M, Achour M, Blumer D (2016) U.S. Patent No. 9359677. Washington, DC: U.S. Pat and Trademark Off

Chapter 3
The Review of National Contingency Plan Towards the Oil Spill Response in Malaysia

Ismila Che Ishak, Aminuddin Md Arof, Md Redzuan Zoolfakar, and Mohd Fairoz Rozali

Abstract This paper reviews the elements involved in the contingency plan regarding the oil spill response activities in Malaysia. The effective elements in the contingency plan cover the definition of terms, effects of the oil spill, tier levels, stockpiles equipment, the roles of the main related government agencies, response times, response techniques, emergency response life cycle, and the influencing factors in the preparedness and response toward the oil spill incidents. The review of the oil spill contingency plan in the oil spill in Malaysia has also highlighted the number of response team members based on The National Oil Spill Contingency Plan (NOSCP) 2014, occupied in responding toward the oil spill incidents in the economic exclusive zone (EEZ) of Malaysia and including Malaysian territorial water (MTW).

Keywords Oil spill preparedness and response · Economic exclusive zone · Contingency plan

3.1 Introduction

The oil spill is known as per a discharge of fuel hydrocarbons hooked on the natural atmosphere, either aquatic or land, unintentionally or operationally, whenever the oil

I. Che Ishak (✉) · A. Md Arof
Maritime Management Section, Universiti Kuala Lumpur Malaysian Institute Marine Engineering and Technology, 32200 Lumut, Perak, Malaysia
e-mail: ismila@unikl.edu.my

A. Md Arof
e-mail: aminuddin@unikl.edu.my

M. R. Zoolfakar
Marine & Electrical Engineering Technology, Universiti Kuala Lumpur Malaysian Institute Marine Engineering and Technology, 32200 Lumut, Perak, Malaysia
e-mail: redzuan@unikl.edu.my

M. F. Rozali
Bahagian Maritim, Kementerian Pengangkutan Malaysia, Wilayah Persekutuan Putrajaya, Malaysia
e-mail: fairoz@marine.gov.my

A. Ismail et al. (eds.), *Design in Maritime Engineering*, Advanced Structured Materials 167, https://doi.org/10.1007/978-3-030-89988-2_3

is generated or transferred [2, 47] and is a cause of contamination that is encountering waterbodies universal. The incidents of oil spills occurred due to numerous sources. The potential sources could be such as unintended leakage from vessels, cargo, and offshore oil platforms from the ship-to-ship in MARPOL 73/78 and OPRC, 1990 [24]. In Malaysia, statistically, 119 oil spill incidents occurred since 2014–2020 [3, 45]. From the statistic, it displays around 40 oil spill cases happened due to unknown sources, 15 cases due to discharging the oil to the ocean, 6 oil spill incidents due to ship collision, 4 oil spill incidents due to pipeline leakage, 3 oil spill incidents from shipwreck, and one incident each due to load and unload movement, oil digging, cutting activity of vessel, and incidents at port terminal [3, 45]. The oil spill incidents to some extent may not form from a marine casualty. But, it was also occurred due to illegal discharge from unauthorized bunkering activity which has not to notify the related authority as required [14]. Once the oil spill has occurred, it has affected several ecosystems and species-rich areas [7]. Oil spills are a worldwide occurrence together with influences that slash throughout socioeconomic, well-being, and ecological elements of the shoreline environment [4]. It has caused excessive economic loss and costs and destructive marine ecological life [42]. These oil spill incidents could produce a negative effect on the marine environment and have a poor impact on the maritime community. The effects such produce effects on marine life, public health, human life, and loss of tourism and aquaculture activities [20]. To overcome this effect, our country is highlighting the requirement of contamination management concerning coastline and maritime environments and is required to encounter the needs of the quality natural environment as obliged with the Environmental Quality Act (EQA), 1974 [30]. This research paper highlights the overview of the oil spill preparation and response contingency plan activity in Malaysia.

3.2 Literature Review

3.2.1 The Effects of the Oil Spill Cases

When the oil spill happened, it produced several effects particularly to environmental, social, economic, human lives, marine lives, investors, fishery, amount of safety and ecotourism, and the country [16]. The oil spill cases generate causes marine pollution and carrying enormous economic damage to humanity, maritime environmental, directs to the harms in environmental stability, leads to the devastation of the environment and the creature [15]. The effects of the oil spill could be explained as follows.

3.2.2 *The Effects on the Ecological*

The effect of the oil spill has acute impact on at the ecosystems and the environment once the oil spills have been released. It has affected the maritime environment, seafood, maritime life, plankton, and coral [52]. The oil spill disaster affects the oceans and produces marine life destruction [15]. It has also affected the flora and fauna, fish life which has been trapped with oil [11, 25]. Besides, it also harms nearby society [2]. In addition, the effects also have reduced nutrients and proteins in seafood which could harm humans and animals by eating the polluted seafood [35]. The oil spills do not only destroy the marine life and wetland but also disturb the estuarial habitat as mutual [10, 39].

3.2.3 *The Effects on the Economy*

The cleaning-up process after the oil spill incidents has occurred, needed a lot of costs and manpower [54]. During the cleaning-up activities, the working people who are participating could affect their health condition from the toxic hazardous exposure and could instigate skin inflammation and respiratory difficulties [29]. This effect could exist longer and affected the coastal areas which could disrupt camping, fishing, and swimming. It also has reduced the income and caused the rural populations to relocate to urban areas to acquire an additional resource of income in maintaining life. After the oil spill incidents happened, it may similarly resume time to recover. As a result, it may affect the industry of hotels, restaurant vendors, tenant boats, dive tour operators, and others who have income from the entertaining activities in the coastline zone, and they are experiencing a substantial financial loss [21].

3.2.4 *Definition of Marine Pollution*

When the oil spill occurred, it generates harmful substances, which have penetrated the ecosystem's human deeds and affect the natural cycle [8]. The marine oil spill incidents have happened near to coastline settings, influencing enormous spread of flora, beach, aquatic, and birds [44]. In various components of the globe, a significant proportion of oil spill influences on coastline environments is associated to the inaccessibility of dependable decision support systems (DSS) which can precisely describe the susceptible coastline reserves for quick interference [5, 7, 50]. A common definition of pollution is any objects which have been thrown into the water and affected direct and indirect harm and generate bad effects on humans, marine life, ecosystem, social, and economics to the affected countries. Meanwhile, according to UNCLOS 1982, Article 1 defines pollution as the introduction by man, directly or indirectly, of substances or energy into the marine environment, including

Table 3.1 Definition of marine pollution

Authors	Explanations
[2]	Such as a contaminated element that penetrates an ecosystem that harms humans and wildlife
[23]	Ås the introduction by man, directly or indirectly, of substances or energy into the marine environment, including estuaries, which results or is likely to result in such deleterious effects as harm to living resources and marine life, hazards to human health, a hindrance to marine activities, including fishing and other legitimate uses of the sea, impartment of quality for use of seawater, and reduction of amenities

estuaries, which results or is likely to result in such deleterious effects as harm to living resources and marine life, hazards to human health, a hindrance to marine activities, including fishing and other legitimate uses of the sea, impartment of quality for use of seawater, and reduction of amenities (UNCLOS 1982, Article 1 (4) [22, 33] (Table 3.1).

3.2.5 *Definition of Oil Spill*

Generally, contamination is defined as any arrangement of contamination in an ecosystem with a harmful influence upon the organisms by changing the progress rate and the reproduction of plants or animal species by intervening with human being conveniences, reassurance, well-being, or estate values [12]. Oil spills are substantial forms of hydrocarbons that go into the collecting marine ecosystem and an effective approach to eliminating hydrocarbons from seawater properties. Table 3.2 explains several definitions of the oil spill from different authors.

3.2.6 *Definition of Oil Spill Preparedness and Response*

The oil spills will remain to arise if the population hangs on petroleum and its products [16, 17]. Generally, preparedness refers to activities prepared by the community to respond when a disaster occurs [31]. Consequently, sufficient collision avoidance methods and oil spill preparedness planning are recommended to enrich marine security and maritime conservation safety [46]. The acknowledgment through ratification of several IMO conventions is as follows [23, 24].

(i) Marine pollution convention—MARPOL 73/78
(ii) Preparedness and response—OPRC 1990 and OPRC HNS 2000
(iii) Liability and compensation regime—CLC, Fund 1992, and BCC 2001.

Table 3.2 Explanation of oil spill

Authors	Explanation
[12]	Oil slicks are the enormous sources of hydrocarbons that come into the receiving maritime situation
[28]	The elimination of hydrocarbons in the maritime natural environment is triggered by human being movement, and environmental catastrophes can trigger a lot of consequences on both the natural environment and wealth
[37]	The existence of oil spills appeared to be on the increase because of the release of excess or mishaps of the crash of crude oil tankers, as well as spills of refined petroleum products; petrol, diesel, offshore platforms, drilling rigs and wells, and the practice of weightier oils by big vessels, the bunker oil, or the spill of any oily waste or excess oil
[2]	Oil is a common phrase utilized to signify liquid petroleum products which primarily comprise hydrocarbons. The release of oil into the environment is labeled an oil spill
[38]	Any sort of type of pollution that is discovered in an ecosystem causing in a damaging destructive effect on the organisms in the ecosystem and disrupting the expansion rate and plant or animal breeding by obstructing human contentment, well-being, property values, and conveniences
[41]	Any kind of oil that has been issued into the natural environment and pollution from land-based sources, mainly after the impact of industrialization and urbanization alongside coastal areas, has been seen as the major contributor to maritime contamination
[1]	A collision that requires oil spills would be devastating for the environment varying on exactly how terrible it is and subject matter to other considerations such as oil management and deterrence of the oil scattered actions

Responsibility and accountabilities for dealing with this major spillage of oil and other hazardous substances that have been thrown into the sea from shipping operations and activities that could intimidate Malaysia's interest. These responsibilities are implemented through the collaboration between the Department of Environment, Marine Department, and other organizations that are dealing in corresponding oil spill emergencies in Malaysian seawaters containing the exclusive economic zone or EEZ [40]. The regional contingency plan acclaims the representative to cultivate an own national contingency plan and endangered the reserves mandatory to react to marine oil spills dilemma. Additionally, it challenged its state approach for oil spill preparation and response and increased the necessity for a provincial and global organization to improve cope with the oil spill concern [9]. It is also suggested to proactively participate in the organizational collaborative efforts to establish a healthier natural environment for shareholder commitment in boosting preparation and response procedures. Meanwhile, the main objective of preparedness includes the advancement of suitable preparation procedures and systems for detecting and describing the presence of a contamination case inside the regions of responsibility of the concerned group. It is also important in supplying the means to introduce trigger actions to constrain the additional disperse of oil and distributing the method by which sufficient properties may be utilized to react to oil contamination occurrences [43].

Once the oil spill has occurred, it is essential to concentrate on methods to avoid the oil spill. Appropriate and feasible approaches for monitoring the oil spill from being dispersed to an additional area and cleaning the oil spill must also be developed. Normally, the application of a unified approach of contingency plans and response options could speed up and enhance response to an oil spill and expressively lessen the environmental effect and harshness of the spill [17–19]. In summary, the oil spill response encircling all actions intricate in comprising and cleaning up oil for several reasons such as to maintain the safety of human life, to alleviate a condition to prevent it from deteriorating, and to diminish adversative environmental and socioeconomic impacts by coordinative all restraint and elimination actions to transmit out a timely, effective response [45].

3.2.7 The National Oil Spill Control Agency

In the occurrence of oil spill cases, a collaboration among the related government agencies is significant and interrelated. In Malaysia, 17 government agencies have been appointed as the national oil spill control committee as shown in Table 3.3 [22].

These 17 government agencies are known as flat organizations. To date, few government agencies have been consolidated into one department. As such, Marine Park Department is no longer exists, where the function has been integrated into the Department of Fisheries Malaysia [14]. At the same time, some of the appointed national committees will only function once receive a command from DOE as the custodian and do not develop any specific task force toward the oil spill preparation and response.

Table 3.3 Response committee members {22]

Parties	Parties
1. Department of Environment (DOE)	9. Royal Malaysian Navy
2. Ministry of Foreign Affairs	10. Royal Malaysian Air Force
3. Marine Department	11. Fire and Rescue Department of Malaysia
4. Malaysian Maritime Enforcement Agency (MMEA)	12. Malaysia Meteorology Department
5. Royal Malaysian Police Air Operation Force	13. Department of Fisheries Malaysia
6. Petroliam Nasional Berhad (PETRONAS)	14. Department of Marine Park Malaysia
7. Petroleum Industry of Malaysia Mutual Aid Group (PIMMAG)	15. National Security Council
8. Royal Malaysian Police Marine Operations Force	16. Royal Malaysian Customs Department
	17. Immigration Department of Malaysia

Meanwhile, in the proposed NOSCP, 2021, [14] has differentiated the control committees into primary and secondary functions. Tier 1 response will involve the control committee at primary, and the secondary committees will react once Malaysia is facing Tier 3 incidents. For example, the role of the Royal Malaysian Customs Department and Immigration Department of Malaysia is needed to collaborate toward the oil spill incidents by making the process of releasing the equipment and manpower accordingly effectively [14].

In practice, for instance, the Royal Malaysian Customs Department and Immigration Department of Malaysia has never been involved in any of the oil spill incidents in Malaysia at Tier 1 or Tier 2 [14]. These two agencies will get activated at Tier 3 or oil spill transboundary requirements. At this time, when the current national assets are insufficient or not available. It requires the country to import expertise, assets, and diplomatic requirement from other related neighborhood countries [14, 15]. The Royal Malaysian Customs Department will process and release immediately any duty tax on the imported stockpiles purchased. In addition, it is also depending on the location of Tier 3 of oil spill incidents which requires a standardization from the Ministry of Foreign Affair, i.e., communication either from Department of Maritime Affairs, ASEAN Desk, or International Desk. The diplomatic clearance letter will be issued to the imported expertise or assistance needed by the country from the neighborhood countries [14].

3.2.8 The Roles of Marine Department Toward the Oil Spill Occurrences

In the event of oil spill cases in Malaysia, there is one main government agency, the Marine Department, which will involve. The desired roles of the Marine Department are as follows [13]

- (i) On-scene commander (OSC),
- (ii) Maintain and operationalize stockpiles of the oil spill incidents and oil craft response boat,
- (iii) Run simulation of oil spill trajectory and to determine the needs for the place of refuge,
- (iv) Salvage and wreck activities,
- (v) Protect and safety of ships and safety of navigation,
- (vi) Issuance of Malaysia Shipping Notice to shipping community,
- (vii) Custody the international liability convention,
- (viii) Claims and compensation processes from merchant ships.

Besides that, the main activities played by the Marine Department in the event of the oil spill incidents are as follows [13]

- (i) Involve in Tier 1 of the oil spill cases, if required

(ii) Respond accordingly to the requirements of the oil spill incidents at Tier 2 and Tier 3,
(iii) Responsible as a technical specialist agent in responding and preventing the issue of oil spill cases,
(iv) Use stockpile facilities available throughout the country,
(v) Responsible for responding within 24 h after the oil spill cases have happened for Tier 1 and Tier 2 in the peninsular Malaysia and within 48 h in East Malaysia including Sabah and Sarawak,
(vi) Appointed as a focal agency and custodian under IMO, and
(vii) Serve as a liaison body and supervising all international matters abbreviated shipping sectors.

3.2.9 Response Time

In responding to the oil spill incidents, it requires time to respond. The response time is explained as the length of time needed for a person or system to respond to a conferred incentive or event [36]. In oil spill incidents, the response time is necessary and crucial. In the issue of oil spill occurrences in Malaysia, there is a disparity in the response time between Malaysia and other countries. The response time differs, such as 30 min in Japan, 24 h in Mexico, within 4 h in the UK [6] and the US, within 12 h in Canada as for less than 1000 ton of spilled oil, within 6 h for less than 150 ton spilled oil [26].

Meanwhile, as for Malaysia, the response time is within 24 h following the oil spill events that have happened at Tier 1 and Tier 2 in the peninsular Malaysia. While, the response time is within 48 h in East Malaysia. However, in terms of a benchmarking time of responding toward the oil spill incidents, Malaysia does not relate to any benchmarking country as a benchmark in the matter of the oil spill response hour [13, 32]. Among the other contributing factors regarding the oil spill incidents, they are the location of the stockpiles, and readiness of the stockpiles contributes to the oil spill incidents either at Tier 1 or Tier 2.

3.2.10 Types of Oil Spill Response Equipment

The reaction apparatus is necessarily used for counteracting the oil spill incidents. The response equipment for Malaysia is provided by the Marine Department. The equipment categories are divided as follows [22]

(i) Containment equipment: consists of oil booms and beach booms,
(ii) Recovery equipment: consists of oil skimmers, oil nets, vacuum pumps,
(iii) Storage equipment: consists of temporary storage tank (RO tank), storage tanks (troiltanks),

Table 3.4 Equipment categories stockpile base and equipment for the Marine Department [22]

Equipment	Central region (Wilayah Tengah)	Eastern region (Wilayah Timur)	Southern region (Wilayah Selatan)	Northern region (Wilayah Utara)	F.T Labuan and South China Sea (Wilayah Persekutuan Labuan and South China Sea)
Containment equipment	4 sets	1 set	5 sets	4 sets	2 sets
Recovery equipment	6 sets	2 sets	12 sets	5 sets	3 sets
Dispersant sprayer				1 set	1 set
Storage equipment	5 sets		4 sets	3 sets	3 sets
Support equipment	30 sets		30 sets	15 sets	15 sets

(iv) Support equipment: consists of anchors, marker buoys, hydrocarbon detectors, chemical suits, gas-tight suits, communication equipment.

The Marine Department equipment is allocated based on the regions as shown in Table 3.4, which have been divided as central region, east region, south region, north region, Federal Territory of Labuan and South China Sea. In addition, the practice of in situ burning is not allowed to be applied in Malaysia. However, the Malaysian policy at the Ministry of Transport will issue penalty that does not allow the international business to be performed at any economic exclusive zone (EEZ) and including Malaysian territorial waters (MTW) [14].

3.2.11 Techniques of Response

The response technique depicts several techniques which can be used to respond to oil spill incidents [16]. Superficial and remote sensing technology could be utilized to discover and trace the oil slicks. Moreover, separate tools and techniques could be applied to remove the oil slick from the sea. Booms and ancillary equipment are applied to contain oil on water. Meanwhile, skimmers, sorbents, and manual recuperation are applied to recover the oil from the water surface. The consumption of pumps, in situ burning, and chemical agents, such as dispersants, are used for oil elimination. The oil spill could contaminate the shorelines and effects the numerous ecologies and the numerous creatures within the water. Meanwhile, [53] recommended that it is required to assess the mechanical improvement capability of skimmers as an essential part of the whole oil spill response and preparedness capacity. As a result of using this new application of skimmer, methodology reveals as more accurate and useful for capacity preparing and real oil spill reaction procedure.

3.2.12 Emergency Response Management Life cycle

A life cycle approach conveys a comprehensive and organized point of view of activities concerning disaster reaction administration [51]. The emergency response framework is altered to each one of the phases in the life cycle, and it can be imagined in terms of different sets of events on the timeline range which include actions taken as,

(a) before an incident which usually deals with preparedness issues such as planning and training,
(b) for the duration of the event or modification, and
(c) following the case or the response and improvement phase. The life cycle is generally specified in conditions of four stages such as mitigation, preparedness, response, and recovery [48] as shown in Fig. 3.1.

The response life cycle includes several processes such as

(a) mitigation as the application of actions to avert the onset of a disaster or to reduce the impacts should an oil spill occur,
(b) preparedness as an activity to organize the society to react when a catastrophe arises,
(c) response as the employment of resources and emergency actions which is driven by policies to protect life, property, the environment, and the social, economic, and political composition of the society, and
(d) recovery includes the long-term actions after the instant impact of the disaster to ensure peace to the community and to reinstate some presence of routine [44].

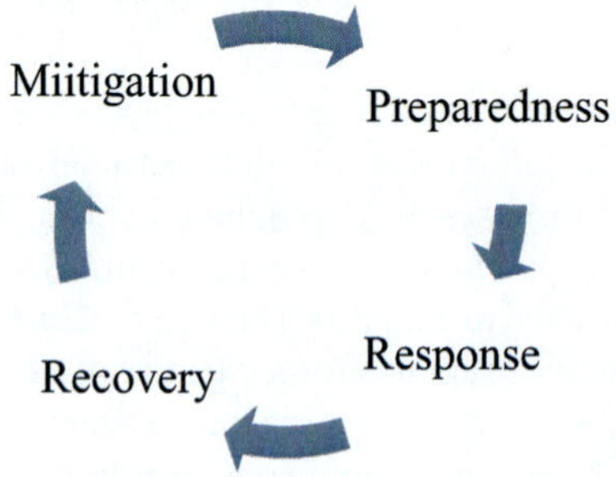

Fig. 3.1 Stages in response life cycle

3.2.13 Malaysian National Oil Spill Contingency Plan

Meanwhile, NOSCP edition 2014 [22] has endorsed and authorized by *Arahan Majlis Keselamatan Negara* (MKN) 20 in March 2012 in which the oil spill incident is treated as one of the maritime disasters. Therefore, under Article 7 Instruction 20, the National Safety Council and Department of Environment have been assigned as authorized agents using the National Oil Spill Contingency Plan (NOSCP), while the Marine Department has been assigned as a technical agent to oversee an oil spill incident or accident that leads to oil pollution. Besides that, for maritime incident or emergency which comprises human life and assets, the authorized agent is Maritime Enforcement Agency (MMEA) using the National Search and Rescue Manual, International Aeronautical and Maritime Search and Rescue (IAMSAR).

The National Oil Spill Contingency Plan 2014 [22] has been developed to switch oil spill incidents happening inside Malaysian seawaters or waters encompasses along Malaysian shorelines and the EEZ situated inside the Straits of Malacca and the South China Sea, as well as Brunei Bay, Sabah, and Sarawak seawaters, and the Celebes Sea [49]. The National Contingency Plan has been activated by referring to the three types of tiers which include Tier 1, Tier 2, and Tier 3 established on the subsequent factors

(a) location of the spill,
(b) quantity of spill, and
(c) capability and competency to respond.

3.2.14 The Tier Response

Oil risk and response are classified corresponding to the magnitude of the spill and its vicinity to a corporation's operational capacity. It directs to the concept of a tiered response to the oil spill. The contingency plan would include each tier and be exactly associated with the corporation's possible spill consequences [22]. Figure 3.2 shows the tiered reaction with the amount of spill. Generally, Tier 1 refers to a small spill, Tier 2 refers to a medium spill, and Tier 3 refers to a large spill.

3.2.15 Tier 1—Local Response

The clean-up activities in Tier 1 have been carried out by local ownership in which the oil spill occurs in own plants, buildings, premises, ports, terminals, or offshore by using own capacity and equipment as of its activities. The oil spill in Tier 1 generally a small volume and the response is based on the local response contingency plan. An individual company would habitually offer resources for response at this tier [22].

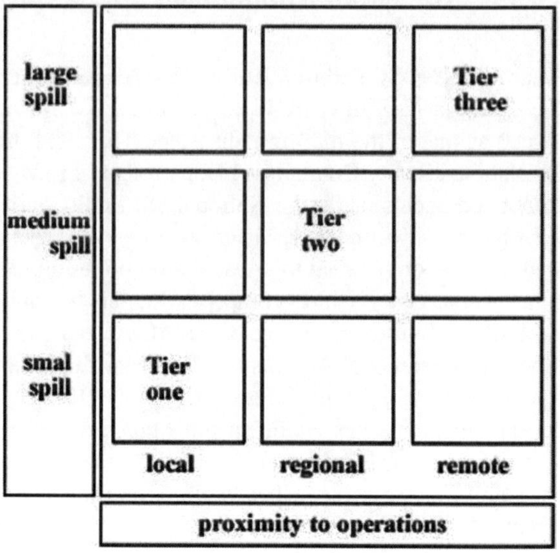

Fig. 3.2 Tier response

3.2.16 *Tier 2—National Response*

Tier 2 is executed when parties in Tier 1 are incapable and incompetent to react to an oil spill that occurs beyond the scope of Tier 1. It is known as a bigger spill in the area of corporation amenities where properties from other companies, industries, and possible government agencies could be called in, on a shared aid base. Tier 2 is run by the State Operation Committee (SOC) chaired by the Director of District Officer. The SOC needs to convene to discuss the issues in the oil spill, the response in the oil spill, and, indeed, to certify the readiness and adequacy of equipment stockpiles, manpower, and financial capacity [22].

3.2.17 *Tier 3—International Response*

Meanwhile, Tier 3 is activated to react when the oil spill incidents have been spread out to the other neighborhood shorelines. It is a larger spill that requires substantial additional resources where assistance from national or global collaborative stockpiles might be essential. This response has been designed by the National Committee of Oil Spill Control chaired by the Director General of Environment. Tier 3, as shown by Fig. 3.2, involves the response to regulate oil spill from spreading out to other neighborhood territories, in which it requires resources and expert corporation from the affected neighborhood.

Table 3.5 Range of discharged oil and tier levels [13, 32]

Tier level	Amount of spilled oils
Tier 1	Discharged less than 50 ton
Tier 2	Discharged more than 50 ton
Tier 3	To be announced

The standard of the international response is led by the National Committee of Oil Spill Control based on the regionally developed contingency procedures which include the following [22]

(a) ASEAN-Oil Spill Response Action Plan (ASEAN-OSRAP),
(b) Standard Operating Procedure for Joint Oil Spill in The Straits of Malacca and the Straits of Singapore,
(c) Standard Operating Procedure for Malaysia and Brunei Darussalam, and
(d) Cooperative Network for Oil Spill Countermeasures in the Lombok/Makassar Straits and the Celebes Sea.

3.2.18 Range of Ton of Oil Spill and Category of Tier Indication

Based on the information received from the Marine Department and Department of Environment (DOE) on the oil spill incidents in Malaysia, the tier announcement could be developed as referring to this range as shown in Table 3.5. This is consistent with the IMO MEPC.1 Circle 318 which purpose is to deliver IMO with a synopsis of emissions (Article 2(3)) not allowed below the requirements of MARPOL 73/78 and contamination anticipated to fatalities to vessels [22–24].

3.2.19 Influencing Factors in Tiered Preparedness and Response

Conventionally, the tiered preparation and response is functioning established on the magnitude and position of the possible oil spill. However, there are influencing factors that could vary accordingly to locations, operations, and perception of the operators, government authorities, agencies, and other stakeholders. There are four major categories of factors, each affecting the way response's capabilities such as; operational factors, setting factors, response capability factors, and legislative factors as shown in Fig. 3.3.

In responding to the oil spill, it is to certify that all possible hazards are minimal and possible in considering actions to alleviate the enduring effects. The incidents which could take the lead to the oil spill may be discovered in which indicative circumstances can be established and classified in terms of probability of happening

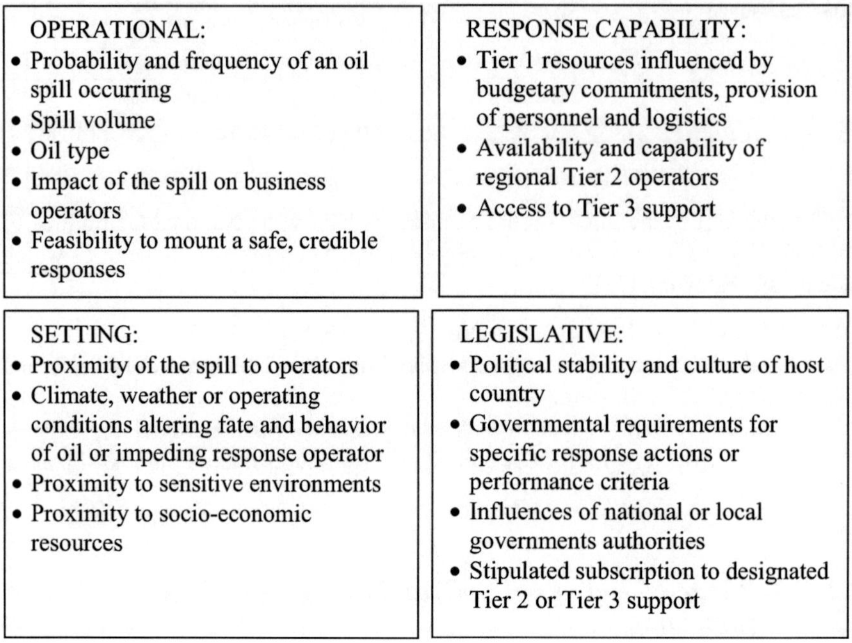

Fig. 3.3 Influencing factors in oil spill response capability needed

and possible effect by getting into consideration operating and establishing considerations. Besides that, it is also useful to utilize with any of the related integration of the oil spill modelling with the coastal cause resource information for protecting the environment from the oil spill incidents [27].

3.3 Conclusion

As usual, once the oil spill has occurred, it requires high cost and attempts to clean up the disturbed areas. Thus, reviewing the effectiveness of the preparedness and reaction concerning the oil spill incidents is crucial to reduce the cost after the incidents. In Malaysia, the two main primary important government agencies intricate in the oil spill occurrences are the Department of Environment and the Marine Department concerning the oil spill, with the main private key player such as PIMMAG and associated response teams to ensure the successful command and control management of oil spill. The DOE and the Marine Department together with another key player will collaborate and associate as a team in combating the oil spill incidents in Malaysia [13]. The oil spill occurrences not necessarily occurred from the oil, but also occurred from chemicals as the other substances that exist the threat to the environment. Thus, the power of planning for hazard mitigation encourages to revise the related oil spill federal policies to help build local understanding of the oil spill risk, commitment to hazard mitigation and continue support for the oil spill response planning [20]. Finally, with good collaboration among the oil and gas industry is may help to encourage in responding to the specific challenge in sustainability development of the human rights promotion in responding to the oil sill incidents [34].

Acknowledgements This paper was written on the collaboration obtained from the associated government agencies associated with the contingency plan regarding the oil spill preparation and response in Malaysia. This study was financially supported by the Universiti Kuala Lumpur Malaysian Institute of Marine Engineering Technology (UniKL MIMET). Undoubtedly, with no cooperation conferred on or after all members especially from En Mohd Fairoz Rozali, the paper could not be productively proficient.

References

1. Al-Majed AA (2012) Adebayo A R & Hossain M E. A sustainable approach to controlling oil spills. J Environ Manage 113:213–227
2. Ajide O M (2013) An Assessment of the Physical Impact of Oil Spillage Using GIS and Remote Sensing Technologies: Empirical Evidence from Jesse Town, Delta State, Nigeria. 12(II): 235–252
3. Azila A (2021) Statistic Oil Spill, Department of Environment (DOE), Putrajaya. https://www.doe.gov.my/portalv1/en/. Accessed 23 June 2021
4. Balogun A L, Matori A N & Kiak K W (2018) Developing An Emergency Response Model For Offshore Oil Spill Disaster Management Using Spatial Decision Support System (SDSS). ISPRS J. Photogramm. Remote Sens. & J. Spat. Inf. Sci. 4(3): 21–27
5. Canu DM, Solidoro C, Bandelj V, Quattrocchi G, Sorgente R, Olita A, Cucco A (2015) Assessment of oil slick hazard and risk at vulnerable coastal sites. Mar Pollut Bull 94(1–2):84–95
6. Chow P, Wat B, Molumo M, Law A & Walker S (2017) Environmental Factors of BC Wildfires, University of British Columbia
7. Craveiro N, De A A R V, Da S J M, Vasconcelos E, Alves-Junior D A F & Rosa F J S (2021) Immediate effects of the 2019 oil spill on the macrobenthic fauna associated with macroalgae on the tropical coast of Brazil. Mar. Pollut. Bull. 165: 112107
8. Commoner B (2020) The closing circle: nature, man, and technology. Courier Dover Publications
9. Chung SY, Lee G (2016) Combating oil spill accidents in Northeast Asia: A case of the NOWPAP and Hebei Spirit oil spill. Mar Policy 72:14–20. https://doi.org/10.1016/j.marpol.2016.06.014
10. DeLeo DM, Ruiz-Ramos DV, Baums IB, Cordes EE (2016) Response of deep-water corals to oil and chemical dispersant exposure. Deep-Sea Res Pt II(129):137–147
11. Demopoulos A W & Strom D G (2012) Benthic community structure and composition in sediment from the northern Gulf of Mexico shoreline, Texas to Florida. US Department of the Interior, US Geological Survey
12. Doshi B, Repo E, Heiskanen JP, Sirviö JA, Sillanpää M (2017) Effectiveness of N, O-carboxymethyl chitosan on destabilization of Marine Diesel, Diesel, and Marine-2T oil for oil spill treatment. Carbohydr Polym 167:326–336
13. Fairoz R (2019) Perkongsian Maklumat Dengan Pihak Jabatan Alam Sekitar. https://www.marine.gov.my/. Accessed 25 June 2021
14. Fairoz R (2021) Perkongsian Maklumat Dengan Pihak Jabatan Alam Sekitar. https://www.marine.gov.my/. Accessed 25 June 2021

15. Farrington J W (2014) Oil pollution in the marine environment II: fates and effects of oil spills. Environment: Science and Policy for Sustainable Development, 56(4): 16–31
16. Fingas M (2013) The basics of oil spill cleanup. CRC Press
17. Fingas M & Brown C (2012) Oil Spill oil spill Remote Sensing oil spill remote sensing. Encyclopedia of Sustainability Science and Technology: 7491–7527
18. Fingas M (2011) Oil Spill Dispersants: A Technical Summary. Elsevier Inc., Oil Spill Science and Technology. https://doi.org/10.1016/B978-1-85617-943-0.10015-2.Accessed23June2021
19. Fingas M, Fieldhouse B (2012) Studies on water-in-oil products from crude oils and petroleum products. Mar Pollut Bull 64(2):272–283
20. Godschalk D R (1991) Disaster mitigation and hazard management. Emergency management: Principles and practice for local government: 131–160
21. Gill DA, Picou JS, Ritchie LA (2012) The Exxon Valdez and BP oil spills: A comparison of initial social and psychological impacts. Am Behav Sci 56(1):3–23
22. Halimah Hassan (2014) Rancangan Kontingensi Kebangsaan Kawalan Tumpahan Minyak (RKKKTM), Jabatan Alam Sekitar, Putrajaya. https://enviro2.doe.gov.my/ekmc/wp-content/uploads/2016/08/1422257438-RKKKTM%202014-1.pdf. Accessed 23 June 2021
23. International Maritime Organization (IMO) (2012) International Shipping Facts and Figures. Information resources on trade, safety, security, environment
24. International Maritime Organization (2021). https://www.imo.org/. Accessed 23 June 2021
25. Ishak I C, Ab R W M H W, Ismail S B & Mazlan N (2020) A Study of Oil Spill at Marine Companies: Factors and Effects. In Advancement in Emerging Technologies and Engineering Applications: 1–12
26. Kanjilal B (2015) Enhanced Marine Oil Spill Response Regime for Southern British Columbia. Canada. Aquat. Procedia 3:74–84
27. Kankara RS, Arockiaraj S, Prabhu K (2016) Environmental sensitivity mapping and risk assessment for oil spill along the Chennai Coast in India. Mar Pollut Bull 106(1–2):95–103
28. Lin Q, Mendelssohn IA, Graham SA, Hou A, Fleeger JW, Deis DR (2016) The response of salt marshes to oiling from the Deepwater Horizon spill: Implications for plant growth, soil surface-erosion, and shoreline sustainability. Sci Total Environ 557:369–377
29. Montewka J, Weckström M, Kujala PA (2013) probabilistic model estimating oil spill clean-up costs–case research for the Gulf of Finland. Mar Pollut Bull 76(1–2):61–71
30. Mustafa M (2019) Environmental law in Malaysia. Kluwer Law International BV
31. Narimannejad A, Bidhendi GN, Hoveidi H (2015) Emergency response management native model to reduce environmental impacts using by analytic hierarchy process (AHP). Bulg Chem Commun 47:250–258
32. Norazaimah A (2019) Statistic Oil Spill, Department of Environment (DOE). Putrajaya. https://www.doe.gov.my/. Accessed 22 June 2021
33. Nordquist M, Nandan S N & Kraska J (1995) UNCLOS 1982 commentary
34. Owens J, Sykes R (2005) The International Petroleum Industry Environmental Conservation Association social responsibility working group and human rights. Int Soc Sci J 57:131–141
35. Ordinioha B, Brisibe S (2013) The human health implications of crude oil spills in the Niger delta, Nigeria: An interpretation of published studies. Niger J Med 54(1):10
36. Oxford University Press (2019) Oxford Dictionaries at http://oxforddictionaries.com. Accessed 22 June 2021
37. Oyebamiji M, Mba C (2014) Effects of oil spillage on community development in the Niger Delta region: Implications for the eradication of poverty and hunger (Millennium Development Goal One) in Nigeria. World 1(1):27–36
38. Potters G (2013) Marine pollution. Bookboon
39. Sagerup K, Nahrgang J, Frantzen M, Larsen LH, Geraudie P (2016) Biological effects of marine diesel oil exposure in red king crab (Paralithodes camtschaticus) were assessed through water and foodborne exposure experiment. Mar Environ Res 119:126–135
40. Sativale, M A. (2015) Malaysian Maritime Zones. https://www.sativale.com.my/Practice_Areas/Malaysian_Maritime_Zones.php. Accessed 23 June 2021

41. Sholeye O, Salako A, Ayankoya S (2012) Oil spills and community health: Implications for resource limited settings. J. Toxicol. Environ. Health Sci. 4(9):145–150
42. Shi X, Wang Y, Luo M, Zhang C (2019) Assessing the feasibility of marine oil spill contingency plans from an information perspective. Saf Sci 112:38–47
43. Sydnes AK, Sydnes M (2013) Norwegian-Russian cooperation on the oil-spill response in the Barents Sea. Mar Policy 39(1):257–264. https://doi.org/10.1016/j.marpol.2012.12.001
44. Trustees D N (2016) Deepwater Horizon oil spill: Final programmatic damage assessment and restoration plan and final programmatic EIS. Deepwater Horizon
45. Tuler S, Seager T P, Kay R, Linkov I & Figueira J R (2007) Defining and selecting objectives and performance metrics for oil spill response assessment: A process design integrating analysis and deliberation. Durham, NH: Coastal Response Research Centre : 1–68
46. Van Dorp JR, Merrick JR (2011) On a risk management analysis of oil spill risk using maritime transportation system simulation. Ann Oper Res 187(1):249–277
47. Wahi R, Chuah LA, Choong TSY, Ngaini Z, Nourouzi MM (2013) Oil removal from the aqueous state by natural fibrous sorbent: an overview. Sep Purif Technol 113:51–63
48. Hy W L, W R J (1990) Handbook of emergency management: programs and policies dealing with major hazards and disasters. Greenwood Publishing
49. Wing C K (2005) Oil spill response management and transboundary issues in Malaysia. In SPE Asia Pacific Health, Safety and Environment Conference and Exhibition. Society of Petroleum Engineers. https://onepetro.org/SPEAPHS/proceedings-abstract/05APHS/All-05APHS/SPE-96480-MS/89385 Accessed 23 June 2021
50. Yekeen ST, Balogun AL, Yusof KBW (2020) A novel deep learning instance segmentation model for automated marine oil spill detection. ISPRS J Photogramm Remote Sens 167:190–200
51. Yuan Y, Detlor B (2005) Intelligent mobile crisis response systems. Commun ACM 48(2):95–98
52. Zengel S, Pennings S C, Silliman B, Montague C, Weaver J, Deis D R & Nixon Z (2016) Deepwater Horizon oil spill impacts on salt marsh fiddler crabs (Uca spp.). Estuaries Coast Copy, 39(4): 1154–1163
53. Zhang C, Han L, Shi X (2015) Modified Assessment Methodology for Mechanical Recovery Capacity for Oil Spill Response at Sea. Aquat. Procedia 3:29–34. https://doi.org/10.1016/j.aqpro.2015.02.224
54. Zock JP, Rodríguez-Trigo G, Rodríguez-Rodríguez E, Espinosa A, Pozo-Rodríguez F, Gómez F, Barberà JA (2012) Persistent respiratory symptoms in clean-up workers 5 years after the Prestige oil spill. Occup Environ Med 69(7):508–513

Chapter 4
Simulation of Three-point Bending Sandwich Composite Panels Through Finite Element Analysis

Zulzamri Salleh and Goh Wei Kee

Abstract In marine industry, the usage of sandwich composite structures has been increased due to these structure can provide strong properties that the manufacturer desires. The useful properties of these structure are not only useful in marine industry but also in other industries such as aviation, automotive and aerospace industries. The main propose of this study is to provide useful simulation testing results for the future references by comparing the results of three-point bending tests through experimental and finite element analyses of sandwich composite structures. This research is conducted with fiberglass and a core material which is carbon from pine wood with different weight percentage (10, 20,30, 40 and 50 wt%). It was found that the sandwich panel with higher content of pine wood carbon 50 wt% has achieved the maximum stress 25.268 kPa and strain of 3.184×10^{-7} mm/mm. When the percentage of the pine wood-activated carbon increased or the core's diameter increased, all the values of maximum stress, maximum strain and maximum displacement also have been increased. This has also shown that specimens with higher percentage of the pine wood-activated carbon or the core's diameter can withstand more stress before the specimen rupture.

Keywords Three-point bending · Pine wood · Stress–strain

4.1 Introduction

According to American Society of Testing and Materials (ASTM Composite Standard), the sandwich composite structure concept is explained as follows [1]:

> A structural sandwich is a special form of composite comprising of a combination of different materials that are bonded to each other so as to utilize the properties of each separate component to the structural advantage of the whole assembly.

Z. Salleh (✉) · G. W. Kee
Universiti Kuala Lumpur Malaysian Institute of Marine Engineering Technology, Lumut, Perak, Malaysia
e-mail: zulzamri@unikl.edu.my

A. Ismail et al. (eds.), *Design in Maritime Engineering*, Advanced Structured Materials 167, https://doi.org/10.1007/978-3-030-89988-2_4

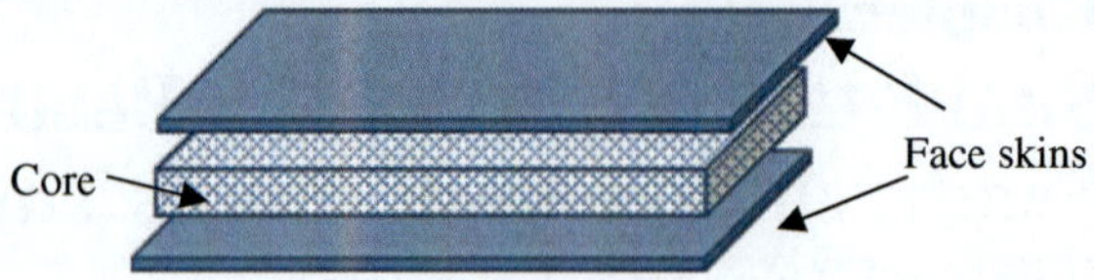

Fig. 4.1 Sandwich composite structure [1]

A sandwich composite structure usually consists of two thin layers known as the facing sheets and a core between the layers, so that the structure can have a maximum mechanical properties (see Fig. 4.1). In order to complete a sandwich structure, the core must be glued to the two facing sheets and forming two additional glued sides. To form a good quality of sandwich composite structure, the facing sheets must be able to withstand all the applied loads and bending moment and the core must provide the shear strength needed to the structure. So, thin facing sheets are enough to obtain great stiffness, and a lightweight core which is able to withstand the shear forces and able to make the two facing sheets that attached to it being stable.

By identifying the type of facing sheets and core used, the types of sandwich composite structures can be classified. For the type of core used, sandwich composite structures can be classified into three types, that is, the honeycomb core sandwich composite, foam-type core sandwich composite and corrugated core sandwiches. The facing sheets can be divided into two types, metallic and non-metallic sheets. The metallic facing sheets are made of, for example, aluminum, stainless steel and titanium. Non-metallic facing sheets usually use fiber-reinforced plastics (FRP) with different resins like epoxy and polyester. The types of core and facing sheets chosen to use will depend on the application of the sandwich composite structure [1].

Nowadays, the demand of sandwich composite structures in the industry becomes higher (see Fig. 4.2). This requires the quality of the sandwich composite structure also being high enough. Thus, the core materials must have low density, sufficient thermal insulation, sufficient thickness, sufficient shear stiffness, sufficient stiffness to prevent local buckling and sufficient shear strength to prevent core shear failure. The facing materials must have high tensile and compressive strength, good impact resistance, good water resistance, good environmental resistance, good surface finish and high stiffness and high flexural rigidity [1].

4.2 Methodology

The three-point bending test is a test that is performed on the materials or products to obtain their ultimate flexural load, which is the maximum transverse force that the materials or products can tolerate without permanent damage. This test is commonly carried out by a universal testing machine (UTM) with three-point or four-point bend fixture. The three-point bending test is the one of the most common procedures for products testing (see Fig. 4.3). The main objective to perform this bending testing is to gain the following data [3]

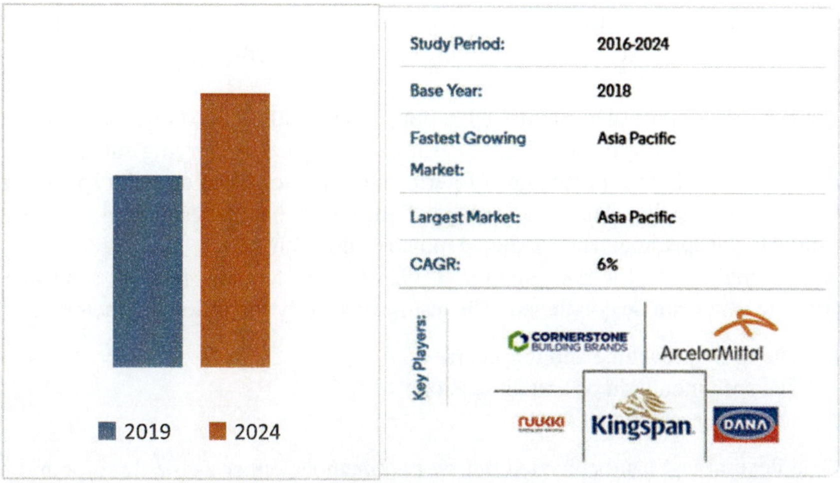

Fig. 4.2 Sandwich panels market—growth, trends and forecast (2019–2024) [2]

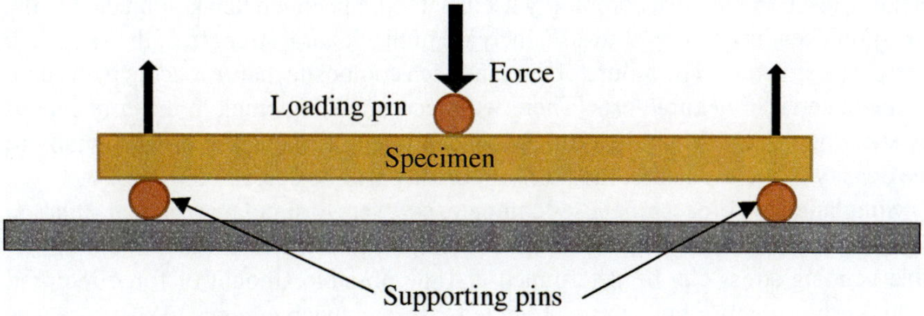

Fig. 4.3 Basic set-up for three-point bending test

(a) Flexural strength: Maximum load a material can withstand before its failure.
(b) Flexural modulus: Degree of slope of stress or strain curve. This is to indicate the material's stiffness.
(c) Yield point: The point where a material no longer displays elastic behavior. The material has become plastic, and from this point ahead, the further increase of loading will cause permanent defect on the material.

As stated by Hansdah, Tata Steel (2012), in engineering mechanics, materials that are used for flexural tests are [4]:

(a) Composite sandwich materials (e.g., carbon fiber for car's spoiler).
(b) Woods (e.g., walnut for propeller).
(c) Metals (e.g., aluminum 7075 in aircraft's wing).
(d) Brittle materials (e.g., ceramic for construction).

A specimen with rectangular shape and flat cross-section is placed on two parallel supporting pins in this test. A mean loading pin is applied in the middle of the loading force (see Fig. 4.3).

Values for modulus of elasticity in bending, flexural stress, flexural strain and the flexural stress–strain response of the material can be gained by carrying out the three-point bending test. The advantages of performing this test is to ease the specimen preparation and testing. The disadvantage of this test is that the result gained is very sensitive to the specimen and loading geometry and strain rate.

According to ASTM Designation C393-00, the flexural test of sandwich composite panel can be conducted. The test must specify the following factors [5]:

(a) The testing machine and its accuracy
(b) The specimen's shape and dimension
(c) Analyzing result's method.

The flexural test can be carried out by two methods either as the three-point or as the four-point bending method. The four-point bending test is due to there is a lack of destructive influence of the pressure stamp on the top part of the specimen tested. Because of this disadvantage of the four-point bending test, the three-point bending test became most commonly used method. Specimen that is subjected to the four-point bending testing shows an increase of the flexural strength of the sandwich composite structure. The failure of the sandwich composite mainly occurs in the core of the material in flexural tests. There will occur a sliding crack in the core caused by shearing forces. When the ultimate shear strength is increased proportionally to the density of the used core, the level of porosity will be decreased.

Simulation is a process that uses computer software to simulate an actual situation by using experimental data. By using simulation, the structure's maximum allowable bending stress can be determined and the possible amount of force required to make the structure fail can be predicted. The sandwich composite structure was designed using a design software that is SpaceClaim with the specification stated in the American Society of Testing and Material Designation C393-00. Then, testing was performed testing by using finite element analysis simulation computer software, that is, CREO simulation. Comparison of results obtained from simulation computer software was made.

The five models were designed by using the SpaceClaim software by following the dimensions that are stated in ASTM C-393(00) (ASTM 2000) (0) which is 10 wt% activated carbon (core: 4.0 mm), 20 wt% activated carbon (core: 4.5 mm), 30 wt% activated carbon (core: 5.0 mm), 40 wt% activated carbon (core: 5.5 mm) and 50 wt% activated carbon (core: 6.0 mm). After completing the design, the design was imported to the CREO software to run the simulation testing. Before running the simulation testing, there are a few steps that need to set. The first step is setting the load and constrains (see Fig. 4.4). The second step is setting for the specimen the material which is CF-FG and soft pine wood (same properties with mangrove wood-activated carbon) (see Fig. 4.5). The third step is to create a refined mesh of all the surface of the specimen (see Fig. 4.6). The fourth step is running the AutoGem

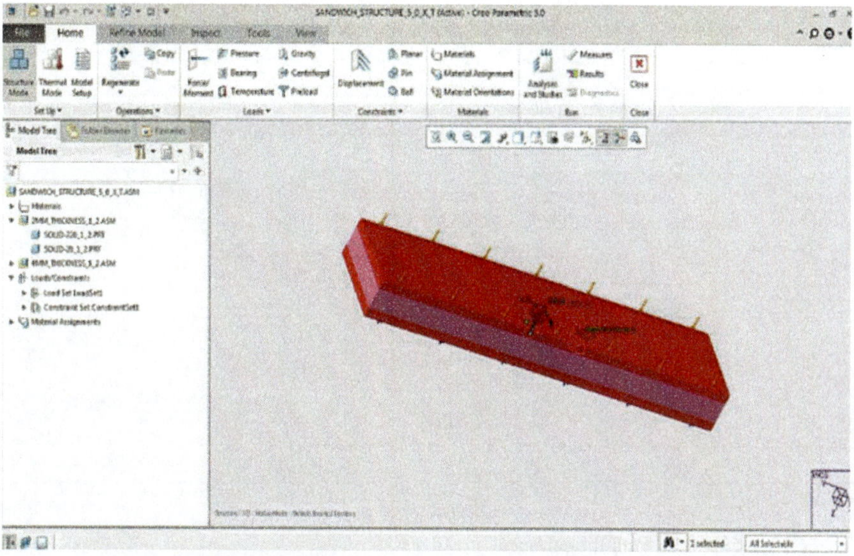

Fig. 4.4 Setting the load and constrains

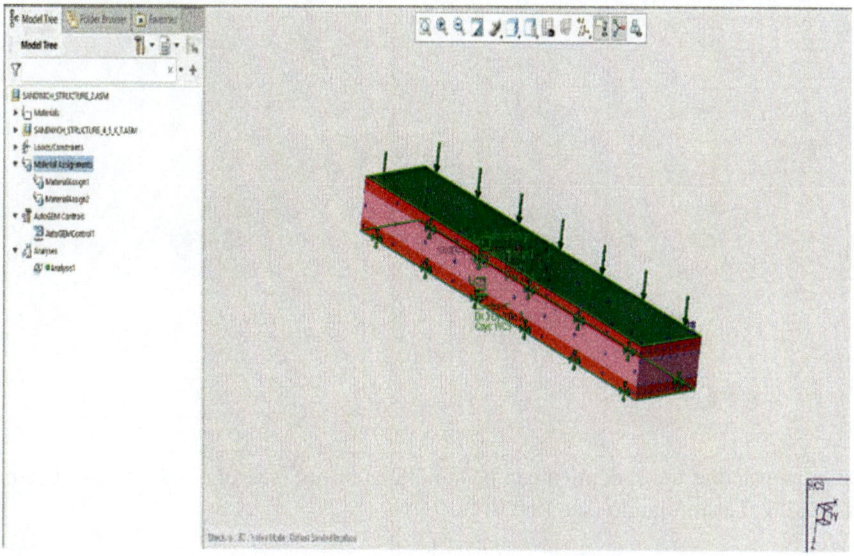

Fig. 4.5 Assigning material properties

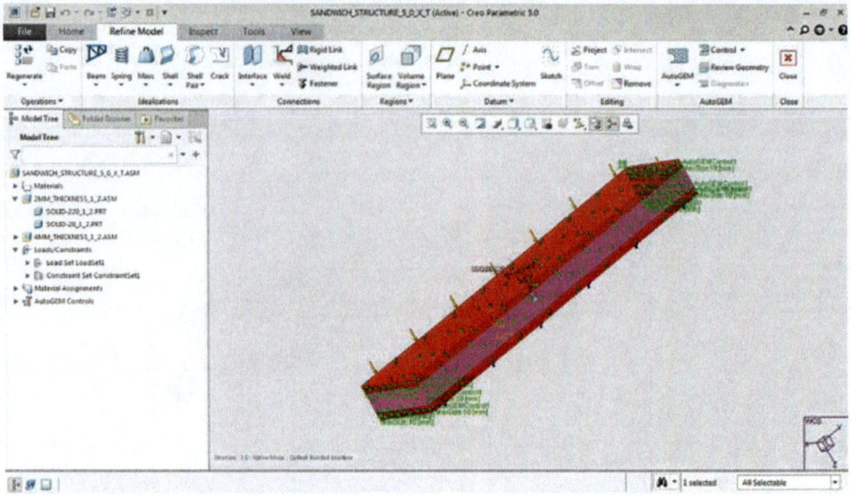

Fig. 4.6 Refined mesh of all the surfaces

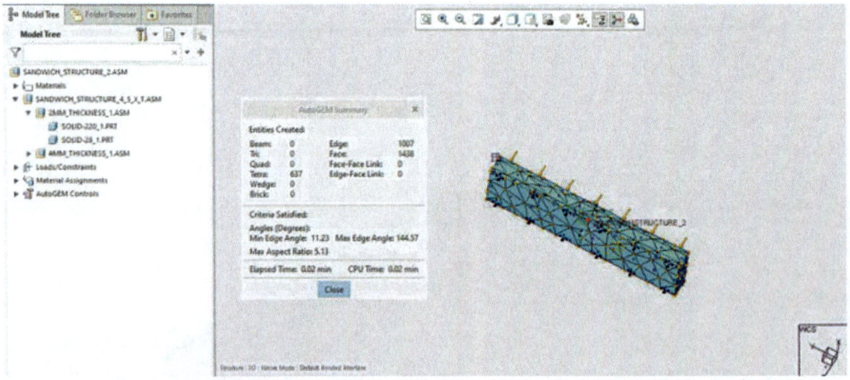

Fig. 4.7 Running the AutoGem

to know whether the specimen can be simulated or not (see Fig. 4.7). The last step is running the simulation (see Fig. 4.8).

4.3 Results and Discussion

In the CREO simulation software, there are three types of data or results that can be obtained which are stress, strain and displacement. All the data or results for each specimen was collected and tabulated.

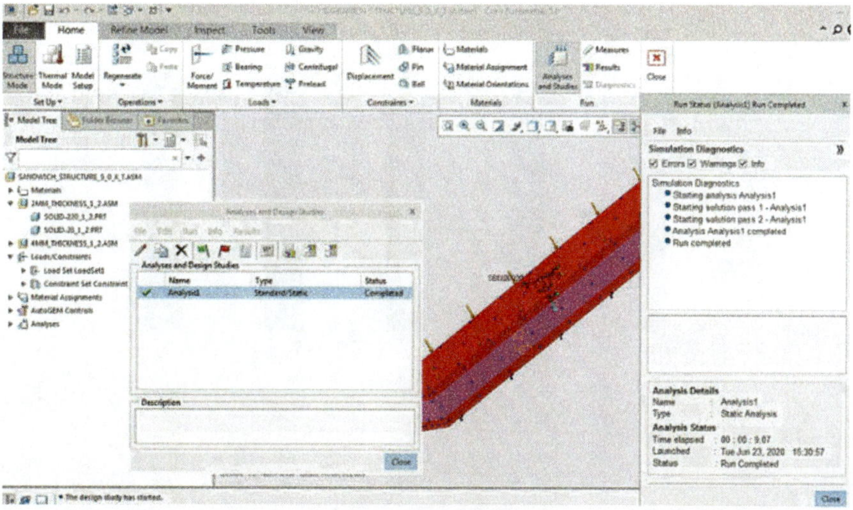

Fig. 4.8 Running simulation

Specimen 1 = core with 10 wt% activated carbon (core: 4.0 mm).
Specimen 2 = core with 20 wt% activated carbon (core: 4.5 mm).
Specimen 3 = core with 30 wt% activated carbon (core: 5.0 mm).
Specimen 4 = core with 40 wt% activated carbon (core: 5.5 mm).
Specimen 5 = core with 50 wt% activated carbon (core: 6.0 mm).

It can be concluded from Figs. 4.9 and 4.10 that there are differences in the values of stress, strain and displacement between all the five specimens. When the percentage of the mangrove wood-activated carbon increases or the core's diameter increases, all the value of maximum stress, maximum strain and maximum displacement

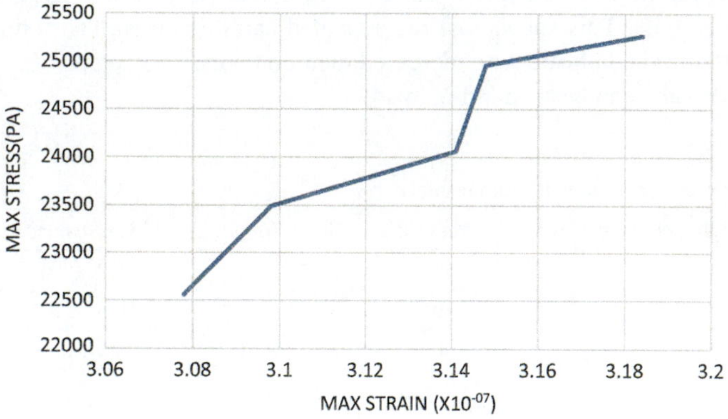

Fig. 4.9 Stress–strain diagram

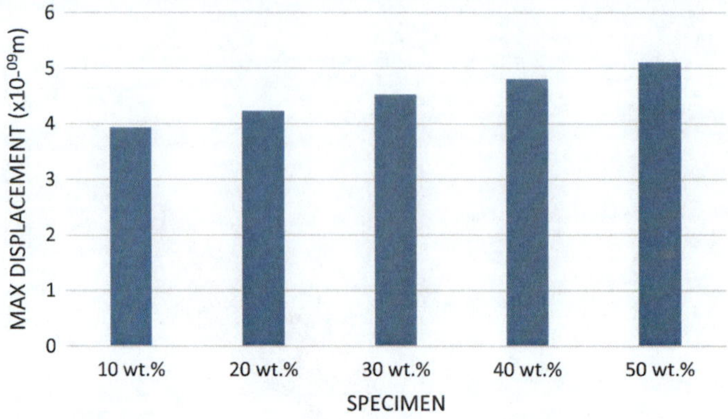

Fig. 4.10 Displacement diagram

also increase. This shows that a specimen with higher percentage of the mangrove wood-activated carbon or the core's diameter, the more stress that the specimen can withstand before the specimen's rupture. Other than that, it can be concluded that by increasing the mangrove wood-activated carbon or the core's diameter, the mechanical properties of the specimen also increase. Since the mechanical properties increase, the specimen with higher mangrove wood-activated carbon or the core's diameter is more suitable for marine industry application. Based on the data gained from the simulation, both with the highest carbon content (6 mm) recorded the highest stress and strain. For the double layer with 50 wt% carbon content, the maximum stress is 25.27 kPa, while for the sample with 10 wt%, the minimum strength is 22.56 kPa. The details of the results are shown at Table 4.1.

The strain the sample with 50 wt% carbon content also recorded the highest which is 3.184×10^{-7}, while for the double layer, it is 3.078×10^{-7}. This data can be simply used for better understanding that the more layers and higher carbon content of the core, the highest maximum point of stress and strain can withstand before the specimen fails. This shows that the activated carbon content can influence the fiberglass-reinforced plastic with increased their contents. The details of the sample simulation results can be seen in Fig. 4.11.

Table 4.1 Details of stress and strain results

Sample weight percentage (wt%)	Stress (kPa)	Strain (mm/mm)	Max. displacement (m)
10	22.56	3.078×10^{-7}	3.913×10^{-9}
20	23.49	3.098×10^{-7}	4.225×10^{-9}
30	24.06	3.141×10^{-7}	4.515×10^{-9}
40	24.96	3.148×10^{-7}	4.799×10^{-9}
50	25.27	3.184×10^{-7}	5.102×10^{-9}

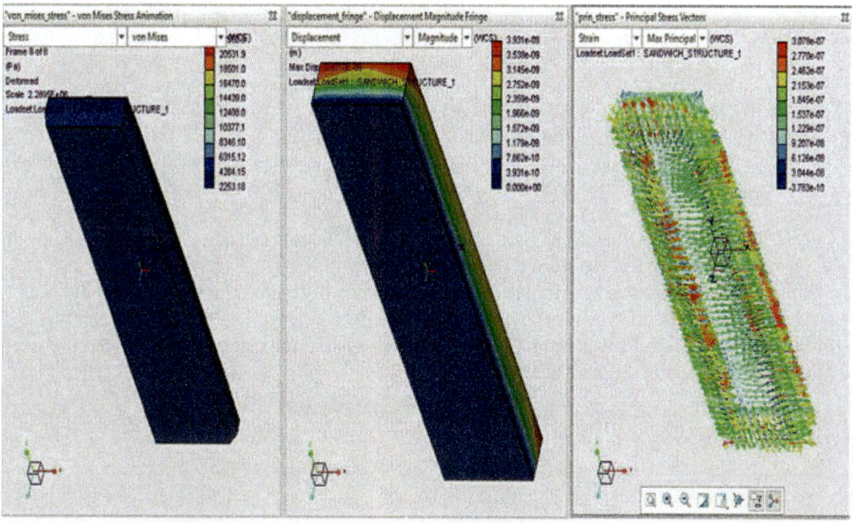

Fig. 4.11 Sample 1 simulation flexural testing result

4.4 Conclusion

In conclusion, this research was carried out to explore the relationship between stress and strain by using finite element analysis. The research was done by FEA method using the PTC CREO software. From the data gained, it will ease the production team to study the properties of the material. Overall, the objectives are achieved while conducting this project. Firstly, the design of the sandwich structure followed the ASTM C393-00 for double layer and triple layer. Second, the analysis of data gained from FEA allows to decide the best mechanical properties.

The mechanical behavior of the sample shows to be enhanced with increasing weight percentage of activated carbon powder. In the experimental study, the specimen with 50 wt% (10 mm) was able to withstand about 25.2682 kPa of stress, while for the sample with 10 wt%, the limit of the stress was 22.56 kPa.

For the displacement, the sample with 50 wt% recorded 5.102×10^{-9} mm the highest, while for sample with 10 wt% showed a displacement of 3.913×10^{-9} mm. However, it still experiences in the limitation and deficiencies such as low of impact strength, low in tensile strength, brittleness and low conductivity and others. The stress experienced on the specimen is different based on the percentage of carbon content in the specimens.

Acknowledgements This research has been conducted during COVID-19 pandemic by our student final year project at UniKL MIMET.

References

1. Nallagula S (2006) Behavior and flexure analysis of Balsa wood core sandwich composites: experimental, analytical and finite element approaches. Eng Struct 31(9):2032–2044
2. Wong YC, Mahyuddin N, Aminuddin AMR (2020) Development of thermal insulation sandwich panels containing end-of-life vehicle (ELV) headlamp and seat waste. Waste Manage 118:402–415
3. Lokesh KS (2018) Synthesis and study on effect of thickness on 3-point bending strength of sandwich composites. Int Res J Eng Tech 5(08):2395–2456
4. Sadhinoch M, Kampczyk M, Mulder R (2019) Statistics of fracture in 3-point bending. Mater Sci Eng C 651(1):012075
5. Freitas DST, Kolstein H, Bijlaard F (2011) Sandwich system for renovation of orthotropic steel bridge decks. J Sandw Struct Mater 13(3):279–301

Chapter 5
The Application of Unmanned Aerial Vehicles (UAV) for Slope Mapping with the Determination of Potential Slope Hazards

Muhammad Farhan Zolkepli, Mohd Fakhrurrazi Ishak, and Mohd Sharulnizam Wahap

Abstract In our modern world, the application of small unmanned aerial vehicles (UAV) for monitoring work or slope mapping expanded and is widely used by people in the construction field and researchers. Slope mapping can be considered challenging when using traditional surveying methods since most slopes especially in forest regions are high and considered risky if monitored by human themselves. Other than that, mapping by using UAV need a lower number of manpower to operate the device itself which is more than enough to be conducted by a single person only. This paper discusses the applications of unmanned aerial vehicles for mapping and also its important parameters including perimeter, area and also volume of certain selected area. With the development of modern technology, the utilization of UAV to gather data for geological mapping is becoming easier as it is quick, reliable, precise, cost-effective and also easy to operate. High imagery quality and high-resolution images are essential for the effectiveness and nature of normal mapping output such as digital elevation model (DEM) and also orthoimages. With the help of established software, the parameters of three selected study areas (stockpile, slope A and slope B) can be determined easily which can be considered as one of the main interest in this study. In addition to that, the horizontal and vertical cross section of every selected area can be obtained which help to determine the highest and lowest point of each area. From this cross section, the slope path profile can be determined. Other than that, from this path profile, the potential slope hazard will be determined based on the slope angle (slope classes) as suggested by the United States Department of Agriculture (USDA). Overall, the application of unmanned aerial vehicles for photogrammetry together with slope mapping and slope hazard monitoring can be considered as a reliable modern technology which ease the work with proper assurance of analysis due

M. F. Zolkepli · M. F. Ishak (✉) · M. S. Wahap
Faculty of Civil Engineering Technology, Universiti Malaysia Pahang, Lebuhraya Tun Razak, 26300 Kuantan, Malaysia
e-mail: fakhrurrazi@ump.edu.my

M. S. Wahap
e-mail: shahrulnizam@ump.edu.my

A. Ismail et al. (eds.), *Design in Maritime Engineering*, Advanced Structured Materials 167, https://doi.org/10.1007/978-3-030-89988-2_5

to its advancement and powerful technology. This modern surveying device helps workers and researchers to simplify and fasten their work.

Keywords Unmanned aerial vehicle (UAV) · Slope mapping · Slope parameters · Slope cross section · Digital elevation model (DEM) · Slope hazard

5.1 Introduction

Unmanned aerial vehicles (UAV) or commercially known as drones were first developed during World War I (WWI) in 1918 [1]. During this time, UAV were constructed based on the creation of small planes equipped with cruise missiles (self-flying aerial torpedo) which aim to attack enemies at short distances. UAV were created by two persons known as Orville Wright and Charles Kettering, an electrical engineer. According to [1], further information is known about UAV during World War I in which UAV were developed to attack target points without return (one way mission).

UAV keep developing in recent years. The development of UAV technology over the last decade helps to obtain data of high-resolution and real-time aerial images for mapping, which was considered challenging before for normal airplanes [2–4]. In our modern world, UAV were used usually in work or research projects. Nowadays, the instruments used for data acquisition in geological topography have been rapidly improved [5].

With the development of modern technology, the equipment used to gather all information related to earth surfaces become cheaper, smaller, accurate and can gather large number of data within a short period of time [6–10]. According to [11], the cost needed to operate an unmanned aerial vehicle is lower than operating a manned aerial vehicle. These devices are light, mobile, easily to operate, completely automated and providing access to almost unavailable study areas. Advances in UAV technology have enabled the acquisition of high-resolution and real-time aerial images for photogrammetry [3, 12]. An unmanned aerial vehicle is normally an aircraft that launches and flies without a human onboard [13]. UAV have significantly developed during the last decades. They operate remotely in the form of small platforms, attached completely with camera and available as small or micro-aircrafts [14].

Most UAV can be considered as environmental-friendly (except UAVs for military application) due to the fact that they are powered by electricity and do not emit greenhouse gases directly [15, 16]. In terms of photogrammetry, the UAV technology has led to several economic and environmental advancements. Even with these advancements, there are both advantages and disadvantages in innovation that kept developed as time passed [17]. Recently, the use of unmanned aerial vehicles in research studies and also commercial terms are ending up progressively normal [6, 7]. According to [18], geophysical surveys in mountains and natural terrains are normally challenging due to the site conditions, which may affect the quality of data acquisition. Unmanned aerial vehicles allow for the effectiveness of monitoring which cover large areas of land and infrastructures within a very short time interval compared to conventional techniques [14].

UAV photogrammetry provides information used for image stitching. A modern autopilot system guarantees the planned flight path, camera triggered auto-control to take a picture at every waypoint [6, 7]. As suggested by [19], since UAV can fly below clouds (low altitude), a high capacity of aerial photography can be obtained compared to airplane photogrammetry flights which usually faces the cloudiness. The greatest amount of information is contained in the video signal [20, 21]. According to [22], the use of UAV with remote sensing systems help to carry a point and shoot digital camera for capturing an aerial photograph located at visible land.

The UAV-based imaging which could take very high-resolution images economically in restricted area such as polar regions compared to satellite and aerial photography will be used in a variety of areas [11, 23]. With the help of UAV, the effectiveness of land monitoring together with existing infrastructures can be conducted within a short time period compared to conventional techniques, especially for urgent cases such as natural disasters [24, 25]. The advantages of this photographic method compared to traditional in situ surveys are that it helps to reduce time for large data collections [11, 26, 27].

As suggested by [28], digital elevation models (DEMs) can be considered as one of the most important spatial information tools used by geomorphological applications. This is due to the fact that the DEM allow for the extraction of several attributes (slope, aspect, curvature of profile together with flow direction). According to [29], there are three main sources which may contribute to obtain elevation data: (1) ground surveys [30, 31]; (2) topographical map that are already existed [32–35]; and (3) remote sensing [36–40]. Even though traditional ground survey techniques are very accurate compared to topographic mapping and remote sensing, the time taken will be much longer for data acquisition when the applications require high-resolution DEMs.

Slope monitoring is very important as an early precaution to detect whether a slope has the potential or not in terms of failure. Slope failures are usually found in man-made slope, known as landslides. These kinds of slopes involved cut and fill activities which often occur at highways, residential area and urbanized areas [41–50]. Several examples of slope failure include slope falls, slope topples, landslides, flows and spread of slopes [51]. Detailed orthophotos are very valuable for mapping and classifying morphological phenomena that occur at that point [52].

The aim of this paper is to obtain the geological map of several selected study areas (stockpile, slope A and slope B) located at Kuantan, Pahang by using an unmanned aerial vehicle or commercially known as drone. The photos collected by the UAV during monitoring work will be converted into a map by using the PIX4D mapping software. This map will be presented into two types of analyses which are the digital orthophoto and the digital elevation model (DEM). Other than that, from the analysis, the parameters of these areas will be determined such as perimeter, area, volume and more which are considered as the main interest in this study. Besides, with the help of established software, the vertical and horizontal cross section for this area can be determined. The view from the cross section will show the path profile and terrain of the slope itself. In addition, another focus of this paper is to determine the potential slope hazard based on the slope angle. The angles of the slopes are divided into several classes according to the United States Department of Agriculture (USDA).

5.2 Methodology

The study area is at Kuantan, Pahang. The site area consists of different terrain profiles. This site is completely free from any distraction and obstacle in the air for UAV to freely move around. Figure 5.1 shows the map of the study area.

5.2.1 *DJI Inspire 2*

The DJI Inspire 2 weights around 3.44 kg and is a powerful and high technology drone. This UAV has a speed of 94 km/h which makes it pretty impressive. The maximum ascent speed is 6 m/s in sport mode and the maximum descent speed is 4 m/s. The length of this UAV is 42.7 cm, with a height of 31.7 cm and a width of 42.5 cm. DJI Inspire 2 has a maximum transmission distance of 7 km and is capable to deliver both 1080p and 720p videos. Figure 5.2 shows the image of the DJI Inspire 2. Table 5.1 shows the specification and features of the DJI Inspire 2.

Fig. 5.1 Location of study area

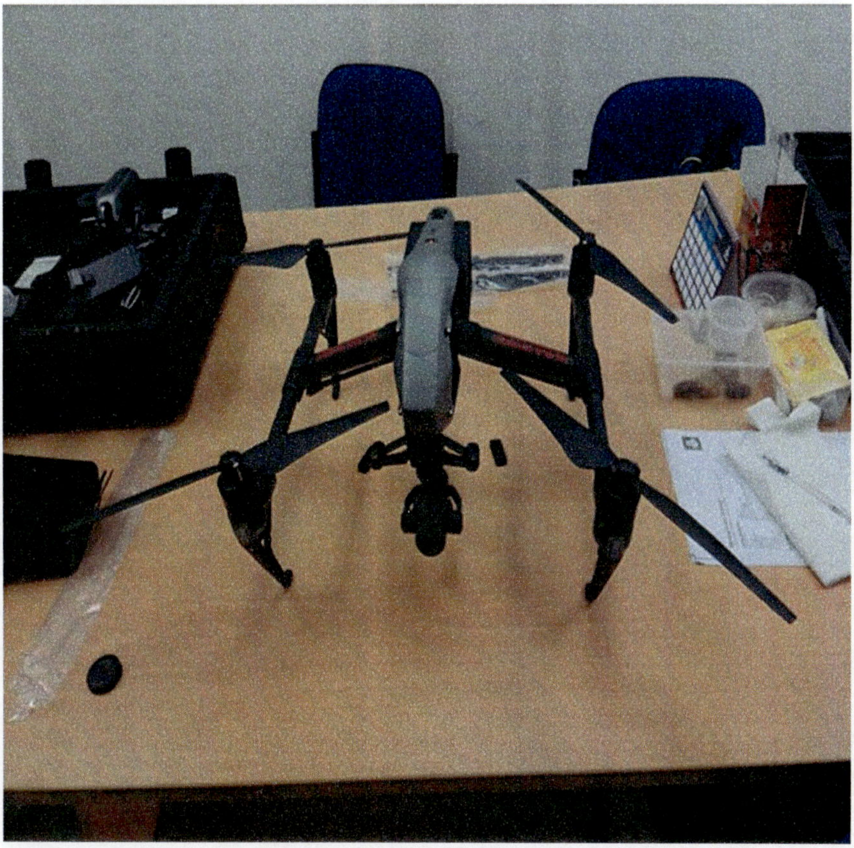

Fig. 5.2 UAV model DJI Inspire 2

Table 5.1 UAV specifications

No.	Parameters	Capability values
1	Flight time (minute)	25–27
2	Speed (km/h)	94
3	Sensory range (m)	30
4	Battery	98 Wh dual battery
5	Raw video recording	Yes
6	Ports	USB and HDMI
7	Control range (km)	7
8	Video resolution	5.2 K and 4 K
9	Live view	1080P
10	Design material	Magnesium aluminium composite shell with carbon fibre arm

5.2.2 *Image Acquisition*

The normal workflow accepted for image acquisition has been used by many researchers and practitioners. Following are the steps for image acquisition as shown

Fig. 5.3 Workflow for data acquisition

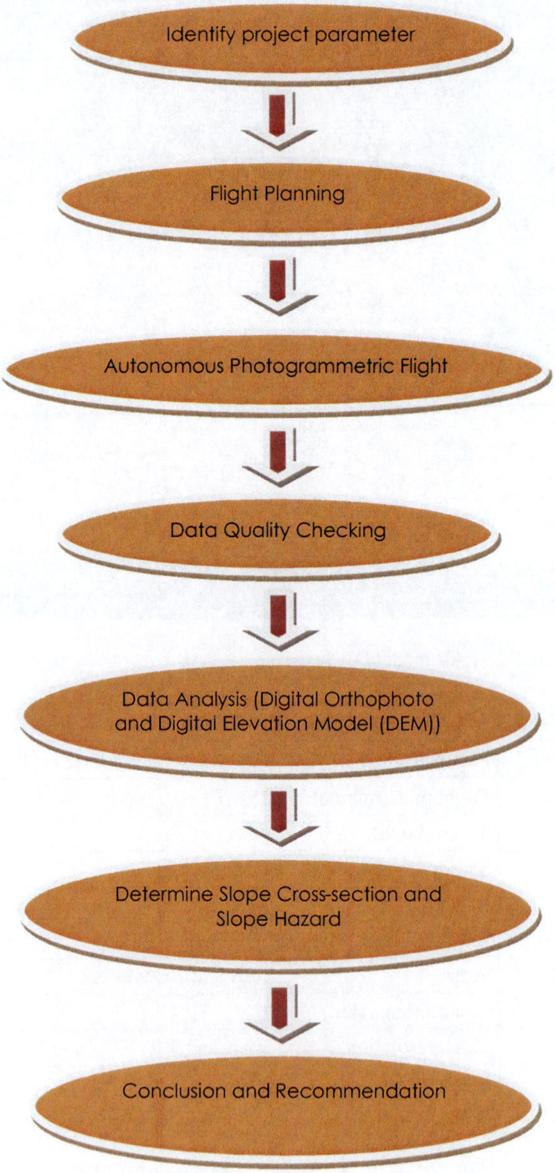

in Fig. 5.3. The results obtained from UAV monitoring will then be transferred into the global mapper software version 18.1 for further analysis.

5.3 Results and Discussion

The results of this study are presented in digital orthophoto and digital elevation model. The ground control points (GCP) are not used in this study as they do not give much error for the results obtained. The independent orthoimages are used to generate the digital orthophoto in the photogrammetric process. Figure 5.4 shows the image of stockpile. Figure 5.5 presents the image of slope A, and Fig. 5.6 shows image of slope B.

5.3.1 Digital Orthophoto and Digital Elevation Model (DEM)

Three slopes were selected (stockpile, slope A and slope B) for further analyses. These slopes were presented in both a digital orthophoto and a digital elevation model.

The application of the unmanned aerial vehicle is not only limited for mapping, but it can also help to determine the important parameters of those areas such as perimeter, area, volume and more. The use of the global mapper software version 18.1 can meet the requirement needed in order to obtain the properties of selected study areas. The measurement of the stockpile, A and B, is shown in Figs. 5.7, 5.8

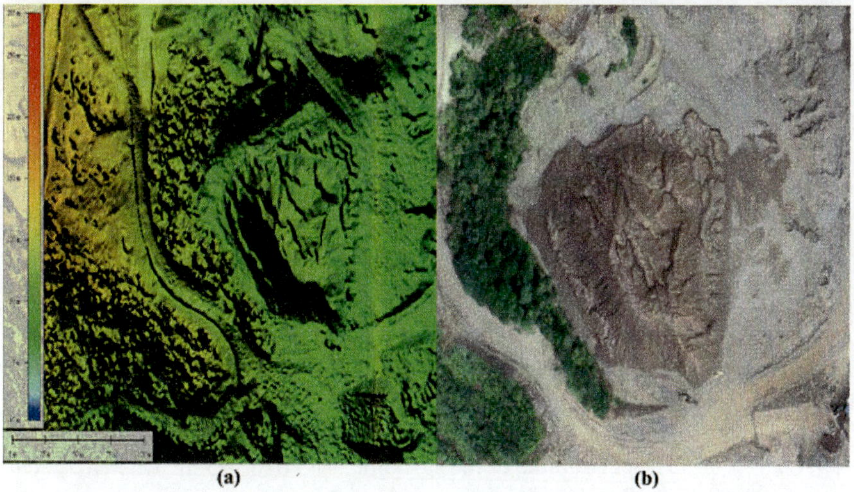

(a) (b)

Fig. 5.4 **a** Digital elevation model (DEM) of stockpile, **b** digital orthophoto of stockpile

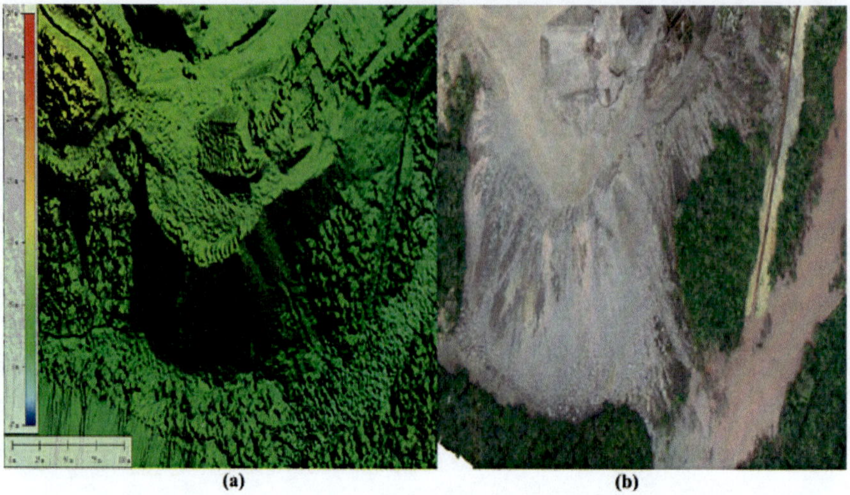

Fig. 5.5 a Digital elevation model (DEM) of slope A, b digital orthophoto of slope A

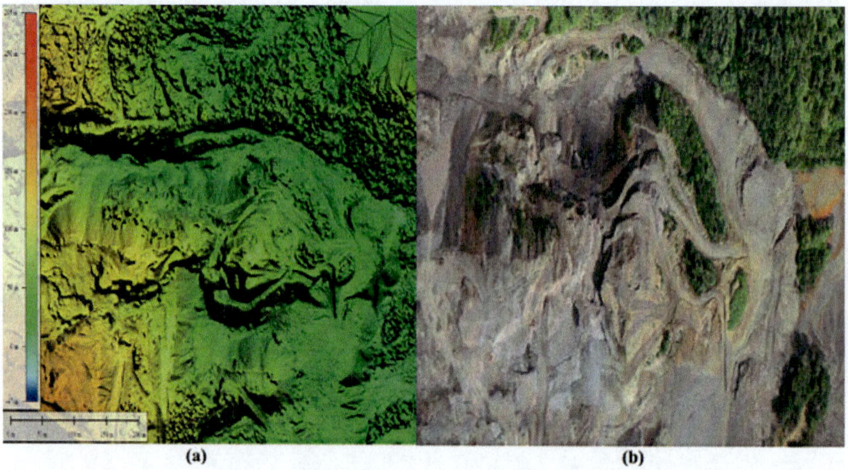

Fig. 5.6 a Digital elevation model (DEM) of slope B, b digital orthophoto of slope B

and 5.9. The dimensions of stockpile, A and B, are presented in Tables 5.2, 5.3 and 5.4. Besides, with the help of the UAV and also the established software, the determination of the profile for each area according to their respective cross section can be done.

From Table 5.2, the total volume of stockpile is 45,799.494 m^3. The net volume is 45,432.792 m^3. The cut volume of the stockpile is 45,616.143 m^3 whereas its cut area is 0.00933 km^2. Besides, the cut area 3D is 0.0108 km^2. The amount of fill volume of this stockpile is 183.35065 m^3 while the fill area is 0.001394 km^2. The

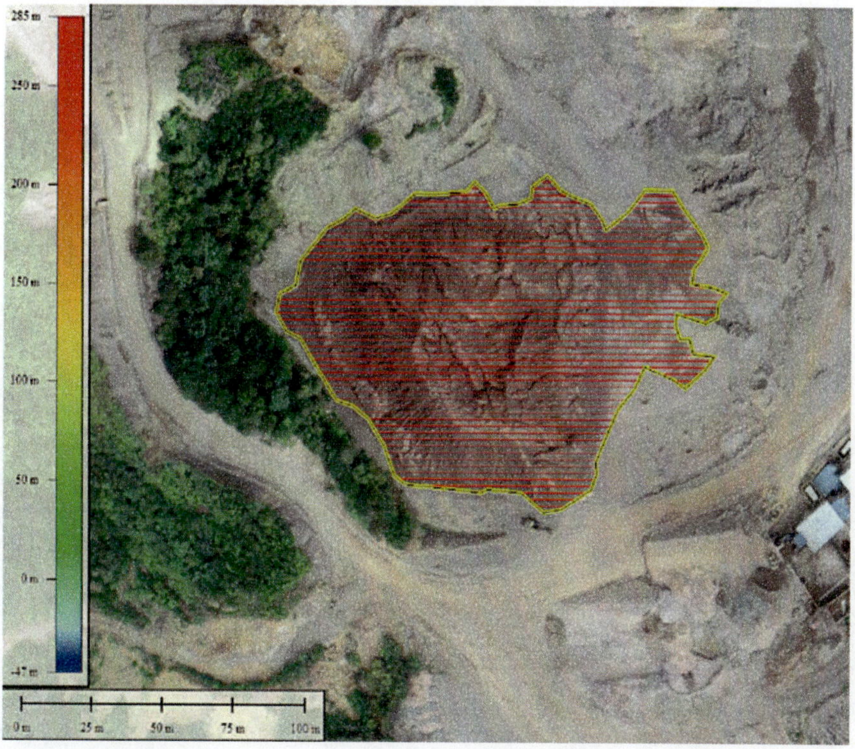

Fig. 5.7 Measurement of stockpile

total fill area 3D is 0.001409 km^2. The enclosed area of stockpile is 0.01069 km^2 and the existing parameter is 499.71 m.

From Table 5.3, the total volume of slope A is 41,028.486 m^3. The net volume is 12,771.773 m^3. The cut volume of the slope A is 14,128.357 m^3 whereas its cut area is 0.00789 km^2. Besides, the cut area 3D is 0.00954 km^2. The amount of fill volume of this slope A is 26,900.129 m^3 while the fill area is 0.00959 km^2. The total fill area 3D is 0.01171 km^2. The enclosed area of slope A is 0.01743 km^2 and the existing parameter is 777.75 m.

From Table 5.4, the total volume of slope B is 465,914.18 m^3. The net volume is 463,347.59 m^3. The cut volume of the slope B is 464,630.89 m^3 whereas its cut area is 0.04129 km^2. Besides, the cut area 3D is 0.057 km^2. The amount of fill volume of this slope B is 1283.2932 m^3 while the fill area is 0.001478 km^2. The total fill area 3D is 0.001916 km^2. The enclosed area of slope B is 0.04264 km^2 and the existing parameter is 1.018 km. Figure 5.10 shows horizontal cross section of stockpile.

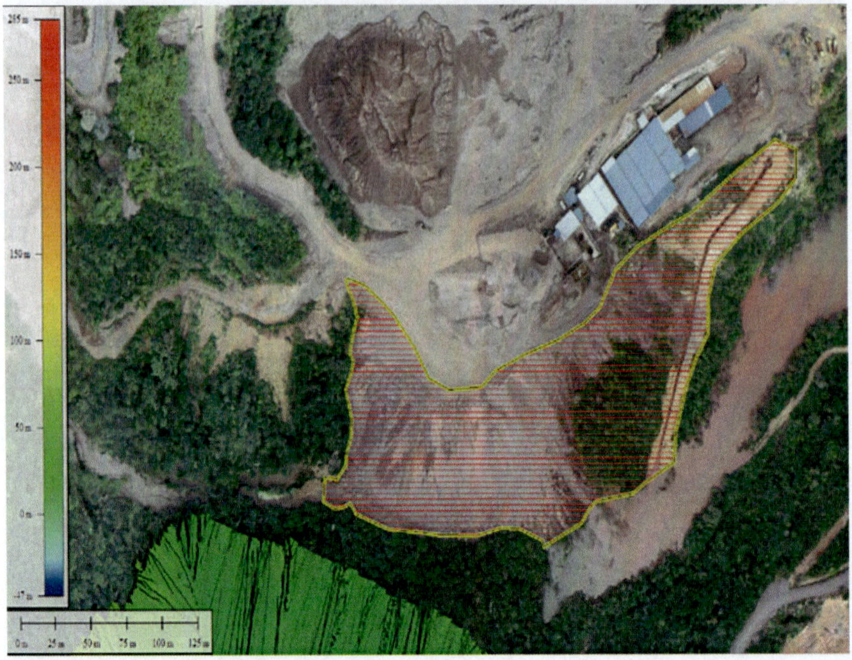

Fig. 5.8 Measurement of slope A

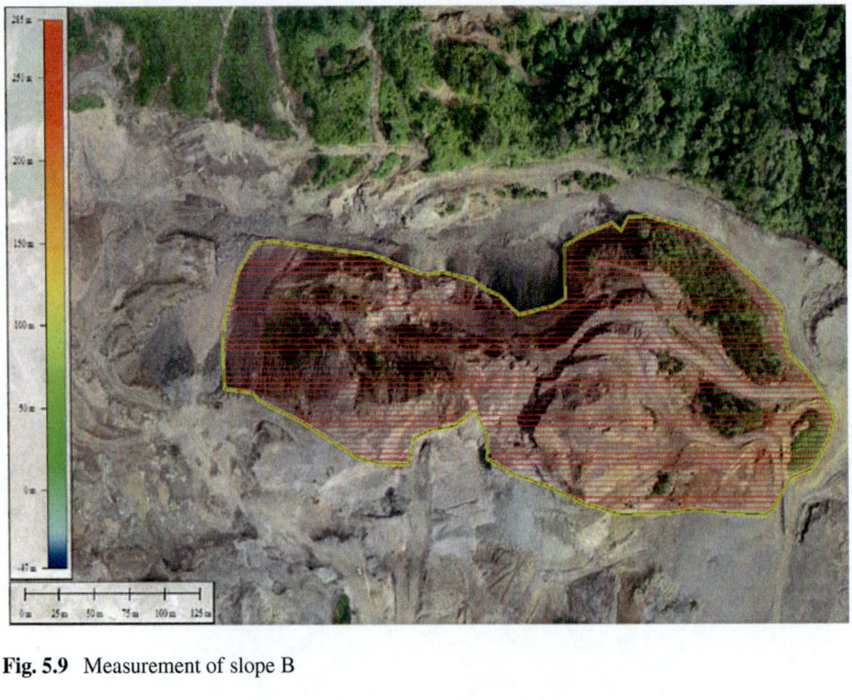

Fig. 5.9 Measurement of slope B

Table 5.2 Dimensions of stockpile

No.	Measurement	Unit
1	Total volume	45,799.494 m^3
2	Net volume	45,432.792 m^3
3	Cut volume	45,616.143 m^3
4	Cut area	0.00933 km^2
5	Cut area 3D	0.0108 km^2
6	Fill volume	183.35065 m^3
7	Fill area	0.001394 km^2
8	Fill area 3D	0.001409 km^2
9	Enclosed area	0.01069 km^2
10	Perimeter	499.71 m

Table 5.3 Dimensions of slope A

No.	Measurement	Unit
1	Total volume	41,028.486 m^3
2	Net volume	12,771.773 m^3
3	Cut volume	14,128.357 m^3
4	Cut area	0.00789 km^2
5	Cut area 3D	0.00954 km^2
6	Fill volume	26,900.129 m^3
7	Fill area	0.00959 km^2
8	Fill area 3D	0.01171 km^2
9	Enclosed area	0.01743 km^2
10	Perimeter	777.75 m

Table 5.4 Dimensions of slope B

No.	Measurement	Unit
1	Total volume	465,914.18 m^3
2	Net volume	463,347.59 m^3
3	Cut volume	464,630.89 m^3
4	Cut area	0.04129 km^2
5	Cut area 3D	0.057 km^2
6	Fill volume	1283.2932 m^3
7	Fill area	0.001478 km^2
8	Fill area 3D	0.001916 km^2
9	Enclosed area	0.04264 km^2
10	Perimeter	1.018 km

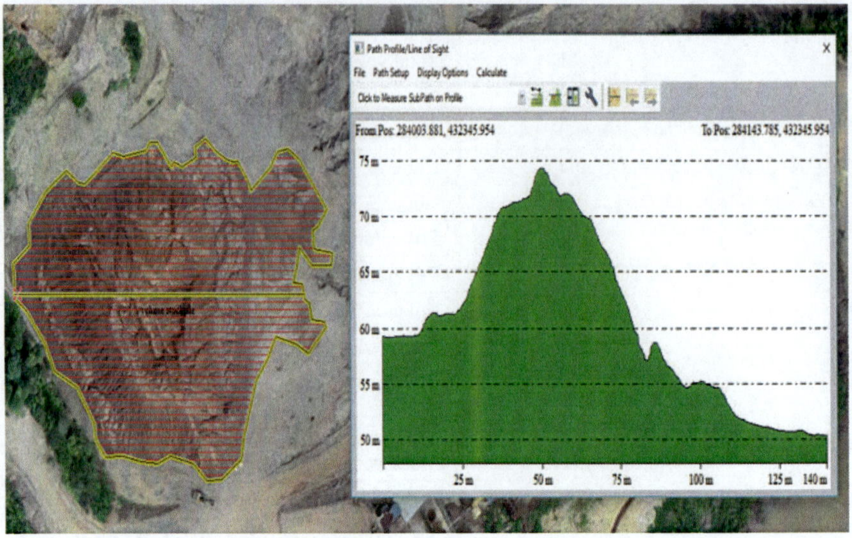

Fig. 5.10 Horizontal cross section of stockpile

5.3.2 *Cross-Sectional Area of Slopes*

Figure 5.10 shows the horizontal cross section (*x*-axis) of the stockpile. It can be seen from this figure that the highest point from this cross section is almost 73 m and the lowest point is 51 m. The total distance is 140 m.

Figure 5.11 shows the vertical cross section (*y*-axis) of the stockpile. It can be seen from this figure that the cross section is almost 71 m and the lowest point is 54 m with the total distance of 88 m.

Figure 5.12 shows the horizontal cross section (*x*-axis) of slope A. It can be seen from this figure that the highest point from this cross section is almost 55 m and the lowest point is 9 m. The total distance is 232 m.

Figure 5.13 shows the vertical cross section (*y*-axis) of slope A. It can be seen from this figure that the highest point from this cross section is almost 51 m and the lowest point is 7 m with the total distance of 103 m.

Figure 5.14 shows the horizontal cross section (*x*-axis) of slope B. It can be seen from this figure that the highest point from this cross section is almost 101 m and the lowest point is 25 m with a total distance of 387 m.

Figure 5.15 shows the vertical cross section (*y*-axis) of slope B. It can be seen from this figure that the highest point from this cross section is almost 74 m and the lowest point is 25 m with the total distance of 140 m.

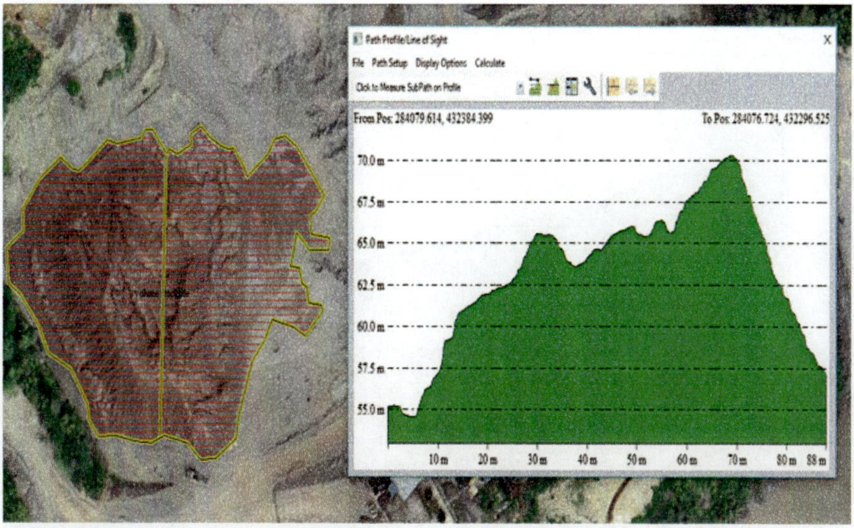

Fig. 5.11 Vertical cross section of stockpile

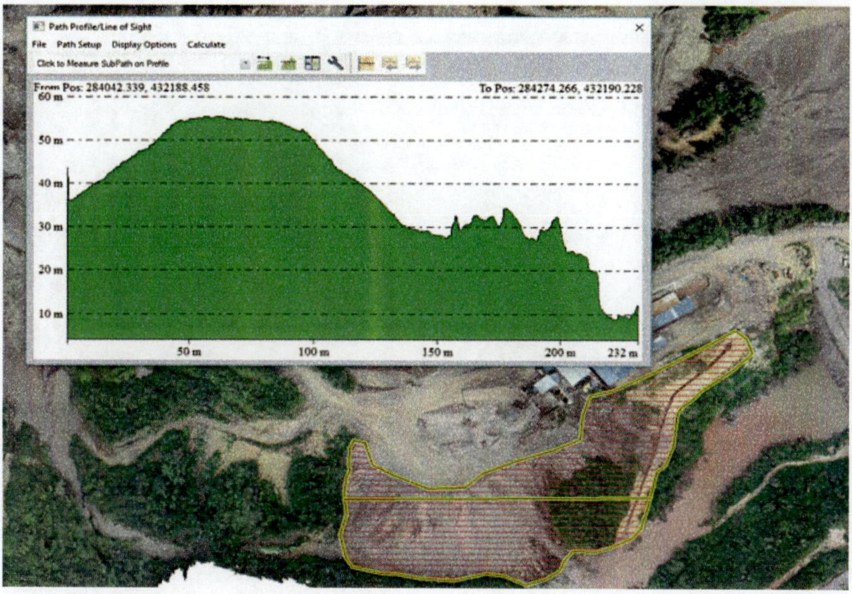

Fig. 5.12 Horizontal cross section of slope A

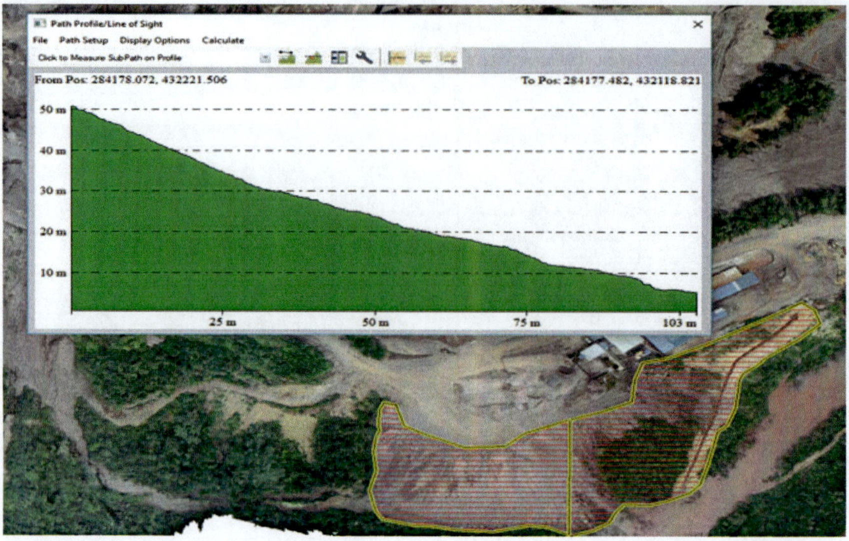

Fig. 5.13 Vertical cross section of slope A

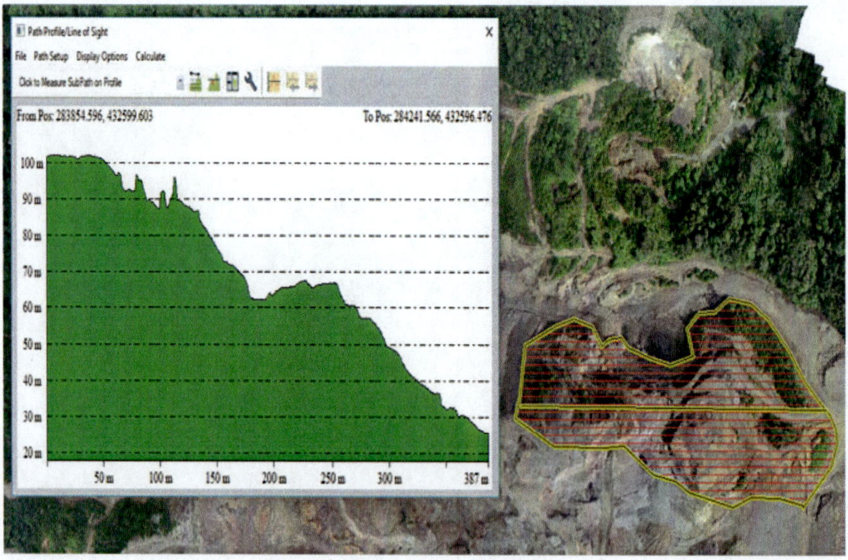

Fig. 5.14 Horizontal cross section of slope B

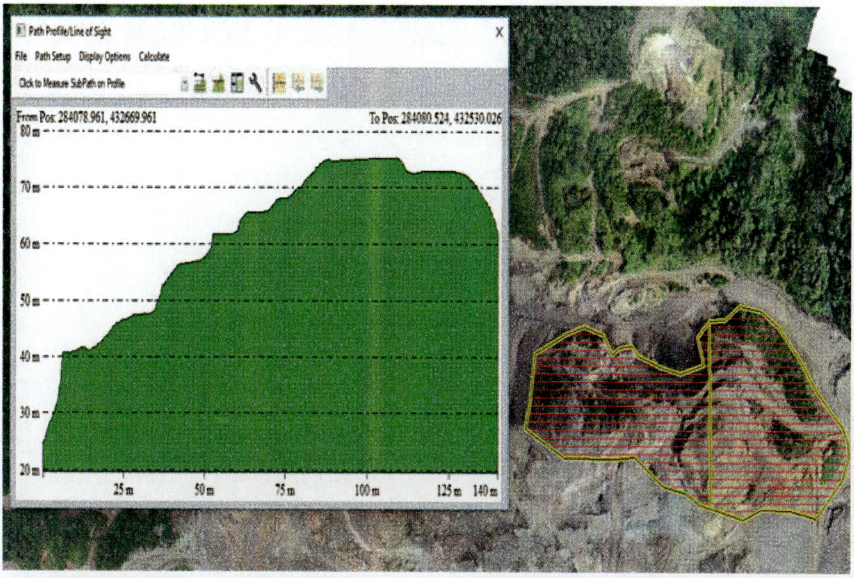

Fig. 5.15 Vertical cross section of slope B

5.3.3 Potential Risks of Slope Hazard

Reference [45] states that the determination of the slope angle is very important before further construction on a slope. By using the UAV to do mapping and with the help of the commercial software for the analysis process, the potential slope hazard can be determined which is based on the slope angle of certain selected areas. In these cases, the determination of the cross-sectional area for stockpile, slope A and slope B will be further extended to determine the potential slope hazard based on the slope angle and this will be discussed further in this section. Table 5.5 shows the slope classes by United States Department of Agriculture.

Figure 5.16 shows horizontal cross section (x-axis) of the stockpile with possible slope hazard. Two selected points from (284,063.407, 432,344.695) to (284,143.528, 432,344.136) were selected. The total length of this cross section is 138 m, and the horizontal distance between these two points is 25.873 m. The height of the first

Table 5.5 Slope classes according to USDA [46]

Angle of slope (°)	Letter designation	Descriptive term
0–2	A	Nearly level
3–7	B	Gently sloping
8–15	C	Sloping
16–25	D	Moderately steep
>26	E, F	Steep

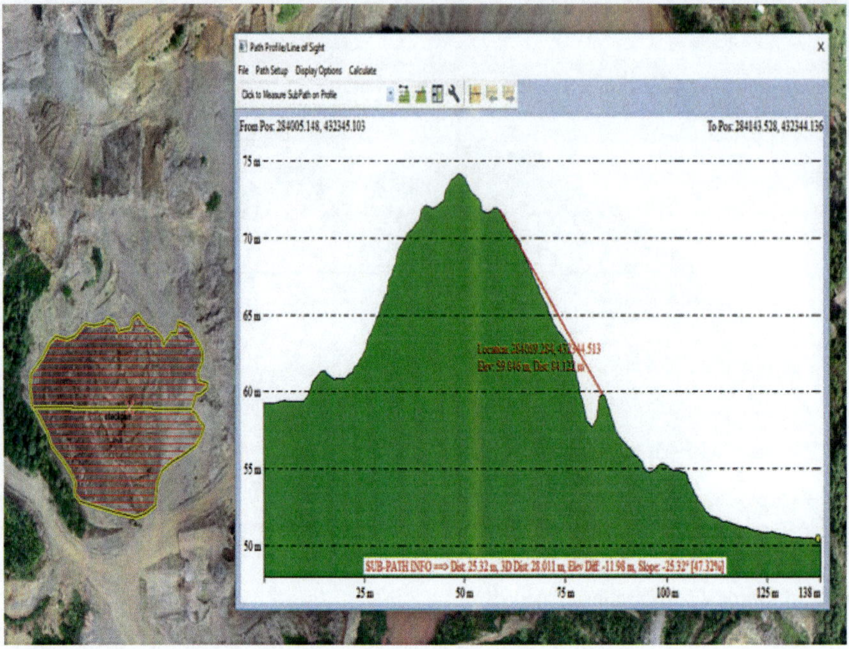

Fig. 5.16 Potential slope hazard of stockpile (horizontal cross section)

point is 71.826 m, and the height of the second point is 59.846 m. The slope angle between these two points is 25.32°. Based on Table 5.5 of slope classes, the slope between these two points can be considered as a steep slope (>than 26°) and can be considered as slope hazard.

Figure 5.17 shows vertical cross section (y-axis) of the stockpile with possible slope hazard. Two selected points from (284,078.327, 432,313.905) to (284,078.327, 432,291.190) were selected. The total length of this cross section is 94 m and the horizontal distance between these two points is 19.085 m. The height of the first point is 69.474 m, and the height of the second point is 55.727 m. The slope angle between these two points is 35.42°. Based on Table 5.5 of slope classes, the slope between these two points can be considered as a steep slope (>than 26°) and can be considered as slope hazard.

Figure 5.18 shows horizontal cross section (x-axis) of slope A with possible slope hazard. Two selected points from (284,138.186, 432,191.820) to (284,195.900, 432,192.063) were selected. The total length of this cross section is 232 m and the horizontal distance between these two points is 57.7 m. The height of the first point is 54.133 m, and the height of the second point is 28.133 m. The slope angle between these two points is 24.26°. Based on Table 5.5 of slope classes, the slope between these two points can be considered as a moderately steep slope (16° to 25°) and can be considered as moderate slope hazard.

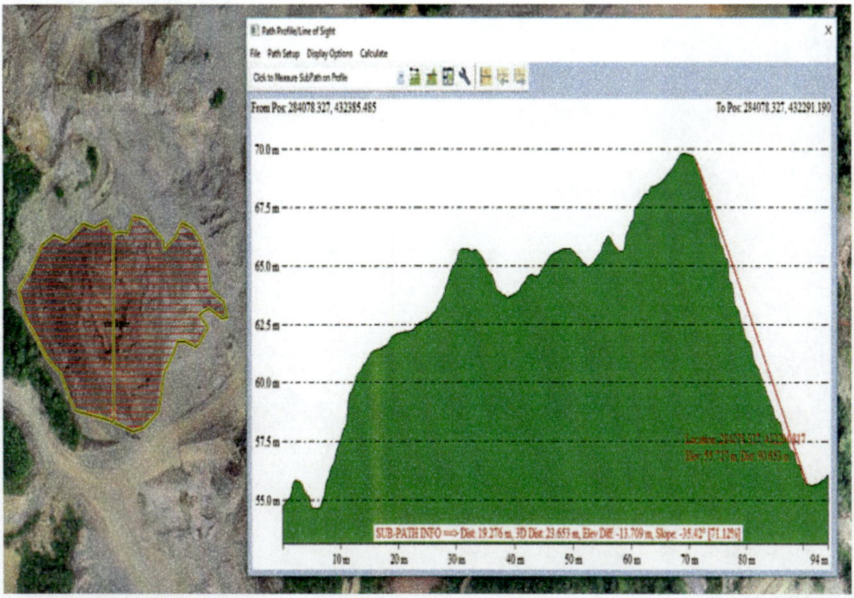

Fig. 5.17 Potential slope hazard of stockpile (vertical cross section)

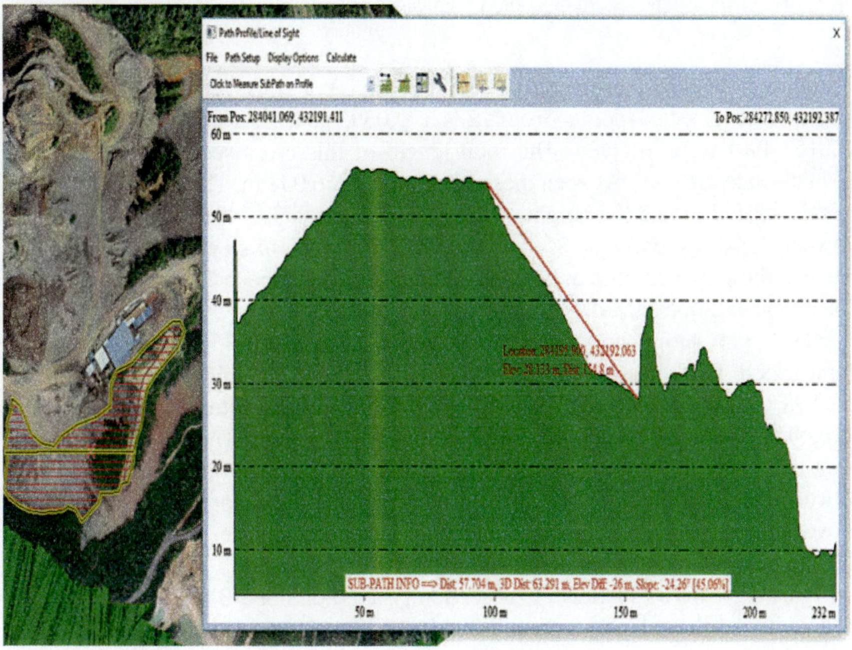

Fig. 5.18 Potential slope hazard of slope A (horizontal cross section)

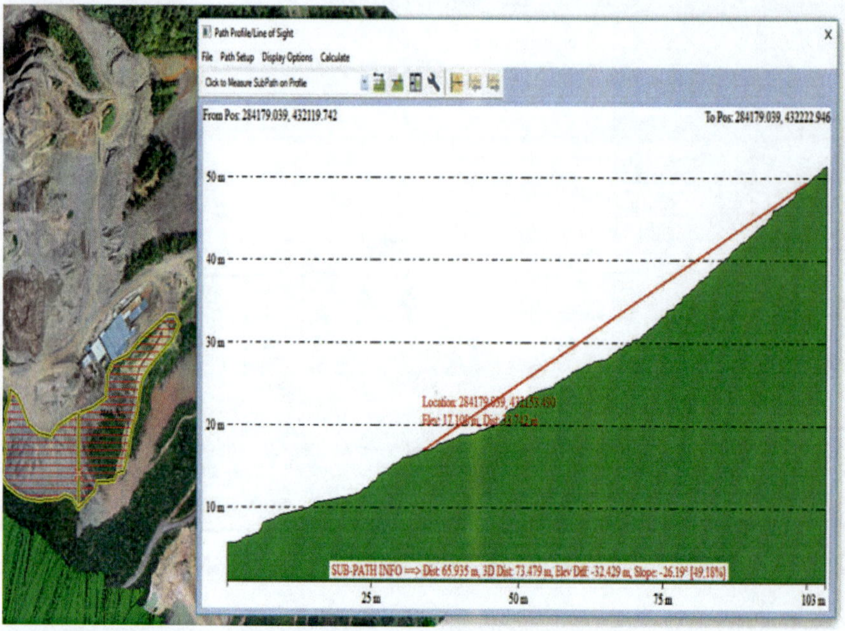

Fig. 5.19 Potential slope hazard of slope A (vertical cross section)

Figure 5.19 shows vertical cross section (*y*-axis) of slope A with possible slope hazard. Two selected points from (284,179.039, 432,219.540) to (284,179.039, 432,153.490) were selected. The total length of this cross section is 103 m and the horizontal distance between these two points is 66.038 m. The height of the first point is 49.422 m and the height of the second point is 33.742 m. The slope angle between these two points is 26.19°. Based on Table 5.5 of slope classes, the slope between these two points can be considered as a steep slope (>than 26°) and can be considered as slope hazard.

Figure 5.20 shows horizontal cross section (*x*-axis) of slope B with possible slope hazard. Two selected points from (284,108.918, 432,597.654) to (284,210.269, 432,598.156) were selected. The total length of this cross section is 394 m and the horizontal distance between these two points is 101.332 m. The height of the first point is 32.628 m and the height of the second point is 66.596 m. The slope angle between these two points is 18.58°. Based on Table 5.5 of slope classes, the slope between these two points can be considered as a moderately steep slope (16° to 25°) and can be considered as moderate slope hazard.

Figure 5.21 shows vertical cross section (*y*-axis) of slope B with possible slope hazard. Two selected points from (284,083.643, 432,664.189) to (284,083.643, 432,672.668) were selected. The total length of this cross section is 177 m and the horizontal distance between these two points is 8.477 m. The height of the first point is 22.412 m and the height of the second point is 41.099 m. The slope angle

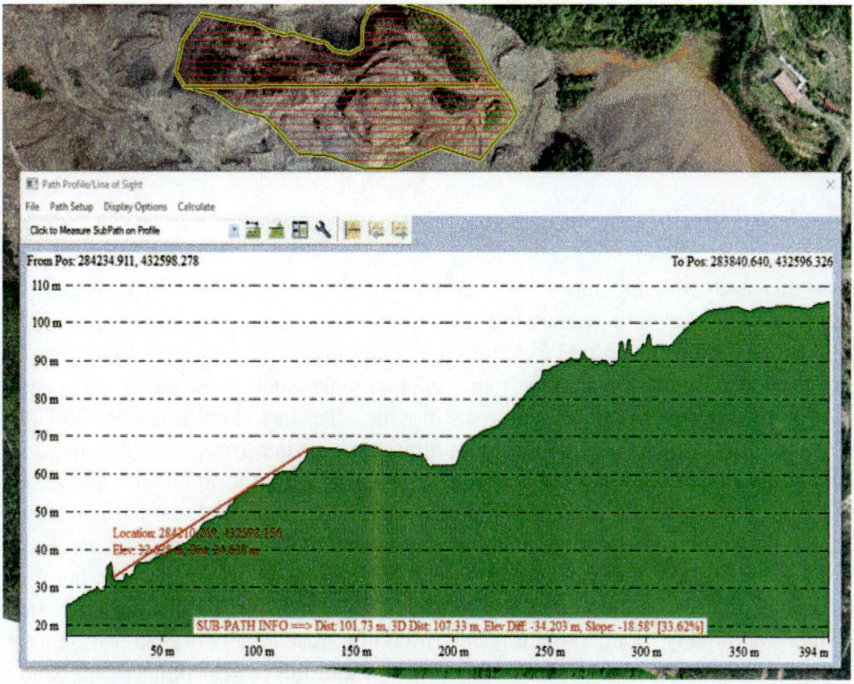

Fig. 5.20 Potential slope hazard of slope B (horizontal cross section)

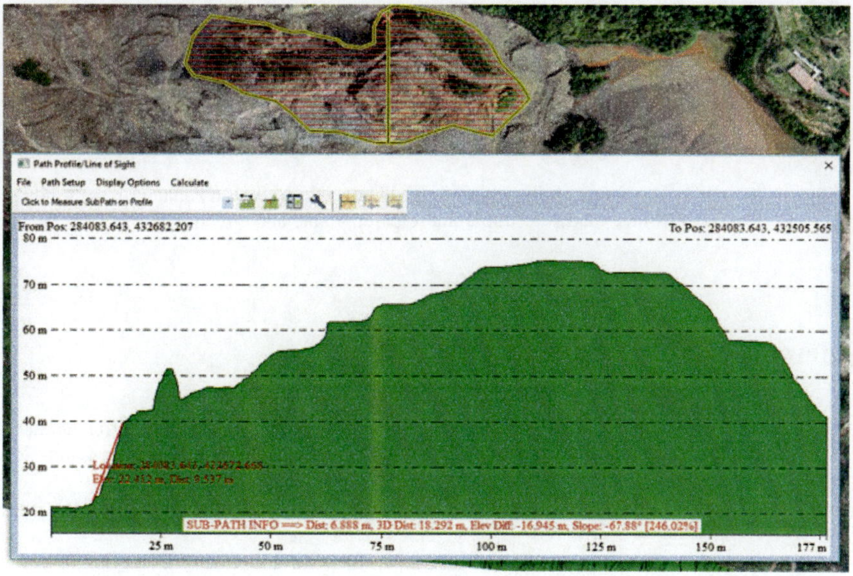

Fig. 5.21 Potential slope hazard of slope B (vertical cross section)

between these two points is 67.88°. Based on Table 5.5 of slope classes, the slope between these two points can be considered as a steep slope (>than 26°) and can be considered as slope hazard.

5.4 Conclusion

From this study, the application of UAV has proven to be very effective because of the low costing, low time consuming, easy to operate and can gather huge amount of data within short time interval for geological mapping. This modern technology will help in research and also commercial works to make such work easier and faster. UAV helps in terms of photogrammetry in which the data is collected by the UAV or drone in term of images. With the combination of data from the UAV and also the established software, researchers can be provided with important parameters and information about the topography of study area. Other than that, the parameters of the study area such as its perimeter, area, volume and more can be obtained precisely. Besides, the elevation of the selected study area can be determined which helps in researches and also commercial works to determine the profile of high area such as slopes and many more. From this determination, researchers may know the total distance of marked areas as well as the highest and the lowest point of the study area. Another main focus in this study was to determine the possible slope hazard based on slope angles. These slope angles can be divided into several classes. This is very important as an early precautionary step to prevent slope failure from occur. Slope monitoring using UAV becomes much easier compared to previous survey techniques due to the fact that slope usually high enough which may cause danger to human lives. This can be very helpful for geotechnical engineers, geologists and researchers to simplify their work in the future.

Acknowledgements The authors fully acknowledged Universiti Malaysia Pahang for the funding of grant PGRS200380 which made this important research viable and effective.

References

1. Estrada MAR, Ndoma A (2019) The uses of unmanned aerial vehicles-UAV's-(or Drones) in social logistic: natural disaster response and humanitarian relief aid. Procedia Comp Sci 149:375–383
2. Wozencraft, JM, Lillycrop WJ (2003) SHOALS airborne coastal mapping past present and future. J Coast Res 207–215
3. Turner IL, Harley MD, Drummond CD (2016) UAVs for coastal surveying. Coast Eng 114:19–24
4. Yeh FH, Huang CJ, Han JY, Ge L (2018) Modeling slope topography using unmanned aerial vehicle image technique. MATEC web of conferences, 147:1–6

5. Ardi ND, Iryanti M, Asmoro CP, Nurhayati N, Agustine E (2018) Mapping landslide potential area using fault fracture density analysis on unmanned aerial vehicle (UAV) image. IOP Conf Ser: Earth Environ Sci 145:012010
6. Kumar NS, Ismail MAM., Sukor NSA, Cheang W (2018) Geohazard reconnaissance mapping for potential rock boulder fall using low altitude UAV photogrammetry. IOP Conf Ser: Mater Sci Eng 352:012033
7. Kumar NS, Ismail MAM, Sukor NSA, Cheang W (2018) Method for the visualization of landform by mapping using low altitude UAV application. IOP Conf Ser: Mater Sci Eng 352:012032
8. Bondarchuk AS (2018) System of technical vision for autonomous unmanned aerial vehicles. IOP Conf Ser: Mater Sci Eng 363:012027
9. Qin F (2014) The market share of power transmission line inspection about CSG and market development strategies for H Company, Beijing University of Technology
10. Du W, Niu J, Wang M, Zhang J, Yang Y, Li Y (2018) An analysis model of helicopter and UAV in overhead power line inspection. IOP Conf Ser: Earth Environ Sci 189:062056
11. Park S, Lee H, Chon J (2019) Sustainable monitoring coverage of unmanned aerial vehicle photogrammetry according to wing type and image resolution. Environ Pollut 247:340–348
12. Zolkepli MF, Rozar NM, Ishak MF, Sidik MH, Ibrahim NAS, Zaini MSI (2021) Slope mapping using unmanned aerial vehicle (UAV). Turk J Comput Math Educ 12(3):1781–1789
13. Eid BM, Chebil J, Albatsh F, Faris WF (2013) Challenges of integrating unmanned aerial vehicles in civil application. IOP Conf Ser: Mater Sci Eng 53:012092
14. Tziavou O, Pytharouli S, Souter J (2018) Unmanned aerial vehicle (UAV) based mapping in engineering geological surveys: considerations for optimum results. Eng Geol 232:12–21
15. Delacourt C, Allemand P, Jaud M, Grandjean P, Deschamps A, Ammann J, Cuq V, Suanez S (2009) DRELIO an unmanned helicopter for imaging coastal areas. J Coast Res, pp 1489–1493
16. Anderson K, Gaston KJ (2013) Lightweight unmanned aerial vehicles will revolutionize spatial ecology. Front Ecol Environ 11(3):138–146
17. Watts AC, Perry JH, Smith SE, Burgess MA, Wilkinson BE, Szantoi Z, Ifju PG, Percival HF (2010) Small unmanned aircraft systems for low-altitude aerial surveys. J Wildl Manag 74(7):1614–1619
18. Ismail MAM, Kumar NS, Abidin MHZ, Madun A (2018) Systemic approach to elevation data acquisition for geophysical survey alignments in hilly terrains using UAVs. J Phys: Conf Ser 995:012104
19. Colomina I, Molina P (2014) Unmanned aerial systems for photogrammetry and remote sensing: a review. ISPRS J Photogrammetry Remote Sens 92:79–97
20. Shashev DV, Shidlovskiy SV, Syriamkin VI, Yurchenko AV (2015) Application of reconfigurable computing environments for image processing in X-ray tomography of materials. IOP Conf Ser: Mater Sci Eng 81:012101
21. Panin SV, Syryamkin VI, Glukhih AI (2003) Avtometriya (4):79–92
22. Rokhmana CA, Utomo S (2016) The low-cost UAV-based remote sensing system capabilities for large scale cadaster mapping. IOP Conf Ser: Earth Environ Sci 47:012005
23. Chi YY, Lee YF, Tsai SE (2016) Study on high accuracy topographic mapping via UAV-based images. IOP Conf Ser: Earth Environ Sci 44:032006
24. Greenwood W, Zekkos D, Lynch J, Bateman J, Clark M, Chamlagain D (2016) UAV based 3-D characterization of rock masses and rock slides in Nepal. In: 50th US Rock mechanics/geomechanics symposium, American Rock Mechanics Association, Houston, TX, 26–29 June
25. Tannant DD, Giordan D, Morgenroth J (2017) Characterization and analysis of a translational rockslide on a stepped planar slip surface. Eng Geol 220:144–151
26. Reimers B, Griffiths CL, Hoffman MT (2014) Repeat photography as a tool for detecting and monitoring historical changes in South African coastal habitats. Afr J Mar Sci 36:387–398
27. Konar B, Iken K (2018) The use of unmanned aerial vehicle imagery in intertidal monitoring. Deep-Sea Res Part II 147:79–86

28. Ouedraogo MM, Degre A, Debouche C, Lisein J (2014) The evaluation of unmanned aerial system-based photogrammetry and terrestrial laser scanning to generate DEMs of agricultural watersheds. Geomorphology 214:339–355
29. Nelson A, Reuter HI, Gessler P (2009) DEM production methods and sources. Geomorphometry: concepts, software, applications. Dev Soil Sci 33:65–85
30. Peucker TK, Fowler RJ, Little JJ, Mark DM (1978) The triangulated irregular network. Proceedings: American Society of Photogrammetry, Digital Terrain Models Symposium, St. Louis, Missouri., pp 516–540
31. Niewinski M (2004) Distributed monte carlo simulation of a dynamic expansion system. Vacuum 73:257–261
32. Zolkepli MF, Ishak MF, Yunus MYM, Zaini MSI, Wahap MS, Yasin AM, Sidik MH, Hezmi MA (2021) Application of unmanned aerial vehicle (UAV) for slope mapping at Pahang Matriculation College, Malaysia. Physics and Chemistry of the Earth (article in press)
33. Henry JB, Malet JP, Maquaire O, Grussenmeyer P (2002) The use of small-format and low-altitude aerial photos for realization of high-resolutions DEMs in mountainous areas: application to the super-sauze earthflow (Alpes-de-Haute-Provence, France). Earth Surf Process Landf 27:1339–1350
34. Fabris M, Pesci A (2005) Automated DEM extraction in digital aerial photogrammetry: precision and validation for mass movement monitoring. Ann Geophys 48:973–988
35. Hladik C, Alber M (2012) Accuracy assessment and correction of a lidar-derived salt marsh digital elevation model. Remote Sens Environ 121:224–235
36. Huising EJ, Pereira LMG (1998) Errors and accuracy estimates of laser data acquired by various laser scanning systems for topographic applications. ISPRS J Photogramm Remote Sens 53:245–261
37. Wehr A, Lohr U (1999) Airborne laser scanning: an introduction and overview. ISPRS J Photogramm Remote Sens 54:68–82
38. Eisenbeiss H, Zhang L (2006) Comparison of DSMs generated from mini UAV imagery and terrestrial laser scanner in a cultural heritage application. Int Arch Photogramm Remote Sens Spat Inf Sci XXXVI:90–97
39. Guarnieri A, Vettore A, Pirotti F, Menenti M, Marani M (2009) Retrieval of small-relief marsh morphology from terrestrial laser scanner, optimal spatial filtering and laser return intensity. Geomorphology 113:12–20
40. Hohle J (2009) Dem generation using a digital large-format frame camera. Photogramm Eng Remote Sens 53:245–261
41. Ishak MF, Zolkepli MF, Yunus MYM, Ali N, Kassim A, Zaini MSI (2021) Verification of tree induced suction with numerical model. Phys Chem Earth 121:102980
42. Ishak MF, Zolkepli MF, Yunus MYM, Ali N, Kassim A, Zaini MSI (2021) The effect of tree water uptake on suction distribution in tropical residualsoil slope. Phys Chem Earth 121:102984
43. Zolkepli MF, Ishak MF, Zaini MSI (2018) Analysis of slope stability on tropical residual soil. Int J Civ Eng Technol (IJCIET). 9(2):402–416
44. Ishak MF, Zolkepli MF (2016) Exploration of methods for slope stability analysis influenced by unsaturated soil. Electron J Geotech Eng 21(17):5627–5641
45. Zaini MSI, Ishak MF, Zolkepli MF (2020) Monitoring soil slope of tropical residual soil by using tree water uptake method. IOP Conf Ser: Mater Sci Eng 736:072018
46. Yue LJ, Ishak MF, Zaini MSI, Zolkepli MF (2019) Rainfall induced residual soil slope instability: building cracked and slope failure. IOP Conf Ser: Mater Sci Eng 669:012004
47. Zolkepli MF, Ishak MF, Zaini MSI (2019) Slope stability analysis using modified Fellenius's and Bishop's method. IOP Conf Ser: Mater Sci Eng 527:012004
48. Goh JR, Ishak MF, Zaini MSI, Zolkepli MF (2020) Stability analysis and improvement evaluation on residual soil slope:building cracked and slope failure. IOP Conf Ser: Mater Sci Eng 736:072017
49. Zaini MSI, Ishak MF, Zolkepli MF (2019) Forensic assessment on landfills leachate through electrical resistivity imaging at Simpang Renggam in Johor, Malaysia. IOP Conf Ser: Mater Sci Eng 669:012005

50. Ishak MF, Zolkepli MF, Affendy M (2017) Tropical residual soil properties on slopes. Int J Eng Technol Sci (IJETS) 8(1):1–9
51. Braathen A, Blikra LH, Berg SS, Karslen F (2004) Rock-slope failures in Norway; type, geometry, deformation mechanisms and stability. Norw J Geol 67–88
52. Giordan D, Manconi A, Tannant DD, Allasia P (2015) UAV: low-cost remote sensing for high-resolution investigation of landslides. Int Geosci Remote Sens Symp (IGARSS), pp 5344–5347

Chapter 6
Tensile Strength Testing of +45° Isotropic FRP Laminate on Different Universal Testing Machines

Roslin Ramli, Mohd Hisbany Mohd Hashim, Anizahyati Alisibramulisi, Suhailah Mohamed Noor, and Mohd Faizal Abdul Razak

Abstract The present work intends to analyze the tensile strength testing of +45° of isotropic fiberglass-reinforced polymer (FRP) laminates on the different universal testing machines. The tensile strength test used two different universal testing machines. The tensile test has one type of specimen which is under +45° angle. The FRP material that had effects on the angle orientation was a woven roving mat. The laminated specimen has used a polymer (resin) which is an unsaturated polyester, and the catalyst was methyl ethyl ketone peroxide (MEKP). The stacking sequence resulted in quasi-isotropic properties. The specimen had nine layers with the same size, i.e., 250 mm × 25 mm × 8 mm which was a flat bar, balanced, and symmetric according to ASTM D3039. This research revealed that the use of the universal testing machine "MFL system, UPD-20" resulted in the highest tensile strength and the highest force.

Keywords Tensile test · FRP · Isotropic · Angle orientation · Stack sequences

---

R. Ramli (✉) · M. F. Abdul Razak
Universiti Kuala Lumpur Malaysian Institute of Marine Engineering Technology, 32200 Lumut, Perak, Malaysia
e-mail: roslin@unikl.edu.my

M. F. Abdul Razak
e-mail: mfaizalar@unikl.edu.my

R. Ramli · M. H. Mohd Hashim · A. Alisibramulisi
Faculty of Civil Engineering, Universiti Teknologi MARA (UiTM), 40450 Shah Alam, Selangor, Malaysia
e-mail: hisbany@uitm.edu.my

S. Mohamed Noor
Faculty of Civil Engineering, Universiti Teknologi MARA (UiTM), 13500 Permatang Pauh, Pulau Pinang, Malaysia
e-mail: suhailahmn@uitm.edu.my

A. Ismail et al. (eds.), *Design in Maritime Engineering*, Advanced Structured Materials 167, https://doi.org/10.1007/978-3-030-89988-2_6

6.1 Introduction

The tensile strength is the resistance of a material to break under tension. Besides that, tensile strength of composite materials is indicative of the strength derived from factors such as fiber strength, fiber length, and bonding [1]. This is a measurement of the force required to elongate something such as a rope, structural bar to the point it breaks. The tensile strength is tested by using universal testing machine (UTM). The standard test method that is used is the ASTM Standard Test Method for Tensile Properties of Polymer Matrix Composites (D 3039). This test method covers the determination of a material's tensile strength, tensile strain, force/load, and modulus of elasticity [2].

The material that is used is a fiber-reinforced polymer (FRP). FRP is a composite material made of fibers with a polymer matrix reinforced [3, 4]. The common types of fibers are glass, carbon, and aramid. The polymers have two categories, i.e., thermoset resins and thermoplastic resins. A thermoset resin is the most common use for structural applications. The types of thermoset resins are polyester, vinyl ester, and epoxy. FRPs are commonly used in the marine, aerospace, automotive, and construction industries [4–7].

The FRP laminate process can be classified as a manufacturing method which can result in isotropic materials. Isotropic materials have the same material properties in all directions [8]. These material properties are independent of angle orientation direction within the material. The orientations have various angles such as 0°, 90°, −45°, or +45° as shown in Fig. 6.1.

The objective of this research paper is to analyze the highest strength and the highest force by using different universal testing machines. Two different testing machines, i.e., a Galdabini Quasar 100 and MFL system, UPD-20, were used.

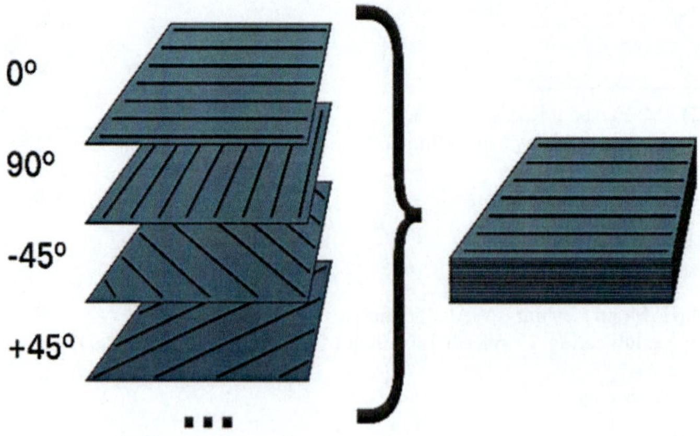

Fig. 6.1 Angle orientation

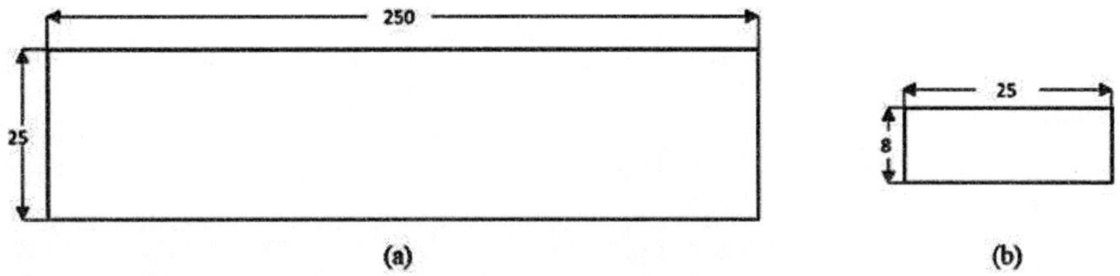

Fig. 6.2 Specimen S + 45° size

6.2 Methodology

This research methodology consists of four parts. The four parts are the parameters of the specimen, specimen preparations, tensile test on an automatic machine, and the last is the tensile test on a manual machine. The specimen is called S + 45°.

6.2.1 Parameter of the Specimen

The parameters of the specimen S + 45° are chosen according to the ASTM D 3039. The size of the specimen S + 45° is 250 mm × 25 mm × 8 mm which is a flat bar, balanced, and symmetric as shown in Fig. 6.2. The specimen S + 45° has nine layers. The layer of the specimen S + 45° laminate consists of the woven roving mat, the tissue mat, and the chopped strand mat. The direction of the angle orientation of the woven roving is +45°. The stack sequence results in quasi-isotropic material behavior.

6.2.2 Specimen Preparation

The specimen S + 45° preparation is according to the ASTM D 3039. The test specimen requires at least five specimens for each test condition unless a useable result can be gained with less specimens. The specimen S + 45° preparation has used the woven roving 600 (WR600), chopped strand mat 450 (CSM 450), and tissue mat (T). The polymer is the unsaturated polyester, and the catalyst was methyl ethyl ketone peroxide (MEKP).

The specimen S + 45° laminate has a sort of isotropic stacking sequence. The stacking sequence is the arrangement of each layer. The lamination sequence is T/CSM450/WR600/CSM450/WR600/CSM450/WR600/CSM450/T, and the angle orientation is [0/0/+45°/0/+45°/0/+45°/0/0] as shown in Table 6.1.

The preparation of one plate of glass (304.8 mm × 304.8 mm) was used to give a smooth surface. Before the process, Mirror Glaze® mold release wax was applied

Table 6.1 S + 45° stacking sequence and angle orientation

Stacking sequence	Angle orientation
T	0°
CSM450	0°
WR600	+45°
CSM450	0°
WR600	+45°
CSM450	0°
T	0°

onto the glass to prevent the unwanted bonding between the materials and the glass surfaces. The specimen + 45° laminate is referred in Table 6.1 in regards the stacking sequence and angle orientation. The fiber mat was impregnated with an unsaturated polyester with 3% methyl ethyl ketone peroxide (MEKP) as a catalyst. The ratio of resin to fiber is fixed at approximately 2:1. The laminated composite was cured at room temperature for 24 h. Figure 6.3 shows the specimen + 45°.

Finally, the laminates (304.8 mm × 304.8 mm) were cut according to the specimen S + 45° size using a tenoning and squaring machine. The prepared specimen size is 25 0 mm × 25 mm × 8 mm, and a total of 10 specimens were prepared. Figure 6.4 shows the sample specimen + 45°.

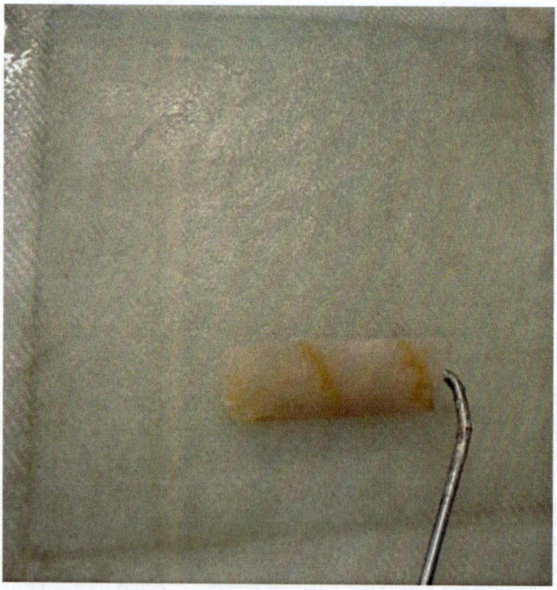

Fig. 6.3 Preparation of the specimen S + 45°

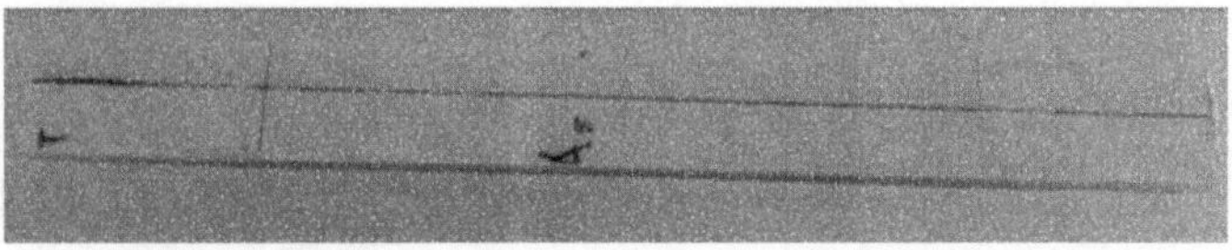

Fig. 6.4 Sample of the specimen S + 45°

6.2.3 Test with Automatic Control

The automatic Galdabini Quasar 100 universal testing machine was used. This UTM has a two-column rigid system and suitable for metals, plastics, composites, and other materials.

The tensile test was conducted at the Materials Laboratory UniKL MSI by using the Galdabini Quasar 100 as shown in Fig. 6.5. Before the experiment was carried out, the specimen thickness and gage length were measured and numbered for identification as shown in Fig. 6.6.

The specimen S + 45° that has been placed accurately was applied with a vertical load of maximum 30 kN as shown in Fig. 6.7. This process was five times repeated, and the related data and observation were recorded.

Fig. 6.5 UTM Galdabini Quasar 100

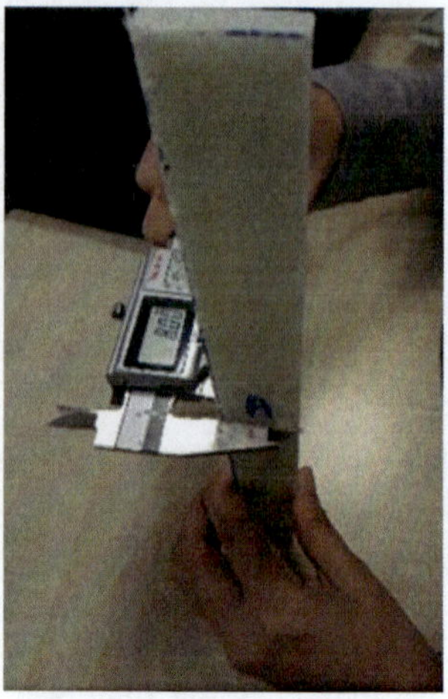

Fig. 6.6 Measurement of the specimen S + 45°

Fig. 6.7 Tensile test at UTM Galdabini Quasar 100

6.2.4 *Test Manual Control*

A manual MFL system, UPD-20, universal testing machine was used.

The UTM has a two-column rigid system and suitable for metals, plastics, composites, and other materials.

The tensile test was conducted at the laboratory "Laboratorium Konstruksi dan Kekuatan Kapal" at Institut Teknologi Sepuluh Nopember (ITS), Indonesia, by using MFL system, UPD-20, as shown in Fig. 6.8. Before the experiment was carried out, the specimen thickness and gage length were measured and numbered for identification, see Fig. 6.6.

The specimen S + 45° that has been placed accurately was applied with a vertical load of maximum 30 kN as shown in Fig. 6.9. The manual UTM must pull and push

Fig. 6.8 UTM MFL system, UPD-20

Fig. 6.9 Tensile test at UTM MFL system, UPD-20

the handle vertically. This process was five times repeated, and the related data and observation were recorded (Figs. 6.10 and Fig. 6.11).

Fig. 6.10 Specimen S + 45° results by automatic control

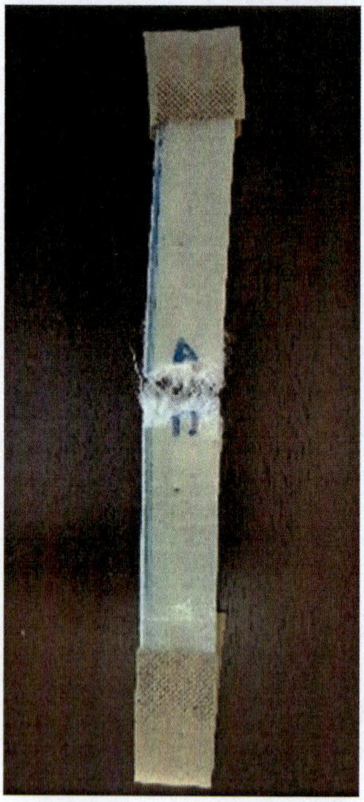

Fig. 6.11 Specimen S + 45° results by UTM MFL system, UPD-20

Table 6.2 Result of S + 45° from UTM Galdabini Quasar 100

Tensile test			
Specimen S + 45°	**Ultimate force (kN)**	**Tensile strength (MPa)**	**Strain**
A1	25.94	134.39	0.15
A2	24.80	126.46	0.14
A3	26.96	137.48	0.14
A4	24.36	125.21	0.15
A5	25.05	130.47	0.15
Average	**25.42**	**130.802**	**0.15**

6.3 Results and Discussion

According to ASTM D3039, FRP fabric properties are tested by the tensile test which gives the result for stress vs. strain a linear elastic behavior up to failure, no yielding, higher ultimate strength, and lower strain at failure. The maximum load for all tests is 30 kN.

6.3.1 *Result of S + 45° UTM Galdabini Quasar 100*

The average results of the maximum load for the specimen S + 45° (Galdabini Quasar 100) are shown in Table 6.2 as a value was 25.40 kN, while the value of stress was 130.80 MPa. Meanwhile, the strain response to the maximum load for the specimen S + 45° is 0.15. The overall result of the tensile test for the automatic control is shown in Table 6.2.

The maximum results of the specimen S + 45° were 26.96 kN, while the value of stress was 137.48 MPa. Meanwhile, the strain response to the maximum load for the specimen S + 45° is 0.14. However, the minimum results of the specimen S + 45° were 24.80 kN, while the value of stress was 126.46 MPa. Meanwhile, the strain response to the minimum load for the specimen S + 45° is 0.14.

6.3.2 *Result of S + 45° UTM MFL System, UPD-20*

The average results of the maximum load for specimen S + 45° (MFL system, UPD-20) are shown in Table 6.2 as a value was 26.50 kN, while the value of stress was 135.362 MPa. Meanwhile, the strain response to the maximum load for the specimen S + 45° is 0.13. The overall result of the tensile test for the automatic control is shown in Table 6.3.

Table 6.3 Result of S + 45° from UTM MFL system, UPD-20

Tensile test			
Specimen S + 45°	**Ultimate force (kN)**	**Tensile strength (MPa)**	**Strain**
A1	29.00	145.61	0.12
A2	26.00	132.3	0.14
A3	30.00	151.78	0.14
A4	26.00	131.32	0.12
A5	21.50	115.8	0.12
Average	**26.50**	**135.362**	**0.13**

The maximum results of the specimen S + 45° were 30 kN, while the value of stress was 151.78 MPa. Meanwhile, the strain response to the maximum load for the specimen S + 45° is 0.14. However, the minimum results of the specimen S + 45° were 21.50 kN, while the value of stress was 115.8 MPa. Meanwhile, the strain response to the minimum load for the specimen S + 45° is 0.12.

6.3.3 Comparison of the Tensile Test with Different Universal Testing Machines

The tensile strength results show that the MFL system, UPD-20 provides the highest value compared to the Galdabini Quasar 100 at which the highest tensile strength is 151.78 MPa and 137.48 MPa. Meanwhile, the highest force shows that the MFL system, UPD-20, provides the highest value compared to the Galdabini Quasar 100 at which the value is 30 kN and 26.96 kN. Both of the tensile test comparison charts are shown in Figs. 6.12 and Fig. 6.13.

The difference in the tensile strength from both universal testing machines is 4.25%; meanwhile, the percentage of the force is 3.37%. This shows that the manual UTM is relevant to use.

6.4 Conclusion

These two types of experiments have done according to ASTM D3039. The FRP fabric properties are tested by the tensile test at which the result for stress vs. strain shows a linear elastic behavior to failure, no yielding, higher tensile strength, and lower strain at failure.

The results of the experiment showed that the difference between the automatic control UTM and the manual UTM is less than 5%. This study has identified that the manual control has the highest tensile stress, highest force, and lower strain.

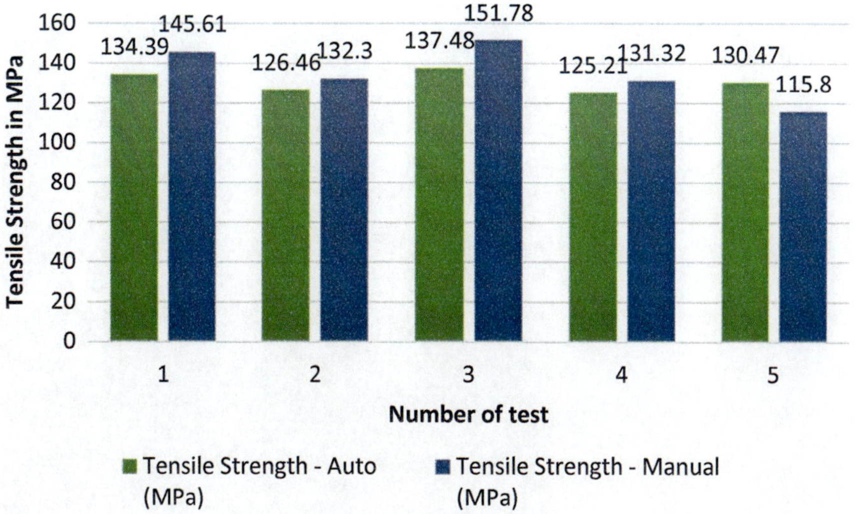

Fig. 6.12 Comparison of the tensile strength results

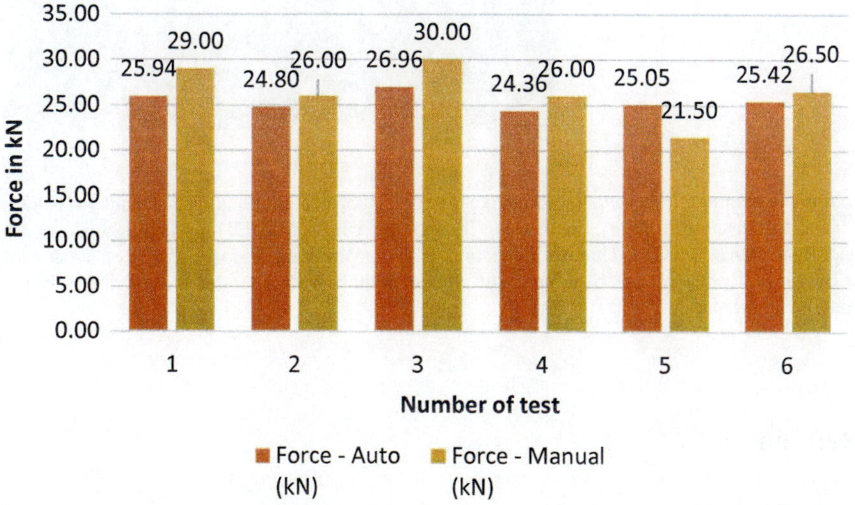

Fig. 6.13 Comparison of the force results

Both experimental works have proven that the average result of the experiment is less than 5% indifference and that the objective of the analyses is achieved. This can be concluded that the manual UTM is still relevant to use, but the result of the graph generated by using a graph paper and ball pen is as shown in Fig. 6.14.

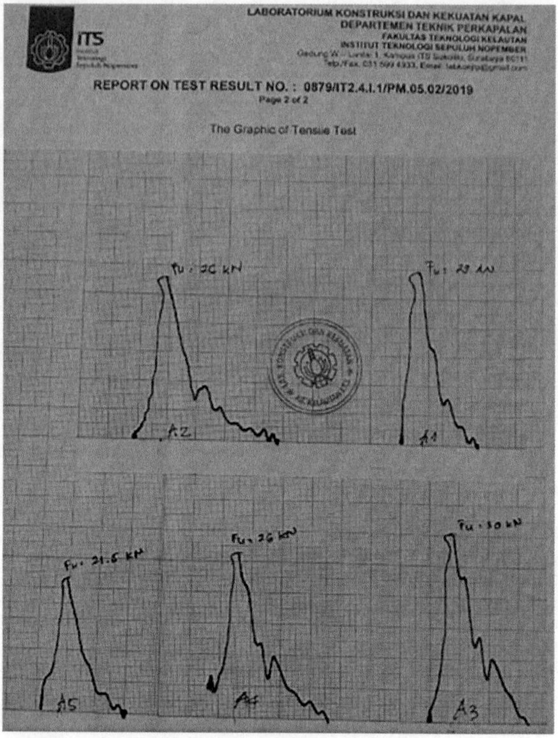

Fig. 6.14 Manual results of UTM MFL system, UPD-20

Acknowledgements The authors would like to express their gratitude and special thanks to UniKL - Malaysian Institute of Marine Engineering Technology (UniKL MIMET) for funding this research paper under the Research and Innovation section (R&I) and to her main supervisor and co-supervisor at UiTM. Lastly, regards and blessing are offered to all those who supported the completion of the paper in any respect.

References

1. Bajpai P (2018) Paper and Its Properties. In: Biermann's Handbook of Pulp and Paper. https://doi.org/10.1016/b978-0-12-814238-7.00002-7
2. ASTM International (2000) ASTM D 3039M—Standard test method for tensile properties of polymer matrix composite materials. Annual Book of ASTM Standards 15(03):1–13
3. Dong PAV, Azzaro-Pantel C, Boix M, Jacquemin L, Domenech S (2015) Modelling of environmental impacts and economic benefits of fibre reinforced polymers composite recycling pathways. In: Computer aided chemical engineering. https://doi.org/10.1016/B978-0-444-63576-1.50029-7g
4. Alberto M (2013) Introduction of fibre-reinforced polymers—polymers and composites: concepts, properties and processes. In: Fiber reinforced polymers-the technology applied for concrete repair. https://doi.org/10.5772/54629

5. Rajak DK, Pagar DD, Menezes PL, Linul E (2019) Fiber-reinforced polymer composites: manufacturing, properties, and applications. Polymers. https://doi.org/10.3390/polym11101667
6. Rubino F, Nisticò A, Tucci F, Carlone P (2020) Marine application of fiber reinforced composites: a review. J Mar Sci Eng. https://doi.org/10.3390/JMSE8010026
7. Yaacob A, Zakaria ZA, Zarina MKP, Koto J, Kidd P (2015) Production process of fiberglass dast interceptor boat. https://isomase.org/JOMAse/Vol.19%20May%202015/19-3.pdf. Accessed 22 June 2021
8. Campbell FC (2010) Structural composite materials. In: Structural composite materials. https://doi.org/10.31399/asm.tb.scm.9781627083140

Chapter 7
Instruments Utilized in Short Sea Shipping Research: A Review

Amayrol Zakaria, Aminuddin Md Arof, and Abdul Khabir

Abstract This study explores the instruments of short sea shipping (SSS) that has been utilized through an inclusive review of papers published in well-known journals over the 2002–2021 period. Systematic exploration shows that maritime policy and management plays a dominant role in publishing short sea shipping research. At the same time, classification and identification of important determinants and barriers for successful SSS has been found as the main research area, followed by attractiveness and competitiveness of SSS; policy and subsidisations in SSS and multimodal transportation network; energy efficiency, emission and environmental issue and protection; ports and transport system efficiency; cargo operations, inventory management, competitive SSS technology and ICT; potential demand, opportunity and competitive advantage; sustainable development and influence of meteorological and weather conditions on SSS operations as popular topics. Since 2002, the use of quantitative and qualitative analysis techniques has progressively increased in SSS in order to help researchers make decisions through selected scientific methods. With this work, present and prospective researchers can understand the contemporary development and popular research topics in SSS. By presenting a review on the common research instruments and techniques used in SSS research, this study is expected to fill the gap in the present literature through the collation of information on the research approaches in contemporary SSS studies.

Keywords Short sea shipping (SSS) · Information communication technology (ICT) · Sustainable development · Quantitative analysis · Qualitative analysis

---

A. Zakaria (✉) · A. Md Arof
Universiti Kuala Lumpur, Malaysian Institute of Marine Engineering Technology, Lumut, Malaysia
e-mail: amayrol@unikl.edu.my

A. Md Arof
e-mail: aminuddin@unikl.edu.my

A. Khabir
Faculty of Business Management, Malaysia Institute of Transport, Universiti Teknologi MARA, Shah Alam, Malaysia
e-mail: abdulkhabir@uitm.edu.my

A. Ismail et al. (eds.), *Design in Maritime Engineering*, Advanced Structured Materials 167, https://doi.org/10.1007/978-3-030-89988-2_7

7.1 Introduction

This paper is organized as follows: Sect. 7.1 identifies the main research area topics in short sea shipping. Section 7.2 presents the collection of relevant papers and a review process. Next, Sect. 7.2 presents volume outline and analysed research methods and data analysis techniques, and Sect. 7.3 presents conclusions and discussion.

Basically, SSS defined as oceanic transport between the seaports of a nation-state as well as between a country's seaport and the seaports of neighbouring countries [13]. This definition includes the type of service, being cabotage or coastal within the ports of a country that may extend the geographic coverage to adjacent countries. In 1992, the European Commission presented its first SSS publication that describes SSS as a carriage of goods and passengers by ocean between ports situated on the mainland of one-member state without calling at islands or services between ports of one-member state where one or more ports are situated on islands and offshore supply services [25]. In 2005, the US Maritime Administration (MARAD) defined SSS as a commercial maritime transportation that does not passage an ocean. It is another method of commercial conveyance that employs coastal waterways and inland to transfer commercial freight from main domestic harbours to its destination [121]. Constant with the MARAD definition, [129] reveals the following basics to the above-mentioned definitions: feedering, intermodalism, inter-regional cargo, border crossing, transhipment, spoke and hub networks and a substitute to road transport for trailers or container [129]. In a more inclusive approach, [5] indicates that SSS generally involves the carriage of cargo, passengers and vehicles by vessels along the coasts, to and from nearby islands, within internal waters such as rivers and lakes but without passage crossing an ocean.

For a nation, SSS not only ensures the transportation of resources needed for production processes but also facilitates the transshipments of vehicles, which accumulates more advantages for the nation. It is argued by [109] that SSS has some advantages which are decreased environmental impact, better utilization of infrastructure, potential cost efficiencies and coastal economic development. Along similar line of argument, [3] reveals that a feasible SSS that alternately connected by road transport will minimize road congestion, reduce road construction and maintenance costs and improve the surrounding environment. On the other hand, since maritime transport offers higher fuel economy and lower emissions of harmful pollutants, SSS has been considered as one of the most sustainable and economically competitive modes of transport [72].

7.2 Research Area in Short Sea Shipping

The SSS sector has traditionally been regarded as an important component of a nation's economic system. Therefore, research on SSS has received a countless deal of attention from scholars worldwide. For instance, an impressive volume of studies have been conducted for classification and identification of important determinants for its successful undertaking [2, 4, 6, 7, 10, 11, 14–18, 36, 54, 63, 65, 72, 73, 79, 84, 86, 88, 88, 89, 104, 106, 112, 113, 126, 138, 141, 143, 147, 151, 164, 166, 167, 169].

Subsequently, the next popular SSS research area is on the attractiveness and competitiveness of SSS in a multimodal transport chain [8, 9, 19–21, 27, 34, 35, 37, 39, 42, 62, 66, 68, 71, 77, 78, 87, 94, 102, 113, 120, 130, 145, 146, 151, 155, 161]. This is followed by research on policy, subsidization, funding and government assistance in SSS and multimodal transportation network as the third-most popular area [12, 44, 51, 59, 60, 81, 82, 92, 111, 113, 131, 136, 139, 148, 153, 156, 157, 164].

The fourth research area involves energy efficiency, emission and environmental issue and protection [23, 40, 47–49, 52, 53, 55, 58, 61, 67, 70, 105, 116, 120, 122, 124, 132, 143, 156, 160, 162, 165, 168].

This is followed by ports and transport system efficiency, cargo operations, inventory management, competitive SSS technology and ICT [1, 22, 26, 28, 31, 33, 41, 45, 56, 57, 63, 76, 83, 94, 97, 99, 103, 106, 107, 113, 128, 135].

The sixth research area covers potential demand, opportunity, competitive advantage and sustainable transport development [32, 38, 50, 58, 59, 69–71, 74, 75, 90, 92–94, 98, 100, 108–110, 119, 124, 125, 127]. Eventually, only few studies can be traced that focus on the influence of meteorological or weather conditions on SSS operations [29, 43].

Basically, such reviews are helpful to understand the contemporary development and popular research topics in SSS. By presenting a review of the major research instruments and techniques used in SSS studies, this study fills the gap in the present literature by collating information on the research approaches of contemporary SSS studies, whether through the qualitative or quantitative techniques.

Consequently, this study reviews the major data analysis techniques used in SSS research. Next, comparing the status of current research in SSS, it is found that excessive commitment has been given to shipping operations, port strategies and SSS developments, but few efforts have been made to research in meteorological or weather conditions, policy and government assistance. The next section will explain the most exciting instruments that have been employed and followed by rare instruments that can be traced employed in SSS study.

Based on thorough observation, there are numerous studies on SSS that have been conducted between 2002 and 2021. Obviously, the Maritime Policy and Management journal plays an important role in publishing SSS studies involving 50 journal articles, followed by Maritime Economics and Logistics with 30 papers, The International Journal of Shipping and Transport Logistics with 25 papers, Journal of Transport Geography with 15 papers, Transportation Research Part A with11 papers, Marine Pollution Bulletin with 8 papers, Transport Reviews with 3 papers, The Asian Journal of Shipping and Logistics with 3 papers, Journal of Navigation with 3 papers, Transport Policy with 3 papers, Transportation Research Part D with 4 papers, Journal of Shipping and Trade with 2 papers, Journal of Maritime Research with 2 papers, Marine Policy with 2 papers, Journal of Shipping and Ocean Engineering with

2 papers, Transport and Telecommunication Journal with 2 papers, International Journal of e-Navigation and Maritime Economy with 2 papers and the last one is the Naval Architect with 2 papers. Impressively, the number of the articles and studies that has been published in SSS keeps increasing for the past 3 years until the present day.

The papers are identified using Scopus, Research Gate, and Google Scholar as research reference and carrying out queries with particular keywords. In order to ensure consistency and homogeneity, all non-indexed journals and dissertations are excluded from the selection due to their more subjective impact on academic literature. This study also presents an academic review on SSS and its significant techniques and method that are utilized by the respective areas. Lastly, a systematic summary of the significant techniques and the gap that this study expects to fill are offered. Apart from that, in order to comprehend the review on the selected instruments, the next section will be deliberated on the recognizable instruments that have been employed in SSS.

7.3 Results and Discussion

7.3.1 Algorithms

Based on Table 7.1, there are several approaches of algorithms that have been used in open literature. Different researcher utilized different levels depending on the area that he or she had explored. For instance, [50] employed the label-setting algorithm in a study to determine the optimal routes for mother and daughter ships, as well as the optimal size of the daughter ships. Authors of [41] employed the pathfinding

Table 7.1 Approaches of algorithms in SSS

Instruments	Scholar	Area
Label-setting algorithm	[50]	Optimal routes optimal size of ships
Pathfinding algorithm	[41]	Ship routing
Algorithm and meteo-oceanographic predications	[40]	Ship routing
Guided algorithm	[22]	Schedules of ships
Multi-objective evolutionary algorithm	[70, 71]	Fleets for the sea motorways
NSGA-ii algorithm	[70]	Optimization of container fleets
Frank–Wolfe hybrid algorithm	[96]	Carbon emission reduction
Algorithms	[52]	Energy efficiency
L-shaped algorithm	[1]	Inventory routing
Label-setting algorithm and path-flow-based formulation	[157]	Short sea shipping cost efficient

algorithm in ship routing and revealed the economic benefits of using ship routing in SSS during energetic wave episodes. Contrarily, authors of [40] employed the algorithm and Meteo-Oceanographic technique to come out with the predictions of ship routing from a European perspective. Within similar research area, [22] employed the guided algorithm and come out with the optimal dynamic rearrangement of schedules of other ships to meet the demand after a vessel's immobilization. From a Spanish and French perspective, [70, 71] employed the multi-objective evolutionary algorithm and revealed that the most suitable fleets for the Sea Motorways are Vigo-St.Nazaire and Gijón-St.Nazaire. Additionally, [70] employed the NSGA-II algorithm to investigate the influence of external costs on the optimization of container fleets and revealed that optimized vessels are able to provide 'greener' multimodal chains than the road. In a hybrid approach, [95] employed the Frank–Wolfe hybrid algorithm to investigate the design of coastal shipping services subject to carbon emission reduction targets and state subsidy levels and proposed a model that was applied to Bohai Bay in China.

In term of energy efficiency in SSS, [52] employed fuel consumption, operational and design specifications algorithms, and the results show that even a conservative estimate of one to four hours of reduced time per port call would lead to a reduction in energy use of 2–8%. Last but not least, from an African perspective, authors of [44] employed the L-shaped algorithm to investigate the inventory routing problem with stochastic sailing and port times. This study presented a computational study based on real-world instances. The following section will discuss the qualitative studies but no specific instruments that have been employed in SSS study.

7.3.2 *Qualitative Research*

Based on Table 7.2, numerous researchers employed a qualitative approach, and it is difficult to identify whether they employed a specific qualitative instrument or techniques to reveal their findings. Among others, [38] conducted a survey to investigate autonomous technologies in SSS and revealed that autonomous technologies are practical to the shipping industry that is facing the problems of crew costs and skill shortage. From the same perspective, [122] employed a survey to investigate an environmental friendliness of SSS towards road transport and presented a new tool to identify the geographical scope where each transport alternative is more environmentally friendly. Similarly, Hamilton [44] employed the same approach to investigate the SSS network and finance model and suggested that proper consultation with stakeholders and design of the implementation can mitigate constraints. [100] also employed the qualitative method when they investigated on slow steaming as part of SECA compliance strategies among Ro-Ro and Ro-Pax shipping companies and conclude that for Ro-Pax and Ro-Ro segment, bunker prices, rigorous competition and most importantly different service quality requirements have significantly restricted the potential implementation of slow steaming.

Table 7.2 Approaches of qualitative research in SSS

Instruments	Scholar	Area
Survey	[38]	Autonomous technologies
Survey	[122]	Environmental
Survey	[44]	SSS network and finance model
Qualitative methodology	[100]	Ro-Ro and Ro-Pax shipping
Qualitative methodology	[51]	Policy implications
Qualitative methodology	[57]	Barriers for SSS
Survey	[85, 89–91, 118]	Current practices of European, China, Japan and Korean SSS
Survey	[21]	Competitiveness of Turkish coaster merchant fleet
Survey	[109]	Food loss reduction in Northeast Asia
Qualitative and observations	[14]	Short sea shipping sustainable
Interviews	[165]	Safety in short sea seafaring
Interviews and triangulated analysis	[167]	Role of policy in supporting SSS

Form both American and European point of view, [94] employed this instrument to investigate the SSS and its prospects. This study concludes that SSS can improve technologically effectiveness for advanced solutions and customized that will more integrate it into the intermodal transportation chain and will advance its image among shippers. From the South American perspective, [51] employed a qualitative approach to investigate potential and policy implications in Brazil and revealed that the cabotage law restricts the coastal and inland water trade to vessels flying its national flag. Additionally, [57] employed the same approach to investigate the barriers and enablers for SSS in Southern Africa and revealed that SSS has the theoretical potential to work in the African continent. Lastly, some other scholars also employed the same approach to investigate SSS in Asia.

For both European and Northeast Asia, [85, 89–91, 118] conducted a survey to investigate the current practices of European, China, Japan and Korean SSS. Similarly, [21] conducted a survey and interview to investigate the competitiveness of the Turkish coaster merchant fleet, and this study revealed that Turkish needs to build low draft, river going and box-type vessels to gain a competitive advantage. In the Northeast Asia, [109] employed the same method to investigate the food loss reduction in emerging economies by exploiting SSS opportunities. This study facilitates the applicants in eliminating both unfruitful options with respect to business effectiveness and food loss reduction.

Table 7.3 Approaches of decision support systems (DSS), discrete event simulation and decision networks in SSS

Instruments	Scholar	Area
Decision networks	[134]	Sustainability of SSS
DDS	[76]	Cascading feeder vessels
Discrete event simulation	[125]	Sustainability of SSS
DSS	[124]	Ballast water risk

7.3.3 Decision Support Systems (DSS), Discrete Event Simulation and Decision Networks

Based on Table 7.3, there are four studies that were traced to have employed these instruments. All the studies are European based. Firstly, [134] employed decision networks to investigate the sustainability of SSS and revealed that the capacity of regasification terminals for LNG, which was under construction, and the modal distribution of inland water cargo transportation are the two most important variables for the decision to implement liquefied natural gas (LNG) as fuel. Second, [76] employed the DDS to investigate the cascading feeder vessels and the rationalization of small container ports. The results of this research show that by 15% of the sub-1000 TEU fleet presently laid up and very limited on order, bigger feeders with deeper drafts seem certain to assist at least nearly of these routes. On the other hand, [125] employed discrete event simulation to investigate sustainability development in SSS. Finally, [124] used DSS to investigate a ballast water risk indication for the North Sea, and they concluded that exemptions from BWM are not recommended for the North Sea area.

7.3.4 Multi-criteria Decision-Making (MCDM), AHP, Delphi, Fuzzy Dematel, Fuzzy Logic

Based on Table 7.4, from the European perspective, [71] employed multi-criteria decision-making to analyse the Motorways of the Sea for France and Spain and come out with the most suitable motorways of the sea options. Authors of [96] also employed the same instrument to explore the logistics network and externalities for short sea transport. This study discloses savings in intermodal transportation costs and a slight decrease in externalities with respect to land transportation. On the other hand, there are pretty numbers of academia that employed the same instruments under MCDM which are Delphi, Fuzzy Dematel and AHP (analytic hierarchy process).

From European perspectives, [63] employed the Delphi technique and presented a transport model using hierarchy process study on key performance indicators of island transport services. With the same perspectives, [88] also employed the same method to investigate the strengths and weaknesses of SSS and proposed a new approach to clarify some concepts in SSS. Using the same technique, [104] explored

Table 7.4 Approaches of multi-criteria decision-making (MCDM), AHP, Delphi, fuzzy Dematel and fuzzy logic in SSS

Instruments	Scholar	Area
Multi-criteria decision-making	[71]	Motorways of the sea
MCDM	[97, 153]	Logistics network and externalities for SSS
Delphi technique	[63]	Indicators of island transport services
Delphi technique	[88]	Strengths and weaknesses of SSS
Delphi technique	[105]	Potential for British coastal shipping
Delphi-fuzzy dematel	[126]	Barriers to coastal shipping
Fuzzy logic	[30]	Economic feasibility study of SSS
Delphi technique	[7]	SSS in Archipelagic Southeast Asia (ASEA)
Delphi-AHP	[6]	Interstate Ro-Ro SSS routes in ASEA sub-region
AHP	[2, 4]	Ro-Ro SSS operations in archipelagic southeast Asia
AHP	[98]	Sustainability on the feeder service improvement in Malaysia
AHP-Delphi	[138]	Ro-Ro operation

the potential for British coastal shipping in a multimodal chain and concluded that managers are in favour of multimodal developments, in particular cooperation between coastal shipping and road haulage. In addition, [126] employed the Delphi with Fuzzy DEMATEL to investigate the barriers to coastal shipping development in India and recommended relaxing cabotage rules to stimulate the inflow of foreign capital in order to grow coastal shipping, improving the current port system through joint efforts of the ports, Indian customs and government, and fostering supply chain collaboration. On the other hand, [30] employed the Fuzzy logic to investigate an economic feasibility study of SSS, and the results show that SSS has impressive prospective for further enlightening its environmental performance by depressing ship emissions at ports. In terms of multi-criteria decision method, [97, 153] employed this method to evaluate the performance of potential alternative ports as intermodal cargo consolidation centre.

In Asia, [7] employed this instrument to investigate the potential benefits and obstacles of interstate SSS in Archipelagic Southeast Asia (ASEA), and this study addresses the gap in the literature by focusing on SSS in ASEA, particularly those involving interstate Ro-Ro operations and extending the usage of the Delphi technique to the realm of interstate SSS. Subsequently, [6, 138] employed this study together with the AHP technique, and this study finalizes with the development of a decision-making model for interstate Ro-Ro SSS routes in ASEA sub-region.

For AHP, [2, 4] employed this instrument to develop a model of Ro-Ro short sea shipping operations in Archipelagic Southeast Asia and come out with the improvement of a decision-making model that was tested on three interstate Ro-Ro SSS routes within the ASEA sub-region. With the same region, [98] employed this instrument to investigate influences of the feeder service expansion in Malaysia concerning its sustainability, and this study revealed that the service sub-criteria is the most important feature on the feeder service improvement in Malaysia.

7.3.5 Econometric Analysis, Cost Benefits Analysis (CBA), Economic Analysis, Cost Model and Monetary Cost, and Costs and Transit Time Model

Based on Table 7.5, in terms of financial analysis, [132] employed the cost benefit analysis (CBA) to investigate the role of regulation and fuel prices, and the results show that there are significant implications of the new sulphur limits to the payback period of emissions abatement investments, particularly following the unexpected drop in fuel prices. From a European perspective, [130] employed the same approach by using economic statistical analysis to investigate in the global logistics trend

Table 7.5 Econometric analysis, cost benefits analysis (CBA), economic analysis, cost model and monetary cost, and costs and transit time model in SSS

Instruments	Scholar	Area
CBA	[132]	Regulation and fuel prices
Economic analysis	[130]	Container and Ro-Ro shipping
Economic analysis	[68]	Analysis of SSS container
Cost and transit time model	[131]	Costs of averting modal shifts
Cost and transit time	[35]	Competitiveness of SSS
Cost model	[71]	Fleets for sea motorways
CBA	[111, 143]	Sustainable transportation
Costs and transit time	[74]	Container transport
Econometrics and spatial shift-share	[110]	SSS services
Costs and transit time	[108]	Demand in SSS
Approximate analysis and assessing external costs	[58–60]	Role of SSS in sustainable development
CBA with SWOT analysis	[80, 81]	Nigerian coastal and inland shipping cabotage policy
CBA	[133]	Ro-Ro SSS network
CBA	[148]	Integrated short sea shipping
Life-cycle cost assessment (LCCA)	[160]	CO_2 emissions and environmental
Life cycle assessment—LCA method	[168]	CO_2 emissions and environmental
Quality function deployment (QFD)	[151]	Quality of high-speed ferry
Visualization analysis	[153]	Short sea shipping
Configurational approach of the space syntax analysis	[159, 161]	Short sea shipping

spillover through container and Ro-Ro shipping in North Europe SSS. This study concludes that SSS in North Europe tends to divide into two different ways, participating in competition as a logistics provider or strengthening their own position as a pure carrier. With the same approach, [68] investigates the analysis of SSS container routes in the Mediterranean and the Black Sea. Their study revealed that the Italian foreign trade is becoming more oriented to high value goods.

From the same perspectives, [131] employed cost and time model to investigate the costs of averting modal shifts in the European SSS sector and come out with the proposed measures that can successfully reduce the negative effects of the regulation. Authors of [35] also employed the monetary cost and transit time to explore in the competitiveness of SSS, and the results show the road option is about 30 and 34% most costly than the best SSS option available for the exportations from Jaén and Southern Catalonia, respectively. Authors of [71] also employed the cost model instruments to explore the optimal fleets for sea motorways. This study concludes that the most suitable fleets for the sea motorways are Vigo-St.Nazaire and Gijón-St.Nazaire. Similarly, [111] also employed CBA to investigate the implementation of the 'ecobonus' project in the Republic of Croatia and contributes to the political initiatives of the European Union for the European maritime space without boundaries and promotes environmentally sustainable transportation. Next, [143] employed the same method to investigate the concept of VT (Vessel Train) to determine the effectiveness of associated cost towards the reduction of crew size. In terms of CO_2 emissions and environmental aspect, [160] revealed that the adoption of green technical and process improvements by Europe's SSS corporations is positively and significantly tied to environmental restrictions as an external institutional driver.

In North America, [74] employed the costs and transit time model to scrutinize the modal choice preferences in short-distance hinterland container transport. The outcomes of the research suggest that, to improve a further modal shift, operators should attempt to deliver daily services at a competitive price, with an emphasis on providing more consistent services than road transport. Authors of [110] furthermore employed the spatial econometrics and spatial shift-share to investigate SSS along the Atlantic Arc. The conclusion of this study shows that along the Atlantic facade of the Iberian Peninsula, there are growing nodes with great potential to enable the increase in their throughput, in terms of inner competitiveness or by establishing new SSS services between neighbouring positions. Authors of [108] also employed the costs and transit time model to investigate a modelled transportation demand in SSS. This model produces the amounts of cargo that could potentially be carried annually through each transport solution, for different freight rates and ship speeds.

On the other hand, [58–60] employed the approximate analysis and assessing external costs to investigate the role of SSS in sustainable development and came out with a model that can be applied to evaluate external benefits of infrastructure investment or a new shipping line. From an African point of view, [80, 81] employed the CBA with SWOT analysis to investigate the benefit of maximizing criteria from the Nigerian coastal and inland shipping cabotage policy. The study reviews government policies affecting investments in ship sizes operating in lakes, rivers canals, inland waters and coastal waters of the maritime state. Last but not least, [133] employed the

CBA to investigate the establishment of a Ro-Ro SSS network connecting Southeast Asian countries. In terms of space syntax analysis, [159, 161] employed this method to determine the relationship between the spaces of the urban grid.

7.3.6 Factor Analysis, Sensitivity Analysis, Statistical Techniques, OLS Regression, Concentration Analysis and Pearson Correlation Analysis

Based on Table 7.6, in terms of statistical technique, there are few studies that were traced to have employed factor analysis as an instrument in SSS. In North America, [127] employed factor analysis to investigate on customer segmentation of freight forwarders and impacts on the competitive positioning of ocean carriers in the Taiwan–Southern China trade lane. This study contributes to a managerial implications and directions for future research. From South America, [20] employed univariate and multivariate statistical techniques to investigate the domestic SSS services in Brazil. This study reveals that cabotage users aim to enhance the integration of logistics between transport modes and to adopt modal shift strategies if better services could be provided, including a real-time information system, shorter transit times and freight offered on a door-to-door basis. In the same research area, [82] employed a statistical technique focusing in regression analysis (OLS Regression)

Table 7.6 Factor analysis, sensitivity analysis, statistical techniques, OLS regression and concentration analysis in SSS

Instruments	Scholar	Area
Factor analysis	[127]	Customer segmentation
Factor-cluster analysis	[166]	The performance of SSS
Univariate and multivariate statistical techniques	[20]	Domestic SSS services in Brazil
Regression analysis (OLS regression)	[82]	Over-exertion in Swedish SSS
Pearson correlation analysis	[169]	Ro-Ro operation
Statistical techniques	[28, 123]	Ro-Ro connectivity
Employed the statistical techniques and secondary data	[103]	Cargoes in tramp shipping
Statistical method and used correlation analysis	[23]	Transportation, production and distribution on costs and environment
Statistical method and used correlation analysis	[36]	Ferry service in SSS
Sensitivity analysis	[24]	Transport mode
Concentration analysis	[117]	Port rationalization and port systems
A numerical method	[164]	Demand on alternative short sea shipping

to explore on administrative burdens and over-exertion in Swedish SSS and revealed that highest levels of exertion were reported by employees in the catering department, positions not generally associated with high administrative burden. Authors of [28] also employed the statistical techniques to explore on port connectivity indices and revealed that for Ro-Ro connectivity, neither the number of links nor the link quality strictly dominate the results of their proposed indicator. Authors of [103] employed the statistical techniques and secondary data to come out with a mixed method to explore in the viability of part cargoes in tramp shipping, and this study stretches a better understanding of potential risks and benefits related to utilization of part cargo operations. In Asia, [23] employed statistical method and used correlation analysis to investigate the impact of restructuring transportation, production and distribution on costs and environment. This study contributes to the Thai rubber industry as a whole.

Additionally, [36] employed these instruments to explore the critical success factor of the ferry transport service in SSS, and this study provides some corresponding reference suggestions about customized service for ferry providers. From Southeast Asia, [24] employed sensitivity analysis to investigate the shippers' choice behaviour in picking transport mode, and this study conclude that shippers potency shift to the SSS mode when the importance weights of CO_2 emission increase, and cost and to trucking mode when the weight of time declines. For concentration analysis, [117] employed this instrument to investigate a port rationalization and the evolution of regional port systems and revealed that constant discussion of reform in Norway concerning regional port governance, analogous to other European nations, may ultimately result in such rationalization. For the numerical method, [164] revealed that factor-cluster analysis can offer policymakers with a preliminary abundance of information that can aid in the development of focused policy for SSS-oriented projects.

7.3.7 *Modal Choice Model, Discrete Choice Models and Preferences of Modal Choice Decision, Employed International Competition Model*

Based on Table 7.7, in Europe, two studies have been traced to have employed these instruments. Authors of [37] employed modal choice model to investigate the European common transport policy and SSS. This study revealed how important the politico-economic evaluation is in a European perspective. Second, [114] employed international competition model to investigate the SSS as an intermodal competitor, and this study concludes that EU as a whole needs to emphasis on ports and transport system effectiveness in order to compete effectively in the freight transport market. Form a North American perspective, [19] employed the modal choice model to investigate the requirements of Atlantic Canadian shippers and proposed a new method of collecting data on how organizations split business in the decisions. In addition, [102]

Table 7.7 Modal choice model, discrete choice models and preferences of modal choice decision, employed international competition model in SSS

Instruments	Scholar	Area
Modal choice model	[37]	Transport policy and SSS
International competition model	[114]	SSS as an intermodal
Modal choice model	[19]	Atlantic Canadian shippers
Discrete choice models	[102]	Competition between Ro–Ro and Lo–Lo services in SSS
Mixed integer linear programming model	[147]	Optimal network configuration for the organization of transport services
Mathematical and calculation models	[156]	Sustainable sources
A mixed multinomial logit model and a latent class choice model	[151]	Inter-urban freight mode choice
Aggregate discrete choice model	[145]	Competitive and sustainable freight transport system
A path-flow model	[136]	Short sea liner shipping network
Transport demand modelling	[146]	Freight transport
Optimization model	[156]	Short sea feeder network

employed discrete choice models to investigate the competition between Ro–Ro and Lo–Lo services in SSS market in Mediterranean countries. This study revealed that the important element that emerges, in general terms, is the segmentation of the market in relation to the distances existing between each couple of countries. For mixed multinomial logit model and a latent class choice model, [151] revealed that freight mode choice is majorly influenced by transit time and frequency of service.

7.3.8 Comparison of Data Collected, Comparative Analysis, Destination (OD) Matrices

Under this category, Table 7.8 shows that [69] employed comparison of data collected instrument to investigate the SSS container routes in the Mediterranean and in the Black Sea. This study concludes that the most important Italian port cluster is the

Table 7.8 Comparison of data collected, comparative analysis, destination (OD) matrices in SSS

Instruments	Scholar	Area
Comparison of data collected	[69]	SSS container routes
Comparison and forecasting calculations	[43]	Vessel speed on bunker cost in SSS
Comparative analysis mixed with destination (OD) matrices	[116]	Emission control in SSS

Ligurian. On the other hand, [43] employed the comparison and forecasting calculations to investigate the impacts of vessel speed on bunker cost in SSS. Their study revealed that the negative economic impacts of the oil price variation can be mitigated to some extent by using lower vessel speeds. Additionally, [116] also employed the comparative analysis mixed with destination (OD) matrices to investigate the environmental effects of emission control area regulations on SSS in Northern Europe focusing on container feeder vessels. This study effectively illustrates how empirical data supports the necessity of stricter SO_x regulations in order for maritime operations to uphold a green image set up against competing transport modes. For Aggregate discrete choice model, [145] employed this method to determine the impacts of green collaboration in land-sea freight transport.

7.3.9 Energy Efficiency Design Index, Analysis of the Sea, Advanced Modelling Approach, SO_2 Emission Calculations

In terms of environmental, Table 7.9 shows that [61] employed the advanced modelling approach to investigate the methodology of power distribution system design for hybrid SSS, and this study presents a modelling methodology for dimensioning marine vessel ES and hybrid PDS components. On the other hand, [120] employed the energy efficiency design index to examine Ro-Pax and passenger ships in Greece and demonstrated that the 'EEDI baseline' calculation is highly influenced by vessel design and operational characteristics, dictating the need for close monitoring of the EEDI effectiveness in this sector. In terms of Port connectivity index, [155] employed this method to study foreland port connectivity disaggregated by destination market. This study revealed that the technique allows for a more specific identification of the major port rivals by destination market, resulting in more detailed data to aid port managers and policymakers in their decision-making.

Table 7.9 Energy efficiency design index, analysis of the Sea, advanced modelling approach, SO_2 emission calculations in SSS

Instruments	Scholar	Area
Advanced modelling approach	[61]	Hybrid SSS
Energy efficiency design index	[120]	Ro-Pax
Port connectivity index (FPCI)	[155]	Container short sea shipping (SSS) services

Table 7.10 Strength–weaknesses–threats–opportunities (SWOT) analysis in SSS

Instruments	Scholar	Area
SWOT analysis	[101]	The rewards of SSS
SWOT analysis	[81]	Benefit maximizing criteria

Table 7.11 Job Demands–resources (JD–R) model in SSS

Instruments	Scholar	Area
JD–R model	[93]	Shipping lines
JD–R model	[92]	Short sea cargo

7.3.10 Strength–Weaknesses–Threats–Opportunities (SWOT) Analysis

From a European perspective, Table 7.10 shows that [101] employed SWOT analysis to analyses the rewards of SSS in Croatia as well as the environmental anxieties related to shipping in universal and SSS particular. This study revealed that strengths and opportunities definitely reduce the implication of threats and weaknesses. From Africa, [81] employed (SWOT) analysis to investigate the benefit maximizing criteria from the Nigerian coastal and inland shipping. The study shows that in the valuation of marine policy, one tool that has been very helpful is the social costing technique.

7.3.11 Job Demands Resources (JD-R) Model

In general perspective, Table 7.11 shows [93] employed a JD-R model to examine the seafarers' perceptions of job demands of SSS shipping lines and their effects on work and life on board. Thus, this study concludes that a good working climate was pivotal in counteracting negative emotions and supporting motivation and collaboration. In another research, [92] investigated effects of job demands and social interactions on fatigue in short sea cargo shipping and highlight the importance of considering social interactions on board to advance our understanding of stressors and strain in seafaring.

7.3.12 Data Envelopment Analysis (DEA)

Based on Table 7.12, [47] employed a data envelopment analysis (DEA) to investigate the environmental demands and the future of the Helsinki – Tallinn freight route. This study revealed that current truck and semi-trailer-based transportation is challenged by containers, irrespective of how they are carried.

Table 7.12 Data envelopment analysis (DEA) in SSS

Instruments	Scholar	Area
DEA	[47]	Freight route

7.3.13 Data Mining Techniques

From the European perspectives, [39] employed the data mining techniques to evaluate liquid natural gas (LNG) utilization in SSS. This study found that the capacity of LNG regasification terminals under construction and modal distribution of cargo transport by inland waters are the two root nodes of the network.

7.3.14 Porte's Five Force Model

The author of [86] employed Porter's five force model to investigate the motorway of the sea port's requirements. The study revealed that the viewpoint of port authorities regarding this matter and suggested a list of 21 pre-requisites that ports can use to assess their potential.

7.3.15 Assessing External Costs

In Europe, [60] employed the assessing external costs instrument to explore the role of SSS in sustainable development of transport, and this study proposed a model that can be applied to evaluate external benefits of infrastructure investment or a new shipping line.

7.3.16 Impact Pathway and Top-Down Approaches

In Northeast Asia, [62] employed this instrument to explore in the management of empty container flows through SSS and regional port systems. This study found that SSS represents a viable tactic in the implementation of a regional port system development strategy.

7.3.17 Longitudinal Analysis

Through this instrument, [46] investigated the growing trade of unitised SSS for Finland and Germany, and this study discovered that growth of unitised maritime cargo flow has faced difficulties after global financial crisis between Germany and Finland.

7.3.18 Formal Concept Analysis (FCA) Method

The authors of [64] employed this instrument to investigate the revitalization of SSS through simplified, slender and standardized designs and indicates that significant fuel and cost savings can be achieved by designing and building slender, simplified and standardized short sea ships. The savings might be of a similar magnitude as the traditional economies of scale benefits which are achievable by doubling the vessel size.

7.3.19 Novel Methodology

The authors of [107] employed novel methodology to examine the methodology for Ro-Ro ship and fleet sizing with application to SSS that allows the identification of the most suitable ship and fleet sizes for different market penetration levels.

7.3.20 Simulation Modelling Method

From the study of [83], the simulation modelling method was employed in an analysis of Ro-Ro terminals and shows that the variable that mostly affect the terminal capacity is the number of trucks arriving to terminals. It has also been utilized by [164] who focused on the Ro-Ro operation.

7.3.21 Two-Phase Hybrid Matheuristic

The authors of [45] employed a two-phase hybrid matheuristic to analyse the multi-product SSS inventory-routing problem and come out with an adaptive large neighbourhood search to solve the resulting ship routing and scheduling problem.

7.3.22 Bi-Objective Optimization, Mathematical Model, a Structural Equation Model, Discretization Method and Mathematical Formulation

In Europe, [99] employed bi-objective optimization and mathematical model in a study of container transport flow of goods and proposed a mathematical model that can minimize the transit time and transportation costs of container imports to Serbia. For the structural equation model, [162] revealed that there is a significant relationship between environmental regulations and the green technological among Europe's SSS companies. The other study by [140, 150] also employed the same method that focused on the cost reductions in ship and port operations.

7.3.23 Theoretical Intermodal Competition Model

Eventually, [115] employed the theoretical intermodal competition model in the topic of SSS as intermodal competitor, and the study concludes that the EU needs to focus on ports and transport system efficiency as a whole in order to compete effectively in the freight transport market. Thus, discussion on the papers that have been review and a conclusion will be simplified for the next section.

7.4 Conclusion

This study reviews a thorough literature review of data analysis techniques used in SSS between 2002 and the middle of 2021. It shows how SSS research has been undertaken to give researchers a superior understanding of the current state of research in the realm of SSS. Throughout the review period, there has been an increasing tendency in the utilization of both the quantitative and qualitative instruments for SSS research. More academic and research publications have also been consistently published throughout the period of consideration. The advance in SSS research is attributable to the fact that SSS has traditionally been regarded as an important element of a nation's economic system. During the review period, the MPM journal has maintained a dominant position in publishing research in the various areas involving SSS. Today, experience is greater than before by expert analysis and informed by research findings made using a variety of quantitative and qualitative instruments, and some of scholars pursued mix methods then become qualitative–quantitative approach to come out with the inspiring research contribution. As a whole, quantitative instruments have been used more often as compared to the qualitative approaches. The popularity of quantitative approaches is associated with the introduction and application of computer science, statistics and mathematics in social science research. Despite the comprehensive approach, this study retains some limitations. For instance, this study considered only journal articles that visibly revealed their instruments because a clear understanding is essential to simplify the exact instruments that have been used by some scholars. Nonetheless, this study is suggestive and offers researchers a deep exploration of the present situation of SSS research.

Acknowledgements The gratitude of the authors is for the University Kuala Lumpur Malaysian Institute of Marine Engineering Technology for providing a constructive environment to present this research.

References

1. Aperte XG, Baird AJ (2013) Motorways of the sea policy in Europe. MPM 40(1):10–26
2. Arof AM (2013) The Challenges in the Development of a Feasible Short Sea Shipping Project in BIMP EAGA Sub-Region. Marine Frontier
3. Arof AM (2018) Decision making model for Ro-Ro short sea shipping operations in Archipelagic Southeast Asia. AJSL 34(1):33–42
4. Arof AM, Nair R (2014) Key factors for a feasible Ro-Ro Short Sea Shipping in BIMP-EAGA sub-region. IMC. Labuan, Malaysia, Universiti Malaysia Sabah, pp 161–178
5. Arof AM, Nair R (2017) The identification of key success factors for interstate Ro-Ro short sea shipping in Brunei-Indonesia-Malaysia-Philippines: a Delphi approach. IJSTL 9(3):261–279
6. Arof AM, Hanafiah RM, Ooi IUJ (2016) A Delphi study on the potential benefits and obstacles of interstate short sea shipping in Archipelagic Southeast Asia. IJME 5:97–110
7. Asia Pacific Economic Cooperation (APEC) (2007) Short Sea Shipping Study: a report on successful SSS models that can improve ports efficiency and security while reducing congestion, fuel costs and pollution. APEC Transportation Working Group & Inha University, Singapore
8. Asian Development Bank (2010) ADB bridges across ocean; initial impact assessment of the Philippines Nautical Highway System and Lesson for Southeast Asia, ADB and Asia Foundation
9. Baindur D, Viegas JM (2011) An agent-based model concept for assessing modal share in inter-regional freight transport markets. JTG 19(6):1093–1105
10. Baindur D, Viegas JM (2012) Success factors for developing viable motorways of the sea projects in Europe. Logist Res 4(3–4):137–145
11. Baird AJ (2007) The economics of Motorways of the Sea. MPM 34:287–310
12. Balduini G (1982) Short Sea Shipping in the economy of inland transport in Europe: Italy. European Council of Ministers of Transport, 1–2
13. Becker JFF, Burgess A, Henstra DA (2004) No need for speed in short sea shipping. MEL 6(3):236–251
14. Bendall HB (2013) Australian flag shipping: have the regulatory changes of 2012 help or hindered opportunities for Australian coastal shipping investor, a paper presented at IAME 2013 conference, July 3–5, Marseilles, France
15. Breuillet A, Harding A, West R (2002) The potential for regional cabotage in Central America. In: IAME Panama 2002 proceedings
16. Briceno GC, Bofinger H, Cubas D, Millan PMF (2015) Connectivity for Caribbean countries: an initial assessment
17. Brooks* MR, Frost JD (2004) Short sea shipping: a Canadian perspective. MPM 31(4):393–407

18. Brooks MR, Trifts V (2008) Short sea shipping in North America: understanding the requirements of Atlantic Canadian shippers. MPM 35(2):145–158
19. Casaca ACP, Galvão CB, Robles LT, Cutrim SS (2017) Domestic short sea shipping services in Brazil: competition by enhancing logistics integration. IJSTL 9(3):280–303
20. Çetin İB, Akgül EF, Koçak E (2018) Competitiveness of Turkish coaster merchant fleet: a qualitative analysis by short sea shipping perspective. TransNav: IJSMSS 12
21. Chainas KA (2017) Dynamic routing system for short sea shipping following ship immobilisation. IJBSR 11(1–2):198–212
22. Chanchaichujit J, Saavedra RJK (2016) Analysing the impact of restructuring transportation, production and distribution on costs and environment–a case from the Thai Rubber industry. IJLRA: 1–17
23. Chang CH, Thai VV (2017) Shippers' choice behaviour in choosing transport mode: the case of South East Asia (SEA) Region. AJSL 33(4):199–210
24. Commission of the European Communities (1992) Council Regulation (EEC) No. 3577/1992, Applying the principle of freedom to provide services to maritime transport within member states (maritime cabotage). Brussels, O J no. L 364
25. Daduna JR, Hunke K, Prause G (2012) Analysis of short sea shipping-based logistics corridors in the Baltic Sea Region. JSOE 2(5)
26. Danesi A, Farina A, Lupi MA (2010) Comparative Analysis of Lo-Lo and Ro-Ro short sea shipping networks in Italy. In: Proceedings of the international conference ICTS: 27–28
27. De LPW, Udenio M, Fransoo JC, Helminen R (2016) Port connectivity indices: an application to European RoRo shipping. JST 1(1):6
28. De OFXM, la CM (2008) Heavy weather in European Short Sea Shipping: its influence on selected routes. J Navig 61(01):165–176
29. Denisis A (2009) An economic feasibility study of short sea shipping including the estimation of externalities with fuzzy logic
30. Dias JQ, Calado JMF, Mendonça MC (2010) The role of European «ro-ro» port terminals in the automotive supply chain management. J Transp Geogr 18(1):116–124
31. Donner P, Johansson T (2018) Sulphur directive, short sea shipping and corporate social responsibility in an EU context. In: Corporate social responsibility in the maritime industry. Springer, Cham, pp 149–166
32. Eriksson K, Wikström K, Hellström M, Levitt RE (2019) Projects in the business ecosystem: the case of short sea shipping and logistics. PMI 50(2):195–209
33. Frémont A, Franc P (2010) Hinterland transportation in Europe: combined transport versus road transport. J Transp Geogr 18(4):548–556
34. Galati A, Siggia D, Crescimanno M, Martín AE, Saurí MS, Morales FP (2016) Competitiveness of short sea shipping: the case of olive oil industry. Br Food J 118(8):1914–1929
35. Gan GY, Chung CC, Lee HS, Wang SW (2017) Exploring the critical success factor to the ferry transport service in short sea shipping. IJSTL 9(3):304–322
36. García ML, Feo VM (2009) European common transport policy and short-sea shipping: empirical evidence based on modal choice models. Transp Rev 29(2):239–259
37. Ghaderi H (2019) Autonomous technologies in short sea shipping: trends, feasibility and implications. Transp Rev 39(1):152–173
38. González CN, Serrano BM, Soler FF (2019) Seaport sustainable: use of artificial intelligence to evaluate liquid natural gas utilization in short sea shipping. Transp J (Pennsylvania State University Press) 58(3):197–221
39. Grifoll M, Castells SM, Martínez DOFX (2016) Enhancement of maritime safety and economic benefits of short sea shipping ship routing. In: Proceedings of SEAHORSE 2016, international conference on maritime safety and human factors
40. Grifoll M, de OFM, Castells M (2018) Potential economic benefits of using a weather ship routing system at short sea shipping. WMU J Marit Aff 17(2):195–211
41. Grosso M, Lynce AR, Silla A, Vaggelas GK (2010) Short sea shipping, intermodality and parameters influencing pricing policies: the Mediterranean case. NETNOMICS 11(1):47–67

42. Hämäläinen E, Inkinen T (2018) Impacts of vessel speed on bunker cost in short sea shipping: a cross-examination. In SHS web of conferences, vol 58, p 01011. EDP Sciences
43. Hamilton T (2018) Short sea shipping network and finance model for the Caribbean (No. IDB-TN-01419). Inter-American Development Bank
44. Hemmati A, Hvattum LM, Christiansen M, Laporte G (2016) An iterative two-phase hybrid matheuristic for a multi-product short sea inventory-routing problem. EJOR 252(3):775–788
45. Hilmola OP, Tolli A (2019) Growing trade, but slowing unitized short sea shipping: analysing Finland and Germany. Transp Telecommun 20(1):82–91
46. Hilmola OP, Lorentz H, Rhoades DL (2015) New environmental demands and the future of the Helsinki–Tallinn freight route. MEL 17(2):198–220
47. Hjelle HM (2011) The double load factor problem of Ro-Ro shipping. MPM 38(3):235–249
48. Hjelle HM, Fridell E (2012) When is short sea shipping environmentally competitive? INTECH Open Access Publisher
49. Holm MB, Medbøen CAB, Fagerholt K, Schütz P (2018) Shortsea liner network design with transhipments at sea: a case study from Western Norway. Flex Serv Manuf J 1–22
50. Iimi A, Rao K, Medina AC (2018) Short Sea Shipping in Brazil: Potential and Policy Implications
51. Johnson H, Styhre L (2015) Increased energy efficiency in short sea shipping through decreased time in port. Transp Res Part A Policy Pract 71:167–178
52. Johnson H, Johansson M, Andersson K (2014) Barriers to improving energy efficiency in short sea shipping: an action research case study. J Clean Prod 66:317–327
53. Kennedy SD (2008) Short sea shipping in the United States-the new marine highways. Tul Mar LJ 33:203
54. Koliousis I, Koliousis P, Papadimitriou S (2013) Estimating the impact of road transport deregulation in short sea shipping: experience from deregulation in the European Union. IJSTL 5(4–5):500–511
55. Konings R, Ludema M (2000) The competitiveness of the river–sea transport system: market perspectives on the United Kingdom-Germany corridor. J Transp Geogr 8(3):221–228
56. Konstantinus A, Zuidgeest M, Christodoulou A, Raza Z, Woxenius J (2019) Barriers and enablers for short sea shipping in the Southern African development community. Sustainability 11(6):1532
57. Kotowska I (2015) The role of Ferry and Ro-Ro shipping in sustainable development of transport. Rev Econ Perspect 15(1):35–48
58. Kotowska I (2016) Method of assessing the role of short sea shipping in sustainable development of transport. IJSTL 8(6):687–704
59. Kotowska I (2016) Policies applied by seaport authorities to create sustainable development in port cities. Transp Res Proc 16:236–243
60. Lana A, Pinomaa A, Peltoniemi P, Lahtinen J, Lindh T, Montonen JH, Pyrhonen OP (2019) Methodology of power distribution system design for hybrid short sea shipping. IEEE Trans Aerosp Electron 66(12):9591–9600. https://doi.org/10.1109/TIE.2019.2892665
61. Le GHD, Griffin MT (2009) Managing empty container flows through short sea shipping and regional port systems. IJSTL 2(1):59–75
62. Lekakou MB, Remoundos G (2015) Restructuring coastal shipping: a participatory experiment. WMU J Marit Aff 14(1):109–122
63. Lindstad E, Eskeland G, Sandaas I, Steen S (2018) Revitalization of short sea shipping through slender, simplified and standardized designs SMC-007-2016
64. Lombardo GA (2004). Short sea shipping: practices, opportunities and challenges. Transportgistics White Paper Series, 24
65. López NMÁ, Ángel MM, María RR, Sánchez J (2011) Accompanied versus unaccompanied transport in short sea shipping between Spain and Italy: an analysis from transport road firms perspective. Transp Rev 31(4):425–444
66. Lu C, Yan X (2015) The break-even distance of road and inland waterway freight transportation systems. MEL 17(2):246–263

67. Lupi M, Farina A, Orsi D, Pratelli A (2017) The capability of motorways of the sea of being competitive against road transport. The case of the Italian mainland and Sicily. J Transp Geogr 58:9–21
68. Lupi M, Pratelli A, Falleni M, Farina A (2019) An analysis of short sea shipping container routes in the Mediterranean and in the Black Sea
69. Martínez LA, Sobrino PC, González MM (2016) Influence of external costs on the optimisation of container fleets by operating under motorways of the sea conditions. IJSTL 8(6):653–686
70. Martínez LA, Sobrino PC, Santos LC (2015) Definition of optimal fleets for sea motorways: the case of France and Spain on the Atlantic coast. IJSTL 7(1):89–113
71. Medda F, Trujillo L (2010) Short-sea shipping: an analysis of its determinants. MPM 37(3):285–303
72. Medina AC, De VNAL, Botter RC Baird AJ (2011) Short sea shipping in Brazil: potential and policy implications, a paper presented at IAME 2011 conference, Santiago De Chile from 25th–28th October 2011
73. Meers D, Macharis C, Vermeiren T, van LT (2017) Modal choice preferences in short-distance hinterland container transport. RTBM 23:46–53
74. Molina SB, González CN, Soler FF (2018) Sustainable short sea shipping: social component analysis through decision networks. Proc Inst Civ Eng-Marit Eng 171(4):135–144
75. Monios J (2017) Cascading feeder vessels and the rationalisation of small container ports. J Transp Geogr 59:88–99
76. Morales FP, Saurí S, Lago A (2012) Potential freight distribution improvements using motorways of the sea. J Transp Geogr 24:1–11
77. Ng AKY (2009) Competitiveness of short sea shipping and the role of port: the case of North Europe. MPM 36(4):337–352
78. Ombo G (2012) Coastal shipping in Africa; prospect and challenges, SAMIC, July 6th, Cape Town
79. Onyemechi, C (2015) Benefit maximizing criteria from the Nigerian Coastal and Inland Shipping Cabotage Policy. RGCI 15(3)
80. Onyemechi C (2015) Benefit maximizing criteria from the Nigerian Coastal and Inland shipping cabotage policy. Revista de Gestão Costeira Integrada- JICZM 15(3):409–416
81. Österman C, Hult C (2016) Administrative burdens and over-exertion in Swedish short sea shipping. MPM 43(5):569–579
82. Özkan ED, Nas S, Güler N (2016) Capacity analysis of Ro-Ro terminals by using simulation modeling method. ASJL 32(3):139–147
83. Paixão CAC, Marlow PB (2005) The competitiveness of short sea shipping in multimodal logistics supply chains: service attributes. MPM 32(4):363–382
84. Paixão CAC, Marlow PB (2009) Logistics strategies for short sea shipping operating as part of multimodal transport chains. MPM 36(1):1–19
85. Paixao CAC (2008) Motorway of the sea port requirements: the viewpoint of port authorities. Int J Logist Res Appl 11(4):279–294
86. Paixao CAC, Marlow PB (2001) A review of the European Union shipping policy. MPM 28(2):187–198
87. Paixao Casaca AC, Marlow PB (2005) The competitiveness of Short Sea Shipping in multimodal logistic chains: service attributes. MPM 32(4):363–382
88. Papadimitriou S, Lyridis DV, Koliousis IG, Tsioumas V, Sdoukopoulos E, Stavroulakis PJ (2018) Strategic planning of short sea shipping within maritime clusters. In: The dynamics of short sea shipping. Palgrave Macmillan, Cham, pp 37–59
89. Papadimitriou S, Lyridis DV, Koliousis IG, Tsioumas V, Sdoukopoulos E, Stavroulakis PJ (2018) Short sea shipping in various regions. In: The dynamics of short sea shipping. Palgrave Macmillan, Cham, pp 163–186
90. Papadimitriou S, Lyridis DV, Koliousis IG, Tsioumas V, Sdoukopoulos E, Stavroulakis PJ (2018) Supply, demand, and major short sea shipping networks in Europe. In The dynamics of short sea shipping. Palgrave Macmillan, Cham, pp 9–36

91. Park YA, Medda FR (2016, May). Cabotage Policy and Development of Short Sea Shipping in Korea, China and Japan. In International Forum on Shipping, Ports and Airports (IFSPA) 2015: empowering excellence in maritime and air logistics: Int J Innov Manag Technol
92. Pauksztat B (2017) 'Only work and sleep': seafarers' perceptions of job demands of short sea cargo shipping lines and their effects on work and life on board. MPM 44(7):899–915. https://doi.org/10.1080/03088839.2017.1371347
93. Pauksztat B (2017) Effects of job demands and social interactions on fatigue in short sea cargo shipping. MPM 44(5):623–640. https://doi.org/10.1080/03088839.2017.1298868
94. Perakis AN, Denisis A (2008) A survey of short sea shipping and its prospects in the USA. MPM 35(6):591–614
95. Chen K, Yang Z, Notteboom T (2014) The design of coastal shipping services subject to carbon emission reduction targets and state subsidy levels. Transp Res E-Log 61:192–211
96. Pérez MJC, Galdeano GE, Andújar JAS (2012) Logistics network and externalities for short sea transport: an analysis of horticultural exports from southeast Spain. Transp Policy 24:188–198
97. Radmilović Z, Zobenica R, Maraš V (2011) River–sea shipping–competitiveness of various transport technologies. J Transp Geogr 19(6):1509–1516
98. Rahmatdin MN, Abdul NSF (2016) The influential factors of the feeder service development in Malaysia towards Its sustainability. JMR 13(3)
99. Rajkovic R, Zrnic N, Stakic D (2016) Application of a mathematical model for container transport flow of goods: from the far East to Serbia. TEH VJESN 23(6)
100. Raza Z, Woxenius J, Finnsgård C (2019) Slow steaming as part of SECA compliance strategies among RoRo and RoPax shipping companies. Sustainability 11(5):1435
101. Runko LL, Ančić I, Šestan A (2013). The viability of short-sea shipping in Croatia. Brodogradnja: Teorija i praksa brodogradnje i pomorske tehnike 64(4):472–481
102. Russo F, Musolino G, Assumma V (2016) Competition between ro–ro and lo–lo services in short sea shipping market: the case of Mediterranean countries. RTBM 19: 27–33
103. Saari J (2016) Viability of part cargoes in tramp shipping company's routing and scheduling decisions
104. Saldanha J, Gray R (2002) The potential for British coastal shipping in a multimodal chain. MPM 29(1):77–92
105. Sambracos E, Maniati M (2012) Competitiveness between short sea shipping and road freight transport in mainland port connections; the case of two Greek ports. MPM 39(3):321–337
106. Sánchez RJ, Wilmsmeier G (2005) Short-sea shipping potentials in Central America to bridge infrastructural gaps. MPM 32(3):227–244
107. Santos TA, Guedes SC (2017) Methodology for ro-ro ship and fleet sizing with application to short sea shipping. MPM 44(7):859–881. https://doi.org/10.1080/03088839.2017.1349349
108. Santos TA, Soares CG (2016) Modeling transportation demand in short sea shipping. MEL
109. Schripsema A, Soethoudt H, Tromp S, Jani B, Vaishnav P, Anit A, Pradhan S (2018) Food loss reduction in emerging economies by exploiting short sea opportunities
110. Seoane MJF, Laxe FG, Montes CP (2017) Short sea shipping in the Atlantic Arc: a spatial shift-share approach. IJSTL 9(3):323–370
111. Smojver Ž, Jugović A, Hadžić AP (2012) Implementation of" Ecobonus" Project In the Republic Of Croatia. ICES J Mar 26(1)
112. Styhre L (2009) Strategies for capacity utilisation in short sea shipping. MEL 11(4):418–437
113. Suárez AA, Campos J, Jiménez JL (2015) The economic competitiveness of short sea shipping: an empirical assessment for Spanish ports. IJSTL 7(1):42–67
114. Suárez AA, Trujillo L, Medda F (2015) Short sea shipping as intermodal competitor: a theoretical analysis of European transport policies. MPM 42(4):317–334
115. Svindland (2016) The environmental effects of emission control area regulations on short sea shipping in Northern Europe: the case of container feeder vessels. Transp Res D: Trans Environ
116. Svindland M (2018) The environmental effects of emission control area regulations on short sea shipping in Northern Europe: the case of container feeder vessels. Transp Res D: Trans Environ 61:423–430. https://doi.org/10.1016/j.trd.2016.11.008

117. Svindland M, Monios J, Hjelle HM (2019) Port rationalization and the evolution of regional port systems: the case of Norway. MPM 1–17
118. Tong J, Watanabe Y (2016) China-Japan Port networks suitable for short sea shipping. JTTE 4:205–220
119. Trujillo L, Medda F, Gonzalez MM (2011) An analysis of short sea shipping as an alternative to freight transport. In Cullinane K (ed), International handbook of maritime economics. Cheltenham, England. Edward Elgar, pp 284–300
120. Tzannatos E, Stournaras L (2015) EEDI analysis of Ro-Pax and passenger ships in Greece. MPM 42(4):305–316
121. MARAD (2008). Short Sea Shipping Brochure
122. Vallejo PJA, Garcia AL, Álvarez FR, Mateo MI (2019) Iso-emission map: a proposal to compare the environmental friendliness of short sea shipping vs road transport. Transp Res D Transp Environ 67:596–609. https://doi.org/10.1016/j.trd.2019.01.015
123. van DBG, Wiegmans B (2018) Short sea shipping: a statistical analysis of influencing factors on SSS in European countries JST 3(1):6
124. van DMR, de BMK, Liebich V, ten HC, Veldhuis M, Ree K (2016) Ballast water risk indication for the North Sea. Coast Manage 44(6):547–568
125. van LT, Caris A, Macharis (2016) Sustainability SI: bundling of outbound freight flows: analyzing the potential of internal horizontal collaboration to improve sustainability. NETS 16(1):277–302
126. Venkatesh VG, Zhang A, Luthra S, Dubey R, Subramanian N, Mangla S (2017) Barriers to coastal shipping development: an Indian perspective. Transp Res D: Trans Environ 52:362–378
127. Wen CH, Lin WW (2016) Customer segmentation of freight forwarders and impacts on the competitive positioning of ocean carriers in the Taiwan–southern China trade lane. MPM 43(4):420–435
128. Woxenius J (2012) Flexibility versus specialisation in ro-ro shipping in the South Baltic Sea. Transport 27(3):250–262
129. Yonge M, Henesey LA (2005) Decision tool for identifying the prospects and opportunities for short sea shipping. JMTL Advisors, LLC, Ft Lauderdale
130. Yoo JY (2018) Global logistics trend spillover through container and RoRo shipping in North Europe
131. Zis T, Psaraftis HN (2019) The costs of averting modal shifts in the European Short Sea Shipping Sector (No. 19–05017)
132. Zis T, Angeloudis P, Bell MG, Psaraftis HN (2016) Payback period for emissions abatement alternatives: role of regulation and fuel prices. Transp Res Rec 2549:37–44
133. JICA (2013) THE MASTER plan & feasibility study on the establishment of an ASEAN RO-RO shipping Network and short sea shipping, ASEAN/JICA
134. Molina SB, González CN, Soler FF (2018, December) Sustainable short sea shipping: social component analysis through decision networks. P I CIVIL ENG-MAR EN. 171(4):135–144
135. Hilmola OP, Tolli A (2019) Growing trade, but slowing unitized short sea shipping: analysing Finland and Germany. Transp Telecommun 20(1):82–91
136. Akbar A, Aasen AK, Msakni MK et al (2021) An economic analysis of introducing autonomous ships in a short-sea liner shipping network. Int Trans Oper Res 28(4):1740–1764
137. Akbar A, Aasen AK, Msakni MK, Fagerholt K, Lindstad E, Meisel F (2021) An economic analysis of introducing autonomous ships in a short-sea liner shipping network. Int Trans Oper Res 28(4):1740–1764
138. Arof AM, Zakaria A (2020) 11 Short sea shipping in the Association of Southeast Asian Nations. Short Sea Shipping in the Age of Sustainable Development and Information Technology, 242
139. Baylon AM, Dragomir C (2020) Maritime and logistics perspectives on short sea shipping sustainable sollutions to road congestion. TransNav 14(1)
140. Chandra S, Christiansen M, Fagerholt K (2020) Analysing the modal shift from road-based to coastal shipping-based distribution–a case study of outbound automotive logistics in India. MPM 47(2):273–286

141. Christodoulou A, Cullinane K (2020) Potential for, and drivers of, private voluntary initiatives for the decarbonisation of short sea shipping: evidence from a Swedish ferry line. MEL, 1–23
142. Christodoulou A, Kappelin H (2020) Determinant factors for the development of maritime supply chains: the case of the Swedish forest industry. Case Stud Transp Policy 8(3):711–720
143. Colling A, Hekkenberg R (2020) Waterborne platooning in the short sea shipping sector. Transp Res Part C Emerg Technol 120:102778
144. Comi A, Polimeni A (2020) Assessing the potential of short sea shipping and the benefits in terms of external costs: application to the Mediterranean Basin. Sustainability 12(13):5383
145. Danloup N, Allaoui H, Gonzales FJ, Goncalves G (2020) Assessing the environmental impacts of green collaboration in land-sea freight transport. In Handbook of research on interdisciplinary approaches to decision making for sustainable supply chains. IGI Global, pp 113–139
146. Elem TR, Ogwude IC, Ibe CC, Nnadi KU, Ejem EA (2020) Decoupling of economic activity and freight transport volume: an evidence for short sea shipping future in the ECOWAS sub-region. JSDTL 5(2):124–134
147. Fadda P, Fancello G, Mancini S, Pani C, Serra P (2020) Design and optimisation of an innovative Two-Hub-and-Spoke network for the Mediterranean Short-Sea-Shipping market. Comput Ind Eng 149:106847
148. Fancello G, Serra P, Carta M, Aramu V, Fadda P (July 2020) Decision-making for maritime networks: evaluating corporate and social profitability of an integrated short sea shipping network in the Upper Tyrrhenian Sea. In: ICCSA. Springer, Cham, pp 83–95
149. Huang ST, Shang KC, Su CM, Chang KY, Tzeng YT (2020) Applying QFD to assess quality of short sea shipping: an empirical study on cross-strait high-speed ferry service between Taiwan and Mainland China. IJSTL 12(4):284–306
150. Jia B, Fagerholt K, Reinhardt LB, Rytter NGM (2020) Stowage planning with optimal ballast water. ICCL. Springer, Cham, pp 84–100
151. Konstantinus A, Zuidgeest M, Hess S, De JG (2020) Assessing inter-urban freight mode choice preference for short-sea shipping in the Southern African Development Community region. J Transp Geogr 88(C)
152. Lee SW, Shin SH, Bae HS (2020) Short Sea Shipping on the West Coast of Korea: Keys to activating the shipping industry in preparation for Korea unification era. JST 18(2):91–105
153. Martínez LA, Trujillo L (2020) 6 Multi-criteria decision method applied to Motorways of the Sea. Short Sea shipping in the age of sustainable development and information technology 141
154. Martínez LA, Romero FA, Chica M (2021) Specific environmental charges to boost Cold Ironing use in the European Short Sea Shipping. Transp Res D: Trans Environ 94:102775
155. Martínez MJ, Feo VM (2020) Measuring foreland container port connectivity disaggregated by destination markets: an index for short sea shipping services in Spanish ports. J Transp Geogr 89:102873
156. Medbøen CAB, Holm MB, Msakni MK, Fagerholt K, Schütz P (2020) Combining optimization and simulation for designing a robust short-sea feeder network. JALGO 13(11):304
157. Msakni MK, Akbar A, Aasen AK, Fagerholt K, Meisel F, Lindstad E (2020) Can autonomous ships help short-sea shipping become more cost-efficient? Oper Res 2019. Springer, Cham, pp 389–395
158. Perčić M, Vladimir N, Fan A (2020) Life-cycle cost assessment of alternative marine fuels to reduce the carbon footprint in short-sea shipping: a case study of Croatia. Appl Energy 279:115848
159. Pratelli A, Rosselli A, Farina A, Lupi M (2021) An analysis between spatial relationships and Short Sea Shipping impacts on Messina's waterfront. Int J Transp Dev Integr 4(3):199–217
160. Raza Z (2020) Effects of regulation-driven green innovations on short sea shippińgs environmental and economic performance. Transp Res D: Trans Environ 84:102340
161. Santos TA, Soares CG (2020) Assessment of transportation demand on alternative short-sea shipping services considering external costs. In Maritime supply chains. Elsevier, pp 13–45

162. Santos TA, Escabelado J, Botter R, Soares CG (2020) 5 Simulating Ro-Ro operations in the context of supply chains. Short sea shipping in the age of sustainable development and information technology, 116
163. Schwartz H, Gustafsson M, Spohr J (2020) Emission abatement in shipping–is it possible to reduce carbon dioxide emissions profitably? J Clean Prod 254:120069
164. Serra P, Fancello G (2020) Performance assessment of alternative SSS networks by combining KPIs and factor-cluster analysis. ETRR 12(1):1–24
165. Shan D, Neis B (2020) Employment-related mobility, regulatory weakness and potential fatigue-related safety concerns in short-sea seafaring on Canada's Great Lakes and St. Lawrence Seaway: Canadian seafarers' experiences. Saf Sci 121:165–176
166. Spoof TK, Niem S (2020) Environmental and economic evaluation of fuel choices for short sea shipping. Clean Technol 2(1):34–52
167. Talbot D, Boiral O (2021) The role of policy in supporting SSS–the case of Quebec. MPM, 1–15
168. Ülker D, Bayırhan İ, Mersin K, Gazioğlu C (2020) A comparative CO_2 emissions analysis and mitigation strategies of short-sea shipping and road transport in the Marmara Region. Carbon Manag 1–12
169. Zakaria A, Arof AM, Ishak IC, Mukti AQ (2020) Ro-Ro Port facilities toward customer satisfaction: evidence from Kuala Perlis Terminal, Perlis, Malaysia. In: Advancement in emerging technologies and engineering applications. Springer, Singapore, pp 299–303

Chapter 8
Issues on Palm Oil Shipment with Regard to the Revised MARPOL Annex II: A Review

Naterah Abdullah Sani, Kanagalingam Selvarasah, and Aminuddin Md Arof

Abstract Issues connected with the discharge of high-viscosity and persistent floating products from tankers and cargo ships have been reported by countries affected by the discharge. This concern prompted the International Maritime Organization (IMO) to amend Annex II of the International Convention for the Prevention of Pollution from Ships (MARPOL 73/78), which came into force on 1st January 2021. This new amendment, however, has a negative impact on the shipping business particularly in the transport of palm oil. The aim of this study is to underline the weaknesses in the policy process of MARPOL Annex II amendment as well as to examine articles and journals about issues involving palm oil, which ended in a misleading view of palm oil among public. The study has involved literature review of published and unpublished materials relating to MARPOL Annex II and vegetable oils, particularly palm oil. The findings revealed flaws in the policy framework of the MARPOL Annex II revision and the perception issues that are working against palm oil are linked with the lack of scientific research, which may serve as proof to indicate that palm oil is not harmful to marine life and living things. The public perception of palm oil and its products has also been proven to be influenced by media.

Keywords MAPROL Annex II · Palm oil · Vegetable oils · GESAMP

N. Abdullah Sani · A. Md Arof (✉)
Universiti Kuala Lumpur Malaysian Institute of Marine Engineering Technology, 32200 Lumut, Perak, Malaysia
e-mail: aminuddin@unikl.edu.my

N. Abdullah Sani
e-mail: naterah.abdullah@s.unikl.edu.my

K. Selvarasah
Marine Department Malaysia, Peti Surat 12, Jalan Limbungan, 42007 Port Klang, Malaysia
e-mail: kanagalingam@marine.gov.my

A. Ismail et al. (eds.), *Design in Maritime Engineering*, Advanced Structured Materials 167, https://doi.org/10.1007/978-3-030-89988-2_8

8.1 Introduction

Each year, large volumes of fully refined or unrefined (slack) vegetable oil such as palm oil, soybean oil, sunflower oil, and olive oil are transported in bulk by tankers and cargo ships around the world. Certain products, like palm oil, must be stored at temperatures above their melting point in order to be filled or discharged in liquid form, and ships are often fitted with cargo heating coils to facilitate this. Following unloading, residuals known as "stripping" frequently stay at the bottom of cargo tanks or crystallise against bulkheads and interior machinery. The crew cleans the tanks manually or automatically with rotary-jet cleaning devices that use steam, hot water or chemical solvents [1]. The residuals can then be treated by port reception facilities or be discharged at sea under certain conditions.

Operational practices for this are regulated by the Annex II of the International Convention for the Prevention of Pollution from Ships (MARPOL 73/78) issued by the International Maritime Organization (IMO), which contains regulations for the control of pollution by noxious liquid substances (NLS) transported in bulk, defining the standards and principles, which must be adopted to discharge harmful substances at sea, as well the standards for controlling such releases. According to the revised MARPOL Annex II, which entered into force in 2007, palm oil that was previously categorised as unrestricted has been classified as "high-viscosity and solidifying substances" that fall within the intermediate pollution category. Category Y of the NLS comprised substances which, if discharged into the sea from tank cleaning or deballasting operations, are deemed to present a hazard to either marine resources or human health or cause harm to amenities or other legitimate uses of the sea. Therefore, it justifies a limitation on the quality and quantity of the discharge into the marine environment [2]. However, chemicals, such as vegetable oils, were still permitted to be discharged into the ocean based on the revised MARPOL Annex II and the International Code for the Construction and Equipment of Ships Carrying Dangerous Chemicals in Bulk (IBC Code) as regulated by the IMO, as long as the shipowner follows the guidelines for the discharge procedure and does not violate the regulation.

In response to years of extensive pollution of the coastlines in Northern Europe as a result of cleaning residues after carriage of vegetable oils and waxes commonly known as persistent floaters, the IMO has amended the IBC Code and MARPOL Annex II on 17 May 2019 under "Cargo residues and tank washings of persistent floating products" that entered into force in North West European Waters, Baltic Sea, Norwegian Sea and Western European Waters on 1 January 2021. According to this amendment, when discharging category Y high-viscosity or solidifying substances (e.g. with a viscosity equal to or greater than 50 mPa s at 20 °C and/or a melting point greater than or equal to 0 °C), the ship tanks should be stripped to the greatest degree feasible, followed by a tank pre-wash operation, with the residue/water combination formed during the pre-wash being discharged to a receiving facility at the port of unloading.

8.2 Background of the Study

This new amendment undoubtedly has negative impacts on the shipping industry, especially in the transport of palm oil. There is a range of concerns to address when dealing with this new provision, such as the shortage of port reception facilities (PRF) to dispose the residue, which is still lacking since not every port has this facility. Inadequate facilities can cause port congestion and will increase demurrage charges as a result of longer waiting times. The commercial washing takes at least one hour, but with certain cargoes, the whole cleaning process that includes flushing and steaming will take up to four to six hours [3]. As the world's most traded vegetable oil, any changes to regulations governing the transport of vegetable oils would have a huge impact on the palm oil business. Shipping palm oil to the Port of Rotterdam now costs between USD 40 and USD 70 per tonne. The cost of a pre-wash procedure for a vessel transporting 35,000 tonnes of palm oil to the Port of Rotterdam will rise to USD 68,600, or USD 1.96 per tonne of palm oil, due to the possibility of a longer wait at the port. The present shipping cost of palm oil from Malaysia to the Port of Rotterdam may rise by 2.5–7.1% as a result of this. Exporting palm oil will cost between USD 16 million and USD 20 million more for exporters. This increased expense would have to be passed on to customers, thus making it less competitive. Apart from the increased costs due to higher charter rate, it has been a major concern that negative perception about palm oil will be heightened. There is also a possibility that further efforts may be in the pipeline to extend this regulation to other geographical areas outside Europe, such as India and China; the two biggest palm oil consumers in the world.

The decision by IMO to stringent condition for vegetable oil through the amendment of MARPOL Annex II may be seen as an attempt to further discredit palm oil to its use and transport. Besides that, it is seen to be another attempt of putting a non-tariff barrier against palm oil. Although other vegetable oils will be impacted by the new law, they will not be as severe as palm oil. Most of the domestically produced vegetable oils such as sunflower oil, rapeseed oil and olive oil are transported internally using land transport such as trains or cargo trucks, and the MARPOL regulations will not apply on such mode of transport. The decision taken at the IMO level to impose stricter requirements on Category Y product is debatable since there is no observational and literature analysis on palm oil to prove that it is hazardous to marine life and living things. Indeed, scientific research has revealed that most of the vegetable oil-including palm oil is biodegradable and non-toxic. This state of fact leads the researchers to wonder why public opinion is polarised in terms of viewpoints that palm oil can threaten marine life and other living things even though it has not been proven so. This is the question that prompted the researchers to embark on this study; to conduct an analytical and literary evaluation of palm oil in order to refresh people's mind but also draw the attention to the characteristics of palm oil that people seem to take less attention to. This paper aims to:

(a) Underline the weaknesses in the policy process of MARPOL Annex II;
(b) Examine articles/journals on issues involving palm oil that ended in a misleading view of palm oil among the public;
(c) Highlight the behaviour and characteristics of palm oil; and
(d) Propose the future field of study, this study might serve as a starting point for further research on the feasibility of shifting the palm oil category in MARPOL Annex II from Y to Z.

The study has involved literature research of published and unpublished materials relating to MARPOL Annex II and vegetable oils, particularly palm oil. This includes an analytical review of the policy process of MARPOL Annex II, people's perception towards palm oil and behaviour and characteristics of palm oil.

8.3 MARPOL Annex II

8.3.1 The Policy Process

As early as 1994, the IMO circulated to all member states guidance and warnings concerning the transport of vegetable oils and their effect on birds after discharge [4]. Despite the lack of particular legislation, the organisation asked governments and port state administrations to bring the information to the attention of all ship operators in order to limit the discharge or emission of such compounds. The IMO's specialists engaged in the creation of the Globally Harmonised System (GHS) have to coordinate with the Organisation for Economic Cooperation and Development (OECD) and United Nations (UN) expert sessions from 1993 to 2003. During this period, scientists collected all safety data for bulk liquids mentioned or derogated in the IMO standard and created particular hazard categorization criteria in accordance with continuing discussions at the OECD and UN levels, using a flexible methodology. For a number of years, however, the hazard assessment of vegetable oils was excluded [5].

In the 1970s and 1980s, scientists published scientific publications regarding environmental concerns before any political or regulatory discussion. In 1989, however, there were complaints from coast guards and beach patrol agents concerning incidences of pollution and oiled birds that sparked national political action. Incidents involving the deaths of dogs and seabirds in 2013 triggered national political action once again [5]. Following this, in 2016, Norway and Sweden have proposed amendments towards MARPOL Annex II to further reduce the impact on the environment of tank washings containing high-viscosity and persistent floating products during Marine Environment Protection Committee (MEPC) meeting under IMO [6]. Among the recommendations made, include requiring the vessel to go through a pre-washing procedure before releasing the cargo residues to the port's reception facilities, as well as revising the definitions of solidifying and high-viscosity chemicals. In addition, Norway and Sweden proposed reclassifying the discharge material as a high-viscosity material: a substance in category X or Y with a viscosity of 50 mPa s or greater at

20 °C; and a solidifying agent is a noxious liquid with a freezing point of 0 °C or higher.

When the discussion at the international level began, the majority of IMO members, including the major vegetable oil importers, agreed towards the proposals and choosing for editorial work on the current treaty [6]. There was a final political debate with just minor resistance orchestrated by Malaysia, the world's top exporter of palm oil. Malaysia, speaking on behalf of certain other vegetable oil producing countries, questioned the hazards of floating oils for some products (such as palm oil), claiming that the liquids would solidify and hence not harm marine life. Furthermore, Malaysia identified a shortage of suitable tanker tonnage, thus leading to a shortage of renewable oils after 2006 [7]. Between the lines of the papers presented throughout the years, the producing agricultural industry made it clear that tighter transportation laws might have a severe impact on the competitive position of these renewable raw products and have an impact on developing nations in the south.

8.3.2 GESAMP Hazard Profile

Alongside the revision of MARPOL Annex II, the marine pollution hazards of thousands of chemicals have been evaluated by the Evaluation of Hazardous Substances (EHS) working group, giving a resultant of the joint Group of Experts on the Scientific Aspects of Marine Environmental Protection (GESAMP) Hazard Profile, which indexes the substance according to its bio-accumulation; bio-degradation; acute toxicity; chronic toxicity; long-term health effects; and effects on marine wildlife and on benthic habitats. The hazard profile provides an alphanumerical fingerprint of each substance. The numerical scales start from 0 (no hazard), while higher numbers reflect increasing hazard. In this way, information on substances evaluated by GESAMP is made available to the widest possible audience in an instantly readable form. Hazard evaluation and categorisation or "classification" are thus the responsibility of separate bodies [8].

GESAMP, through its EHS working group, encourages industry involvement in the preparation of the hazard profiles. Of necessity, the sessions of the GESAMP EHS working group are closed meetings. However, representatives from chemical manufacturers, their branch associations or sector groups, as well as shipping agencies are frequently invited to provide statements or to comment on specific items under discussion. Such contributions are particularly welcomed by GESAMP in cases where the whole groups of substances are being reviewed or re-evaluated [8]. The results of the evaluation of chemical substances are published in the meeting reports of the GESAMP EHS working group and tabled at the next GESAMP session. Following approval by GESAMP, the hazard profiles are published periodically as circulars by IMO and distributed to IMO member states and observer organisations. In addition, a composite list is published annually by IMO containing the hazard profiles of all chemical substances evaluated during the last thirty years. Through MEPC, IMO is responsible for assigning bulk liquid substances to an appropriate pollution category on the basis of the GESAMP hazard profile as per Table 8.1 [8].

Table 8.1 GESAMP hazard profile [9]

GESAMP hazard profile May 2017 Product	Aquatic environment (score: 0–6)				Human health (score: 0–4)						Interference with other uses of the sea		
					Acute mammalian toxicity			Irritation, corrosion & long-term effect					
	A1 Bio-accumulation	A2 Bio-degradation	B1 Acute toxicity	B2 Chronic toxicity	C1 Oral toxicity	C2 Dermal toxicity	C3 Inhalation toxicity	D1 To skin	D2 To eye	D3 Long-term health	E1 Tainting	E2 Wildlife & benthic	E3 (0–3) Coastal amenities
Palm oil	0	R	0	NI	0	(0)	(0)	0	0			Fp	2
Soybean oil	0	R	0	NI	0	(0)	(1)	(0)	1			Fp	2
Sunflower oil	0	R	0	NI	(0)	(0)	(1)	(0)	(1)			Fp	2
Cotton seed oil	0	R	(2)	NI	(0)	(0)	(1)	0	1			Fp	2
Coconut oil	0	R	(1)	NI	0	(0)	(1)	0	(1)			Fp	2
Olive oil	0	R	(2)	NI	(0)	(0)	(1)	1	1			Fp	2

(continued)

Table 8.1 (continued)

GESAMP hazard profile May 2017	Aquatic environment (score: 0–6)				Human health (score: 0–4)						Interference with other uses of the sea		
					Acute mammalian toxicity			Irritation, corrosion & long-term effect					
Product	A1 Bio-accumulation	A2 Bio-degradation	B1 Acute toxicity	B2 Chronic toxicity	C1 Oral toxicity	C2 Dermal toxicity	C3 Inhalation toxicity	D1 To skin	D2 To eye	D3 Long-term health	E1 Tainting	E2 Wildlife & benthic	E3 (0–3) Coastal amenities
Rapeseed oil	0	R	(2)	NI	(0)	(0)	(0)	(1)	(1)			Fp	2

- No concern over bio-accumulation
- Readily biodegradable
- No confirmed chronic toxicity
- No or low acute toxicity
- No acute mammalian toxicity (bracket of C3 valued in vapour form)
- No known long-term health concerns
- May be mildly irritant to eye and skin in direct contact
- Exists as persistent floater at sea
- Moderately affect coastal amenities, possibly lead to the closure of amenity

Based on Table 8.1, it is found that vegetable oil including palm oil is non-hazardous to aquatic environment and human. Thus, it exists as persistent floater at sea and moderately affects coastal amenities. This evaluation process done by International Tanker Owners Pollution Federation (ITOPF) was based on the GESAMP Hazard Evaluation Procedure. This state of fact leads the researchers to wonder why palm oil has been categorised under Category Y and condition under this category has been stringent even though scientific study has proven it to be as non-hazardous. This might be due to the fact that in instances of accidental poisoning, GESAMP will also take into account of human experience [8]. All available information is considered together by the experts, and ratings are given on the basis of the total weight of evidence, in order to evaluate the hazard produced by the substances. However, where experimental data on bio-accumulation or acute aquatic toxicity are not available, then GESAMP generally accepted estimation techniques. In this case, quantitative structure–activity relationships (QSARs) for the chemical group in question is acceptable. Besides that, estimation techniques for biodegradation also could be accepted by GESAMP to show that a substance is not readily biodegradable, in order to avoid further and often pointless testing [8].

8.4 Excerpts from Media About Palm Oil Harmfulness

8.4.1 The Death of Zanzi the Dog

One of the first cases happened in Cornwall, England, near the Strait of Dover, on the dog-friendly beach of Long Rock. This is also, coincidentally, the world's busiest shipping channel. Zanzi, a black-haired mini-schnauzer, went out on a stroll with his owners in October 2013. He had wandered off and was shortly munching on something, which was claimed to be white, waxy, and smelled a bit like diesel. Zanzi became agitated minutes after biting and swallowing the gluey material. The stuff became entangled, and the dog was having difficulty breathing. Zanzi was sent to the nearest vet, Mounts Bay, where he had an emergency operation. Zanzi died that day, despite his best attempts [9]. That year, news of Zanzi's oil-related death spun through the news cycle. It appeared that palm oil, a substance formerly connected with rainforest devastation, had also killed a beloved pet. Following this case, further complaints of the oil sickening and killing dogs began to appear in online news, with the majority of articles going back to Zanzi himself. Zanzi's death has spurring anxiety to owners up and down the coasts. Cornwall Council and other governments scrubbed their beaches of washed-ashore fats, posted "Be wary of palm oil" signs on the sand, and set up a government website as a warning to safeguard local pets [9].

With this bad connotation, palm oil has penetrated the public mind. Back at Mounts Bay, where Zanzi died, it was discovered that there had never been a proven dog fatality caused by palm oil. In reality, despite the fact that this substance kills a dog every day, none of the canine deaths have been verified to be the effect of this component. Long pointed out that no one seems to be completely sure what exactly in the blobs of floating fat it is that was sickening and killing dogs [10]. The stomach contents of one dog that became really unwell after consuming something on the beach were found to contain not just palm oil, but also substantial amounts of germs. Therefore, it is probable that these dogs are becoming seriously ill as a result of bacterial infection. According to her, dogs do not break down fats properly, so if they consume enough of it—even if it is not enough for a person to metabolise—they might get pancreatitis.

Cudmore claimed that the initial Cornishman article that revealed the "deadly palm oil" was confirmed by the national body, Public Health England, as a cause of Zanzi's death was incorrect [11]. According to Cudmore, based on her communication with one of that agency press officer, she has been informed that Zanzi's case was actually not in Public Health's hand but in those of the Maritime Coastguard's. Andrew also responded that they did not analyse the substance, and there had never been proof or cause to believe that palm oil itself is responsible for the dog's death. Apart from this, the Poisons Information Service in UK investigated how 30 surviving dogs who had eaten various "palm oil blobs" on various beaches in the country reacted to the substance. While many dogs exhibited no signs of illness, one did have a minor cough that lasted three days, and another recovered from aspiration pneumonia after seven days. Eleven of them vomited, which was the most prevalent adverse effect, and a few of them also had diarrhoea. Both of these are signs of petroleum toxicity in dogs [11].

It is fully aware that cargo ships or tankers also discharge chemicals, such as diesel fuel residue into the sea as it is permitted by MARPOL Annex II and IBC Code, as long as the shipowner follows the guidelines for the discharge procedure and does not violate the regulation. Therefore, the water is contaminated not only by palm oil waste but also by other chemicals like petroleum. As a result, the ship waste and palm oil residue mix to produce a gelatinous form with a petrol-like odour, attracting dogs to sniff and even swallow the oil. This petrol mixture could cause major health risks to both humans and dogs. See Table 8.2 for the list of articles and journals regarding the death of the dogs.

Table 8.2 Articles and journals regarding the death of the dogs [12–16]

Article/journal	Title
Mail Online https://www.dailymail.co.uk/news/article-2571647/Toxic-waste-killing-dogs-beaches-Poisoned-pets-left-writhing-agony-palm-oil-dumped-sea.html	Toxic waste killing dogs on beaches: poisoned pets left writhing in agony by palm oil dumped at sea (2 March 2014)
Shoreham Herard https://www.shorehamherald.co.uk/news/washed-palm-oil-dangerous-dogs-beach-2301467	Washed-up palm oil dangerous to dogs on beach (24 Feb 2014)
The Westmorland Gazette https://www.thewestmorlandgazette.co.uk/news/14208063.warning-issued-to-dog-walkers-after-dog-eats-deadly-palm-oil-on-beach-at-silverdale/	Warning issued to dog walkers after dog eats deadly palm oil on beach at Silverdale (15 Jan 2016)
Daily Post UK https://www.dailypost.co.uk/news/north-wales-news/toxic-palm-oil-deadly-dogs-13847250	Toxic palm oil deadly to dogs could be washing up from killer wreck again (2 Nov 2017)
Metro https://metro.co.uk/2018/03/23/dog-owners-warned-deadly-palm-oil-washing-uk-beaches-7410402/	Dog owners warned of deadly palm oil washing up on UK beaches (23 March 2018)

8.4.2 The Deaths of Seabirds

In January 2013, it was reported that about 100 birds were found on Chesil Beach in Dorset, UK, 60 a little further west at Brixham and many other individual birds and smaller groups elsewhere along the coast. The birds, mostly guillemots, were covered in the waxy film. Morris stated that the tests by the Environment Agency have established that the problem has been caused by some sort of refined mineral oil, not palm oil as had been suspected [17].

According to [18], forensic specialists determined that the unexplained waxy material coating more than 100 seabirds found up on Southern English beaches was a mineral-based oil. United Kingdom's Maritime and Coastguard Agency (MCA) officials also said the initial findings indicate that the substance is not palm oil, as had been suggested by at least one scientist [19]. That means it might have been anything from hydraulic fluid to Vaseline, but it was clearly not the palm oil that had been linked to the deaths of birds from Cornwall to Sussex. However, the West Sussex Gazette alleged that palm oil was a primary cause of eleven seabird deaths in Worthing [20]. According to senior animal rescue officer at the Worthing, if palm oil makes dogs ill it is certainly not going to be good for birds [21]. More and more articles about palm oil washing up on beaches and killing seabirds have recently appeared on the online news as listed in Table 8.3, causing public distress. Despite

Table 8.3 Articles and journals regarding the death of the seabirds [22–25]

Article/journals	Title
The Free Library https://www.thefreelibrary.com/SLAUGHTER+OF+THE+SEABIRDS%3B+Hundreds+covered+in+palm+oil+washed+up+on+...-a0317122822	SLAUGHTER OF THE SEABIRDS; hundreds covered in palm oil washed up on British beaches (1 Feb 2013)
BBC https://www.bbc.com/news/uk-england-dorset-26313284	Chesil Beach: dead and oil-covered birds still washing up (23 Feb 2014)
West Sussex Gazette https://www.westsussextoday.co.uk/news/seabirds-die-after-ingesting-deadly-palm-oil-1285258	Seabirds die after ingesting deadly palm oil (28 Feb 2014)
Dorset Echo https://www.dorsetecho.co.uk/news/11030832.update-tide-of-dead-seabirds-keeps-rolling-in/	UPDATE: tide of dead seabirds keeps rolling in (25 Feb 2014)

the fact that none of the seabird deaths were caused by this component, such news has heightened the public's negative perception of palm oil.

8.5 Vegetable Oil Behaviour and Characteristics: Palm Oil

8.5.1 Increase in Palm Oil Production in Response to Increased Demand for Vegetable Oils

Vegetable oils have been used by man for many decades in various ways [26]. Vegetable oils' unique characteristics and chemical composition have allowed them to be used as food, lubricants and fuels, as well as in the production of agrochemicals, plasticisers, inks and coatings. Palm oil remains the largest vegetable oil consumed globally for food and industrial use. Higher production and lower prices support expanded global palm oil demand from China, the European Union and many other countries. Despite global production outpacing consumption, ending stocks continue to fall as stock levels recover from 2020/21 consumption exceeding production levels [27]:

i. Indonesia consumption is up 220,000 tonnes to 15.3 million on higher demand for food use.
ii. China consumption is up 400,000 tonnes to 7.2 million mostly on higher demand for food use.
iii. Thailand consumption is up 450,000 tonnes to 2.7 million on higher industrial and food use.

iv. India consumption is down 205,000 tonnes to 8.6 million as consumer preferences shift to other vegetable oils, driving lower food use [28].

The palm oil consumption is increasing since it is the most versatile of all vegetable oils. This is due to the fact that palm oil may be processed to produce a diverse spectrum of products with varying melting points, consistencies and characteristics. There are a couple of reasons why palm oil has been the favoured crop to meet growing demand for vegetable oils. Firstly, it has the lowest production costs. Secondly, its composition means it is versatile and can be used for food and non-food purposes. Some oils are not suited for cosmetic uses such as shampoos and detergents. Third, it gets incredibly high yields [29].

8.5.2 *Oil Behaviour*

Vegetable oils, including palm oil, have melting points above 0 °C that require heating during the voyage and while discharging. The viscosity and melting point of the substance, as well as its characteristics when discharged into the marine environment as tank washings, are the key physical properties at issue for these materials. The viscosity of oil and its displacement on the water surface are inversely proportional. The melting point of various vegetable oil is listed in Table 8.4. While the viscosity of the oil is high, it moves slowly; when the viscosity is low, the oil moves quickly, depending on the water temperature [30].

Vegetable oil such as palm oil, soybean oil, sunflower oil and olive oil are transported in bulk by tankers and cargo ships around the world. Following unloading, residuals frequently stay at the bottom of cargo tanks or crystallise against bulkheads and interior machinery. Tanks are either manually cleaned by the crew or automatically cleaned by rotary-jet cleaning systems that use steam, hot water or chemical

Table 8.4 Vegetable oil melting point temperature

Vegetable oil	Melting point temperature (°C)
Palm oil	30–37
Soybean oil	−20
Sunflower oil	−17
Palm Kernel oil	~24
Peanut oil	0–3
Cottonseed oil	~0
Coconut oil	24–26
Olive oil	−6
Sesame oil	−6
Corn oil	−11
Cocoa butter	34–38

solvents. When discharging tank washings, the majority of the tank washing will be water, with a final component being the cargo residues floating on the tank washing surface. Although the cargo residue may be a viscous liquid while in the slop tank, when it enters the sea, it will become solid because of the ambient sea temperature. Therefore, instead of being spread over a greater distance, it could be concentrated in one area. When cargo residue, which contains oil enters into the seawater, it begins a cycle of processes known as weathering, which alters the properties and behaviour of the oil. The key factors that influence oil behaviour are [30]: physical characteristics of the oil, in particular, specific gravity, viscosity and boiling range; composition and chemical characteristics of the oil; meteorological conditions (sea state, sunlight and air temperatures); and characteristics of the seawater (specific gravity, currents, temperature, presence of bacteria, nutrients and dissolved oxygen and suspended solids). Palm oil is classified into two types: solid fractions called palm stearin (PS) and liquid fractions called palm olein (PO). The separation of this fraction intends to improve the potential application of palm oil, particularly palm stearin, which offers flexible melting properties for a variety of food sectors [31].

Palm stearin, a palm product, solidified at 30 °C and according to a laboratory test conducted by the ITOPF, it is rapidly biodegradable. The result of the test has been evaluated based on GESAMP Hazard Evaluation Procedure. Based on the test result, it shows that palm stearin is non-hazardous, although it is discovered that it may have an impact on coastal amenity usage and wildlife. Table 8.5 shows that palm oil and palm nut have low viscosity in comparison to other vegetable oils [32].

Table 8.5 Bibliographical data of vegetable oil [33]

Oil type	Rapeseed	Palm	Palm nut	Castor
Synonyms	Rape oil, rapeseed oil, canola oil	Palm oil, palm butter	Palm nut oil, palm kernel oil	Castor oil, cosmetol, gold bond, phorbyol, neoloid, ricinus oil
Appearance (state at 20 °C)	Yellow liquid	Orange-red solid, light to dark	Light yellow liquid, hazelnut odour	Colourless/pale yellow liquid, slight odour
Density relative to seawater (at 20 °C)	0.91	0.895–0.95	0.899–0.913	0.96
Solubility in sea water (mg/l)	Insoluble	Insoluble	Insoluble	Insoluble
Viscosity (cSt at 20 °C)	72–82	25–31 (at 50 °C)	17–20 (at 50 °C)	600–1200 (at 840 °C)

8.5.3 Biodegradation at Sea

Vegetable oils are widely found in nature and consist mainly of glycerol esters particularly triacylglycerols (a major form of dietary lipid in fats and oils, whether derived from plants or animals). They are biodegradable, come from renewable sources and have low toxicity [34]. Biodegradation of a compound is often found as a result of the actions of multiple organisms. It is a process by which microorganisms alter or convert (through enzymatic or metabolic action) the chemical structure imported to a polluted environment to improve degradation through a process called bioaugmentation [35]. Several tests that were carried out indicated that vegetable oils, including palm oil undergo about 70–100% biodegradation in a period of 28 days. However, it was discovered that the vegetable oils did not biodegrade on the beach, but rather changed into a gum-like substance known as lump in the case of palm oil. Palm oil lump is formed at the European Union sea because the water temperature is about 20 to −2 °C with heavy waves [6]. Given the state of the ambient waters, palm oil residues do not have ample time to go through the process of biodegradation and have been taken by strong currents to the coastal and marine areas of the European Union.

The oil biodegradation processes in seawater were found to be significantly stimulated by nutrient enrichment and by the presence of mixed microbial consortia found in wastewater. It was also observed that different oils responded in different rates and extents to biodegradation depending on their viscosity, structure and compositions. Auto-oxidation and biodegradation took place in the oil weathering process. The oxidation process either accelerates the biodegradation rates by producing much smaller and easier compounds to be biodegraded or inhibits the microbial attacks by producing antibacterial products [36]. Depending on the sea temperature, palm stearin will easily biodegrade in less than 5 days. The reason why palm oil did not biodegrade on the beach, but instead transformed into a gum-like material known as lump, has attracted researchers' interest. Other chemical residues, which also have been discharged to the sea, may have had a role in why palm oil solidified and did not go through the weathering process.

8.6 Qualitative Content Analysis on Published and Unpublished Materials

The content and meaning of the published and unpublished materials used in this study were examined using qualitative content analysis. The meaning unit has been coded, categorised and organised into themes. Then, graphs were created to provide a graphical representation of the results. According to the graphs in Tables 8.6 and 8.7, the major focus of the materials is on negative propagation of palm oil. Due to a lack of scientific research, palm oil has been propagated as dangerous and harmful

Table 8.6 Total of meaning unit based on theme (The Death of The Dogs)

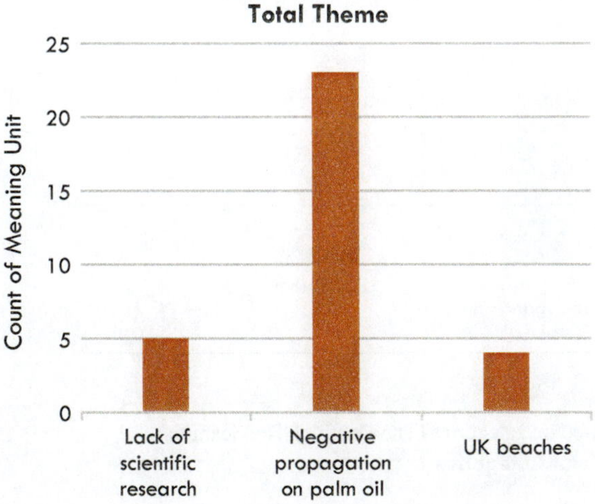

Table 8.7 Total of meaning unit based on categories (The Death of The Dogs)

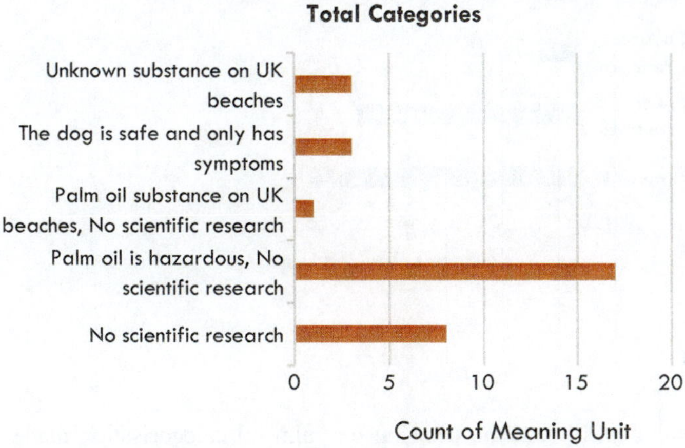

to marine life and living things. Similarly, the content analysis of materials about seabird deaths in Tables 8.8 and 8.9 yielded the same result.

Table 8.8 Total of meaning unit based on theme (The Death of The Seabirds)

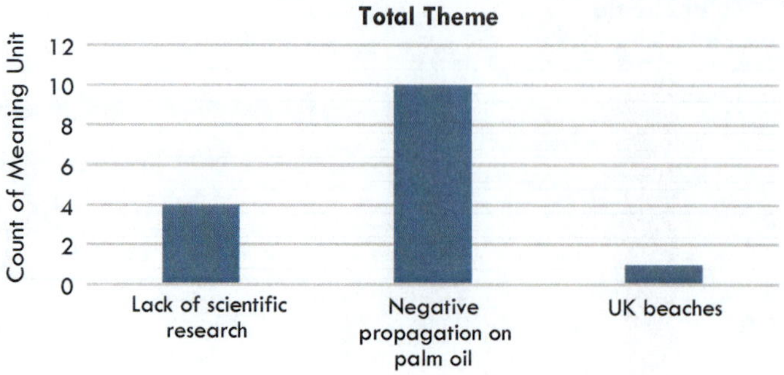

Table 8.9 Total of meaning unit based on categories (The Death of The Seabirds)

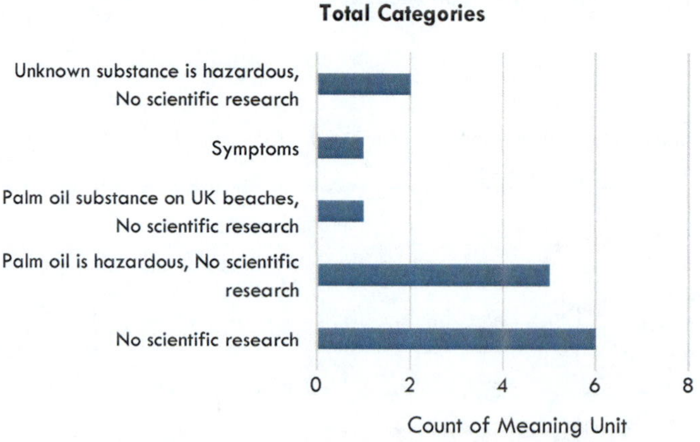

8.7 Discussion

According to the analytical review, the decision on palm oil categorisation made at the IMO forum is debatable. Scientific studies confirming that palm oil is non-hazardous have been recognised as one of the few elements supporting this claim. Furthermore, no scientific analysis has been undertaken to verify that palm oil killed Zanzi or other dogs, as well as the seabirds. Apart from that, human experience and estimating techniques have been acknowledged as flaws in the GESAMP Hazard Evaluation Process, which might lead to inaccurate palm oil categorization at the MEPC at the IMO. This is because data derived from human experience is very subjective and must be supported by solid proofs. However, there have been instances in the past where manufacturers, representatives, and trade organisations have questioned hazard profiles. MEPC has previously advised GESAMP to do research on big chemical groups such as polyether polyols and vegetable/animal oils. GESAMP reviews individual and groupings of chemicals on a regular basis, based on the availability of new data. Therefore, the researchers are of the opinion that there are strong reasons for palm oil categorisation in MARPOL Annex II to be revised. Palm oil major exporters should propose to IMO for the revision of GESAMP Hazard Profile, so that a new categorisation for palm oil can be determined.

The fact that palm oil exists as persistent floater at sea and moderately affects coastal amenities could be among the issues, which should be further analysed. A further research might be needed to find the best solution in order to cater for this issue. Besides that, the qualitative content analysis performed produced several salient observations:

i. The key perception propagated by the media is that palm oil is harmful to the marine ecosystem and living things.
ii. The perception issues that are working against palm oil are linked with the lack of scientific research.
iii. The propagation of negative sentiment towards palm oil started in European countries.
iv. The results of experimental studies and laboratory experiments proved that palm oil's behaviour and characteristics are safe for marine life and other living things.

8.8 Conclusion and Recommendation

News posted on social media has a ripple-like effect, allowing the post to reach a larger audience. This is not the first time that the palm oil industry has faced backlash. In fact, it began in the late 1980s, and misinformation about palm oil continues to exist since then. Palm oil has long been overlooked in the West, despite scientific evidence showing it is both more environmentally friendly and safer than other oils. The perceptions of people and the media impact the acceptability of palm oil and its products. Due to the obvious gap between what the public perceives and what is currently needed for the palm oil industry's long-term survival, all stakeholders must continue to put in efforts to raise their awareness. If there is a biased media narrative, it can be corrected by proactive discussion at international level and should involve political outreach.

The decision by IMO to enforce a more stringent condition for vegetable oil through the amendments of MARPOL Annex II and IBC has negative impacts on the shipping industry, especially in the transport of palm oil. Although other vegetable oils will also be impacted by the new regulation, they will not be as severe as palm oil. Most of the European-produced vegetable oils such as sunflower oil, rapeseed oil and olive oil are transported internally using land transport such as trains or cargo trucks and the MARPOL regulations will not apply on such mode of transport. Besides that, there is still lack of research on assessing the impact of current amendments. Rather than bickering, global industry leaders, particularly major palm oil exporters, should engage one another through dialogues and collaboration to improve the palm oil industry. There is a possibility for a future revision of the palm oil categorization in MARPOL Annex II and IBC. Therefore, a thorough analysis should be undertaken for this reason.

References

1. Sea-Mer A (2017) Industrial paraffin-wax strandings on the eastern coast of the channel, contexts and stakes. Technical report
2. Rahul S (2016) MARPOL Annex II—preserving the marine ecosystem is imperative. Stand Saf (February):2–5
3. Honkanen M, Häkkinen J, Posti A (2012) Tank cleaning in the Baltic Sea—assessment of the ecotoxicity of tank cleaning effluents. Centre of Maritime Studies
4. IMO (1994) Harmful effects on birds from contact with animal and fish oils and fats and vegetable oils (floating lipophilic substances). MEPC/ Circ.274, London
5. Suaria G, Aliani S, Merlino S, Abbate M (2018) The occurrence of paraffin and other petroleum waxes in the marine environment: a review of the current legislative framework and shipping operational practices. Front Mar Sci 5:94. https://doi.org/10.3389/fmars.2018.00094
6. IMO (2017) Review of MARPOL Annex II requirements that have an impact on cargo residues and tank washings of high viscosity and persistent floating products. Unpublished
7. IMO (2004) A study of the impact of reclassification of products relative to the revised MARPOL Annex II. Submitted by Malaysia, MEPC 52/5/5, London
8. GESAMP (IMO/FAO/UNESCO-IOC/WMO/WHO/IAEA/UN/UNEP Joint Group of Experts on the Scientific Aspects of Marine Environmental Protection) (2002) Revised GESAMP hazard evaluation procedure for chemical substances carried by ships. Rep Stud GESAMP No. 64, p 126
9. Zhang AS (2018) Vegetable oil matters. www.itopf.org/fileadmin/data/Documents/Papers/Vegetable_oil_matters. Retrieved 19 Jun 2021
10. Christie L (2017) Are toxic palm oil chunks poisoning our dogs? Article PetCoach
11. Becca C (2017) Are toxic blobs of palm oil poisoning our pets? Article Slate
12. Paul H (2014) Toxic waste killing dogs on beaches: poisoned pets left writhing in agony by palm oil dumped at sea. Article Mail Online, 2 Mar 2014
13. The Newsroom (2014) Washed-up palm oil dangerous to dogs on beach. Article Shoreham Herard, 24 Feb 2014
14. Katie D (2016) Warning issued to dog walkers after dog eats deadly palm oil on beach at Silverdale. Article The Westmorland Gazette, 15 Jan 2016
15. Owen E (2017) Toxic palm oil deadly to dogs could be washing up from killer wreck again. Article Daily Post, UK, 2 Nov 2017
16. Tanveer M (2018) Dog owners warned of deadly palm oil washing up on UK beaches. Article Metro, 23 Mar 2018
17. Steven M (2013) Thousands of seabirds may be harmed by oil off UK coast. Article The Guardian
18. LiveScience Staff (2013) Hundreds of slime-covered seabirds wash Ashore, Puzzle Scientists. Article NBC News, 2 Feb 2013
19. Pilita C (2013) Mineral-based oil blamed for bird deaths. Article Financial Times
20. The Newsroom (2014) Seabirds die after ingesting deadly palm oil. Article West Sussex Gazzete
21. Simencio-Otero RL, Canale LCF, Totten GE (2012) Use of vegetable oils and animal oils as steel quenchants: a historical review. J ASTM Int 9(1):136–195
22. Euan S (2013) Slaughter of the seabirds; hundreds covered in palm oil washed up on British beaches. Article The Free Library, 1 Feb 2013
23. BBC (2014) Chesil Beach: dead and oil-covered birds still washing up. Article BBC News, 23 Feb 2014

24. The Newsroom (2014) Seabirds die after ingesting deadly palm oil. Article West Sussex Gazette, 28 Feb 2014
25. Emma W (2014) Update: Tide of dead seabirds keeps rolling in. Article Dorset Echo, 25 Feb 2014
26. IMO (2005) Manual on oil pollution section IV. IMO Publication
27. Hannah R (2018) Palm oil. Our World Data
28. United States Department of Agriculture Foreign Agricultural Service (2021) Oilseeds: World markets and trade. Global Market Analysis 2021
29. Edy S, Nurannisa RL (2020) The recent application of palm stearin in food industry: a review. Int J Sci Technol Res 9(2):2593–2597
30. Farhan MAF (2013) The impact of maritime oil pollution in the marine environment: case study of maritime oil pollution in the navigational channel of Shatt Al-Arab. World Maritime University Dissertation
31. Edy S, Rizki LN (2020) The recent application of palm stearin in food industry: a review. Int J Sci Technol Res 9(02)
32. Centre of Documentation, Research and Experimentation on Accidental Water Pollution (2014) Vegetable oil spills at sea, operational guide. Cedre
33. International Tanker Owners Pollution Federation (ITOPF) (2018) Vegetable oil matters. ITOPF
34. Ibrahim S, Abd Shukor MY, Yazid NA, Ahmad SA (2018) Microbial degradation of vegetable oils: a review. Malays J Biochem Mole Biol 3(1):45–55
35. Tyagi M, Da Fonseca MMR, De Carvalho CCCR (2011) Bioaugmentation and biostimulation strategies to improve the effectiveness of bioremediation processes. Biodegradation 22:231–241
36. Al-Darbi MM, Saeed NO, Islam MR, Lee K (2003) Biodegradation of natural oils in seawater. Energy Sour 27:19–34
37. US Department of Agriculture; USDA Foreign Agricultural Service (2021) Vegetable oils: global consumption by oil type 2013/14 to 2020/2021. Statista
38. International Tanker Owners Pollution Federation (ITOPF) (2018) A study on palm stearin biodegradation at sea. ITOPF

Chapter 9
Assessing the Efficacy of the Advance Transfer Technique in Calculating the Wheel Over Point Through Simulation Studies

Amir Syawal Kamis, Ahmad Faizal Ahmad Fuad, Aimie Qamarina Anwar, and Sheikh Alif Ali

Abstract Without proper planning, a ship may deviate from the desired course line while changing its course. This is frequently the case when a late course change is made. To ensure the ship stays on the route while changing course, the wheel over point must be indicated on the charted courses to identify the location of the course alteration. The purpose of this research is to conduct a review of the advance transfer technique that is often used to determine and identify the wheel over point. Although this method is regularly used by seafarers aboard, there is a shortage of evidence in the scientific literature to support its efficacy. The cross-track distance data for a ship's movement in different navigation areas was obtained in this research via simulations utilising a ship simulator. After that, the findings were compared to the cross-track limit. As a result, this study discovered the strength and weakness of the technique, which can be used as a foundation for developing an improved mathematical model.

Keywords Wheel over point · Alteration course · Voyage planning · Course keeping · Advance transfer technique

9.1 Introduction

The recent Suez Canal tragedy involving the MV Ever Given demonstrates the critical necessity of ships adhering to their designated route [1]. On 23 March 2021, the ship deviated from its original route, which resulted in the 399.94 m ship blocking the canal for six days. After the incident, the Suez Canal authority detained the vessel, seeking USD 916 million in fines for delaying trade [2].

A. S. Kamis
Rating Programs Department, Akademi Laut Malaysia, Bt 30, Kg Tg Dahan, 78200 Kuala Sungai Baru, Melaka, Malaysia

A. S. Kamis · A. F. Ahmad Fuad (✉) · A. Q. Anwar · S. A. Ali
Universiti Malaysia Terengganu, 21030 Kuala Terengganu, Terengganu, Malaysia

S. A. Ali
e-mail: sheikh_alif@umt.edu.my

A. Ismail et al. (eds.), *Design in Maritime Engineering*, Advanced Structured Materials 167, https://doi.org/10.1007/978-3-030-89988-2_9

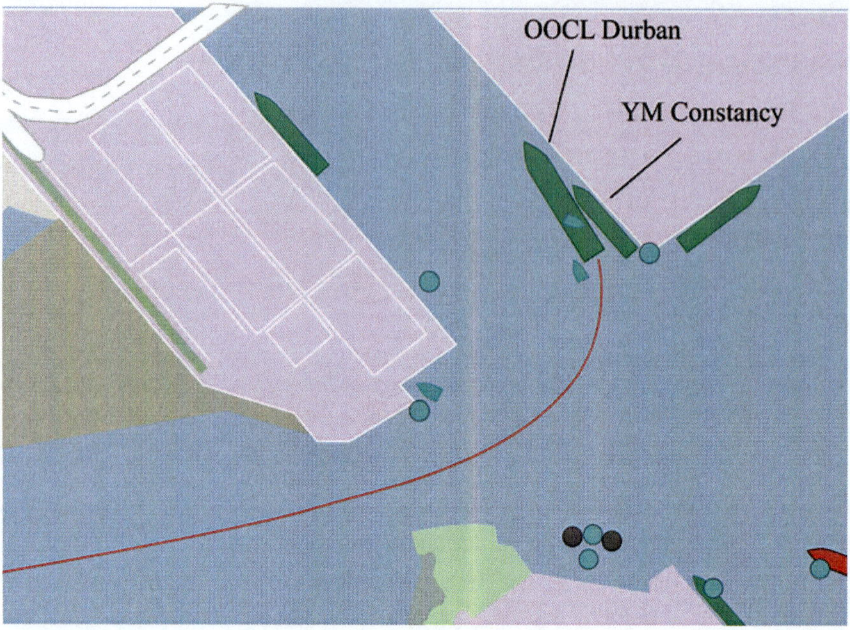

Fig. 9.1 The track of OOCL Durban in the collision with YM constancy

In addition to that, another disaster took place on 3 June 2021 in Kaohsiung, Taiwan, involving the OOCL Durban and YM Constancy, as seen in Fig. 9.1, which appeared to be caused by insufficient turning in restricted waterways. The ship collided immediately with the YM Constancy, before hitting and collapsing two nearby cranes during the incident [3]. The OOCL Durban was nearing berth 66 when it collided with the YM Constancy, which was already alongside berth 70. Nonetheless, the investigation has not finished in its entirety [4].

9.1.1 Voyage Plan

If land transport has a road as a route guide, a ship, on the other hand, has to navigate by carefully monitoring its position obtained from the global navigation satellite system (GNSS) or using fixes from radio detection and ranging (RADAR), terrestrial or celestial observation [5]. Then, at regular intervals, the position or the fixes obtained will be plotted onto the paper navigation chart manually to see whether the ship is still on the charted course [6].

Nowadays, manual plotting has been made easy by the implementation of an electronic chart display information system (ECDIS), where the plotting is done automatically using the position obtained from GNSS [7]. Accordingly, the ship needs to ensure that it is manoeuvring according to the charted route and continues navigating along from one course line to another course line, connected by waypoints (WPT), as shown in Fig. 9.2. When altering a course, the change must be performed at a sufficient distance prior to WPT to prevent the ship from deviating from its intended path [8]. A deviation is denoted by the cross-track distance (XTD), which is the distance measured between a vessel's current position and its planned track, as seen in Fig. 9.2 [9]. A wheel over point (WOP) must be indicated on the planned path to identify the point of alteration [10]. Although the cross-track distance is not a significant issue in open sea since there is adequate sea room for the ship to manoeuvre, it affects the vessel's fuel consumption [11]. However, while navigating in restricted water with little sea room, the cross-track distance is a crucial factor that affects the ship's safety [8] since there were a few documented incidents which happened as a result of ignored or underutilised cross-track alarms [12–17].

Deviation from an intended route in a dense water area may be disastrous if navigators do not grasp the basic principles of hydrodynamics [18, 19]. For instance, if the vessel travels through restricted waterways such as a river, the cross-track distance may cause the ship to approach a riverbank [20]. Hydrodynamic phenomena such as bow cushion and bank suction may develop in this condition, resulting in incidents such as grounding [21]. The shallow water effect makes manoeuvring the ship more difficult and exacerbates the problem [22–25]. When a vessel deviates into

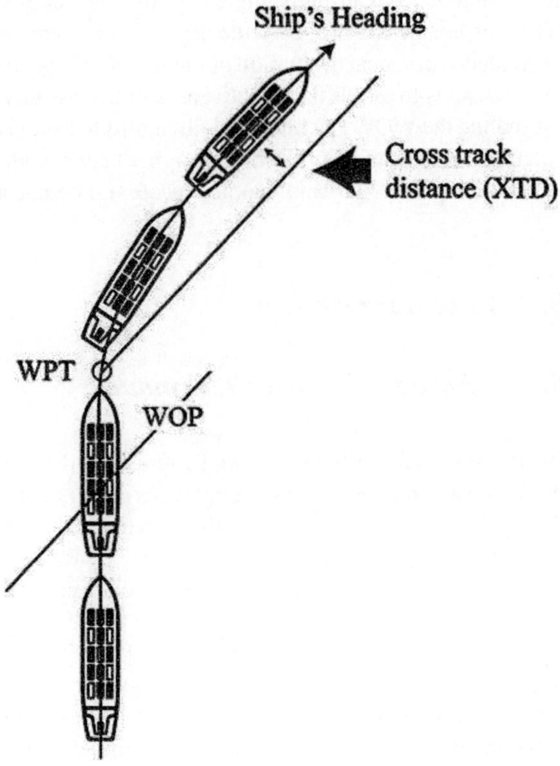

Fig. 9.2 The cross-track distance (XTD) of a vessel due to incorrect calculation of WOP and deviation from a charted course

a particularly narrow channel, it may run aground or crash with a land feature, such as a pier [21, 26].

While the ECDIS aids navigation officers in monitoring the vessels' routes [27], currently, the regulations that require ships to be equipped with such technology is only made mandatory for passenger ships of more than 500 GT and cargo ships of more than 3000 GT [28]. As a result, the remaining navigators without access to such technology continue to depend on traditional navigational skills such as the advance transfer technique (ATT) [29]. Moreover, some particularly old-school seafarers who had a chance to navigate with ECDIS and other state-of-the-art navigation aids still prefer to use the traditional technique. This statement is supported through a study where fourteen out of sixteen seafarers chose to use the traditional technique and avoided over-reliance on the navigation aids [30]. Thus, it is essential to sustain such skill as part of the effort in improving navigation safety.

9.1.2 Problem Statement and Research Aim

As seen in Fig. 9.2, although keeping the ship on the charted course is easy while going in a straight route, the vessel may veer off the desired course line if the right WOP is not established. Currently, there is limited research that evaluates the efficiency of ATT, despite the technique's wide usage by mariners worldwide. Thus, it is essential to conduct a practical evaluation of the methodology to demonstrate its efficacy. This research aims to assess the effectiveness of the advance transfer technique (ATT) for estimating the WOP. The findings will highlight the existing research gap in the WOP calculation and can be utilised to design a better model to assist navigation officers in estimating the WOP and making successful course alterations.

9.2 Literature Review

9.2.1 Advance Transfer Technique

Every ship has its particular turning circle according to its dimension, displacement, draft-to-water ratio and propeller characteristic. A vessel's turning circle is examined during the sea trial, finalised, and then posted in the wheelhouse based on the rudder angle used during a particular manoeuvre [31]. When manoeuvring, fixing the ship's rudder at a certain angle will allow for a constant turning circle [32].

The advance transfer technique (ATT) is a turning method where the rudder is positioned at the maximum rudder angle [33]. It is referring to the use of advance and transfer information related to the manoeuvring characteristic. It is typically used for navigating inland waterways and pilotage due to the fact that the turning will be significantly affected by the rudder angle but less affected by the speed [7, 31, 33].

The speed of a ship can be considered constant in open sea. However, during pilotage in the harbour and confined waterways, the speed will always be adjusted by taking into account traffic density, visibility, water depth, manoeuvrability, navigation hazards, and pilot's advice [34]. Due to the characteristic of the turning circle, which is not significantly affected by the vessel's speed, this method is preferred during pilotage [31]. This method also does not require experienced navigators as the navigators just need to put the rudder at maximum rudder angle without having to worry about the rate of turn [7, 8, 35–37].

As seen in Fig. 9.3, the advance transfer technique involves two values related to the manoeuvring characteristic, namely advance and transfer distances, hence the name. The ship's centre of gravity (CG) is the point at which the ship's weight is assumed to be concentrated. Therefore, it is used as the datum to measure the advance and transfer distances [38]. The advance and transfer distances, generally expressed in nautical miles, are determined from the moment the ship begins to turn by hard over the rudder until the ship's heading varies by 90° from the original heading, where the advance distance is on the $X0$ axis and the transfer distance is on the $Y0$ axis, as seen in Fig. 9.3 [38]. The usage of advance and transfer distances to calculate WOP is as explained below.

The following acronyms are used to describe the formula with reference to Fig. 9.4:

d_{adv} Advance distance measured from CG.
d_{trs} Transfer distance measured from CG.

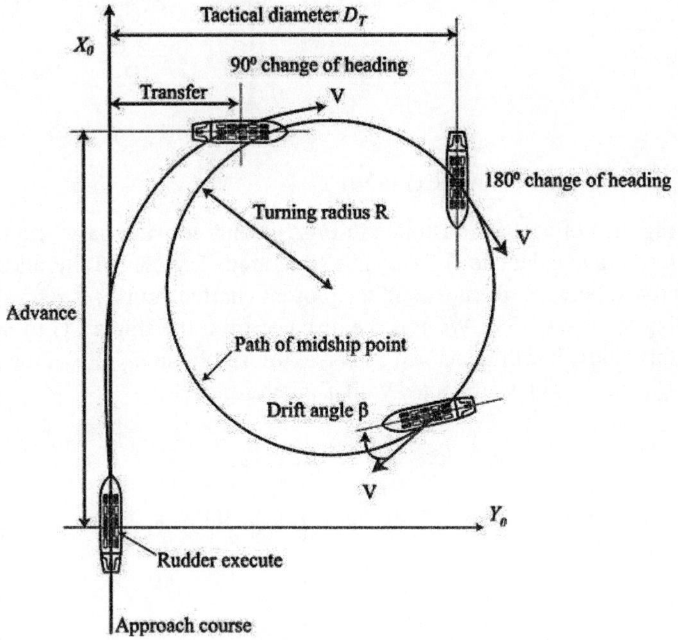

Fig. 9.3 Typical manoeuvring characteristic of a ship

Fig. 9.4 Marking WOP

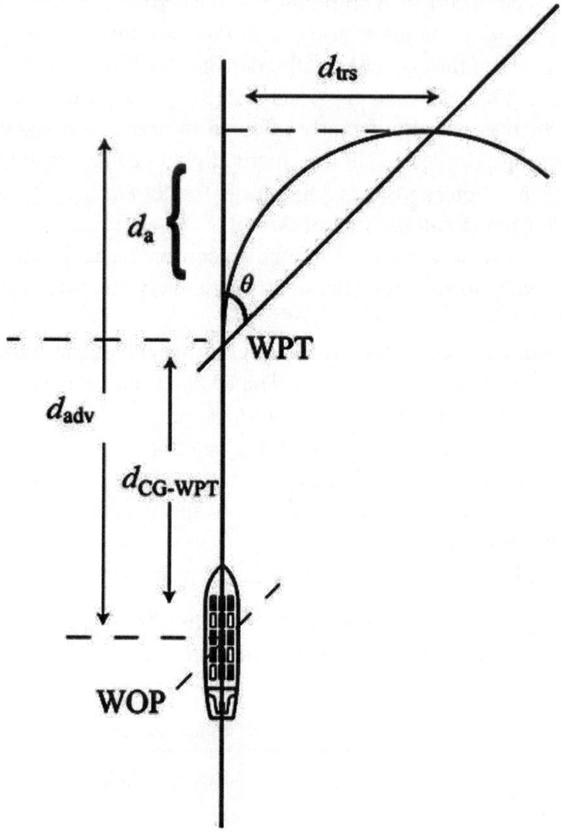

$d_{CG\text{-}WPT}$ Distance from ship's CG to WPT.

Following the concept as illustrated in Fig. 9.4 and utilising basic trigonometry rules, the formula to calculate WOP can be structured. The course alteration, which is the difference between the next and the present charted course, is represented as θ. According to Anwar [33], WOP is the distance from the ship's CG to the WPT. Thus, for this study, it will be named as d_{CG-WPT}. To obtain d_{CG-WPT}, the advance distance, d_{adv}, needs to be subtracted with d_a, therefore:

$$d_{CG-WPT} = d_{adv} - d_a \tag{9.1}$$

d_a can be obtained by utilising the tangent rule as follows:

$$\tan\theta = \frac{d_{trs}}{d_a}$$

$$d_a = \frac{d_{trs}}{\tan\theta}$$

Hence, the formula of advance transfer technique [33] is obtained as below:

$$d_{CG-WPT} = d_{adv} - \frac{d_{trs}}{\tan\theta} \tag{9.2}$$

9.3 Methodology

For this study, the d_{adv} and the d_{trs}, as required in Eq. (9.2), were taken from the ship's particulars from the simulator (Table 9.1). The ship's turning circle is given for four conditions, namely (1) ballast condition in deep water, (2) ballast condition in shallow water, (3) laden condition in deep water, and (4) laden condition in shallow water. The values of d_{adv} and d_{trs} for the selected ship, according to the respective conditions, are given as follows.

Due to the reason that the manoeuvring needed to be carried out in three navigation areas, which were open sea (OS), coastal navigation (CN), and harbour and confined waters (HCW), the arrangement of simulations was made as follows:

i. Manoeuvring in deep water while in ballast condition was carried out in open sea.
ii. Manoeuvring in deep water while in laden condition was carried out in coastal water.
iii. Manoeuvring in shallow water was carried out in harbour and confined water for both ballast and laden conditions.

Using these values, the WOP for nine different turns, where each turn was separated by 10°, were calculated to represent a total turn of 90° in a course, as shown in Tables 9.3 and 9.4. Subsequently, a series of courses representing the three navigation areas were drawn for the simulation in the ECDIS simulator. Using the simulator, a ship was manoeuvred manually by following the prepared course line. The course alteration was executed when the ship arrived at the assigned WOP for each course by steering the rudder to hard starboard angle. The helmsman was instructed to ignore any development of XTD and to not take any additional actions to bring the ship closer to the course line if the ship was deviating away. A total of seventy-two manoeuvres were carried out by the helmsman.

Table 9.1 Values of d_{adv} and the d_{trs} for selected ship in this study

Cond	Ballast				Laden			
Depth	Deep in OS		Shallow in HCW		Deep in CN		Shallow in HCW	
Side	Port	Stbd	Port	Stbd	Port	Stbd	Port	Stbd
d_{adv}	0.464	0.465	0.517	0.519	0.484	0.498	0.619	0.641
d_{trs}	0.203	0.203	0.260	0.259	0.209	0.217	0.285	0.296

Table 9.2 XTL for each navigation area

Navigation area	d_{ZOC} (m)	d_{beam} (m)	d_{pos} (m)	d_{na} (m)	d_{so} (m)	XTL (m)
Harbour and confined waters (HCW)	6.5	22.9	15	50	50.9	145.3
Coastal navigation (CN)	7.5	22.9	15	926	50.9	1022.3
Open sea (OS)	15	22.9	15	1852	50.9	1955.8

For all simulations, the XTD values fro each of the course alterations were recorded and compared to the cross-track limit (XTL). XTL is defined as the maximum limit at which a vessel is allowed to deviated from its planned track as required by the International Maritime Organization (IMO) [39]. Comparison of the XTD against the XTL determined the degree of accuracy of the ATT technique. Therefore, based on the information in [40], the XTL values in Table 9.2 are used for this study.

where d_{ZOC} = zone of confidence accuracy, d_{beam} = half vessel's beam, d_{pos} = own position accuracy, d_{na} = navigational area safety allowance, and d_{so} = (overall length × sin 20°)/2. 20° is the drift angle as in the research work by Kristić et al. [40].

9.4 Results and Discussion

Tables 9.3 and 9.4 show the comparisons between the XTL and XTD values for each condition in the respective areas.

9.4.1 Findings

9.4.1.1 Preparation Before Manoeuvring Analysis

It is important that the navigator chooses the correct values of advance and transfer distances for the WOP calculation based on the draft-to-water depth ratio. The draft can be estimated based on the amount of cargo to be carried for the laden condition, and also for the ballast condition if the ship has nothing but ballast on board. Meanwhile, the depth where the change of course is going to be carried out can be checked from the navigation chart. From there, the navigator will know whether the manoeuvring area is considered deep or shallow. Deep water is defined as a water depth which is more than three times of the ship's draft [23].

In the ATT formula, the information needed are the advance distance, transfer distance and turning angle, which are the variables for the WOP calculation. Thus, the method is suitable for WOP calculation in open sea, coastal navigation, and harbour and confined water. The WOP is difficult to be calculated during pilotage

Table 9.3 Comparison of XTD results against XTL for ballast condition

		Deep water in open sea (OS)					Shallow water in harbour and confined water (HCW)				
Side	θ	d_{adv} (nm)	d_{trs} (nm)	WOP	XTD < XTL (1955.8 m)?		d_{adv}(nm)	d_{trs}(nm)	WOP	XTD < XTL (145.3 m)?	
Starboard	10°	0.465	0.203	−0.686	97	Y	0.519	0.259	−0.950	77	Y
	20°	0.465	0.203	−0.093	185	Y	0.519	0.259	–	170	N
	30°	0.465	0.203	0.113	204	Y	0.519	0.259	0.070	278	N
	40°	0.465	0.203	0.223	172	Y	0.519	0.259	0.210	185	N
	50°	0.465	0.203	0.295	148	Y	0.519	0.259	0.302	143	Y
	60°	0.465	0.203	0.348	185	Y	0.519	0.259	0.369	119	Y
	70°	0.465	0.203	0.391	222	Y	0.519	0.259	0.425	121	Y
	80°	0.465	0.203	0.429	185	Y	0.519	0.259	0.473	127	Y
	90°	0.465	0.203	0.465	222	Y	0.519	0.259	0.519	158	N
Port	10°	0.464	0.203	−0.687	88	Y	0.517	0.260	−0.958	92	Y
	20°	0.464	0.203	−0.094	176	Y	0.517	0.260	−0.197	182	N
	30°	0.464	0.203	0.112	185	Y	0.517	0.260	0.067	278	N
	40°	0.464	0.203	0.222	156	Y	0.517	0.260	0.207	204	N
	50°	0.464	0.203	0.294	126	Y	0.517	0.260	0.299	176	N
	60°	0.464	0.203	0.347	166	Y	0.517	0.260	0.367	162	N
	70°	0.464	0.203	0.390	204	Y	0.517	0.260	0.422	144	Y
	80°	0.464	0.203	0.428	165	Y	0.517	0.260	0.471	152	N
	90°	0.464	0.203	0.464	185	Y	0.517	0.260	0.517	173	N
Compliance to XTL by %					100%					39%	

Y—Yes, *N*—No

Table 9.4 Comparison of XTD results against XTL for laden condition

		Deep water in coastal navigation (CN)					Shallow water in harbour and confined water (HCW)				
Side	θ	d_{adv} (nm)	d_{trs} (nm)	WOP	XTD < XTL (1022.3 m)?		d_{adv} (nm)	d_{trs} (nm)	WOP	XTD < XTL (145.3 m)?	
Starboard	10°	0.498	0.217	−0.733	99	Y	0.641	0.296	−1.038	92	Y
	20°	0.498	0.217	−0.098	222	Y	0.641	0.296	−0.172	222	N
	30°	0.498	0.217	0.122	241	Y	0.641	0.296	0.128	278	N
	40°	0.498	0.217	0.239	311	Y	0.641	0.296	0.288	222	N
	50°	0.498	0.217	0.316	251	Y	0.641	0.296	0.393	204	N
	60°	0.498	0.217	0.373	217	Y	0.641	0.296	0.470	204	N
	70°	0.498	0.217	0.419	215	Y	0.641	0.296	0.533	222	N
	80°	0.498	0.217	0.460	278	Y	0.641	0.296	0.589	222	N
	90°	0.498	0.217	0.498	333	Y	0.641	0.296	0.641	259	N
Port	10°	0.484	0.209	−0.701	85	Y	0.619	0.285	−0.997	112	Y
	20°	0.484	0.209	−0.090	185	Y	0.619	0.285	−0.164	278	N
	30°	0.484	0.209	0.122	222	Y	0.619	0.285	0.125	333	N
	40°	0.484	0.209	0.235	301	Y	0.619	0.285	0.279	352	N
	50°	0.484	0.209	0.309	231	Y	0.619	0.285	0.380	352	N
	60°	0.484	0.209	0.363	211	Y	0.619	0.285	0.454	389	N
	70°	0.484	0.209	0.408	204	Y	0.619	0.285	0.515	407	N
	80°	0.484	0.209	0.447	241	Y	0.619	0.285	0.569	463	N
	90°	0.484	0.209	0.484	259	Y	0.619	0.285	0.619	518	N
Compliance to XTL by %					100%					11%	

Y—Yes, *N*—No

in a harbour and confined water using other techniques such as constant rate of turn and constant radius turn [41]. This is due to the vessel's speed and the ROT, which are often adjusted according to the pilot's discretion [10]. However, since these two factors are not included as variables in the ATT formula, the WOP for courses in the harbour and confined water can be calculated during the planning stage of the voyage plan.

9.4.1.2 Compliances to XTL with Regards to the Navigation Area

In this study, the XTD was compared to the XTL according to the navigation areas, namely (1) harbour and confined water, (2) coastal navigation, and (3) open sea.

The compliances to XTL for the navigation in harbour and confined water were only 39% and 11% while in ballast and laden condition, respectively. In contrast, the compliances to XTL in coastal navigation and the open sea cases were 100% in each area. The results suggest that the ATT technique is more suitable for use in coastal and open sea navigation. However, according to Rawson (2001), rudder forces increased as the vessel's speed increased. Since the vessel was practically kept at full sea speed and at high RPM while in open sea, this technique is not suitable to be utilised. When the rudder was steered at maximum rudder angle while moving at high speed, as required by this technique, it would cause a tremendous amount of pressure on the rudder [42]. For this reason, this technique is best used for coastal navigation while manoeuvring at a slow speed, so that the rudder forces are reduced.

9.4.1.3 Negative Value for Alteration Less Than 20°

For turns of 10° and 20°, the computed WOP values for course alteration using the ATT were negative as seen in Tables 9.3 and 9.4. This implies that the turn must be completed immediately after the WPT, as seen in Fig. 9.5, which is illogical because it implies that the ship has already overshot. As a result, WOP with a negative value was executed precisely at WPT during the manoeuvring analysis in this study. This implies that the ATT has an issue when course alterations of less than 20° are made. As a result, the ATT is inefficient to be used on an autonomous ship, as it causes the ship to overshoot from its intended course.

9.4.1.4 The Final Heading of the Ship Does not Match the Charted Course

The working principle of the technique is as shown in Fig. 9.6. It can be observed that the final heading of the ship is 090°T, which in contrast with the 045°T desired course. The formula would be more relevant if the final heading of the ship matches the desired charted course so that the second overshoot would be avoided.

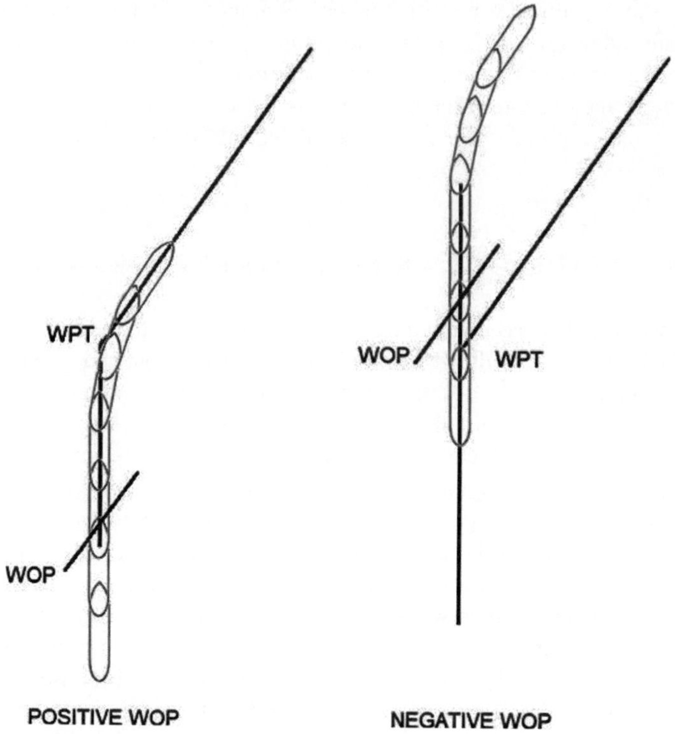

Fig. 9.5 Comparison between positive and negative WOP

9.5 Conclusion

It is critical to adhere to the intended navigation track in ship navigation for fuel economy and navigation safety reasons. The advance transfer technique (ATT) is one of the ways for ensuring that the ship remains on track when altering course. While the use of ECDIS on ships has decreased sailors' reliance on conventional methods such as ATT, not all vessels are equipped with ECDIS, which implies that the ATT approach remains essential to most seafarers. Due to the scarcity of data on the efficiency of ATT, this study used a ship simulator to undertake a technical assessment of the methodology.

9.5.1 Contribution of the Study

Based on the 'preparation before manoeuvring analysis' section that is elaborated in 4.1.1, the ATT is suitable for determining the WOP for harbour and confined water, coastal navigation, and open sea. However, considering the XTD compliance

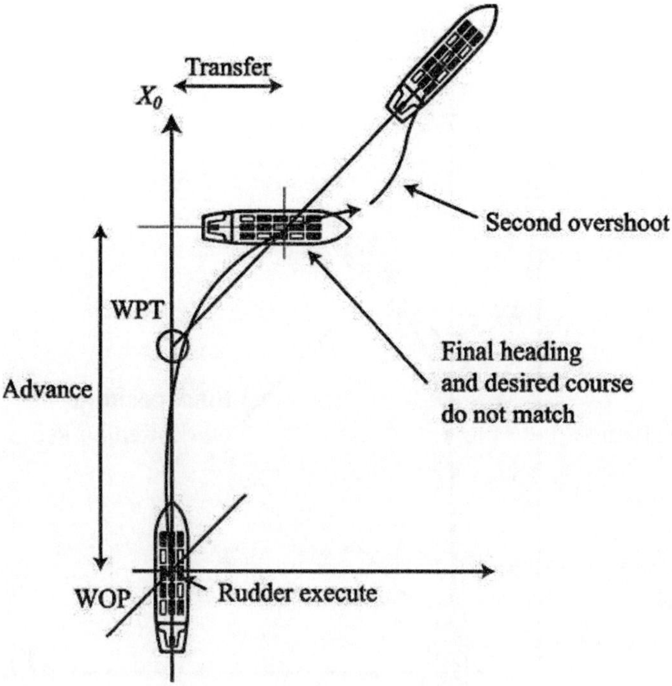

Fig. 9.6 Advance transfer technique principle

to XTL, ATT is only suitable for coastal and open sea navigation. On the other hand, considering the effect on the ship's structural stress, ATT is not ideal for use in open sea due to the rudder forces acting on the rudder depending on the ship's speed. This is because the speed is normally kept at 'full sea ahead' in open sea, so huge rudder application while moving at high speed is not recommended. For this reason, it can be concluded that the study proved the effectiveness of the ATT. However, it is only suitable for coastal navigation, provided the ship's speed has been reduced.

9.5.2 Suggestion for Future Research

9.5.2.1 Suggestion for Improving the Advance Transfer Mathematical Model

After assessing the ATT, the initial hypothesis for future research is that the large XTD found in this study occurred because the final heading did not match the final course line, as seen in Fig. 9.6. Therefore, theoretically, if the turning circle application is adjusted so that the ship's final heading matches the final charted course, as shown in Fig. 9.7, the XTD can be reduced, allowing this technique to be used in harbours and confined water.

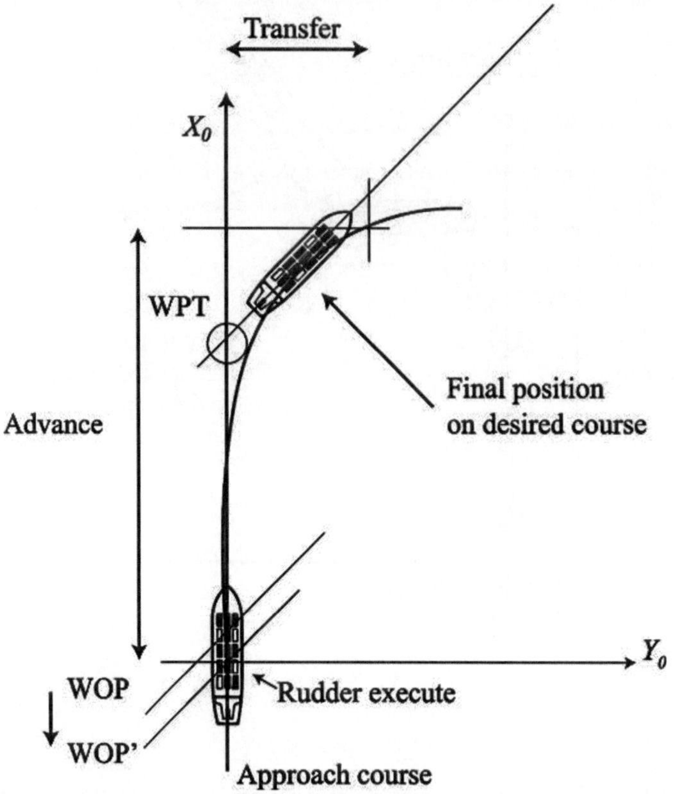

Fig. 9.7 Suggestion for development of an improved model

9.5.2.2 Alternative Method to Determine WOP that Complies with All Three Navigation Areas

Further research on other alternative methods to determine the WOP, such as the constant rate of turn and constant radius turn [41], can be carried out. In addition, a similar evaluation of the methods' effectiveness through practical assessment and validation against existing regulations, as demonstrated in this study, can be attempted.

References

1. Boom DV, Keane S (2021) Ever given seized in Egypt after blocking Suez Canal: everything to know—CNET. https://www.cnet.com/news/ever-given-seized-in-egypt-after-blocking-suez-canal-everything-to-know/. Accessed 23 May 2021
2. Reuters (2021) Ever given owner says Suez Canal authority at fault for grounding. Suez

canal. The Guardian. https://www.theguardian.com/world/2021/may/22/ever-given-owner-says-suez-canal-authority-at-fault-for-grounding. Accessed 23 May 2021
3. Ang I, Corbett A (2021) OOCL containership causes double crane collapse at Kaohsiung port. TradeWinds. https://www.tradewindsnews.com/casualties/oocl-containership-causes-double-crane-collapse-at-kaohsiung-port/2-1-1019995. Accessed 4 Jun 2021
4. Voytenko M (2021) OOCL container ship contacted 2 cranes wreaking havoc, 2 trapped inside VIDEOS. OOCL DURBAN—FleetMon Maritime News. https://www.fleetmon.com/maritime-news/2021/34017/oocl-container-ship-contacted-2-cranes-wreaking-ha/. Accessed 4 Jun 2021
5. Swift AJ (2018) Bridge team management: a practical guide, 2nd edn. Nautical Institute, London
6. Skora K, Wolski A (2016) Voyage planning. Sci J Silesian Univ Technol Ser Transp 92:123–128. https://doi.org/10.20858/sjsutst.2016.92.12
7. Lušić Z, Kos S, Galić S (2014) Standardisation of plotting courses and selecting turn points in maritime navigation. PROMET—Traffic Transp 26:313–322. https://doi.org/10.7307/ptt.v26i4.1437
8. Vujičić S, Mohović R, Tomaš ID (2018) Methodology for controlling the ship's path during the turn in confined waterways. Pomorstvo 32:28–35. https://doi.org/10.31217/p.32.1.2
9. Lekkas AM, Fossen TI (2014) Minimization of cross-track and along-track errors for path tracking of marine underactuated vehicles. In: 2014 European Control Conference ECC, pp 3004–3010. https://doi.org/10.1109/ECC.2014.6862594
10. Georgiana S, Stefan G (2010) Planning and execution of blind pilotage and anchorage. Constanta Marit Univ Ann 14(2):35–40
11. Reid RE (1978) Improvement of ship steering control for merchant ships—phase IIA. National Maritime Research Center, New York
12. MAIB (2017) Grounding of MUROS Haisborough Sand North Sea—MAIB. https://www.gov.uk/maib-reports/grounding-of-bulk-carrier-muros. Accessed 25 Sept 2020
13. Steamship Mutual (2014) The importance of ECDIS training and good watch-keeping practices—Steamship Mutual. https://www.steamshipmutual.com/Downloads/Risk-Alerts/RA49TheImportanceECDIS_TrainingGoodWatch-keepingPractices.pdf. Accessed 8 Jan 2021
14. MAIB (2015) Report on the investigation of the grounding and flooding of the ro-ro ferry Commodore Clipper in the approaches to St Peter Port, Guernsey on 14 July 2014—MAIB. https://www.gov.uk/maib-reports/grounding-and-flooding-of-ro-ro-ferry-commodore-clipper. Accessed 25 Jan 2021
15. Gale H, Patraiko D (2007) Improving navigational safety—Seaways. https://www.nautinst.org/uploads/assets/uploaded/b311f375-f2da-4c3c-aacfa6df9f604b50.pdf. Accessed 17 Dec 2020
16. Marine Insight (2017) Real life accident: improper bridge procedures and ECDIS use causes grounding of ship. In: MARS Rep. https://www.marineinsight.com/case-studies/mproper-bridge-procedures-ecdis-use-caused-grounding/. Accessed 25 Aug 2020
17. Marine Insight (2020) Real life accident: insufficient passage planning leads to bottom damage of vessel. In: MARS Rep. https://www.marineinsight.com/case-studies/real-life-accident-scraping-the-bottom/. Accessed 25 Aug 2020
18. Molland AF (2008) The maritime engineering reference book: a guide to ship design, construction and operation. Butterworth-Heinemann, Amsterdam
19. Uğurlu Ö, Köse E, Yıldırım U, Yüksekyıldız E (2015) Marine accident analysis for collision and grounding in oil tanker using FTA method. Marit Policy Manag 42:163–185. https://doi.org/10.1080/03088839.2013.856524
20. Meyers SD, Luther ME (2020) The impact of sea level rise on maritime navigation within a large, channelized estuary. Marit Policy Manag 47:920–936. https://doi.org/10.1080/03088839.2020.1723810
21. Du P, Ouahsine A, Sergent P (2018) Influences of the separation distance, ship speed and channel dimension on ship maneuverability in a confined waterway. Comptes Rendus—Mec 346:390–401. https://doi.org/10.1016/j.crme.2018.01.005

22. Briggs MJ, Borgman LE, Bratteland E (2003) Probability assessment for deep-draft navigation channel design. Coast Eng 48:29–50. https://doi.org/10.1016/S0378-3839(02)00159-X
23. Masumi Y, Nikseresht AH (2019) 2 DOF numerical investigation of a planing vessel in head sea waves in deep and shallow water conditions. Appl Ocean Res 82:41–51. https://doi.org/10.1016/j.apor.2018.10.017
24. Paffett JAH (1973) Ship manoeuvring characteristics. J Navig 26:113–122. https://doi.org/10.1017/S0373463300022943
25. Mucha P, Dettmann T, Ferrari V, el-Moctar O (2019) Experimental investigation of free-running ship manoeuvers under extreme shallow water conditions. Appl Ocean Res 83:155–162. https://doi.org/10.1016/j.apor.2018.09.008
26. Molland AF (2008) Chapter 8—Manoeuvring. Marit Eng Ref B 19:578–635. https://doi.org/10.1016/B978-0-7506-8987-8.00008-1
27. Rutkowski G (2018) ECDIS limitations, data reliability, alarm management and safety settings recommended for passage planning and route monitoring on VLCC tankers. TransNav Int J Mar Navig Saf Sea Transp 12:483–490. https://doi.org/10.12716/1001.12.03.06
28. IMO (2019) Electronic nautical charts (ENC) and electronic chart display and information systems (ECDIS)—IMO. https://www.imo.org/en/OurWork/Safety/Pages/ElectronicCharts.aspx. Accessed 11 Jun 2021
29. Marine Insight (2019) understanding different types of manoeuvres of a vessel—Marine Insight. https://www.marineinsight.com/naval-architecture/different-types-of-manoeuvres-of-a-vessel/. Accessed 21 May 2021
30. Wu J, Thorne-Large J, Zhang P (2021) Safety first: the risk of over-reliance on technology in navigation. J Transp Saf Secur 1–28. https://doi.org/10.1080/19439962.2021.1909681
31. Kim MS, Shin HO, Kang KM, Kim MS (2005) Variation of the turning circle by the rudder angle and the ship's speed-mainly on the training ship KAYA-. Bull Korean Soc Fish Technol 41:156–164. https://doi.org/10.3796/KSFT.2005.41.2.156
32. Drachev VN (2012) Calculating wheel-over point. Asia-Pacific J Mar Sci 2:27–46
33. Anwar N (2015) Navigation advanced mates/masters, 2nd edn. Weatherby Seamanship International, Division of Witherbys Publishing Group Limited, London
34. TTEG (2008) Guidelines on voluntary pilotage services in the straits of Malacca and Singapore - TTEG. http://www.marine.gov.my/jlmv4/sites/default/files/ANNEX-U_13.1-ANNEX_2_Guidelines_on_VPS_in_SOMS.pdf. Accessed 13 Nov 2020
35. Van HMJ, Wolkenfelt PHM (2000) The rate of turn required for geographically fixed turns: a formula and fast-time simulations. J Navig 53:146–155. https://doi.org/10.1017/S0373463399008590
36. Inoue K, Okazaki T, Murai K, Hayashi Y (2013) Fundamental study of evaluation at berthing training for pilot trainees using a ship maneuvering simulator. TransNav Int J Mar Navig Saf Sea Transp 7:135–141. https://doi.org/10.12716/1001.07.01.18
37. Kornacki J (2011) Ship's Turning in the Navigational Practice. Navig Syst Simulators 5:187–193. https://doi.org/10.1201/b11343-33
38. ITTC (2002) Full scale measurements manoeuvrability full scale manoeuvring trials procedure—ITTC. https://www.ittc.info/media/8179/75-04-02-01.pdf. Accessed 15 Jan 2021
39. IMO MSC (2006) Adoption of the revised performance standards for electronic chart display and information systems (ECDIS) MSC 82/24/Add.2—IMO. https://www.register-iri.com/wp-content/uploads/MSC_Resolution_23282.pdf. Accessed 12 Jan 2021
40. Kristić M, Žuškin S, Brčić D, Valčić S (2020) Zone of confidence impact on cross track limit determination in ECDIS passage planning. J Mar Sci Eng 8. https://doi.org/10.3390/JMSE8080566
41. Jithin (2019) Constant radius turn—knowledge of sea. https://knowledgeofsea.com/constant-radius-turn/. Accessed 20 May 2021
42. Rawson KJ (2001) Basic ship theory, 5th edn. Butterworth-Heinemann, pp 523–573. ISBN 9780750653985, https://doi.org/10.1016/B978-075065398-5/50016-9

Chapter 10
Polysulfone/Cellulose Acetate Phthalate/Polyvinylpyrrolidone (PSf/CAP/PVP) Blend Membranes: Effect of Evaporation Time on Blend Membrane Characteristics

Asmadi Ali, Rosli Mohd Yunos, Mohamad Awang, Sofiah Hamzah, Mohammad Hakim Che Harun, Fazureen Azaman, and Muhammad Abbas Ahmad Zaini

Abstract Recently, researchers proved that the evaporation time during the membrane fabrication process has a significant effect on the characteristics and performance of membranes. In this study, flat sheet asymmetric polysulfone/cellulose acetate phthalate/polyvinylpyrrolidone (PSf/CAP/PVP) blend membranes were fabricated at different evaporation time in the range of 0–20 s to investigate the effect of the evaporation time on characteristics of the blend membranes. The PSf/CAP/PVP blend membranes were characterized in terms of water content, porosity, pure water flux and permeability coefficient. The results showed that an increase of evaporation time from 0 to 20 s has resulted in decreasing of water content and porosity of the PSf/CAP/PVP blend membranes. The permeation water flux of the blend membrane

A. Ali (✉) · M. Awang · S. Hamzah · M. H. Che Harun · F. Azaman
Environmental Sustainable Materials Research Interest Group, Faculty of School of Ocean Engineering Technology and Informatics, Universiti Malaysia Terengganu, 21030 Kuala Nerus, Terengganu, Malaysia
e-mail: asmadi@umt.edu.my

M. Awang
e-mail: mohamada@umt.edu.my

S. Hamzah
e-mail: sofiah@umt.edu.my

M. H. Che Harun
e-mail: m.hakim@umt.edu.my

R. Mohd Yunos
Faculty of Chemical and Natural Resources Engineering, Universiti Malaysia Pahang, Lebuhraya Tun Razak, Kuantan, Pahang, Malaysia
e-mail: rmy@ump.edu.my

M. A. Ahmad Zaini
Centre of Lipid Engineering and Applied Research (CLEAR), Universiti Teknologi Malaysia, UTM Johor Bahru, 81310 Skudai, Johor, Malaysia
e-mail: abbas@cheme.utm.my

A. Ismail et al. (eds.), *Design in Maritime Engineering*, Advanced Structured Materials 167, https://doi.org/10.1007/978-3-030-89988-2_10

was decreased with the increment of evaporation time. This indicated that an increase of evaporation time would lead to a decrease of the membrane permeability coefficient and hence increased the membrane resistance to permeation of water through the membranes.

Keywords Ultrafiltration · Evaporation time · Blend membrane · Flux

10.1 Introduction

In recent years, membrane technology has been accepted as one of the best available alternative technologies that can be offered to treat the contaminated water from pollutants. This technology is being employed at water treatment plants for producing a high quality of drinking water to comply with a specified standard before it can be supplied to the customer. Ultrafiltration (UF) is one of the promising membrane technology processes that have been received much attention in water treatment process as an effective technology in the removal of suspended solids, colloidal material, inorganic particulates and fatal microorganisms. Most of the commercial membranes used in the UF membrane technology process is an asymmetric membrane and this type of membrane can be fabricated to suit with their suitable application in the water treatment process.

The invention of the asymmetric membranes by Loeb and Sourirajan since 50 years ago has made a great impact on the growth of membrane science and technology. Their breakthrough has put a milestone in the history of membrane technology progress. This remarkable finding has opened the door to commercialize the membrane technology from laboratory-scale membrane application turn to large-scale commercial [1]. Nowadays, most of the commercial membranes have been fabricated by using the phase inversion method. In the phase inversion process, membrane casting solution is changed to a solid thin membrane film by the dry–wet phase inversion technique. During this technique, the membrane casting solution which contains a polymer and a solvent is poured on a plate and cast by a casting knife to produce a thin membrane film [2]. The film is introduced to dry air or nitrogen gas to evaporate the solvent from the membrane film solution for a certain period of time that is known as the evaporation time. Then, the partial solid of thin membrane film is immersed into a coagulation bath to complete the membrane fabrication process [3].

Nowadays, a few of studies by researchers proved that the evaporation time, as one of the important membrane fabrication parameters affects the asymmetric membrane characteristics, morphology and performance of polyethersulfone (PES), polyaniline (PAni) and BDTA-TDI/MDI co-polyimide (P84) membranes in gas separation process [4–6]. In the ultrafiltration (UF) membrane separation process, the separation performance of the UF membrane is solely related to the characteristics and structural morphology of the UF membranes. Hence, in this study flat sheets of PSf/CAP/PVP blend UF membranes were fabricated at different evaporation times to alter their characteristics of the blend membranes. The UF blend membranes were characterized in terms of water content, porosity, pure water flux and permeability coefficient.

10.2 Methodology

10.2.1 Materials

All materials used were of analytical grade. The PSf/CAP/PVP blend membranes were fabricated from ternary casting solutions which consist of PSf (supplied by Amoco Chemical (USA) S. A.) as the membrane backbone polymer, CAP (purchased from Sigma-Aldrich Co.) as the hydrophilic polymer, *N*-Methyl-2-Pyrrolidone (NMP) from MERCK Schuchard OHG (Germany) was used as solvent and polyvinylpyrrolidone (PVP) K15 was purchased from Fluka employed as an organic additive. Distilled water was used as the coagulation bath medium.

10.2.2 Membrane Preparation

The casting solutions of asymmetric PSf/CAP/PVP blend membranes were prepared consisting of 17 wt% of polymer composition (PSf/CAP), 3 wt% of PVP additive and 80 wt% of NMP solvent in the total membrane casting solution. CAP contained 10 wt% in total polymer composition as explained by Ali et al. [1]. The casting solutions were poured onto a stainless steel plate, and then, they were cast by using an automatic casting machine. The cast polymer solutions were introduced with a convective inert stream (nitrogen) for a certain period of time which is known as evaporation time in the dry-phase inversion process.

After that, the cast polymers were immersed in a coagulation bath in order to complete the formation of solid flat sheet membranes. The prepared membranes were stored in distilled water prior usage. In this study, the evaporation time was studied in the range of 0, 5, 10, 15 and 20 s and the produced blend membranes were marked as PCE-0, PCE-5, PCE-10, PC-15 and PCE-20 membranes, respectively.

10.2.3 Membrane Characterization

10.2.3.1 Water Content and Porosity

The water content of the PSf/CAP/PVP blend membranes was evaluated for water absorption capacity and calculated by Eq. (10.1):

$$A = \frac{W_{\text{wet}} - W_{\text{dry}}}{W_{\text{wet}}}, \tag{10.1}$$

where A is the water content (wt%), W_{wet} is the wet weight of membrane (mg), and W_{dry} is the dry weight of membrane (mg). The water content of the membranes was determined by soaking the membranes in water for 24 h at room temperature. The weight of the wet PSf/CAP/PVP blend membranes was first measured after mopping the membranes with a blotting paper and then dried in an oven at 75 °C for 48 h. The porosity of PSf and PSf/CAP blend membranes was evaluated using the expression below:

$$\text{porosity} = \frac{(W_1 - W_2)/d_{\text{water}}}{V}, \tag{10.2}$$

where W_1 and W_2 are the mass of membrane in wet and dry states (mg), d_{water} is the density of water at room temperature (ml/mg), and V *is* the volume of the membrane in wet state (ml).

10.2.3.2 Pure Water Flux and Permeability Coefficient

Pure water flux is important for the determination of the membrane stability and hydraulic properties. Distilled water was used to determine the pure water flux of each PSf/CAP/PVP blend membranes using a dead-end filtration cell. Membranes were then subjected to a pure water flux test with varying operating pressure in the range of 1–5 bar. The membrane permeability coefficient of PSf/CAP/PVP blend membranes can be determined by subjecting these membranes at various pressures towards its pure water fluxes. The effect of the evaporation time on the hydraulic permeability coefficient of PSf/CAP/PVP blend membranes is determined from the slope of the linear line of pure water fluxes.

10.3 Results and Discussion

10.3.1 Water Content and Porosity

Table 10.1 shows the water content and porosity of the PSf/CAP/PVP membranes which were introduced to evaporation time during the membrane fabrication process. It was observed that the water content and porosity of the PSf/CAP/PVP membrane in the absence of evaporation time shows the highest value compared to the blend membranes introduced to the evaporation time. The PSf/CAP/PVP membranes introduced at low evaporation time have higher water content and porosity compared to the blend membrane fabricated at high evaporation time. It was observed that an

Table 10.1 Effects of evaporation time on water content and porosity of PSf/CAP/PVP blend membranes

Membrane	Evaporation time (s)	Water content (%)	Porosity (%)	Permeability coefficient (l/m^2 h bar)
PCE-0	0	78.13	74.59	66.14
PCE-5	5	75.20	69.73	56.64
PCE-10	10	74.93	68.98	47.55
PCE-15	15	72.78	64.22	34.32
PCE-20	20	70.44	62.84	27.31

increase in evaporation time significantly decreased the water content and porosity of the PSf/CAP/PVP blend membranes as given in Table 10.1.

Ohya et al. [7] and Sabri et al. [8] studied the effect of evaporation time on asymmetric aromatic polyimide membranes. They found that shortening the evaporation time will decrease the evaporated quantity of solvent and consequently change the polymer concentration at the polyimide surface which in turn produced membranes with big pore size and high porosity. Ali et al. [6] found the formation of larger pores in blend membranes increased the porosity of the PSf/CAP blend membranes. Based on the results from Table 10.1, it was postulated that a decrease in water content and porosity of PSf/CAP/PVP blend membranes which were fabricated at high evaporation time due to the formation of small pore size at the membrane surfaces. Generally, membranes with small pore size have a high resistance and lower porosity to water to absorb through the membrane structures which in turn decreased the water content of UF membranes.

10.3.2 Pure Water Flux and Membrane Permeability Coefficient

The measurements of pure water permeation as a function of applied pressure were used to investigate the stability and hydraulic properties of UF membranes. Figure 10.1 represents the influence of the evaporation time on the pure water flux of the PSf/CAP/PVP blend membranes measured at various operating pressures in the range of 1–5 bar. As shown in Fig. 10.1, pure water flux was a linear function of the applied pressure. The PSf/CAP/PVP blend membrane in the absence of evaporation time (PCE-0), i.e., the casting solution film which was immediately immersed into a coagulation bath exhibited the highest range of pure water flux in the range of 72.63–318.90 l/m^2 h.

As shown in Fig. 10.1, the trend of the linear line of the pure water flux was decreased by increasing the evaporation time introduced to the casting solution films. The PSf/CAP/PVP membrane introduced to 20 s of evaporation time (PCE-20) shows the lowest pure water flux in the range of 25.02–140.39 l/m^2 h. It was observed that the decrease in pure flux was proportional to the evaporation time. The results revealed

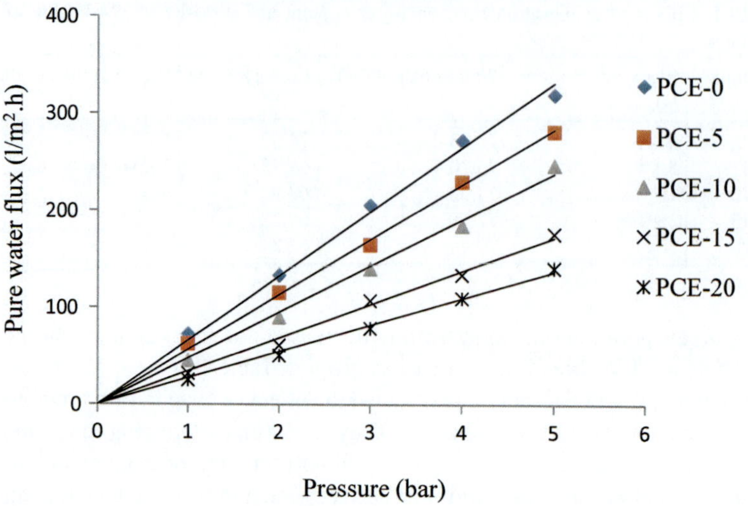

Fig. 10.1 Pure water flux of PSf/CAP/PVP blend membranes fabricated at different evaporation time at the operating pressure of 1–5 bar

that the order of pure water flux of PSf/CAP/PVP blend membranes was decreased as to be PCE-0 > PCE-5 > PCE-10 > PCE-15 > PES-20. The permeability coefficient of five tested membranes of PCE-0, PCE-5, PCE-10, PCE-15 and PCE-20, respectively, can be obtained from the slope of the straight line in Fig. 10.1 and provided in Table 10.1.

Membrane in the absence of evaporation time (PCE-0) exhibited the highest membrane permeability coefficient of 66.14 l/m^2 h bar. However, an increase in the evaporation time from 5 to 20 s gradually decreased the permeability coefficient of PSf/CAP/PVP blend membranes from 56.64 to 27.31 l/m^2 h bar. Noticeably, the permeability coefficient of membranes decreased in the following sequence: PCE-0 > PCE-5 > PCE-10 > PCE-15 > PCE-20. This indicated that an increase of evaporation time would lead to a decrease in the membrane permeability. In other words, an increase in the evaporation time of membrane produced membranes with high membrane hydraulic resistance characteristics and low porosity which in turn decreased the water permeation flux.

According to Huang and Feng [9], the increment in evaporation time removed the most volatile solvent from the membrane surface and led to form more a concentrated nascent skin layer. During the evaporation process, the thickness of the skin layer was a function of time and rate of solvent evaporation. An increasing in the evaporation time induced high evaporation rate of the most volatile solvent and consequently increased the polymer concentration at the nascent skin membrane which in turn formed a thick skin layer. Hasbullah et al. [4] reported that the skin layer thickness was increased with the increase of evaporation time. Jami'an et al. [10] also revealed that an increase of evaporation time produced a membrane with a compact sublayer which contributes to the high resistance to gas molecules to penetrate through membrane layer.

Based on Fig. 10.1 and Table 10.1, it was proven that the pure water flux depends on the membrane permeability coefficient and porosity. It was suggested that the PSf/CAP/PVP blend membranes prepared at high evaporation time produced membranes with thicker skin layer due to higher polymer concentration at the membrane skin layer than the membranes fabricated at low evaporation time. It was due to the increase in the polymer concentrations at the nascent skin layer which tend to increase thickness and dense of the membrane, which consequently results in a low hydraulic permeability, high hydraulic membrane resistance and low porosity. Ismail's et al. [3] and Benhabiles's et al. [11] observed the same results for PES membranes. It was also postulated the sublayer of the membrane also appears in compact structure due to decrease in solvent evaporation rate.

10.4 Conclusion

The results showed that the effects of evaporation time significantly changed the characteristics of the PSf/CAP/PVP blend membranes in terms of water content, porosity, pure water flux and permeability coefficient. The results revealed that an increase of evaporation time has decreased the water content and porosity of the blend membranes. It was postulated that the membranes fabricated at high evaporation time formed membranes with small pores which in turn resist water to absorb through the pores and consequently reduced the water absorption and porosity. It was also observed that the decrease in pure water flux was proportionate to the evaporation time due to decrease in permeability coefficient and the increment of hydraulic membrane resistance of the PSf/CAP/PVP membranes. This study revealed that the fabrication of different UF membrane characteristics was successfully developed by manipulating the evaporation time and these membranes are potentially used for a wide range of water treatment process.

Acknowledgements The authors wish to express high gratitude to Faculty of School of Ocean Engineering Technology and Informatics, Universiti Malaysia Terengganu and Faculty of Chemical & Natural Resources Engineering, Universiti Malaysia Pahang for the contribution and support for this study.

References

1. Ali A, Yunus RM, Awang M et al (2014) Effect of cellulose acetate phthalate (CAP) on characteristics and morphology of polysulfone/cellulose acetate phthalate (PSf/CAP) blend membranes. Appl Mech Mater 493:640–644
2. Ali A, Yu CLM, Sani NAM et al (2018) Preparation, characterization and performance of polyvinylidene fluoride/tetraoctyl phosphonium bromide nanocomposite ultrafiltration membrane. Malays J Anal Sci 22:514–521
3. Ismail AF, Norida R, Abdul RWAW et al (2011) Preparation and characterization of hyperthin-skinned and high performances asymmetric polyethersulfone membrane for gas separation. Desalination 273:93–104
4. Hasbullah H, Kumbharkar S, Ismail AF, Li K (2011) Preparation of polyaniline asymmetric hollow fiber membranes and investigation towards gas separation performance. J Memb Sci 366:116–124
5. Wang R, Chung TS (2001) Determination of pore sizes and surface porosity and the effect of shear stress within a spinneret on asymmetric hollow fiber membranes. J Memb Sci 188:29–37
6. Ali A, Awang M, Mat R et al (2014) Influence of hydrophilic polymer on pure water permeation, permeability coefficient, and porosity of polysulfone blend membranes. Adv Mater Res 931–932:168–172
7. Ohya H, Okazaki I, Aihara M et al (1997) Study on molecular weight cut-off performance of asymmetric aromatic polyimide membrane. J Memb Sci 123:143–147
8. Sabri NSM, Hasbullah H, Tohid MS et al (2020) Effect of solvent evaporation time of polysulfone incorporated copper oxide nanoparticles incorporated polysulfone ultrafiltration membrane on protein removal. J Appl Membr Sci Technol 24:1–12
9. Huang RYM, Feng X (1995) Studies on solvent evaporation and polymer precipitation pertinent to the formation of asymmetric polyetherimide membranes. J Appl Polym Sci 57:613–621
10. Jami'an WNR, Hasbullah H, Mohamed F et al (2016) Effect of evaporation time on cellulose acetate membrane for gas separation. In: 6 IOP conference series: earth and environmental science. IOP Publishing, 12008
11. Benhabiles O, Galiano F, Marino T et al (2019) Preparation and characterization of TiO_2-PVDF/PMMA blend membranes using an alternative non-toxic solvent for UF/MF and photocatalytic application. Molecules 24:724

Chapter 11
Influence of Deadrise on the Dynamic Instability of a 14 Meters Custom Boat in Regular Waves

Hamdan Nuruddin, Aqil Azrai Razali, Muhammad Nasuha Mansor, and Iwan Mustaffa Kamal

Abstract Stability assessment is one of the important criteria for small crafts. Its relationship with safety and seaworthiness has made it an essential guideline to be examined from the early stage of the design process. This paper will focus on the transverse dynamic instability of small crafts that have been performed by the authors for a custom boat with three different deadrise angles. The effects of longitudinal center of gravity, LCG positions, and deadrise angles on the dynamics and hydrodynamics of the vessel in different trims are done. The major parameters taken into accounts in this investigation were the projected area of the planing bottom between the transom and chine, volume displaced, chine length, maximum breadth over chines, centroid of projected area, longitudinal center of gravity, loading coefficient, and length-to-beam ratio in a tabular form for each design. The probability of dynamic instability after reaching planing speed was compared with the guidelines from Blount and Codega. The conditions in which the deadrise angles failed to meet the guidelines were identified.

Keywords Dynamic stability guidelines · Transverse instability · Deadrise angle · Custom boat

11.1 Introduction

Stability is one of the important factors in determining a ship's safety and seaworthiness while in operation. It is a mandatory requirement that the ship designers or shipbuilders submit the stability assessment report (stability booklet) to the related

H. Nuruddin (✉) · A. A. Razali · M. N. Mansor · I. M. Kamal
Maritime Engineering Technology Section, Universiti Kuala Lumpur Malaysian Institute of Marine Engineering Technology, Lumut, Perak, Malaysia
e-mail: hamdann@unikl.edu.my

M. N. Mansor
e-mail: mnasuha@unikl.edu.my

I. M. Kamal
e-mail: iwanzamil@unikl.edu.my

A. Ismail et al. (eds.), *Design in Maritime Engineering*, Advanced Structured Materials 167, https://doi.org/10.1007/978-3-030-89988-2_11

regulatory bodies such as classification society and marine department, well before the construction begins [1]. In a static condition, when a ship is heeled to one side by internal or external forces, the ship will return to its original upright position. This ability is usually expressed as the righting moment, which is the product of force (hydrostatic) and righting arm. Internal forces such as loading, unloading, cargo, and passenger/crew movement can be considered as static forces, whereas external forces such as wind, wave, and high-speed turning during operation should be considered as dynamic forces [1].

Any floating body stability assessment must meet both static and/or dynamic stability requirements. To protect the safety of crews and passengers, the ship must stay afloat and maintain a minimum level of stability while in operation. For large vessels, the hydrostatic stability assessment is adequate due to its large reserve of buoyancy but for small craft this is inadequate. As a result, dynamic stability analysis must be performed early in the design process. Although the static stability criteria have taken into accounts the requirements for high-speed boats, they are still solely based on the hydrostatic assessment [1]. This has in turn created many dedicated works in trying to create the hydrodynamic assessment or dynamic stability assessment to enhance the understanding of ship's dynamic behavior without using the time domain simulation; as a result, there are few guidelines on dynamic stability assessment [2]. One of the hull form design modifications is to change the deadrise angle that will affect the resistance and stability of the ship. It is important at this point to introduce some definitions and basic concepts related to the influence of the transverse dynamic stability to the deadrise angle. For many years, and still the standard practice today, the stability of a ship (static and dynamic) is assessed statically as reflected in many stability codes provided by the classification societies. However, until a more comprehensive approach of assessing the dynamic stability of a ship is established, the designer and naval architect must maneuver within the present guidelines [1].

The ratio of length to beam, the relationship between hull size and gross weight, and the longitudinal location of the center of gravity are three of the most critical parameters impacting the performance of planing hulls [3]. If hulls with various length–beam ratios are evaluated based on similar $A_p/\nabla^{2/3}$, the comparison will be based on nearly similar values of hull area, hull volume, and hull structural weight, as stated in [4]. The distance between the LCG and the centroid of the region A_p, given as a percentage of the length L_p, is known as longitudinal CG location. This will allow the investigation on the impacts of different loadings and LCG locations.

Very little is known about the fundamental causes, and no guidelines presently exist to ensure adequate dynamic stability [2]. The buttocks of a vee-bottom hull are shaped like two airfoil sections that are joined at the keel. The pressure distribution is altered by any asymmetric port and starboard wetted surface or change in trim produced by a movement in weight or sea condition. The equivalent of an airfoil with a high thickness-to-chord ratio, a boat with highly curved buttocks is more prone to developing local low-pressure areas that may cause instabilities at planing speed.

A low-pressure area acting on the mid-body will decrease the transverse stability. The effect of the LCG on dynamic stability is that the trim curve produces the characteristic inflection point when the LCG is shifted forward and the running trim is decreased [2]. The hull loading in relation to the hull dimensions, as well as the position of the LCG, has a significant impact on the possible instability. Porpoising has been successfully predicted and avoided using the guidelines [1]. Most problem boats are that heavily loaded which is $A_p/\nabla^{2/3}$ more than 5.8, and LCG are no more than 3% of the centroid of planing area, CA_p [2].

Samian and Malik [1] found out that the current stability assessment HSC code does not give any guidance for ensuring acceptable dynamic stability. Dynamic instability is a complicated phenomenon that is influenced by a variety of factors, including speed, displacement, weight distribution, hull shape, and appendage design and placement. The relationship between each of the above factors cannot be discussed in depth since the precise relationship is unknown with great uncertainty. Nonetheless, this has led to the recommended design guideline presented in [2] to offer the designer with some tool for analyzing the dynamic stability of a fast boat. Negative trim (bow down) and inflection point might occur if the LCG is located near the centroid of the projected chine waterline area.

The position of the boat's center of gravity is thought to be a major cause that poses serious consideration. To avoid dynamic transverse instability with an $A_p/\nabla^{2/3}$ ratio of more than 5.8, the percentage of $(CA_p\text{-}LCG)/L_p$ should be more than 3%. Blount and Codega found that hulls with a high L_p/B_{px} ratio are more likely to have a reverse slope or inflection, which is more likely to cause instability than hulls with a low L_p/B_{px} ratio. To avoid the saddle point region, $A_p/\nabla^{2/3}$ must be greater than $0.39\,(L_p/B_{px}) + 4.52$ [2].

The Webb Institute of Naval Architecture evaluated a variety of prismatic model hulls with varied deadrise angles. Towing a small model aft while gradually moving the LCG aft till porpoising occurred was the procedure. The discovery that running trim angle, LCG location, deadrise angle, and speed were all factors in the onset of porpoising [5].

All the types had a prismatic aft body, which meant that the deadrise between midships and stern was kept constant. The relative weight or loading factor ($A_p/\nabla^{2/3}$) and the longitudinal position of the center of gravity were modified in the series. It became clear that the deadrise angle was a significant factor in enhancing the seakeeping behavior of these fast-planing boats. However, a greater deadrise has a significant impact on the boat's calm water resistance [2]. In general, a greater deadrise means a higher resistance. To achieve an even better fit over the entire range of all the deadrise angles used in actual designs and in particular because of the fact that a considerable amount of hard chine planing hulls were designed around the 20°–25° of deadrise range, it was decided in 1996 that the Delft Systematic Deadrise Series (DSDS) database was to be extended along the original lines with a similar series but now with 19° of deadrise to better "fill the gap" between 12.5° and 25° of deadrise. The DSDS has been under development for decades by now and consists of a large family of systematically varied hard chine planing monohulls, based on the original research by Clement and Blount, which have all been tested in the same speed range, changing the same parameters and using the same experimental setup [6].

11.2 Methodology

Figure 11.1 shows the flowchart of the research. The project started as early as the previous hulls study including all the necessary data such as design parameters and guidelines requirement, as shown in Table 11.1. The study was initiated by generating a hull form using AutoCAD, based on a planing craft types from one of the Malaysia's enforcement agencies, and then was developed further using Bentley

Fig. 11.1 Project flowchart

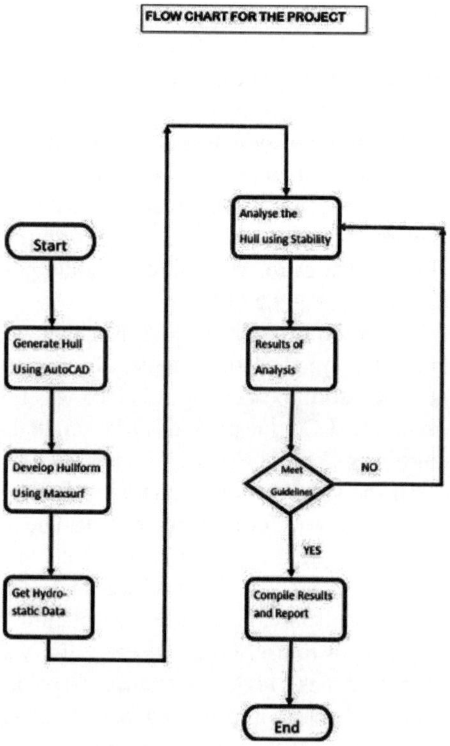

Table 11.1 Transverse dynamic stability guidelines [2]

Dynamic stability criteria	Guidelines/requirement
$(CA_p\text{-}LCG)/L_p$ (%)	>3%
$A_p/\nabla^{2/3}$	$>0.39\ (L_p/B_{px}) + 4.52$

Table 11.2 Main dimension of 14 m custom boat with variations of deadrise design

Design No	LOA (m)	Beam (m)	Depth (m)	Mid-chine deadrise (deg)	Forebody deadrise (deg)	Draft (m)
Design 1	14	4	2.4	12.5°	13°	0.402
Design 2	14	4	2.4	12.5°	19°	0.591
Design 3	14	4	2.4	12.5°	25°	0.813

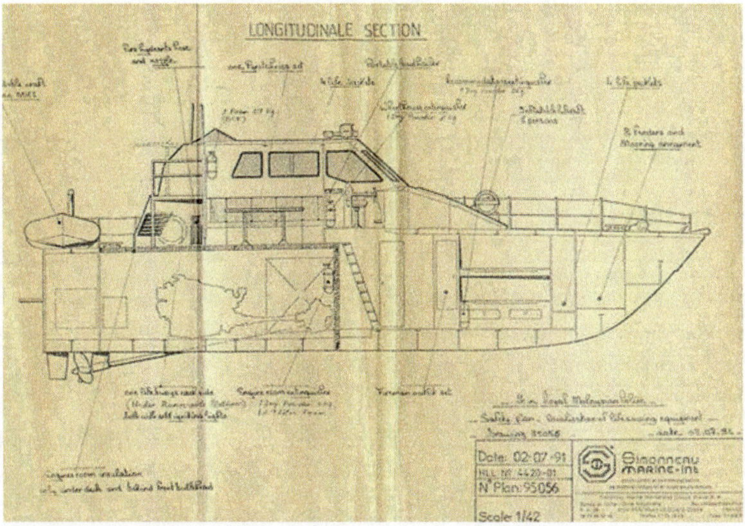

Fig. 11.2 Profile view of the 14 m custom boat

MAXSURF Modeler. Three similar designs were produced with three variations of deadrise angles, i.e., 13°, 19°, and 25° as shown in Table 11.2. On completion of the hull model, the dynamic stability analysis is performed by using Bentley Stability software for all three designs. All parameters were calculated and analyzed in determining which design was dynamically unstable. The result and assessment of each design were based on Blount and Codega guidelines. Figure 11.2 shows the profile of the boat used in the study.

The hull lines that were developed using MAXSURF Modeler are shown in Fig. 11.3a–c.

i. *Development of Hull Form*

   Below are the three designs with different deadrise angles.

ii. *Hull Parameters and Hydrostatics Data*

   Three variations of hullform designs have been modeled by using the MAXSURF Modeler module as shown in Fig. 11.4a–c. The parameters and hydrostatics data from the models were used in the analysis of dynamic stability which included the waterplane area, m^2 (A_w), volume displaced, m^3 (∇), chine

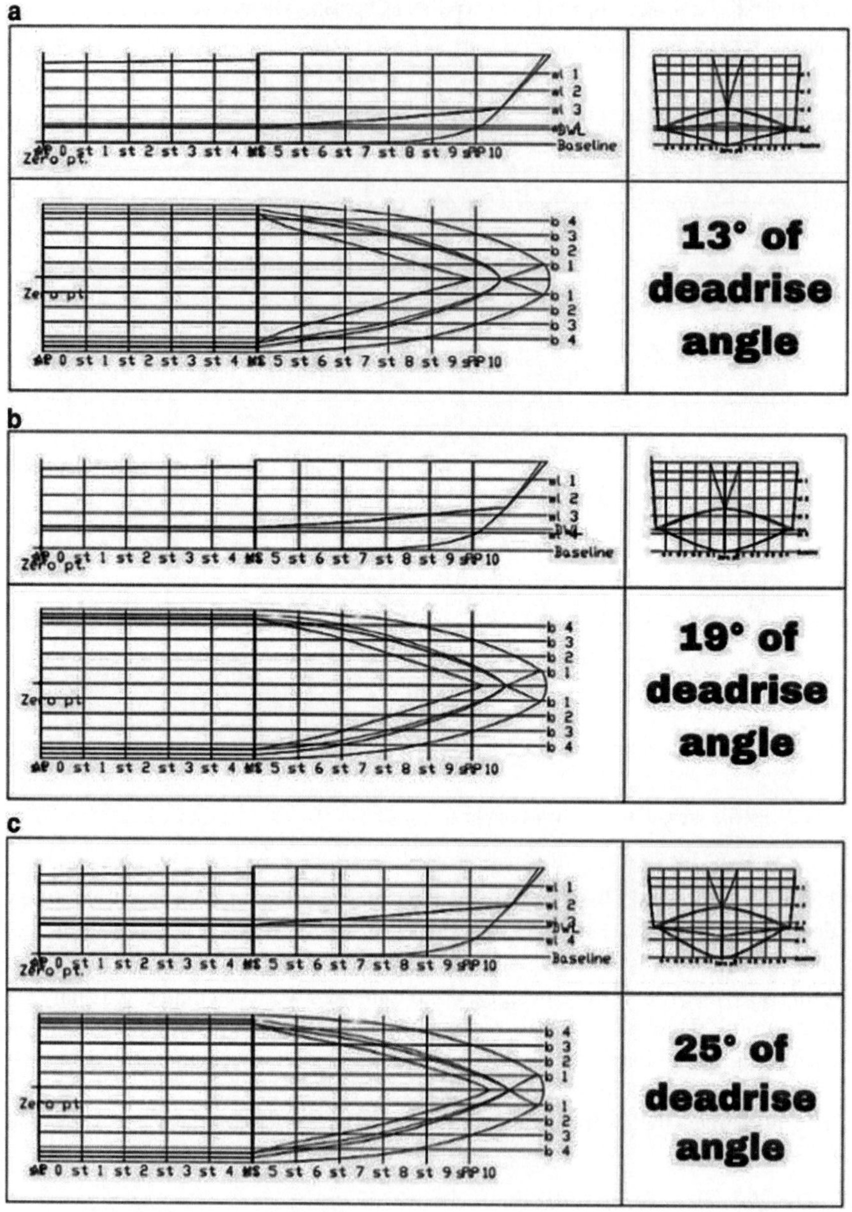

Fig. 11.3 **a** Lines plan of 14 m of custom boat with 13° forebody deadrise angle. **b** Lines plan of 14 m of custom boat with 19° forebody deadrise angle. **c** Lines plan of 14 m of custom boat with 25° forebody deadrise angle

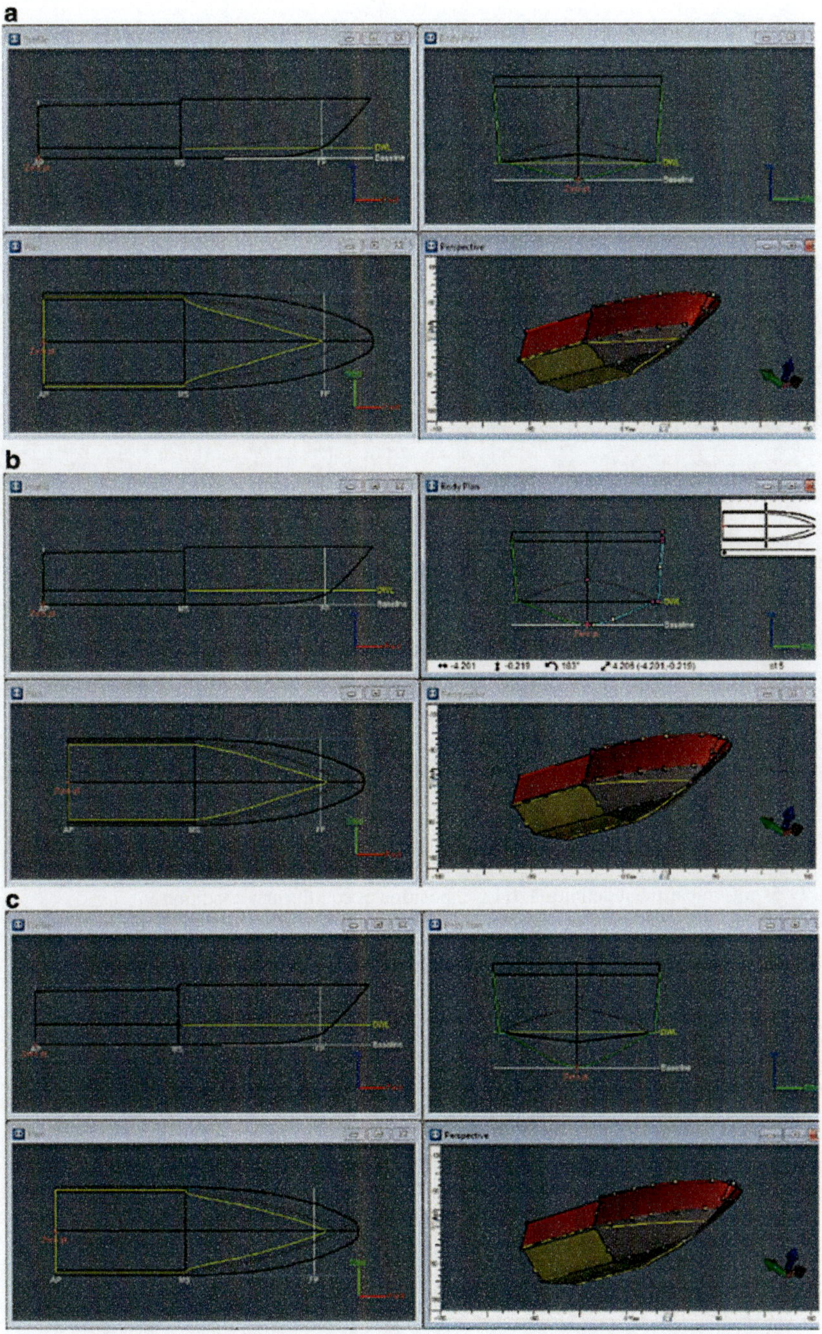

Fig. 11.4 **a** 3D model of 14 m of custom boat with 13° forebody deadrise angle. **b** 3D model of 14 m of custom boat with 19° forebody deadrise angle. **c** 3D model of 14 m of custom boat with 25° forebody deadrise angle

length, m (L_p), maximum breadth over chines, m (B_{px}), projected area between chine and transom (A_p), and longitudinal center of projected area CA_p.

iii. *Longitudinal Center of Gravity Estimation*

Estimating the longitudinal center of gravity (LCG) location of the boat was calculated by using the Bentley Stability module. Weight distribution and general arrangement drawing are among the essential references for the outcome. All the locations of tanks and compartments together with the loading and the locations of respective LCGs in static condition are shown in Fig. 11.5a, b. The location of compartments and tanks was shown in Fig. 11.6

a

	Name	Type	Intact Perm. %	Damaged Perm. %	Specific gravity	Fluid type	Boundary Surfaces	Aft m	Fore m	F.Port m	F.Stbd. m	F.Top m	F.Bott. m
1	SGR	Compartm	100	100			none	0.000	1.200	-2.100	2.100	2.200	0.000
2	Eng Room	Compartm	100	100			none	1.200	6.000	-2.300	2.300	2.700	0.000
3	FOT(S)	Tank	100	100	1.0679	Heel deg	none	4.700	6.000	1.000	2.300	1.900	0.400
4	FOT(P)	Tank	100	100	1.0679	Heel deg	none	4.700	6.000	-2.300	-1.000	1.900	0.400
5	Head/Toilet	Compartm	100	100			none	6.000	8.200	0.500	2.300	2.700	0.000
6	Galley	Compartm	100	100			none	6.000	8.200	-2.300	-0.500	2.700	0.000
7	Alleyway	Compartm	100	100			none	6.000	8.200	-0.500	0.500	2.700	0.500
8	Crew Accom	Compartm	100	100			none	8.200	11.300	-2.100	2.100	2.700	0.500
9	FWT(S)	Tank	100	100	6.2898	Displacemen	none	8.200	10.000	0.000	2.100	0.500	0.000
10	FWT(P)	Tank	100	100	6.2898	Displacemen	none	8.200	10.000	-2.100	0.000	0.500	0.000
11	Store	Compartm	100	100			none	11.300	12.300	-2.100	2.100	2.700	0.000
12	Chain Locker	Compartm	100	100			none	12.300	14.000	-2.100	2.100	2.700	0.000
13	Eng Room (F	Linked Ne	100	100			none	4.700	6.000	1.000	2.300	1.900	0.400
14	Eng Room (F	Linked Ne	100	100			none	4.700	6.000	-2.300	-1.000	1.900	0.400

b

	Item Name	Quantity	Unit Mass tonne	Total Mass tonne	Unit Volume m^3	Total Volume m^3	Long. Arm m	Trans. Arm m	Vert. Arm m	Total FSM tonne.m	FSM Type
1	Lightship	1	14.000	14.000			3.100	0.000	2.000	0.000	User Specifi
2	FOT(S)	100%	1.811	1.811	1.696	1.696	5.350	1.447	1.201	0.000	Maximum
3	FOT(P)	100%	1.811	1.811	1.696	1.696	5.350	-1.447	1.201	0.000	Maximum
4	FWT(S)	80%	2.484	1.987	0.395	0.316	8.982	0.278	0.300	0.701	Maximum
5	FWT(P)	80%	2.484	1.987	0.395	0.316	8.982	-0.278	0.300	0.701	Maximum
6	Total Loadcas			21.596	4.182	4.024	4.560	0.000	1.563	1.401	
7	FS correction								0.065		
8	VCG fluid								1.618		

Fig. 11.5 **a** Compartment definition. **b** LCG position at equilibrium condition

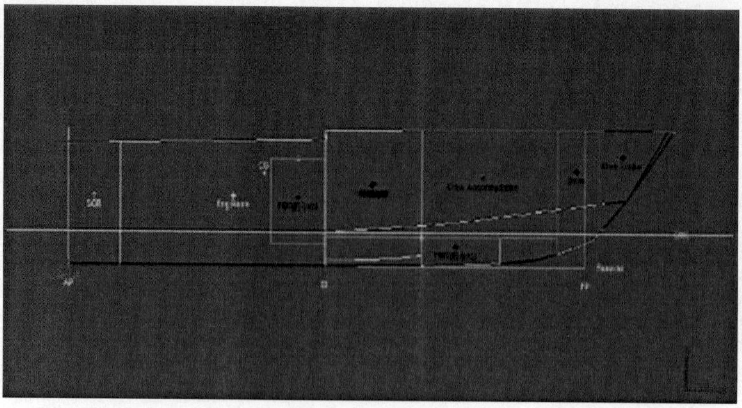

Fig. 11.6 Profile view after room definition and LCG location

11.3 Results and Discussion

Essential parameters were obtained from the MAXSURF Modeler and Stability modules such as projected area of planing bottom between the chine and transom, m^2 (A_p), volume displaced at rest, m^3 (∇), chine length, m (L_p), and maximum beam over chines, m (B_{px}), hydrostatic data at DWL, position of longitudinal center of gravity, m (LCG) and centroid of planing area, m (CA_p). Table 11.2 shows all the parameters taken for each of the various deadrise angles (β) of 13°, 19°, and 25°. The variation of deadrise angles led to different parameters except for the maximum beam value where the value was constant due to the characteristic of the design being changed only at the deadrise, but not at the beam. All these parameters were the main consideration and input for dynamic stability assessment (Table 11.3).

Once the required parameters have been collected, each value is used to calculate the loading coefficient, $A_p/\nabla^{2/3}$, and length-to-beam ratio, L_p/B_{px}, as shown in Table 11.4. These data were required for use in the transverse stability requirement suggested by Blount and Codega as explained earlier.

For the second guideline, the required parameters are the value of centroid of planing area (CA_p), and longitudinal center of gravity (LCG). The parameters were gained in equilibrium condition calculation. Table 11.5 shows the parameter values collected for each design of deadrise angles. In compliance with the requirements

Table 11.3 Main hydrostatics data for different deadrise angle, $\boldsymbol{\beta}$

β	A_p	∇	L_p	B_{px}	LCG	CA_p
13°	32.04	6.13	12.67	3.74	3.0	4.58
19°	33.85	10.32	12.89	3.74	3.1	4.72
25°	34.92	15.33	13.07	3.74	3.5	4.87

Table 11.4 Loading coefficient and length–beam ratio for each design

	$A_p/\nabla^{2/3} > 0.39\ (L_p/B_{px}) + 4.52$		
Deadrise angle	$A_p/\nabla^{(2/3)}$	$0.39\ (L_p/B_{px}) + 4.52$	L_p/B_{px}
13°	9.566	5.841	3.387
19°	7.138	5.864	3.447
25°	5.658	5.883	3.495

Table 11.5 Percentage of $(CA_p\text{-}LCG)/L_p$ for each design

	$(CA_p\text{-}LCG)/L_p$, % > 3%		
Deadrise angle	CA_p/L_p (%)	LCG/L_p (%)	$(CA_p\text{-}LCG)/L_p$ (%)
13°	36.149	23.683	12.465
19°	36.573	24.046	12.527
25°	37.294	26.780	10.513

Table 11.6 Compliance status of dynamic stability for 13° deadrise angle design according to Blount and Codega

Design 1: 13° deadrise angle		
Dynamic stability criteria	Guidelines/requirements	Status
(CA_p-LCG)/L_p (%)	>3%	12.47% (comply)
$A_p/\nabla^{2/3}$	$>0.39(L_p/B_{px}) + 4.52$	9.57 > 5.84 (comply)

proposed by Blount and Codega, (CA_p-LCG)/L_p > 3%, the percentage difference between the centroid of planing area and longitudinal center of gravity divided by chine length must be more than 3%.

Table 11.6 is a summary of all calculated values for design 1, 13° of deadrise angle design and a comparison with the transverse stability guideline proposed by Blount and Codega. The percentage of (CA_p-LCG)/L_p is 12.47%, more than 3% as required. The value of the loading coefficient is 9.57, more than 5.84 as calculated according to the guidelines given. Both requirements show the design of 13° deadrise angle for 14 m custom boat complied with the dynamic stability required by Blount and Codega criteria.

Referring to Table 11.7, for design 2 of 19° deadrise angle case, both dynamic stability requirements also met the criteria proposed by Blount and Codega. The percentage of (CA_p-LCG)/L_p is 12.53%, more than 3% as required. The value of loading coefficient is 7.14, more than 5.87 as calculated according to the guidelines given.

Table 11.8 shows a summary of calculated values for design 3, 25° deadrise angle, and a comparison with the transverse stability guideline proposed by Blount and Codega. For this case, only percentage of (CA_p-LCG)/L_p is complied, which is

Table 11.7 Compliance status of dynamic stability for 19° deadrise angle design according to Blount and Codega

Design 2: 19° deadrise angle		
Dynamic stability criteria	Guidelines/requirements	Status
(CA_p-LCG)/L_p (%)	>3%	12.53% (comply)
$A_p/\nabla^{2/3}$	$>0.39(L_p/B_{px}) + 4.52$	7.14 > 5.87 (comply)

Table 11.8 Compliance status of dynamic stability for 25° deadrise angle design according to Blount and Codega

Design 3: 25° deadrise angle		
Dynamic stability criteria	Guidelines/requirements	Status
(CA_p-LCG)/L_p (%)	>3%	10.51% (comply)
$A_p/\nabla^{2/3}$	$>0.39(L_p/B_{px}) + 4.52$	5.66 < 5.88 (not comply)

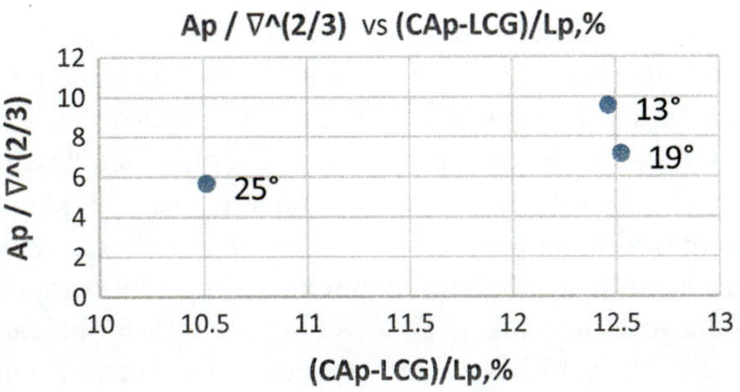

Fig. 11.7 Loading coefficient versus percentage of (CA$_p$-LCG)/L_p

10.51% more than 3% as required. The value of loading coefficient is 5.66, slightly less than 5.88 as required by the criteria.

Figure 11.7 interprets the relationships between loading coefficients and percentage of difference between centroid of planing area and LCG divided by chine length. It seems reasonable to simplify that the hypothetical problem boats were more lightly loaded than $A_p/\nabla^{2/3} = 5.8$ and LCGs are more than 3% L_p aft of the centroid of $A_{p.}$ The hull loading parameter has a small numerical value for a proportionally heavy boat. The proposed design criteria can be seen in Fig. 11.7 to include a bandwidth for margin and offered an engineering approach to avoid non-oscillatory instabilities.

This study was limited to the collection and analysis of data contained in Fig. 11.7 where the design that has good characteristics of transverse stability will be in the same regions, while the design that does not show good transverse stability characteristics will be out of the region, i.e., outlier. Design 3 does not comply with the guideline as stated by Blount and Codega.

Furthermore, the criteria indicated that the curves of high L_p/B_{px} hulls are more likely to have reverse slope/inflection points than low L_p/B_{px} hulls.

Figure 11.8 shows the loading coefficient versus length-to-beam ratio. The graph can be formed from linear regression, ($y = -36.13x + 131.6$). The graph of design 3 did not show good transverse stability characteristics. The trend relates well with paper presented by [1].

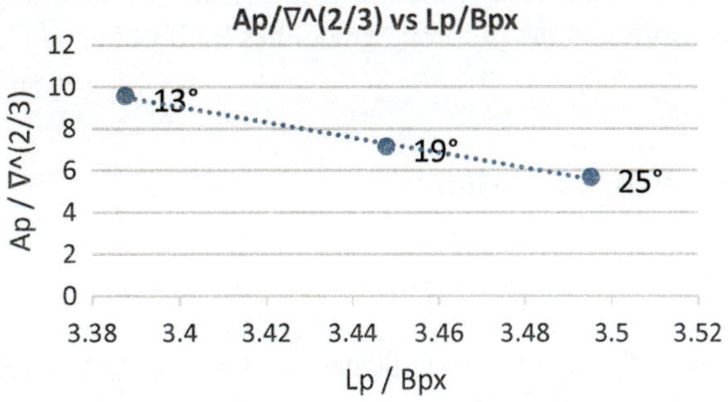

Fig. 11.8 Loading coefficient versus length–beam ratio

From a series of tests on planing hull, it is apparent that hull loading relative to hull dimension and LCG location has an important influenced on potential transverse instability. The development of the proposed design guideline necessitated a dimensionless hull loading parameter and a dimensionless LCG parameter, the two most frequently used hull loading parameters used in planing technology. For a planing boat having ratio of $A_p/\nabla^{2/3}$ more than 5.8, the percentage of $(CA_p\text{-}LCG)/L_p$ should be more than 3%, both requirements must be complied in order to avoid dynamic transverse instability. The result of study also indicated that high L_p/B_{px} hulls are more likely to have a reverse slope or inflection points which will likely exhibit instability than low L_p/B_{px}. A simple guideline in order to avoid the region of saddle points is to ensure that $A_p/\nabla^{2/3} > 0.39\ (L_p/B_{px}) + 4.52$.

From the data obtained, some relationships can be concluded such as

1. Different hull form characteristics generated different parameters, even with similar main dimensions. The higher the deadrise angle, the higher the value of projected area (A_p) of planing bottom and the chine length.
2. The longer the chine length, the further the position of centroid of planing area and LCG move to forward of the boat.
3. High length-to-beam ratio are more likely to have reverse slope or inflection points which will likely exhibit instability than low length-to-beam ratio.
4. The higher the deadrise angle, the higher the potential for transverse instability.
5. The change of parameters will affect the change in loading coefficient, and length–beam ratio which will affect the stability in terms of transverse stability according to the proposed guidelines by Blount and Codega.
6. Moving the LCG forward to improve speed performance reduces running trim angle and thus brings the wetting of the forward curved buttocks into play leading to a situation in which suction can develop [7].

The effect of changing the deadrise angles design will provide a clear behavior change in terms of stability for a boat. This change in deadrise will affect the change of important parameters in the calculation of stability such as waterplane area, volume displaced, chine length, centroid of planing area, and longitudinal center of gravity. The higher the deadrise angle, the higher the parameter value that will be collected. Changes in these parameters will also affect the change in loading coefficient, and length–beam ratio which will affect the stability in terms of transverse stability according to the proposed guidelines by Blount and Codega.

11.4 Conclusion

Although the current stability criteria have taken into accounts the requirements for high-speed vessels, they are still solely based on the hydrostatic assessment. The result of the study also indicated that high L_p/B_{px} hulls are more likely to have reverse slope or inflection points which will likely exhibit instability than low L_p/B_{px}.

Design 3 exhibits unstable characteristics as chine position exhibits instability due to length–beam ratio is too high. The transverse instability is sensitive to LCG, i.e., trim of the vessel. The further forward the LCG the most likely for the vessel to exhibit transverse instability. This is due to the concavity of the hull plate as it progresses forward. Blount et al. discovered that in the course of their work that adding hull wedges forward may improve course keeping but will cause low dynamic hull pressures at the bow which are the source of the problem. Kazemi et al. found out that as the vessel speed increases the leading edge of the wetted surface moved forward due to decreasing in trim angle of the vessel.

This study had succeeded in showing that the guidelines proposed by previous researchers can be used. The need of small boat stability assessments to assure the safety and seaworthiness of its crews, passengers, cargo, and the boat itself cannot be overlooked.

It is recommendation that to prove the usability of the guideline, the time domain approach should be used since it will provide a wider and more conclusive assessment of the dynamic instability. The main advantage of the time domain simulation approach is that it allows the vessel's dynamic behavior to be obtained in a simulated environment in steps of time because dynamic stability can be known more accurately with the analysis related to speed and external forces.

Journal article
Hamburger, C.: Quasimonotonicity, regularity and duality for nonlinear systems of partial differential equations. Ann. Mat. Pura. Appl. **169**, 321–354 (1995)
Journal article only by DOI
Sajti, C.L., Georgio, S., Khodorkovsky, V., Marine, W.: New nanohybrid materials for biophotonics. Appl. Phys. A (2007). https://doi.org/10.1007/s00339-007-4137-z

References

1. Samian Y, Malik AMA (2018) Static dynamic stability assessment of small craft. Department of Marine Technology, Faculty of Mechanical Engineering, Universiti Teknologi Malaysia
2. Blount DL, Codega LT (1992) Dynamic stability of planing boats. Mar Technol 29(1):4–12
3. Kazami H, Salari M (2017) Effects of loading conditions on hydrodynamics of a hard-chine planing vessel using CFD and a dynamic model. Int J Mar Technol 7:11–18
4. Clement EP, Blount DL (1963) Resistance tests of a systematic series of planing hull forms. SNAME Trans 71:491–579
5. Thornhill E, Veitch B, Bose N (2000) Dynamic instability of a high-speed planing boat model. Mar Technol 37(3):146–152
6. Keuning L, Hillege W (2017) The results of the delft systematic deadrise series. In: Proceedings of 14th international conference on fast sea transportation (FAST 2017): innovative materials, pp 97–106
7. Blount D.L. Schleicher D. M.: Correcting dynamic rolls instability. Profesional Boatbuilder. https://dlba-inc.com/wp-content/uploads/2020/07/Correcting-Dynamic-Roll-Instability.pdf. Accessed 24 Jun 2021

Chapter 12
The Implication of the Container Floating Terminal on the Efficiency of Port Klang's Terminal Operations and Domestic Freight Forwarding Industry

Nur Amalia Azmi and Aminuddin Md Arof

Abstract Orders for mega vessels with a capacity of more than 20,000 TEUs have bolstered the sector since 2014. Despite the fact that these mega vessels serve to reduce sea transportation costs and increase global trade, they, on the other hand, also need some port infrastructure and container handling equipment adjustments. This will likely produce higher peaks in port container traffic, which will have far-reaching consequences. A number of seaport and terminal industry players have taken the initiative to establish a floating terminal to cater for the demand, since the development of seaport terminals to meet the call of mega vessels has become such a pressing issue for the sector. The container floating terminal (CFT) is a floating infrastructure that has been developed with facilities to transport containers from larger container ships or load other cargoes on a short-stay platform and then transport them to smaller ships for shipment to end-users. Thus, the purpose of this research is to look into the impact of the CFT on the efficiency of Port Klang's terminal operations and the local freight forwarding business. Using a questionnaire survey that was distributed among Port Klang's employees and local freight forwarding companies, this study found that there is a relationship between the development of the CFT and the improvement of Port Klang's terminal operations efficiency and the local freight forwarding industry.

Keywords Mega vessels · Container floating terminal (CFT) · Port efficiency · Freight forwarding industry

N. A. Azmi · A. Md Arof (✉)
Universiti Kuala Lumpur Malaysian Institute of Marine Engineering Technology, 32200 Lumut, Perak, Malaysia
e-mail: aminuddin@unikl.edu.my

N. A. Azmi
e-mail: amalia.azmi27@s.unikl.edu.my

A. Ismail et al. (eds.), *Design in Maritime Engineering*, Advanced Structured Materials 167, https://doi.org/10.1007/978-3-030-89988-2_12

12.1 Introduction

Ships have grown bigger in recent years, which has attracted the public's attention. The industry is transitioning to a new era of mega vessels, with the most recent attention on the HMM Algeciras, the world's largest container vessel. The season for megaship announcements appears to have begun, with various shipping lines declaring orders for even larger ships almost every two weeks, creating a mathematical sequence that has left everyone guessing when and who will possess the next megaship. Mega vessels are not a new phenomenon in the container shipping sector. The search for economies of scale, as well as the global trade mechanism including non-bulk cargo, has driven motivations for shipping companies to purchase mega container ships, which have risen in size at a rapid pace over the past decade. Furthermore, one of the causes driving most shipping alliances to focus on offering the most efficient yet low-cost maritime logistics services is the tremendous improvement in trade between Europe and Asia.

Mega-ships have established such a large peak in terms of ports and hinterland transit. This type of behaviour is not new, but it is never been witnessed on such a huge scale before. It is fair to say that most ports struggle to deal with these surges, especially when they arise suddenly, such as due to a shipping line's lack of dependability. Larger vessels are being deployed, needing more complex and automated handling procedures, yet fewer ports are being visited at the same time, resulting in increased competition among hub ports. Ports have long struggled to adjust to the massive container ships that container lines have been deploying in recent years [1]. On the other hand, the port terminal's facilities must change. Unless the port interface is updated, an ultra-large container ship (ULCS) is not particularly useful. Large container ships demand large container port terminals that can handle cargo and adhere to port traffic laws.

The mega-ships put a lot of pressure on ports. The annual rate of transport costs associated with mega-ships could be around US$0.4 billion, with roughly a third of the additional costs related to equipment, a third to dredging, and a third to the cost of port infrastructure and port hinterland [2]. Mega-ships, in addition to requiring massive investments, cause increased congestion and intensify competition between ports. A higher cargo density paired with fewer vessels results in higher peak hours for massive numbers of containers, placing a significant burden on ports. Ports require more cranes, as well as more highly qualified workers to operate them efficiently, more yard space, and the capacity to deliver containers inland using more trucks, railcars and barges. Parties participating in terminal operations would have to cooperate more closely together to avoid long turnaround times, congestion and expensive demurrage and detention charges.

Handling all vessels, cargo, and now the varied sizes of mega vessels from all over the world generates such a "stress point" for each government to deliver the greatest version of the port of call to its clients. Providing customers with a fully functional port will almost certainly prompt port developers to seek for a solution. This is where the container floating terminal concept is being established. Because of the available water depth, the development of this floating or offshore terminal inside the port area would aid in the settlement of these challenges, as this facility would be more adapted to receive and handle large vessels.

12.2 Aim

There are very limited studies that are focused specifically on the implementation of the CFT. The majority of the studies currently discussed the impact of mega vessels on the seaport industry. There are numerous studies that focused on solutions for port industries to deal with demand issues by expanding their seaports. However, there is little discussion of the CFT as one of the solutions. Therefore, the aim of this study is to see into the implication of the CFT on the efficiency of Port Klang's terminal operations and the domestic freight forwarding industry that is currently supporting the logistics operations in Port Klang.

12.3 Literature Review

12.3.1 Port Efficiency

A port is a facility that connects maritime and hinterland transportation while also providing warehouses and other value-added services that attract both industrial players and investors [3]. Port output is the performance of any single part of the operation that occurred at the terminal. It could be calculated according to the total time taken to handle the cargo, quality of the performance and capacity to cope with any problem that occurs in the terminal's handling area [3]. Additionally, Rahman et al. stated that port efficiency and output could be measured according to their service performance consisting of the transit period and also the amount of tonnages handled by ship on a regular basis when maintaining the port's asset efficiency and financial results [4]. In this regard, both papers have touched on and addressed port efficiency issues as it is a very significant part that has led to a global economy and international trade's productivity and growth.

Jeevan and Roso said that the seaports are currently competing in terms of the vessels size, which has resulted in an extraordinary operational constraint as well as diseconomies for the seaport [5]. Both of the next papers written by [6] and [7] also mentioned that, seaports are currently facing with great pressure especially in expanding their infrastructure to accommodate the current larger size of the merchant fleet. Ismail et al. stated that in addition to reducing port efficiency, the present huge vessels are affecting numerous sectors of the transportation chain, particularly the multimodal transportation system connecting seaports with the hinterlands [6]. On the other hand, Jeevan et al. stated that despite offering significant benefits to traders and the environment, the introduction of mega vessels has significant implications for seaports, imposing new conditions on seaports, affecting land-side operations, and putting a strain on the entire container logistics chain [8]. Baik also stated that ports spend a lot of money improving their facilities and vying for vessel calls, but it is difficult to handle such demand spikes [1].

In comparison, the study by Rahman et al. is much more relatable with this study as the finding of this paper has shown that the port efficiency could be achieved through the instalments of a complete port infrastructures, logistics superstructures and also great packages of the value-added services [4]. As for the studies by [5–8], the researcher could understand more on the impact of the current picture of the enlargement of the vessel size that has given a big impact towards the performance of the seaports.

12.3.2 Floating Terminal

A floating container storage terminal (FCST) is recognised as a floating facility, which is located near the main terminal on the open sea [9]. Apart from being the most effective and efficient solution for the containerisation industry, this offshore terminal was first developed to cope with the problems arise from oil and gas industry. In 2001, El Paso had been tasked to build a new vessel based on available technology for the Gulf Gateway [10]. Due to this development, the players in the port and terminal industry had come out with an idea to build up an offshore terminal as a backup to the expansion of the inland terminal. The study by Maletic et al. stated that the installation of this floating infrastructure can definitely help in increasing the efficiency of the onshore port performances as it could contribute in reducing the turnover time for both containers and vessels [9]. Meanwhile, Giranza and Bergmann listed out that from their study, it is found out that the investment on floating terminal is more profitable as compared to the onshore terminal [11]. Next, Songhurst (2017) argued that the floating facility is such a bless to the industry as this floating mega structure offers such a lower development cost, faster and flexible schedule and also reusable asset feature [10]. The preceding studies showed how fast the floating terminal industry has grown due to the total investment of the floating infrastructures was much lower than investment for onshore terminals.

On the other hand, the paper written by Souravlias et al. stated that the other benefits offered by this floating terminal is as per followed: (1) it could serve as a disaster relief hub, particularly in coastal areas where land access is difficult or time-consuming, and (2) it would be beneficial not only for port purposes but also for offshore energy production, aqua farming and possibly as a future living location.

As overall, all the papers did provide with the same useful information as regards with the installation of the floating terminal but the most relatable are the papers written by Maletic et al. and Souravlias et al. as both of the papers definitely focused on all the criteria of the floating terminal. The weakness of all the papers selected is on the methodology part as none of the methods used is relevant to be used for this study.

12.3.3 Domestic Freight Development

Arip et al. stated that an international freight transport terminal is defined as the main or hub facility where all the services including the transportation, logistics and goods distribution either for the national or international trade are being connected [12]. As the number of TEUs continues to increase especially in Port Klang, where the container volume has increased to 10.9 million TEUs in 2014 compared to 496,326 TEUs in 1990 [8], the development and expansion of the seaport area is becoming a crucial issue for the players involved in the industry. From primary data collected, Arip et al. discovered that the physical aspects of the freight transport terminal are the most important factor in determining the goods distribution system performance of the intermodal freight transport terminal in Malaysia. This factor is followed by both the core-on site activities and core-on site service elements [12]. The study also discovered that Malaysia was still left behind in the development of the intermodal freight transport terminal [12].

A study by Nasir et al. that covers on the local intermodal transportation services in Malaysia found that there were three major problems when using these intermodal transportation services. The problems included operational efficiency, management and cost. Apart from the problems, it is also stated that the operational efficiency is made up of the total transit time and reliability, the inland terminal operations and also dry port operations [13]. The development of the ports could be such a benefit facility towards the stakeholders as it will help in reducing the waiting time for ships, increasing the efficiency in the supply chain as well as decreasing in the total freight cost [8]. It is also found that the connectivity between the seaport and the hinterland infrastructures including the dry port is very crucial to help providing more numbers of trade volume. Next, [13] stated in their study that the main challenges to overcome in order to increase the usage of intermodal in Malaysia are the efficiency, management issues as well as the cost factors.

A study by Chudasama found that the development of an economy and the development of a port are inextricably linked [14]. This was reinforced by the finding that, whilst, on the one hand, economic growth necessitates port development as part of infrastructure development, and on the other hand, port development facilitates import–export and draws industries to its hinterland, resulting in forward and backward linkages with the rest of the economy. The paper came to a key conclusion, specifically, that economic growth in the hinterland economy would have an effect on cargo traffic at ports making them more competitive [14].

In summary, the three research papers at [8, 12, 13] discussed in the preceding paragraphs are relevant to this research as they were focusing on the development of the freight transportation and the domestic freight terminal in Malaysia. They also discussed in detail on how the pattern of the local freight terminal could affect the business of the local seaports. The methodology used by some of the writers, which is the qualitative method can also be considered as one of the methods for this and subsequent research.

12.4 Methodology

12.4.1 Participants

In this study, the questionnaire was distributed online using a Google Form. The online questionnaire was completed by 55 respondents using the purposive sampling method. Employees of the terminals in Port Klang and representatives from selected freight forwarding companies that operate in the area around the port make up the population of this study. From the total respondents, male accounted for 57.1% of the respondents, while female respondents accounted for 42.9%. More than half of the respondents, i.e. 65.45% or 36 people, are aware of the existence of the CFT, while the remaining 34.55% had never heard of it. According to the data gathered, 100% of respondents agreed that the construction to modernise the port is important in order to handle a variety of difficulties that most seaports face. The majority of those working in the port and adjacent sectors agreed that the creation of this future structure might contribute to in resolving the issues of draught constraint and land for horizontal expansion.

12.4.2 Measure

All the data collected via the online questionnaire was then being analysed using the Pearson coefficient analysis. The Pearson coefficient, r, is a sort of correlation coefficient that represents the relationship between two variables on the same interval or ratio scale.

Table 12.1 Correlation analysis between Port Klang's terminal operations efficiency and the development of CFT

Correlations			
		Port Klang's efficiency	Container floating terminal
Port Klang's efficiency	Pearson correlation	1	0.853**
	Sig. (2-tailed)		0.000
	N	55	55
Container floating terminal	Pearson correlation	0.853**	1
	Sig. (2-tailed)	0.000	
	N	55	55

**Correlation is significant at the 0.01 level (2-tailed)

12.5 Results and Discussion

12.5.1 Correlation Analysis Between Port Klang's Terminal Operations Efficiency and the Development of CFT

Table 12.1 indicates a Pearson's coefficient of correlation of $r = 0.853$, indicating a strong positive linear relationship between Port Klang's terminal operations efficiency and CFT development.

12.5.2 Correlation Analysis Between Freight Forwarding Industry and the Development of CFT

Table 12.2 reveals a strong positive linear relationship between the freight forwarding industry and the development of the CFT, with a Pearson's coefficient of correlation of $r = 1.000$.

12.5.3 Implication of CFT on Port Klang's Terminal Operations Efficiency

The building of the CFT has been determined to have an impact on Port Klang's terminal operations efficiency, according to Pearson's coefficient of correlation, $r = 0.853$. This finding is in line with the previous study's findings, which said that port efficiency is achieved by combining the right mix of port infrastructure, logistics superstructure and related value-added services [3]. In this study, the researcher

Table 12.2 Correlation analysis between freight forwarding industry and the development of CFT

Correlations			
		Freight forwarding industry	Container floating terminal
Freight forwarding industry	Pearson correlation	1	1.000**
	Sig. (2-tailed)		0.000
	N	55	55
Container floating terminal	Pearson correlation	1.000**	1
	Sig. (2-tailed)	0.000	
	N	55	55

**Correlation is significant at the 0.01 level (2-tailed)

discovered a number of essential aspects that help to enhance the efficiency of ports and terminals. According to the data analysed, one of the most important factors for seaports to maintain their efficiency is vessel turnaround time, which can be achieved through good port and terminal infrastructure. According to a research at [1], one of the major challenges facing most seaports is the increase in demand for turnaround time, as the trend of giant container ships has produced new levels of congestion and activities that have hurt the efficiency of most seaports.

On the other hand, the flexibility to expand capacity to meet the current trend of container vessel expansion is the most important attribute for Port Klang to improve its efficiency level. According to the study, the CFT is one of the futuristic alternatives for coping with the container market's growth and the development of mega-sized vessels. This is supported by [15], which says that globalisation has led in a rise in vessel size, necessitating the expansion of seaport container handling capacity. In order to solve the issue of port growth and extension, the study suggests the building and development of a floating terminal, particularly in the case of land limitation. The creation of this CFT could help improve cargo handling, encourage strong competitiveness for seaports and increase the number of TEUs and mega vessel calls which could have a positive impact on Port Klang's terminal operations efficiency in this scenario. The arrival of mega vessels brings such significant effects for the majority of seaports around the world. Therefore, in order to stay competitive, ports must spend a large amount of money to improve their facilities and capacity. As a result, when it comes to the relationship between the development of the CFT and the efficiency of Port Klang's terminal operations, this research shows that the development of this futuristic structure has a significant impact on Port Klang's terminal operations efficiency.

12.5.4 Implication of CFT on Local Freight Forwarding Industry

According to the findings, the development of the CFT has a direct impact on the local freight forwarding industry. As noted in the previous study, the CFT has a strong positive linear association ($r = 1.000$) with the domestic freight forwarding industry. This relationship may also be supported by [14], in which the author stated that the interaction between the port and the hinterland, which includes freight forwarding and other logistics components, is generated by the port's better-operating facilities. The development of this CFT, according to the findings of this study, will have an impact on Port Klang's productivity, resulting in a smooth supply chain and hinterland link. According to the survey findings, 71.4% of respondents agree with the statement that in order to respond to contemporary industrial changes, seaports must engage with their inland connectivity to boost their flexibility as espoused by [5], particularly in terms of seaport infrastructures. The effectiveness of the seaport, as well as the supply chain from the hinterland, will suffer if this is not been sufficiently addressed.

On the other hand, the construction of this CFT, based on the findings, will increase the present use of domestic freight forwarding services. This is verified by a research published by [14], in which the author states that the operational efficiency of the services is linked to the time and reliability of the services, which is linked to the operation at the seaports. As a result of the good development and infrastructure, particularly in dealing with the present trend of giant vessels, the time for cargo handling could be managed effectively, affecting the smooth flow of the freight forwarding business as well as increasing the usage for the services.

As a consequence, in terms of the relationship between the growth of the CFT and the domestic freight forwarding sector, this study demonstrates that the development of the seaport can have a favourable impact on the performance of the local freight forwarding services industry.

12.6 Research Implications

12.6.1 Implication for Academia and Future Research

As this research is exploratory and interpretive, it opens up numerous areas for future research, both in terms of theory development and concept validation. More research will be required and can be done in the future to further elaborate the findings of this study. To start with, based on the facts given above, the researcher discovered the possible impact of the CFT on the efficiency of Port Klang's terminal operations and the local freight forwarding industry. According to the data collected, the CFT's construction could help to improve vessel turnaround time as well as cargo handling efficiency, resulting in improved port performance. As a result, the supply route to the hinterland is becoming more efficient, helping freight forwarding enterprises in the port area to perform much better. The focus of this study was on the general consequences of the CFT. Instead of focusing on generalisation as it appears to have done here, future studies may be expanded to learn more about this CFT. Finally, this research will add to our understanding of this infrastructure. Given how fresh this CFT is and how little attention it has received thus far, conducting additional research on this brand-new development may be a good idea for the future researchers.

12.6.2 Implication for Industry

The contribution of this study in the industry context is the formation of a perspective, particularly for those working in the port industry, on how the CFT could help in resolving various difficulties in the area. People might recognise the impact that this infrastructure could have on the port business based on the literature review that has been conducted. For example, according to [16] on Tunisian seaports' study, most of their seaports are experiencing expansion challenges, where the ability to increase the space was limited to less than 10 m. As a result, it is suggested that industry leaders devise a futuristic strategy to deal with such a situation. As a result, this study may lead industry players to explore developing a CFT as a defence against such issues. This study, as well as other subsequent studies that may develop from this idea, may have an impact on investors or organisations making decisions about how to build and operate a future-ready port.

12.7 Conclusion

According to the findings and the discussion previously stated, the development of the CFT had an impact on Port Klang's terminal operations efficiency and the domestic freight forwarding industry. When determining the port's level of efficiency, there are several crucial aspects to consider. These included reducing vessel turnaround times, increasing seaport capacity, enhancing cargo handling efficiency, increasing the amount of TEUs, and increasing the number of mega vessels calling at Port Klang. These crucial aspects can be achieved by solid long-term growth and technological advancement at the ports, which in this case is the CFT construction. The CFT's construction appears to provide a major impact on the local freight forwarding industry, which transports goods around Port Klang. As a way of dealing with the huge vessel trend, the factors indicated above, such as a smooth supply chain and freight forwarding connectivity, may be established with the good development of seaports. It is believed that the high level of efficiency at the seaport will contribute to the increased efficiency of the freight forwarding business and will also help to strengthen the demand for these services. All these could be achieved through a good development at the seaport and how fast development of the maritime logistics industry are effectively managed. In conclusion, this study will provide a better understanding of the CFT and how it may be considered as one of the finest options for the seaport in dealing with the current challenge.

References

1. Baik JS (2017) The study on impacts of mega container ships on ports. J Supply Chain Manag 22–40
2. Frese F (2019) How mega ships impact terminals, ports and the environment. https://container-xchange.com. Accessed 18 July 2021
3. Hanjra M, Bhatti OK, Niazi S (2017) Understanding port efficiency: a CPEC perspective. J Manag Stud 4(1):90–101
4. Rahman N, Ismail A, Roslin M, Lun V (2018) Decision making technique for analysing performance of Malaysian secondary ports. Int J Shipp Transp Logist 10(4):468–496
5. Jeevan J, Roso V (2019) Exploring seaport dry ports dyadic integration to meet the increase in container vessels size. J Ship Trade 4(8):1–18
6. Ismail N, Mohd N, Jeevan J (2019) Emergence of mega vessels and their influence on future Malaysian seaport expansion requirements. J Undergrad Sci 1(1):58–67
7. Kutin N, Nguyen TT, Vallee T (2017) Relative efficiencies of ASEAN container ports based on data envelopment analysis. Asian J Shipp Logist 33(2):67–77
8. Jeevan J, Menhat M, Anuar D (2019) An empirical analysis on container vessels enlargement: exploring causal factors from the perspective of Malaysian maritime trade system. Int J Recent Technol 8(1S):366–373
9. Maletic J, Prigoda L, Cekerevac Z (2018) Technical solutions and assessment of economic effects of construction of an offshore terminal. MTC-Aj.Com. https://mtc-aj.com/technical-solutions-and-assessment-of-economic-effects-of-construction-of-an-offshore-terminal.1597.en.htm. Accessed 1 July 2021
10. Songhurst B (2017) The outlook for floating storage regasification units (FSRUs). Oxford Institute for Energy Studies. https://ora.ox.ac.uk/objects/uuid:cd2f183a-24c0-4c03-a697. Accessed 1 July 2021
11. Giranza MJ, Bergmann A (2018) An economic evaluation of onshore and floating liquefied natural gas receiving terminals: the case study of Indonesia. In: IOP conference series: environment earth science, vol 150, no. 1, p 012026
12. Arip M, Zainol H, Rashid K et al (2018) Intermodal freight transport terminal planning in Malaysia: assessing goods transportation and distribution system performance. Int J Acad Res Bus Soc Sci 8(11):1758–1771
13. Nasir S, Muhammad A, Jaafar H (2018) Factors influencing the increased usage of intermodal for container movement in Malaysia. Adv Transp Logist Res. https://proceedings.itltrisakti.ac.id/index.php/ATLR/article/view/55/55. Accessed 1 July 2021
14. Chudasama K M (2020) Port developments: relationship between port traffic of Gujarat and its Hinterland. Gujarat University Publication. https://www.researchgate.net/publication/346823111_port_development_relationship_between_port_traffic_of_gujarat_and_its_hinterland. Accessed 1 July 2021
15. Souravlias D, Dafnomilis I et al (2020) Design framework for a modular floating container terminal. Front Mar Sci 7(545637):1–17
16. Hlali A (2018) Tunisian seaport and globalization: the challenges of the first generation ports. Int J Econ Manag Strategy 14(1):61–69

Chapter 13
The Optical Properties of Polyvinyl Alcohol (PVA), Phosphorylated Polyvinyl Alcohol (PPVA), and Phosphorylated Polyvinyl Alcohol—Aluminum Phosphate (PPVA-$AlPO_4$) Nanocomposites: Effect of Phosphate Groups

Asmalina Mohamed Saat, Syarmela Alaauldin, Md Salim Kamil, Fatin Zawani Zainal Azaim, and Mohd Rafie Johan

Abstract Partially phosphorylated polyvinyl alcohol–aluminum phosphate (PPVA-$AlPO_4$) nanocomposites were synthesized through continuous stirring and condensation at 80 °C. The optical properties of all samples were examined using UV–visible and photoluminescence (PL) spectroscopy. The absorption peaks are blue-shifted with the addition of phosphate groups. The addition of phosphate produced a broad peak and increased the optical band gap to 5.75 eV in PPVA; however, the optical band gap was reduced after combination in PPVA-$AlPO_4$ nanocomposite. The PPVA-$AlPO_4$ nanocomposites produced a strong single peak at 208 nm with a 5.5 eV optical bandgap. The excitation at 250 nm of PL spectra shows the presence of characteristic peaks syndiotactic and isotactic configuration of the PPVA-$AlPO_4$ nanocom-

A. Mohamed Saat (✉) · S. Alaauldin · M. S. Kamil · F. Z. Zainal Azaim
Universiti Kuala Lumpur Malaysian Institute of Marine Engineering Technology, Lumut, Perak, Malaysia
e-mail: asmalina@unikl.edu.my

S. Alaauldin
e-mail: syarmela@unikl.edu.my

M. S. Kamil
e-mail: md.salim@unikl.edu.my

F. Z. Zainal Azaim
e-mail: fatinzawani@unikl.edu.my

M. R. Johan
Nanotechology and Catalyst Research Center, University of Malaya, Kuala Lumpur, Malaysia
e-mail: mrafiej@um.edu.my

A. Ismail et al. (eds.), *Design in Maritime Engineering*, Advanced Structured Materials 167, https://doi.org/10.1007/978-3-030-89988-2_13

posite. An intense broad PL band observed at 306–370 nm shows i-PPVA-$AlPO_4$ configuration proves strong interaction of phosphate groups in the PPVA-$AlPO_4$ nanocomposite.

Keywords PVA · PPVA · Phosphate · $AlPO_4$ · Optical properties

13.1 Introduction

Composite materials have been greatly explored due to the necessity to achieve a new material with specific properties. This can be done by selecting suitable materials with desired attributes and combining them into one composite material. These materials are heterogeneous at least on a microscopic scale. Polymer provides a convenient route to prepare composites in which salts are dispersed to a high degree of uniformity and fineness. The usage of composites as polymer fillers, so-called polymer nanocomposites combine size-dependent properties of nanoparticles with desired properties of host polymers. Poly (vinyl alcohol) (PVA) is frequently used as a matrix for a variety of nanoparticles due to its optical clarity, to enable the investigation of optical properties of nanoparticles. PVA is reported to have huge potential in developing new materials for the application of optical communications, optoelectronic devices, microelectronics, organic cells, biosensors, and biological [1]. Partially phosphorylated poly (vinyl alcohol) (PPVA) has attracted considerable interest because of its enhanced properties. PPVA is synthesized by reacting PVA with phosphoric acid [2]. The reaction produced a mixed ester that contained single, double and triple bound phosphorus which gave either a water-soluble (hydrophilic) or a partially soluble (hydrophobic) product. PPVA is used in various applications such as textile [3], coating material/corrosion [4, 5], fire-retardant materials [6, 7], electrolyte [8, 9], membranes [10, 11], metal chelating [12, 13], paper making [14], sensors [15, 16], synthetic bones/teeth [17] as well as nanoparticle/nanocomposite [18–20]. Metal polymer nanocomposites are potentially used for a variety of applications such as optical devices, color filters, sensors, magnetic data storage nano-system, and others. In this case, nanocomposites of aluminum phosphate were reported in various publications [21–24]. Lots of applications of $AlPO_4$ are due to the capacity of water loss by condensation process during the thermal treatment which confers structural improvements to the materials. Earlier optical studies of PPVA show a weak absorption band that changed the color from transparent to brown [25]. Annealing of PPVA at 70 °C for 20 h produced an absorption peak at 350 nm. The peak becomes intense as phosphoric acid increases [26]. The variations of peak intensity are due to the interaction between phosphate ions and OH groups [27]. The optical band gap of PVA slightly increased with the increasing concentration of phosphoric acid due to the decrease in the formation defect within the PVA. Meanwhile, photoluminescence (PL) spectra for PPVA shows five distinct bands at 391, 429, 468, and 481 after excitation at five different wavelengths (272, 300, 350, 400, and 500 nm) which attribute to the $n \leftarrow \pi^*$ transition of the free OH groups in syndiotactic (s-), atactic (a-),

Table 13.1 Formulation of PPVA-$AlPO_4$ nanocomposites

Ratio of Al:P	PPVA	Mole P in PPVA	Al $(NO_3)_3.9H_2O$	
			Sample	mole Al
1.0:3	0.2	0.01091	AP 1.0	0.00364
1.1:3	0.2	0.01091	AP 1.1	0.00401
1.2:3	0.2	0.01091	AP 1.2	0.00437
1.3:3	0.2	0.01091	AP 1.3	0.00472
1.4:3	0.2	0.01091	AP 1.4	0.00508
1.5:3	0.2	0.01091	AP 1.5	0.00575

and isotactic (i-) PPVA configurations. The interactions between OH and phosphate groups modify the bridging in the s-PPVA which changes to a-PPVA configuration at higher phosphoric acid concentrations. In this paper, the influence of phosphoric acid and aluminum phosphate on PVA, PPVA, and PPVA-$AlPO_4$ has been studied. The optical properties of PVA, PPVA, and PPVA-$AlPO_4$ nanocomposites are compared and discussed.

13.2 Methodology

Partially hydrolyzed polyvinyl alcohol (PVA, 86.7%), phosphoric acid/orthophosphoric acid (PA, 85%), and aluminum nitrate (Al $(NO_3)_3.9H_2O$) were sourced from R&M Chemicals. The synthesis method of PPVA was reported earlier [28]. The PPVA-$AlPO_4$ nanocomposite was synthesized from the addition of aluminum nitrate to PPVA. The samples were continuously stirred for 1 h at 90 °C. Then, samples were put at pH 10 and heated at 120 °C. The obtained solution was poured into a Petri dish, dried at room temperature for 3 days, and kept in an oven at 50 °C for 2 h before being sealed in a desiccator for further analysis. Table 13.1 summarizes the details of the formation of PPVA-$AlPO_4$ nanocomposite samples.

The absorption spectra of nanocomposite samples were recorded using a Cary 50 Uv–Visible spectrophotometer. The wavenumber used is 200–900 nm at a scan speed of 60 nm/min and conducted in a 3 mm quartz cell. The photoluminescence (PL) spectra for PPVA-$AlPO_4$ nanocomposite samples were conducted using the Perkin Elmer LS 55 luminescence spectroscopy.

13.3 Results and Discussion

Figure 13.1 shows the absorption spectra for PVA, PPVA, and PPVA-$AlPO_4$ samples. The absorption band for pure PVA has been observed at 204, 277, and 324 nm. Meanwhile, the PPVA absorption band is observed at 274 nm. The absorption band for

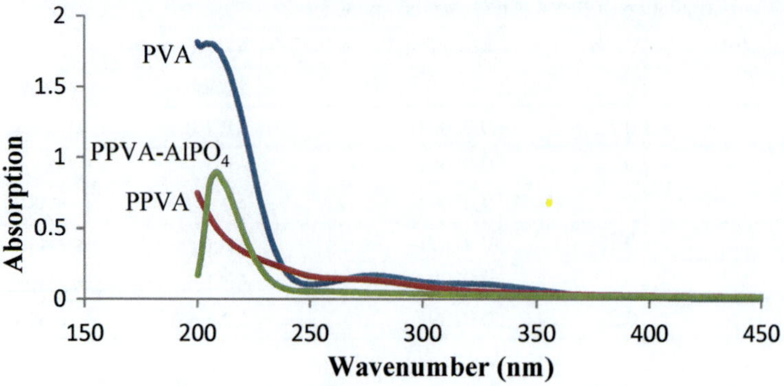

Fig. 13.1 Absorption spectra of PVA, PPVA, and PPVA-$AlPO_4$ samples

the PPVA-$AlPO_4$ sample has appeared at 208 nm. The absorption band for PPVA is shifted to lower wavenumber (274–208 nm) in PPVA-$AlPO_4$ composite due to the interaction of PPVA with the aluminum group. Trunkhin et al. [29] show the incorporation of aluminum and phosphate group produced band at 210 nm indicating the presence of orthophosphate in $AlPO_4$, $ScPO_4$, and $GaPO_4$. This is in good agreement with this finding. Figure 13.2 shows the optical band gap for PVA, PPVA, and PPVA-$AlPO_4$ samples, and their values are presented in Table 13.2. The optical band gap for the PPVA-$AlPO_4$ sample was 5.55 eV. This is lower than PPVA but higher than PVA. The optical band gap for PPVA-$AlPO_4$ composite is almost similar to Devamani [22] which used aluminum phosphate nanoparticles. The decrease in optical band gap as compared to PPVA shows that the degree of disorder was improved with the addition of aluminum. $AlPO_4$ has a structure with repetition of Al and P tetrahedrons. They observed a self-trapped exciton (STE) like silicon dioxide α-quartz. The absorption band for STE is in the visible range. The STE is related to Al–O bond rupture with the creation of a bond between non-bridging oxygen and bonding oxygen on the opposite channel. However, in the PPVA-$AlPO_4$ sample, the result suggests that the UV band is related to complex phosphate–oxygen ions.

Figure 13.3 shows the PL spectra for PVA, PPVA, and PPVA-$AlPO_4$ samples at 250 nm excitation wavelength. The PPVA-$AlPO_4$ sample produces the maximum peak at 318 nm. Meanwhile, two peaks emerged at 521 and 620 nm due to the hydrogen-related species. This finding was in good agreement with Yang et al. [30] which reported that strong PL peaks are observed at 533, 582, 649, and 688 nm in Al/SiO_2 nanocomposite. The PL spectra show the presence of isotactic (i), syndiotactic (s) and atactic (a) configurations of PVA. Table 13.3 summarizes the PL data comparison for the PVA, PPVA, and PPVA-$AlPO_4$ nanocomposites. Figure 13.4 shows the energy level diagrams of non-radiative emission conducted at 250 nm excitation. The 250 nm excitation resonates the excited electronic states in s-PVA at 412 nm shows the lowest intensity of the PL band spectra. As phosphate increased in PPVA, the energy level also increased as observed on i-PPVA. The reaction

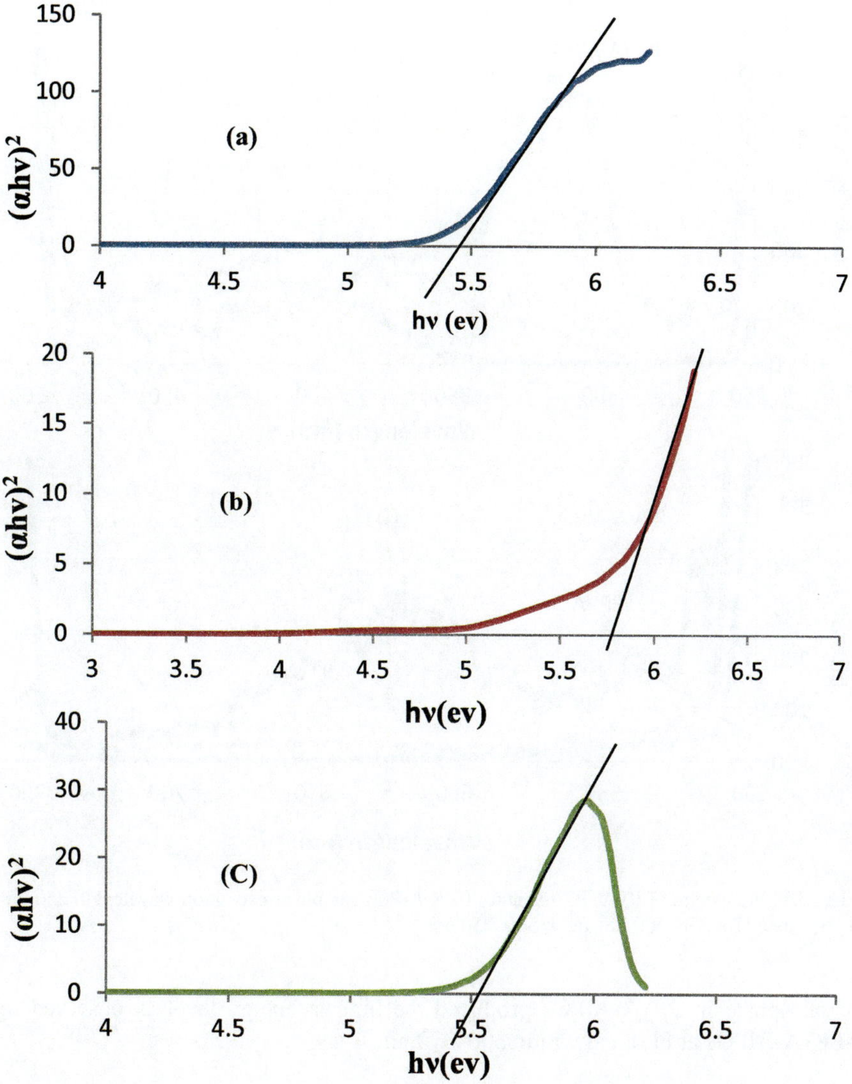

Fig. 13.2 Graph of $(\alpha h\nu)^2$ against photon energy (hν) for **a** pure PVA; **b** PPVA; and **c** PPVA-$AlPO_4$

Table 13.2 Optical parameters for pure PVA, PPVA, and PPVA-$AlPO_4$ samples

Samples	Absorption peak (nm)	Optical band gap, E_g (eV)
Pure PVA	204, 277, 324	5.45
PPVA	274	5.75
PPVA-$AlPO_4$ nanocomposite	208	5.55

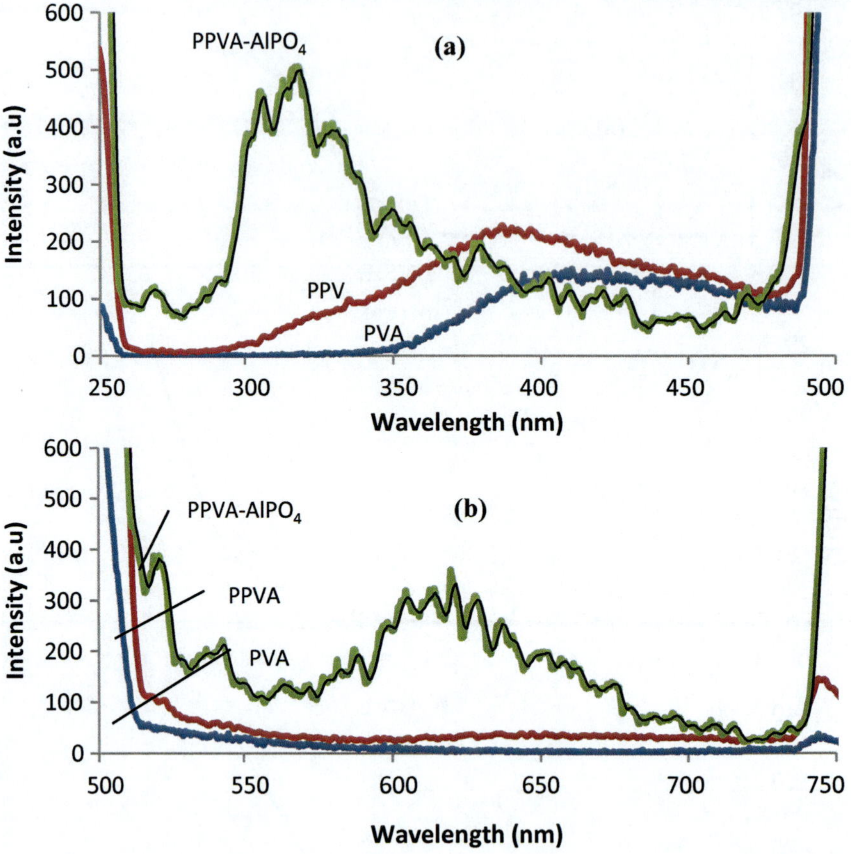

Fig. 13.3 PL spectra of PVA, PPVA, and PPVA-$AlPO_4$ sample at excitation wavelength 250 nm in the range of **a** 250–500 nm and **b** 500–750 nm

of phosphate in PPVA-$AlPO_4$ produced the highest energy level as observed in i-PPVA-$AlPO_4$ at PL band within 306–370 nm.

13.4 Conclusion

PPVA-$AlPO_4$ nanocomposite samples were successfully synthesized. UV–Vis spectroscopy shows that the PPVA-$AlPO_4$ nanocomposite produces a strong single peak at 208 nm. The optical band gap of PPVA-$AlPO_4$ nanocomposite (5.5 eV) was in between pure PVA and PPVA. The shifting of PPVA-$AlPO_4$ nanocomposite absorption peak to lower wavelength confirms the strong interaction of PPVA and aluminum phosphate. The excitation at 250 nm produced comparable PL band spectra between PVA, PPVA, and PPVA-$AlPO_4$. Meanwhile, 300, 350, and 400 nm, the PPVA-$AlPO_4$

Table 13.3 Photoluminescence data in various excitation wavelengths for all samples

Excitation wavelength λ_{exc} (nm)	Samples (mole ratio)	Band position (nm)	Intensity	Assignments
250	Pure PVA	412	70	s-PVA
	PPVA	396	230	i-PPVA
	PPVA-$AlPO_4$ nanocomposite	270	120	
		306, 318, 330	460, 500, 400	i- PPVA-$AlPO_4$
		350, 370	260, 200	i- PPVA-$AlPO_4$
		403, 410, 416, 421, 429	140, 120, 120, 100	s- PPVA-$AlPO_4$
		462, 470	80, 100	i- PPVA-$AlPO_4$
		521, 543	380, 220	i- PPVA-$AlPO_4$
		563, 588	140, 200	
		605, 614, 620, 628, 637	320, 320, 360, 300, 260	
		650, 675, 698	200, 140, 80	
		710, 716	60, 60	
300	Pure PVA	412	70	s-PVA
	PPVA	396	230	i-PPVA
	PPVA-$AlPO_4$ nanocomposite	No spectra		
350	Pure PVA	418, 440	148, 148	s-PVA, a-PVA
	PPVA	421, 444	235, 225	s-PPVA, a-PPVA
	PPVA-$AlPO_4$ nanocomposite	No spectra		
400	Pure PVA	462, 500	135,150	i-PVA
	PPVA	460, 485	125,115	i-PPVA
	PPVA-$AlPO_4$ nanocomposite	No spectra		

nanocomposite shows no PL band spectra. It is observed intense broadband within 306–370 nm after excitation at 250 nm wavelength. The PL band is attributed to the $n \leftarrow \pi^*$ transition of the free OH groups in s- and i-PPVA-$AlPO_4$ configurations. The interaction between the phosphate and OH groups confirms the modified bridging in the s-PPVA which changes to s and i-PPVA-$AlPO_4$ which is more sensitive to H bonding. In conclusion, this finding contributes to the understanding of the change in optical properties of modified PVA and PPVA which will assist for potential optical applications.

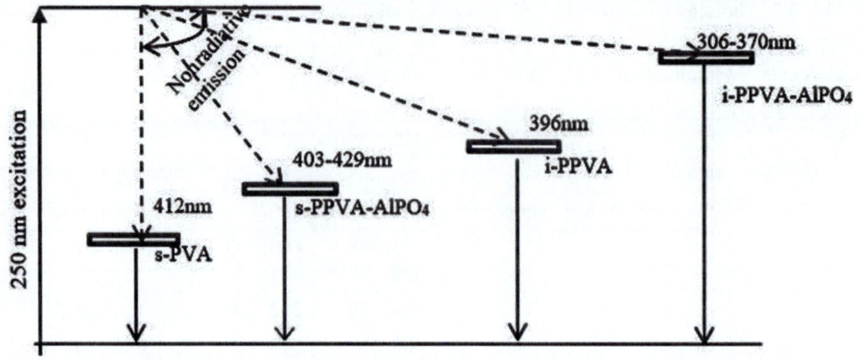

Fig. 13.4 Energy level diagram which shows PL emission and non-radiative process occurring simultaneously in s-PVA, s-PPVA-$AlPO_4$, i-PPVA, and i-PPVA-$AlPO_4$

Acknowledgements This work was financially supported by UM PPP Grant (PS115-2010B & PV129/2012A), UM High Impact Research Grant (UM. C/625/1/HIR-Eng-12). The author also appreciates allowance and leave by University Kuala Lumpur. The author gratefully acknowledges both financial support.

References

1. Aslam M, Kalyar MA, Raza ZA (2018) Polyvinyl alcohol: a review of research status and use of polyvinyl alcohol based nanocomposites. Polym Eng Sci 58(12):2119–2132. https://doi.org/10.1002/pen.24855
2. Saat AM, Johan MR (2014) The surface structure and thermal properties of novel polymer composite films based on partially phosphorylated poly (vinyl alcohol) with aluminum phosphate. Sci World J 1–7
3. Daul GC, Reid JD, Robert MR (1954) Cation exchange materials from cotton and polyvinyl phosphate. Ind Eng Chem 46(5):1042–1045
4. Narayanan TSNS (2005) Surface pretreatment by phosphate conversion coatings—a review. Rev Adv Mater Sci 9:130–177
5. Kaseem M, Hussain T, Baek SH et al (2020) Formation of stable coral reef-like structures via self-assembly of functionalized polyvinyl alcohol for superior corrosion performance of AZ31 Mg alloy. Mater Des 193(May):108823. https://doi.org/10.1016/j.matdes.2020.108823
6. Joseph P, Tretsiakova-Mcnally S (2011) Reactive modification of some chain- and step-growth polymer with phosphorus-containing compounds: effects on flame retardance—a review. Polym Adv Technol 22:395–406
7. Wang DL, Liu Y, Wang DY et al (2007) A novel intumescent flame-retardant system containing metal chelates for polyvinyl alcohol. Polym Degrad Stab 92:1555–1564. https://doi.org/10.1016/j.polymdegradstab.2007.05.001
8. Prajapati GK, Roshan R, Gupta PN (2010) Effect of plasticizer on ionic transport and dielectric properties of PVA–H_3PO_4 proton-conducting polymeric electrolytes. J Phys Chem Solids 71(12):1717–1723. https://doi.org/10.1016/j.jpcs.2010.08.023
9. Gupta PN (2008) Thermal properties and electrochromic behaviour of PVA complexed electrolytes using PEG as plasticizer. Indian J Pure Appl Phys 46(September):657–659

10. Thanganathan U, Parrondo J, Rambabu B (2011) Nanocomposite hybrid membranes containing polyvinyl alcohol or poly(tetramethylene oxide) for fuel cell applications. J Appl Electrochem 41(5):617–622. https://doi.org/10.1007/s10800-011-0270-7
11. Ahmad F, Sheha E (2013) Preparation and physical properties of (PVA)0.7(NaBr)0.3(H_3PO_4)xM solid acid membrane for phosphoric acid fuel cells. J Adv Res 4(2):155–161
12. Bayer E, Grathwohl PA, Geckeler K (1983) Poly (vinyl alcohol) as polymeric support for metal chelating phosphoric acid and derivatives. Macromol Mater Eng 113:141–152
13. Shirai H, Koyama T, Ying A et al (1996) Metal complexes of partially phosphorylated poly(vinyl alcohol)—function and applications. Macromol Symp 222(105):217–222
14. Chen X, Huang R, Pelton R (2005) The reinforcement of calcium carbonate filled papers with phosphorus-containing polymers. Ind Eng Chem Res 44(7):2078–2085. https://doi.org/10.1021/ie048877k
15. Somani P, Amalnerkar DP, Radhakrishnan S (2000) Effect of moisture (in solid polymer electrolyte) on the photosensitivity of conducting polypyrrole sensitized by Prussian blue in solid-state photocells. Synth Met 110:181–187. https://doi.org/10.1016/S0379-6779(99)00279-9
16. Somani PR, Viswanath AK, Aiyer RC et al (2001) Charge transfer complex-forming dyes incorporated in solid polymer electrolyte for optical humidity sensing. Sensors Actuators B Chem 80(2):141–148. https://doi.org/10.1016/S0925-4005(01)00907-8
17. Greish YE, Brown PW (2001) Chemically formed HAp-Ca poly(vinyl phosphonate) composites. Biomaterials 22(July):807–816. https://doi.org/10.1016/S0142-9612(00)00243
18. Pramanik N, Mohapatra S, Alam S, Pramanik P (2008) Synthesis of hydroxyapatite/poly(vinyl alcohol phosphate) nanocomposite and its characterization. Polym. Compos. 29(4):429–436. https://doi.org/10.1002/pc.20410
19. Rajkumar M, Sundaram NM, Rajendran V (2010) In-situ preparation of hydroxyapatite nanorod embedded poly (vinyl alcohol) composite and its characterization. Int J Eng Sci 2(6):2437–2444
20. Mohapatra S, Pramanik N, Ghosh SK, Pramanik P (2006) Synthesis and characterization of ultrafine poly (vinyl alcohol phosphate) coated magnetite nanoparticles. J Nanosci Nanotechnol 6(3):823–829
21. Palacios E, Leret P, Mata MJDL et al (2013) Influence of the pH and ageing time on the acid aluminum phosphate synthesized by precipitation. CrystEngComm 15(17):3359. https://doi.org/10.1039/c3ce00011g
22. Devamani RHP, Alagar M (2012) Synthesis and characterization of aluminium phosphate nanoparticles. Int J Appl Sci Eng Res 1(6):769–775. https://doi.org/10.6088/ijaser.0020101078
23. Yang CS, Kau KY (2005) Synthesis of morphology processable a-$AlPO_4$ nanoparticles, nanowires and multi-strand nano-ropes. J Chinese Chem Soc 52:477–487
24. Chung DDL (2003) Review acid aluminum phosphate for the binding and coating of materials. J Mater Sci 38:2785–2791
25. Somani PR, Marimuthu R, Viswanath A et al (2003) Thermal degradation properties of solid polymer electrolyte (poly(vinyl alcohol)+phosphoric acid)/methylene blue composites. Polym Degrad Stab 79(1):77–83. https://doi.org/10.1016/S0141-3910(02)00240-9
26. Iribarren A, Marzo AL, Lemmetyinen H (2009) Absorption in polyvinyl alcohol phosphoric acids films under different processing conditions. Kinetic study. Rev Cuba Quim 21:3–9
27. Saat AM, Johan MR (2014) Effect of phosphoric acid concentration on the optical properties of partially phosphorylated PVA complexes. Int J Polym Sci 2014(495875):1–8
28. Saat AM, Latiff NA, Yaakup S, Johan MR (2014) Synthesis and characterisation of composite partially phosphorylated polyvinyl alcohol-aluminium phosphate as a protective coating. Mater Res Innov 18(S6):310–313. https://doi.org/10.1179/1432891714Z.000000000974
29. Trukhin AN, Shmits K, Jansons JL et al (2013) Ultraviolet luminescence of $ScPO_4$, $AlPO_4$ and $GaPO_4$ crystals. J Phys Condens Matter 25(38):385502. https://doi.org/10.1088/0953-8984/25/38/385502
30. Yang H, Yao X, Huang D (2007) Sol-gel synthesis and photoluminescence of AlP nanocrystals embedded in silica glasses. Opt Mater (Amst) 29(7):747–752. https://doi.org/10.1016/j.optmat.2005.11.029

Chapter 14
Tensile and Corrosion Resistance Studies of MXenes/Nanocomposites: A Review

Mohd Shahneel Saharudin, Nur Ahza Che Nasir, and Syafawati Hasbi

Abstract MXenes are a relatively new and interesting class of two-dimensional materials with diverse compositions and outstanding characteristics such as dispersibility and metallic conductivity. MXenes appear to be promising fillers for polymer nanocomposites, and data from several studies suggest that this promising material could significantly improve the tensile strength and modulus by 314% and 89%, respectively, when incorporated into a polymer matrix. Corrosion, on the other hand, is a significant issue in numerous industries worldwide, including automotive, defence, aerospace and biomedical. There is a growing body of the literature that recognises MXenes as high-performance corrosion inhibitors. In this review, recent research on the corrosion resistance properties of MXenes-reinforced polymeric composites is also discussed.

Keywords MXenes · Tensile properties · Corrosion inhibitors · Nanocomposites · Coatings

14.1 Introduction

Because of their distinct characteristics, such as low cost, low density and ease of processing, polymeric materials have been used in numerous applications [1]. On the other hand, polymeric materials have limited applicability due to their weak mechanical and tribological properties [2–4]. Many of these polymeric coatings are used owing to their high corrosion resistance properties. Even though, under

M. S. Saharudin (✉) · N. A. Che Nasir
Universiti Kuala Lumpur Malaysia Italy Design Institute, 119 Jalan 7/91, 56100 Kuala Lumpur, Cheras, Malaysia
e-mail: mshahneel@unikl.edu.my

N. A. Che Nasir
e-mail: ahza.nasir@s.unikl.edu.my

S. Hasbi
Department of Mechanical Engineering, National Defense University of Malaysia, 57000 Kuala Lumpur, Malaysia
e-mail: syafawati@upnm.edu.my

A. Ismail et al. (eds.), *Design in Maritime Engineering*, Advanced Structured Materials 167, https://doi.org/10.1007/978-3-030-89988-2_14

tribological conditions, they have low wear resistance there are several methods of enhancing polymeric composite mechanical and tribological performance [5].

Corrosion has always been a major concern in the oil and gas industries, with consequences comparable to natural disasters. Corrosion occurs typically in oil and gas pipelines. These pipelines are constantly threatened by corrosion from the time they are commissioned until they are withdrawn. According to a new report, the entire annual cost of corrosion is estimated to be \$1.372 billion [6]. Corrosion is a severe problem because it wastes thousands of tonnes of iron each year and costs much money to fix or replace. Corrosion inhibitors are one of the effective methods used in the petroleum industry to reduce corrosion. The inhibitors must be added above a specified minimum concentration to achieve optimal inhibition. There are several strategies for fighting corrosion, such as cathodic protection, organic coatings and the use of high-quality corrosion-resistant alloys [7]. Film-forming inhibitors are still known to be the unrivalled means of defence for mild steel in an acidic environment [6]. In industries, film-forming inhibitors are used to establish a molecular layer on the steel surface and an aliphatic tail as a second layer in the hydrocarbon to prevent water from reaching the steel surface and corroding it.

This review paper assesses the most recent publications of MXenes as corrosion inhibitors. The review is expected to lay the groundwork for more study into MXenes as corrosion inhibitors in the future. MXenes, which were discovered in 2011, are one of the fastest-growing 2D materials that have received a lot of scientific attention. MXenes are a novel type of hydrophilic and conductive two-dimensional (2D) nanomaterial. They are 2D sheets of transition metal carbides, nitrides, or carbonitrides derived from the layered ternary $Mn_{+1}AXn$, or MAX, phases which only can be selectively etched to form MXenes. This terminology refers to their dual role as MAX phases and 2D. M-X bonds are supposedly stronger than M-A bonds. A mix of functional groups, good mechanical properties, self-lubricating and high thermal conductivity result in MXenes being formed by the selective etching of the A elements. Figure 14.1 shows the total journal articles published from 2010 to 2020. It can be seen that the research of MXenes only started in 2012 where ten research articles were initially published. The interest in the MXenes' research increased every year. In 2020, around 1585 articles were published, an increase of 60% compared to 2019. The number is expected to rise in 2021 because of the quickly growing interest in this family of materials.

14.2 Synthesis of MXenes

Wet chemical acidic etching of MAX phases yields MXenes in either the colloidal suspension of a few flakes or ML stacks of MXene sheets. Following etching, the post-processing techniques used can have an impact on how each of these forms is formed with the intended quality. As previously reported, Ti_3C_2 nanosheets were created by etching the Ti_3AlC_2 phase with LiF/HCl. Initially, 1 g LiF was dissolved in 20 mL 9 M HCl, and the mixture was well-mixed at room temperature for several

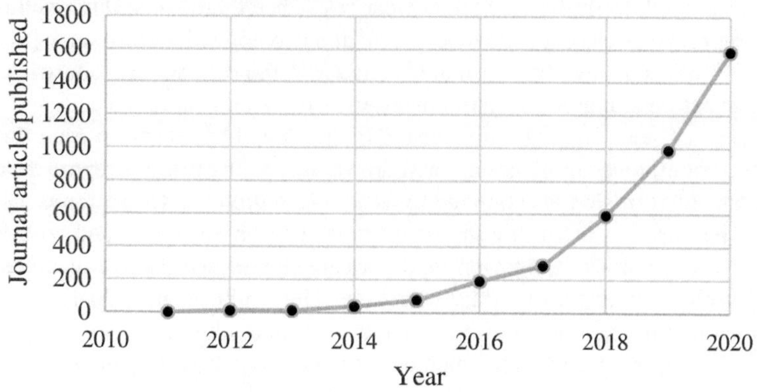

Fig. 14.1 Journal article published from 2011 to 2020 using "MXenes" as keyword obtained from Lens.org on 23rd May 2021

minutes. The solution was prepared by slowly adding 1 g Ti_3AlC_2 into it. Then, the reaction was stirred continuously for 24 h at 35 °C. The etched powder was rinsed with deionised water and left to dry at room temperature. Sonication with Ar resulted in Ti_3C_2 aqueous solutions that were then centrifuged for one hour at 3500 rpm.

Deposition techniques include CVD, ALD, and photo-deposition. CVD has multiple volatile precursors that cause a film to be formed on the substrate's surface. A semiconductor, nanocomposite, alloy or metal can be used as the deposited material. ALD splits the CVD reaction into two halves, ensuring that the precursor materials do not come into contact during the reaction. ALD film growth is constrained and self-dependent on surface reactions, allowing for atomic-scale deposition control. By keeping the precursors separated during the coating process, a monolayer as fine as 0.1 can be fabricated. ALD has various advantages over other processes, including the formation of pinhole-free, chemically linked and substrate conformal layers.

Zhang et al. have successfully employed ALD to develop Pt–TBA–$Ti_3C_2T_x$, which was later used in hydrogen evolution reactions [8]. Electrodeposition processes could be used to create MXene hybrids incorporating C-based materials, transition metal phosphides, oxides and metals. The photo-deposition can be used to apply metallic NPs, such as Cu and Pt, over MXene surface. The deposition methods are regulated and more encapsulating for the fabrication, but the higher cost of production is not preferred.

Solution processing is one of the most frequent processing procedures for synthesising MXene-reinforced polymer composites due to the hydrophilic character of MXene nanosheets created by the presence of a significant number of functional groups.

Premodified MXene and polymer nanoparticles are typically dispersed in polar fluids such as dimethylsulfoxide (DMSO), water and N, N-dimethylformamide to improve dispersibility. The major components are blended together to create a homogeneous slurry. So far, MXenes have been successfully combined with polyurethane, cellulose, polyethylene oxide and a few other materials. This approach can be used to hybridise a wide range of inorganic materials with MXenes. Poor mechanical properties, the creation of substantial amounts of environmental waste associated with the resulting composites and the difficulty in removing solvents using evaporation, which usually limits the use of solution mixing, are all significant disadvantages of this method.

The hydrothermal and solvothermal processes take advantage of the connection between the mineraliser, liquid solvent and precursor molecule, particularly when temperature and pressure are enhanced. It provides the simplest and most cost-effective methods for synthesising diverse secondary materials with different morphologies following the processing conditions. The procedures for solvothermal and hydrothermal are similar apart from for the solvents used in the synthesis process. The chemical reaction takes place in a sealed autoclave above the boiling point in a suitable solvent containing the reaction mixture. As the temperature and pressure rise above the critical point of the solvent, a supercritical fluid phase form is formed. The supercritical fluid phase incorporates both gas and liquid specificities, and there is no surface tension at the solid–supercritical fluid interface. Supercritical fluids had higher viscosities and dissolved a chemical compound that is insoluble at room temperature. One of the method's significant drawbacks is the aggressive corrosion caused by water molecules or hydroxyl units at high temperature and pressure settings, which results in unexpected modifications. MXenes were hybridised with a variety of inorganic compounds, including transition metal oxides, nitrides, phosphides, perovskites and chalcogenides.

14.3 Applications of MXenes

MXenes' versatile and desirable properties make them a potential choice in a wide range of applications. The unique morphological features and high Young's modulus make it attractive in composite formation. On the other hand, these properties drive them as a strong candidate for many applications like catalysis, sensors and energy storage. MXenes performs similarly to or better than most currently utilised products in various instances, such as electromagnetic interference shielding. They are also found in several other fields, for example opto/spintronic, environmental, tribologic and biomaterial. Figure 14.2 shows a 3D model of crystal structures Ti_3AlC_2 max phase, presented isometric and front view. Recently, MXenes have emerged as corrosion inhibitors in coating [7, 9]. Figure 14.3 shows the major industrial applications of MXene composites. They are widely used in energy storage, tribology, biomedical, sensors, catalysis and EMI shielding.

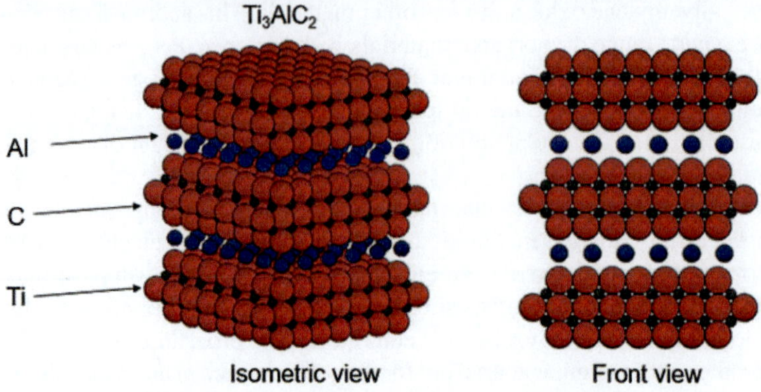

Fig. 14.2 3D model of crystal structures Ti_3AlC_2 max phase (not drawn to scale)

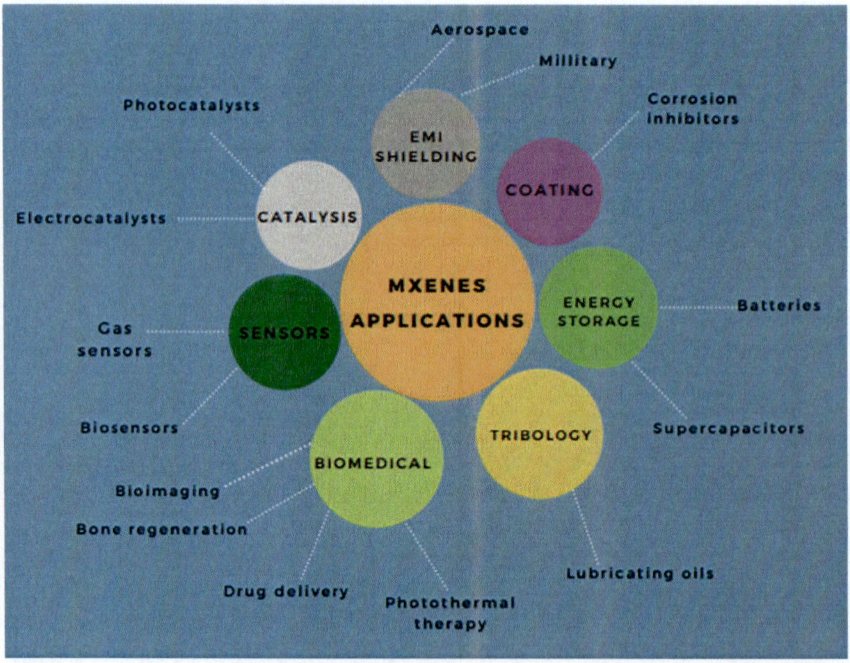

Fig. 14.3 Applications of MXenes from literature

14.4 Tensile Properties of MXenes/Nanocomposites

Figure 14.4 shows the increase in the percentage of the tensile strength for MXenes/nanocomposites obtained from the literature. Zhi et al. studied the MXene-filled polyurethane nanocomposites prepared via an emulsion method [10]. An

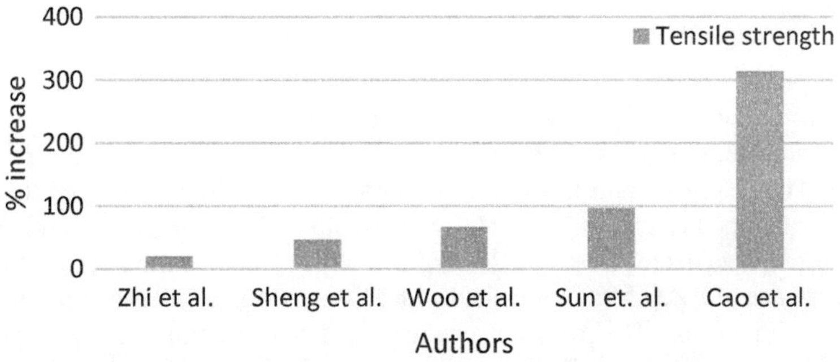

Fig. 14.4 Percentage increase in tensile strength of MXenes/nanocomposites [8, 10–13]

increase of 20% was observed with the addition of 0.5 wt% MXenes into polyurethane. Sheng et al. [8], Woo et al. [11] and Sun et al. [12] reported an increase of tensile properties of 39%, 49% and 89%, respectively. The highest tensile strength of 314% was observed by Cao et al. in the case of MXene/cellulose nanofiber composite [13]. Their works demonstrate that by incorporating MXene into polymers, MXenes significantly improve tensile strength's mechanical properties at a low mass content.

Figure 14.5 shows the percentage increase in elastic modulus of MXenes/nanocomposites. It can be observed that a minimum increase (35%) in elastic modulus was reported by Liu et al. [14]. Sheng et al. [8] reported 40% increase in elastic modulus in their research. Liu et al. in their new study have reported an improvement of 46% in elastic modulus [15]. On the other hand, Saharudin et al. [4] stated that a 49% of the increase in elastic modulus was reported. Last but not least, Wei et al. [5] reported the highest elastic modulus where an 89% increase was observed in their study.

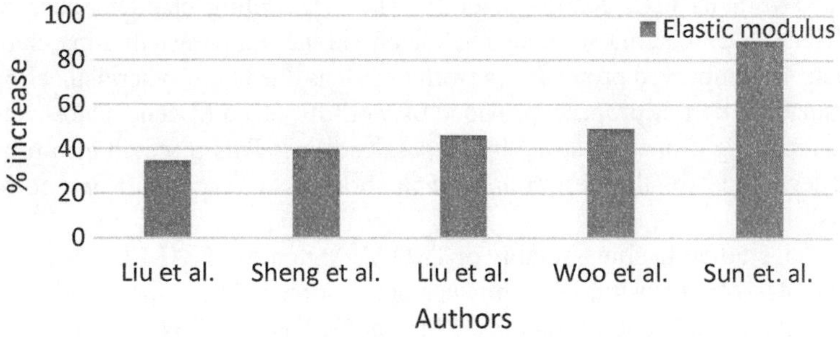

Fig. 14.5 Percentage increase in elastic modulus of MXenes/nanocomposites [8, 11, 12, 14, 15]

14.5 Corrosion Properties of MXenes/Nanocomposites

Yan et al. studied [16] anti-corrosion behaviour of an inorganic–organic multilayer protection system composed of nitriding layer and epoxy coating with Ti_3C_2Tx MXene. The results demonstrated that two orders of magnitude enhanced the electrochemical impedance during 4 weeks of immersion in 3.5 wt% NaCl solution, and the wear rate was decreased by 96.26%. Wu et al. [17] studied the improved corrosion resistance of AZ31 Mg alloy coated with MXenes/MgAl-LDHs composite layer modified with yttrium. The corrosion tests indicated that the composite coating modified with yttrium had an excellent anti-corrosion performance ($E_{corr} = -0.36$ VSCE, $I_{corr} = 9.12 \times 10^{-9}$ A cm^{-2}) and self-healing ability.

Yan et al. [18] studied the Ti_3C_2 MXene nanosheets towards high-performance corrosion inhibitor for epoxy coating. The results reveal that epoxy coatings with Ti_3C_2 nanosheets have strong corrosion resistance due to the intrinsic properties of 2D Ti_3C_2 nanosheets and the establishment of a barrier against corrosive media. They achieved a protective efficiency up to 99%. Therefore, the Ti_3C_2 can enhance corrosion protective performance. Nie et al. studied the MXene-hybridised silane films for metal anti-corrosion and antibacterial applications. They have reported that, after three days of NaCl spray time on aluminium alloy, the corrosion area for Al coated with MXene was 0% compared to chromium and silane films [19].

Shen et al. studied GO-Ti_3C_2 two-dimensional heterojunction nanomaterial for anti-corrosion enhancement of epoxy zinc-rich coatings [7]. Hybride additives of GO-Ti_3C_2 were incorporated into the ZRC coating to improve the utilisation rate of zinc particles. Both GO and Ti_3C_2 provided a barrier for corrosion inhibition. At the end of immersion, R_c value of ZRC/GO-Ti_3C_2 coating was 3.047×10^4 Ω cm^2, it was one order of magnitude better than ZRC coating.

Zhao et al. investigated air-stable titanium carbide MXene nanosheets for corrosion protection [20]. It was discovered that the $Z_{f=0.01\ Hz}$ values of the IL@MXene$_{0.5}$ coating remained nearly unchanged after ten days of immersion. This can be associated with its high coating stability. The self-healing characteristics of the IL@MXene-WEP coating were also validated via the scanning vibrating electrode approach. The improved protective performance was due to the synergistic effects of the outstanding barrier property provided by well-dispersed MXene nanosheets and the self-repairing property caused by IL passive films. This research also provides a good technique for the design and preparation of MXene smart anti-corrosion coatings.

Cai et al. studied in situ assembly of $Ti_3C_2T_x$ MXene@MgAl-LDH heterostructure towards anti-corrosion and antiwear application [21]. As-prepared Ti_3C_2Tx MXene@MgAl-LDH/epoxy coating (C-MXene@LDH) exhibits satisfactory corrosion/wear protection and certain self-healing performance. The corrosion resistance of $Ti_3C_2T_x$ MXene@MgAl-LDH can be attributed to the synergy of good dispersibility, barrier effect and corrosion inhibitor release.

Song et al. studied the thermal and corrosion behaviour of MXenes/Cu composites [22]. The results show that Ti_3C_2/Cu composites have better corrosion resistance than TiC/Cu composites. This can be associated with the outstanding electrical conductivity and easily oxidised property of Ti_3C_2. However, the thermal conductivity of Ti_3C_2/Cu composite with the content of 2 wt% Ti_3C_2 improves about 15% compared to TiC/Cu composites even with the same content, resulting from the low inherent thermal conductivity of filler. Besides that, at the interfacial contact resistance measurement the electric resistance of Ti_3C_2/Cu composites increases around 100% compared to pure Cu, with the content of Ti_3C_2 at 2 wt%. Meanwhile, the anti-corrosion performance of the Ti_3C_2/Cu composites was improved over pure Cu. This work will broaden the application field of Ti_3C_2 and lay the foundation for future research. The E_{corr} of 2.0 wt% Ti_3C_2/Cu composite is 0.048 V higher than that of 2.0 wt% TiC/Cu composite, verifying that Ti_3C_2/Cu composite has better corrosion resistance compared to TiC/Cu composite. Additionally, Ti_3C_2/Cu composites have improved anti-corrosion performance compared to pure Cu.

Sheng et al. reported with 0.4 wt% MXene, the WPU/ Ti_3C_2 MXene composite coatings reach the lowest corrosion current of 2.143×10^{-6} A/cm^2 [23]. A decrease of one order of magnitude compared with blank WPU (1.599×10^{-5} A/cm^2) was achieved, and it has an excellent UV-blocking property [23]. Moreover, measurement against corrosion shows that the WPU Ti_3C2@Si has an exceptional resistance. It was found that the 0.1%Ti_3C_2@Si/ WPU exhibits the lowest corrosion current density (2.67×10^{-9} A/cm^2) and the largest corrosion resistance (3.05×10^6 Ω). After 42 days' immersion, the lowest frequency impedance of 0.1%Ti_3C_2@Si/WPU composite coating was 3.68×10^6 Ω cm^2 [24]. This research is also important for future anti-corrosion applications (Table 14.1).

Table 14.1 Recent literature of corrosion studies in NaCl environment

Ref	Year	Metal type	Exposure	Findings
[16]	2021	Al alloy	28 days	Wear rate decrease 96.26%
[17]	2021	Mg alloy	15 days	$E_{corr} = -0.36$ V SCE, $I_{corr} = 9.12 \times 10^{-9}$ A cm^{-2}
[7]	2021	Steel	50 days	Rc value of ZRC/GO-Ti_3C_2 coating was 3.047×10^4 Ω cm^2
[20]	2021	Steel	10 days	$Z_{f=0.01\ Hz}$ value remains unchanged
[21]	2021	Q345 steel	21 days	The wear rate of C-MXene is 0.0546 μm^3/N μm, reduced by 41.35% compared to that of EP
[23]	2021	Q235 mild steel	Not available	MXene composite coatings reach the lowest corrosion current of 2.143×10^{-6} A/cm^2
[24]	2021	Q235 mild steel	42 days	Largest corrosion resistance (3.05 × 106 Ω)
[22]	2020	Copper	Not available	E_{corr} of 2.0 wt% Ti_3C_2/Cu composite is 0.048 V
[19]	2020	Al alloy	3 days	Corrosion area 0%
[18]	2019	Q345 steel	15 days	Protective efficiency 99%

14.6 Conclusion

This work has provided an overview of applications, tensile properties and corrosion resistance studies of MXenes/nanocomposites. This review also introduces MXene as potential corrosion inhibitors. The composite coating with MXenes exhibits the highest corrosion potential and the lowest corrosion current density, and superior polarisation resistance, which is attributed to the intrinsic properties of Ti_3C_2 and its strong barrier effect. 2D MXene with good anti-corrosion function was proven. Recent research into MXenes as corrosion inhibitors has paved the way for MXene to be used as a high-performance anti-corrosion additive in the real world.

Acknowledgements The authors wish to thank the Centre of Research and Innovation (CoRI), Universiti Kuala Lumpur, for the Research Cluster grant awarded to Composites & Simulation Cluster (CSC) and UniKL MIDI for the research facilities provided.

References

1. Saharudin MS, Wei J, Shyha I et al (2016) The degradation of mechanical properties in halloysite nanoclay-polyester nanocomposites exposed in seawater environment. J Nanomater 2016:1–12
2. Shyha I, Fo G, Huo D et al (2018) Micro-machining of nano-polymer composites reinforced with graphene and nano-clay fillers. Key Eng Mater 786:197–205
3. Wei J, Saharudin MS, Vo T et al (2018) Effects of surfactants on the properties of epoxy/graphene nanocomposites. J Reinf Plast Compos 37:960–967
4. Saharudin MS, Atif R, Inam F (2017) Effect of Short-term water exposure on the mechanical properties of halloysite nanotube-multi layer graphene reinforced polyester nanocomposites. Polymers (Basel) 9:1–27
5. Wei J, Saharudin MS, Vo T et al (2017) N, N-Dimethylformamide (DMF) usage in epoxy/graphene nanocomposites: problems associated with reaggregation. Polymers (Basel) 9:193
6. Tamalmani K, Husin H (2020) Review on corrosion inhibitors for oil and gas corrosion issues. Appl Sci 10:1–16
7. Shen L, Zhao W, Wang K et al (2021) GO-Ti3C2 Two-dimensional heterojunction nanomaterial for anticorrosion enhancement of epoxy zinc-rich coatings. J Hazard Mater 417:126048
8. Sheng X, Zhao Y, Zhang L et al (2019) Properties of two-dimensional Ti3C2 Mxene/thermoplastic polyurethane nanocomposites with effective reinforcement via melt blending. Compos Sci Technol 181:107710
9. Aghamohammadi H, Amousa N, Eslami-Farsani R (2021) Recent advances in developing the MXene/polymer nanocomposites with multiple properties: a review study. Synth Met 273:116695
10. Zhi W, Xiang S, Bian R et al (2018) Study of MXene-filled polyurethane nanocomposites prepared via an emulsion method. Compos Sci Technol 168:404–411
11. Woo JH, Kim NH, Kim SII et al (2020) Effects of the addition of boric acid on the physical properties of MXene/polyvinyl alcohol (PVA) nanocomposite. Compos Part B Eng 99:108205
12. Sun Y, Ding R, Hong SY et al (2021) Mxene-Xanthan nanocomposite films with layered microstructure for electromagnetic interference shielding and joule heating. Chem Eng 410:128348

13. Cao W, Chen F, Zhu Y et al (2018) Binary strengthening and toughening of mxene/cellulose nanofiber composite paper with nacre-inspired structure and superior electromagnetic interference shielding properties. ACS Nano 12:4583–4593
14. Liu L, Ying G, Wen D et al (2021) Aqueous solution-processed Mxene (Ti3C2TX) for non-hydrophilic epoxy resin-based composites with enhanced mechanical and physical properties. Mater Des 197:1–11
15. Liu L, Ying G, Sun C et al (2021) MXene (Ti3C2Tx) Functionalized short carbon fibers as a cross-scale mechanical reinforcement for epoxy composites. Polymers (Basel) 13:1–13
16. Yan H, Cai M, Wang J et al (2021) Insight into anticorrosion/antiwear behavior of inorganic-organic multilayer protection system composed of nitriding layer and epoxy coating with Ti3C2Tx Mxene. Appl Surf Sci 536:147974
17. Wu Y, Wu L, Yao W et al (2021) Improved corrosion resistance of AZ31 Mg alloy coated with Mxenes/Mgal-Ldhs composite layer modified with yttrium. Electrochim Acta 374:137913
18. Yan H, Li W, Li H et al (2019) Ti3C2 Mxene nanosheets toward high-performance corrosion inhibitor for epoxy coating. Prog Org Coatings 135:156–167
19. Nie Y, Huang J, Ma S et al (2020) Mxene-hybridized silane films for metal anticorrosion and antibacterial applications. Appl Sur Sci 527:146915
20. Zhao H, Ding J, Zhou M et al (2021) Air-stable titanium carbide MXene nanosheets for corrosion protection. ACS Appl Nano Mater 4:3075–3086
21. Cai M, Fan X, Yan H et al (2021) In situ assemble Ti3C2Tx Mxene@Mgal-LDH heterostructure towards anticorrosion and antiwear application. Chem Eng J 419:130050
22. Song G, Deng Q, Wang B et al (2020) Thermal and corrosion behavior of Ti3C2/copper composites. Compos Commun 22:1–6
23. Sheng X, Li S, Huang H et al (2021) Anticorrosive and UV-blocking waterborne polyurethane composite coating containing novel two-dimensional Ti3C2 mxene nanosheets. J Mater Sci 56:4212–4224
24. Zhang F, Liu W, Wang S et al (2021) Surface functionalization of Ti3C2Tx and its application in aqueous polymer nanocomposites for reinforcing corrosion protection. Compos Part B Eng 217:108900

Chapter 15
Effect of Nanofillers on the Mechanical Properties of Epoxy Nanocomposites

Nur Ahza Che Nasir, Mohd Shahneel Saharudin, Wan Nursheila Wan Jusoh, and Ong Siew Kooi

Abstract In this research, various types of nanofillers were prepared to investigate the impact of nanofillers on the mechanical properties of the epoxy matrix. Tensile testing, dynamic mechanical analysis (DMA) and scanning electron microscopy (SEM) have been used to compare the efficacy of four distinct nanofillers: MXene, graphene (GNP), carbon nanotubes (CNTs) and halloysite nanotubes (HNTs). Final results indicate that MXene/epoxy nanocomposite lead to a significantly improved tensile strength and elastic modulus of up to 66.57% and 22.65%, respectively, compared to neat epoxy. The homogenous dispersion, size and shape of nanoparticles are the major elements that contribute to the final properties of the nanocomposites.

Keywords Nanofillers · Nanocomposite · Mechanical properties · Dispersion

15.1 Introduction

Epoxy has become one of the most desired engineering applications due to its extraordinary physical and chemical properties, good thermal stability, solvent resistance, ease of processing and superior mechanical properties. Therefore, the number of investigations related to polymer nanocomposites has shown a significant interest

N. A. Che Nasir (✉) · M. S. Saharudin
Universiti Kuala Lumpur Malaysia Italy Design Institute, Taman Shamelin Perkasa, 119 Jalan 7/91, 56100 Cheras, Kuala Lumpur, Malaysia
e-mail: ahza.nasir@s.unikl.edu.my

M. S. Saharudin
e-mail: mshahneel@unikl.edu.my

W. N. Wan Jusoh
Universiti Kuala Lumpur Malaysian Institute of Aviation Technology, Lot 2891, Jalan Jenderam Hulu, 43800 Dengkil, Selangor, Malaysia
e-mail: wannursheila@unikl.edu.my

O. S. Kooi
Universiti Kuala Lumpur Malaysian Institute of Chemical and Bio-Engineering Technology, Lot 1988, Kawasan Perindustrian Bandar Vendor, 78000 Alor Gajah, Malacca, Malaysia
e-mail: skong@unikl.edu.my

A. Ismail et al. (eds.), *Design in Maritime Engineering*, Advanced Structured Materials 167, https://doi.org/10.1007/978-3-030-89988-2_15

to researchers over the last decades due to the development of advanced materials for potential applications [1, 2]. Epoxy and its composites are widely used in many industrial fields, e.g. automotive [2–6], aerospace [2, 5–7], electrical and electronic devices [6, 8–10], biomedical devices, sensors and other general consumer goods [4, 6, 11–13].

Nonetheless, considerable research is being conducted to improve the epoxy's mechanical properties, extreme environmental resistance, damage resistance and high-temperature performance. Second phase materials can be dispersed in epoxy resin systems for improving the performances [14]. With the addition of nanoparticles, i.e. the nanofillers can improve mechanical properties, epoxy coatings' barrier and anti-corrosion properties [1, 15–17]. The nanocomposite is characterized by (1) the matrix type where nanoparticles are dispersed in, and (2) the type of nanofillers. Various nanoparticles have recently been used as fillers in epoxy resins to optimize their properties and characteristics [18]. Nanofillers are materials with at least one dimension between 1 and 100 nm that are manufactured as zero-dimensional (0D), one-dimensional (1D), two-dimensional (2D) and three-dimensional (3D) nanomaterials [1].

MXene is a 2D nanomaterial with the transition group of metal carbides, nitrides and carbonitrides. MXene has attracted much attention due to its excellent electrical and mechanical properties, making it a great filler in polymer composites [19]. Besides, graphene can significantly enhance the matrix's physical and chemical properties at relatively low loadings; nonetheless, this improvement is only possible when the filler is distributed uniformly throughout the matrix [5, 20, 21]. Graphene that is poorly dispersed may lead to increased stress concentrations, hence lowering the mechanical properties [2]. Many sources claimed that CNTs are more robust than steel and carbon fibre, conduct electricity better than silver and platinum and carry a higher current density than copper [22]. Polymeric systems of nanofillers such as HNTs and CNTs can significantly raise the strength and modulus of composites, whilst using lower filler loading. HNTs possess excellent thermal stability, biocompatibility, mechanical strength, and is a promising nanofillers for a wide range of applications [23]. The purpose of this study is to examine the implication of nanofillers towards mechanical properties in an epoxy matrix and to compare the efficacy of four distinct nanofillers: MXene, graphene (GNP), carbon nanotubes (CNTs) and halloysite nanotubes (HNTs).

15.2 Methodology

15.2.1 Materials

15.2.1.1 Epoxy Matrix System

Asasin 3525 bisphenol A based liquid epoxy and Asahard 3525 B hardener were purchased from Asachem (M) Sdn Bhd in Kajang, Malaysia. Liquid epoxy and hardener have viscosities of 12.0–15.0 cps and 10.0–12.0 cps, respectively, at 25 °C. To prepare the epoxy mixture, the mixing ratio (by weight) of epoxy to hardener was 2:1.

15.2.1.2 Nanofillers

Four different nanofillers were used in this study, namely MXene, graphene (GNP), carbon nanotubes (CNTs) and halloysite nanotubes (HNTs). Graphene, CNTs and HNTs were purchased from Sigma-Aldrich, UK, whilst MXene was purchased from Sunway University, Malaysia. The particle size of the graphene nanoplatelets was 2 mμ with a specific surface area of 300m^2/g and 99.2% purity. As per the manufacturer's technical data sheet, the length and diameter of CNTs were 0.5–10 mμ and 7–15 nm, respectively. The HNTs were 1–4 mμ in length and 30–70 nm in diameter.

15.2.1.3 Sample Preparation

Nanofiller samples of 0.3 wt.% were weighed in the analytical balance and dispersed in the hardener by hand mixing for 5 s gently and then sonicated for 30 min at room temperature using a bath sonicator. Then, the suspension was cooled to room temperature and then mixed with liquid epoxy by the ratio of epoxy to hardener of 2:1. After another 5 min of thorough hand mixing, the mixtures were cast in moulds and cured at room temperature for 7 days before being post-cured at 90 °C for 5 h to achieve a complete cross-linking. Making the silicone mould was relatively simple, and the gelation time was about 30 min. The silicone mould materials were purchased from Portal Trading Sdn Bhd, Penang, Malaysia. The mixing ratio of silicone rubber to curing agent of 40:1 was poured into the acrylic mould and been left 24 h for curing. The schematic illustration of the sample preparation is illustrated in Fig. 15.1.

15.2.2 Characterization

The dynamic storage modulus (E') and loss modulus (E'') of the sample was determined using a DMA 8000 dynamic mechanical analyzer, PerkinElmer. The loss

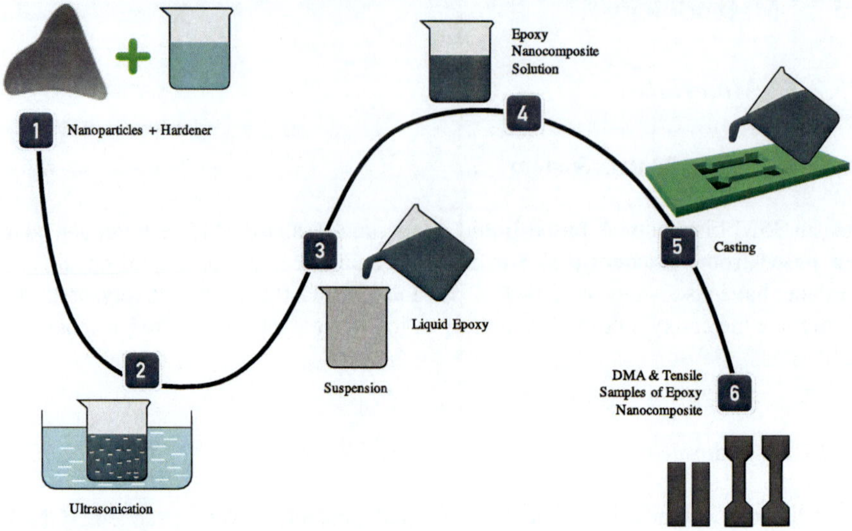

Fig. 15.1 Schematic illustration of sample preparation

factor of the tan delta was calculated on the basis of E''/E'. The rectangular test specimens with a dimension of 35 × 10 × 3 mm were prepared. All tests were run at a constant frequency of 1 Hz using the temperature sweep method (temperature ramp of 30–130 °C at 5 °C/min). ASTM D638 (type V geometry) standard was used to measure tensile properties on a specimen with a thickness of 3 mm. For all tests, the crosshead speed was retained constant at 1 mm/min. For each set of conditions, a minimum of five specimens was tested.

15.3 Results and Discussion

15.3.1 Mechanical Properties of Nanocomposites

Based on the tensile test results, the average value of UTS of neat epoxy was found to be the lowest UTS with 34.80 MPa. The minimum UTS improvement of 5.37% was observed for HNTs/epoxy nanocomposites. As for CNTs and graphene, the tensile strength was increased up to 47.44% and 53.94%, respectively. The UTS recorded a significant improvement of 66.57% for MXene/epoxy nanocomposites. The average value of elastic modulus of neat epoxy was 498 MPa. Similarly, the result indicates stiffness increment for HNTs, CNTs, graphene and MXene by 6.01%, 11.62%, 14.43% and 22.65%, respectively. However, the tensile strain result shows the opposite. Nanocomposite with CNTs filler showed a 5.81% decrement compared to neat epoxy. However, other nanocomposites showed an increment of 2.75%, 11.86% and 18.2% for MXene, HNTs and graphene, respectively.

Comparing the mechanical properties result with the SEM images shows that poor dispersion could contribute to the lower tensile strength, elastic modulus, and tensile strain. In general, homogenous dispersion of nanofillers resulted in higher nanocomposite properties. Characteristically, agglomeration can be avoided if using a technique that improves the dispersion throughout the matrix. In that case, each nanocomposite should see a significant improvement in mechanical properties [24]. Overall, the tensile and elastic properties of nanocomposites were improved compared to neat epoxy. These increments indicate that nanofillers have a significant reinforcement effect in the epoxy matrix. These findings strongly imply that the incorporation of nanofillers had a significant impact on the mechanical properties of the nanocomposites. Results are summarized in Fig. 15.2.

15.3.2 DMA Results of Nanocomposites

In order to evaluate the role of nanocomposites under stress, the storage modulus (*E*') and tan delta values with regards to temperature were analyzed. As shown in Table 15.1, the incorporation of nanofillers in the epoxy matrix affects the nanocomposites' storage modulus value at room temperature. However, the nanocomposite with HNTs filler shows the lowest storage modulus of *E*' = 24.12 GPa, whilst the neat epoxy shows *E*' = 26.84 GPa, which indicates 10.13% depletion. This was due to poor dispersion, which performs as a stress raiser and affects the stress concentration, resulting in micro-cracks within the sample. The value reached its peak at the MXene/epoxy nanocomposites with a 26.27% increment.

The temperature dependence of the tan delta values of epoxy/nanofiller nanocomposites is shown in Fig. 3b. The maximum peak of tan delta signifies the material's glass transition temperature (T_g), as stated in Table 15.1. The T_g value of nanocomposites increased from 58.20 °C of neat epoxy to 58.85 °C of MXene. For other nanofillers, a decrement in the T_g value was recorded. As reported in the literature [25–27], this is due to the presence of more significant aggregation spots in the nanocomposites that impede curing and reduce cross-linking density [27, 28].

15.3.3 SEM Images of Nanocomposites

The nanofillers dispersion in the epoxy matrix was carried out using SEM in this section, Fig. 15.4. The observation shows a smooth fracture surface in Fig. 15.4a, which indicates the typical crack in brittle failure of epoxy resin. Besides, aggregates and minor voids were sparsely located on the surface of MXene/epoxy nanocomposites in Fig. 15.4b. The presence of this agglomeration and voids inevitably reduced the contact area of the epoxy matrix with MXene. It served as crack initiation sites,

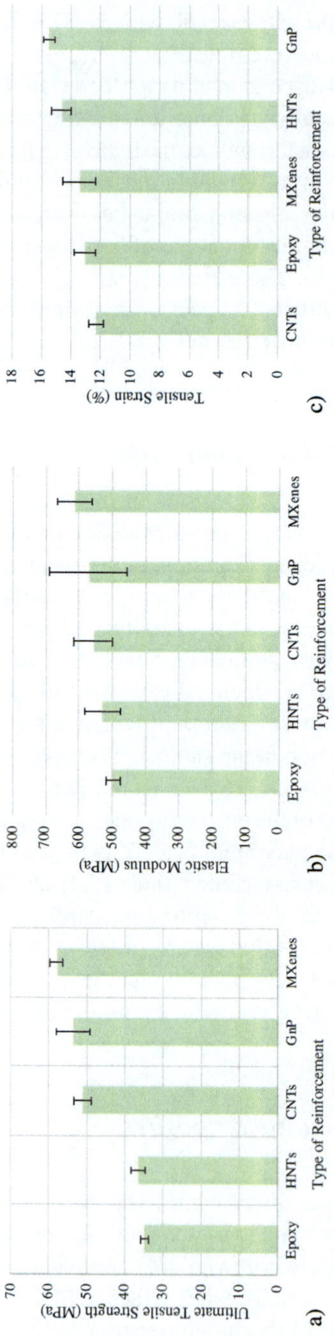

Fig. 15.2 Mechanical properties of epoxy nanocomposite: **a** ultimate tensile strength, **b** elastic modulus, and **c** tensile strain

Table 15.1 Thermal, mechanical properties of epoxy nanocomposites

Sample	Storage modulus (GPa)	T*g* (°C)
Epoxy	26.84	58.20
MXene	33.89	58.85
Graphene	27.87	55.05
CNTs	26.92	48.57
HNTs	24.12	49.98

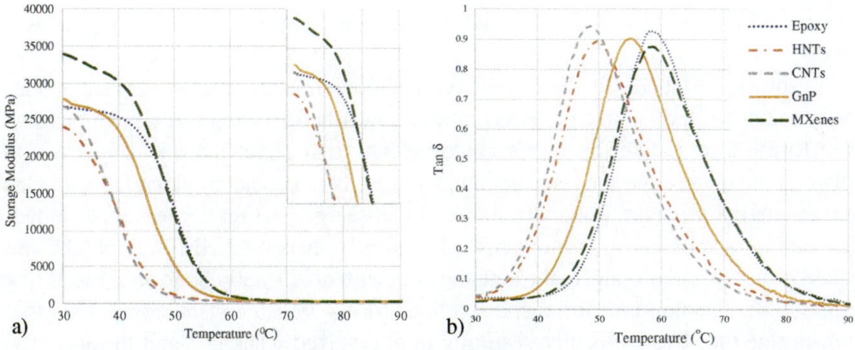

Fig. 15.3 Dynamic mechanical analysis (DMA) of epoxy nanocomposites **a** storage modulus and **b** tan delta

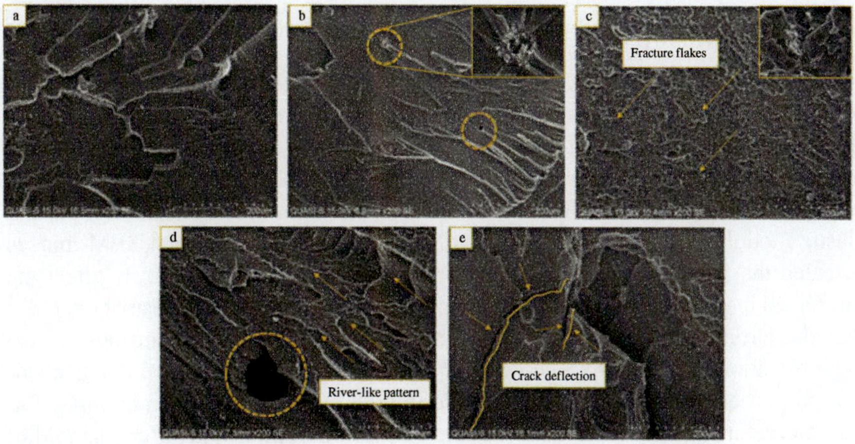

Fig. 15.4 SEM images of fracture surfaces of **a** neat epoxy, **b** MXene, **c** graphene, **d** CNTs and **e** HNTs sample

hence hindering the ability of MXene to reinforce the matrix. These poorly dispersed surfaces indicate a low interfacial interaction between the epoxy matrix and its filler.

A relatively rougher and coarser fracture surface can be seen in Fig. 15.4c, which indicates the toughening mechanism. The fracture flakes obviously can be observed on the surface of nanocomposites. In examining the SEM image of graphene nanocomposites, a large agglomeration was also observed. It suggests the irregular dispersion and distribution of graphene within the epoxy matrix. As reported in the literature [6, 25, 29–34], graphene has a high tendency to re-aggregate in the matrix due to the pi stacking and strong van der Waals force interaction between graphene sheets.

As shown in Fig. 15.4d, the CNTs/epoxy nanocomposite fracture surface is relatively rough, with river-like patterns oriented towards fracturing force and a massive void forming on the surface. This could be due to stronger interfacial interaction between the matrix and the reinforcement since CNTs can act as extended tentacles that entangled with the polymer chains. As observed in Fig. 15.4e, crack deflection can be seen clearly. It indicates some poorly dispersed HNTs as cracks can quickly initiate and propagate through the area with no virtually bond with the epoxy matrix [24]. The non-uniform dispersed filler forms defects in nanocomposites that concentrate the stresses locally, resulting in a localized weakness and lowering the nanocomposites' properties [5].

15.4 Conclusion

Nanofillers dispersion in the matrix plays a vital role in nanocomposite performance. It has also been reported that the degree of dispersion is directly proportional to the material's mechanical properties [24, 35–37]. Four distinct types of nanocomposites were successfully produced. The maximum increment in ultimate tensile strength and elastic modulus were observed in the MXene/epoxy nanocomposite. SEM images revealed that the MXene has excellent properties as defects not severely affect the interfacial interaction and reinforcement of the nanocomposite. The result concludes that the strong van der Waals of the graphene and CNTs will contribute to re-agglomeration in the matrix. Incorporating, nanofillers would improve the mechanical properties of nanocomposite, considering the uniform dispersion is important. Therefore, further research needs to be carried out, as a result would differ if nanofiller dispersion increases and reduces the morphological defects; agglomeration, void, cracks of the nanocomposite.

Acknowledgements The authors wish to express their gratitude to the Centre of Research and Innovation UniKL, and Composites & Simulation Cluster of UniKL MIDI. This paper and its research would not have been possible without the provision of the funding and research facilities.

References

1. Frigione M, Lettieri M (2020) Recent advances and trends of nanofilled/nanostructured epoxies. Materials (Basel) 13:3415–3438
2. Saharudin MS, Atif R, Hasbi S et al (2019) Synergistic effects of halloysite and carbon nanotubes (HNTs + CNTs) on the mechanical properties of epoxy nanocomposites. AIMS Mater Sci 6:900–910
3. Subadra P, Yousef S, Griškevičius P et al (2020) High-performance fiberglass/epoxy reinforced by functionalized CNTs for vehicle applications with less fuel consumption and greenhouse gas emissions. Polym Test 86:106480–106489
4. Campo M, Redondo O, Prolongo SG (2020) Barrier properties of thermal and electrical conductive hydrophobic multigraphitic/epoxy coatings. J Appl Polym Sci 137:49281–49288
5. Wei J, Saharudin MS, Vo T et al (2017) Dichlorobenzene: an effective solvent for epoxy/graphene nanocomposites preparation. R Soc Open Sci 4:170778–170787
6. Wei J, Saharudin MS, Vo T et al (2017) N, N-Dimethylformamide (DMF) usage in epoxy/graphene nanocomposites: Problems associated with reaggregation. Polymers (Basel) 9:193–203
7. de Oliveira JB, Guerrini LM, dos Conejo L, S, et al (2019) Viscoelastic evaluation of epoxy nanocomposite based on carbon nanofiber obtained from electrospinning processing. Polym Bull 76:6063–6076
8. Rafiei HR, Ranjbar Z, Yari H (2018) Modeling of electrical conductive graphene filled epoxy coatings. Prog Org Coatings 125:411–419
9. Wu Y, Zhang X, Negi A et al (2020) Synergistic effects of boron nitride (BN) nanosheets and silver (Ag) nanoparticles on thermal conductivity and electrical properties of epoxy nanocomposites. Polymers (Basel) 12:426–438
10. Yuan Y, Qu Z, Wang Q et al (2020) Reversible nonlinear I-V behavior of ZnO-decorated graphene nanoplatelets/epoxy resin composites. Polymers (Basel) 12:951–963
11. Glaskova-Kuzmina T, Aniskevich A, Zotti A et al (2020) Flexural properties of the epoxy resin filled with single and hybrid carbon nanofillers. J Phys Conf Ser 1431:12001–12006
12. Sanli A, Müller C, Kanoun O et al (2016) Piezoresistive characterization of multi-walled carbon nanotube-epoxy based flexible strain sensitive films by impedance spectroscopy. Compos Sci Technol 122:18–26
13. Spinelli G, Lamberti P, Tucci V et al (2020) Damage monitoring of structural resins loaded with carbon fillers: experimental and theoretical study. Nanomaterials 10:434–449
14. Balguri PK, Samuel DGH, Thumu U (2021) A review on mechanical properties of epoxy nanocomposites. Mater Today Proc 44:346–355
15. Kausar A (2020) Performance of corrosion protective epoxy blend-based nanocomposite coatings: a review. Polym Technol Mater 59:658–673
16. Jing Y, Wang P, Yang Q et al (2020) Molybdenum disulfide with poly(dopamine) and epoxy groups as an efficiently anticorrosive reinforcers in epoxy coating. Synth Met 259:116249–116256
17. Saharudin MS, Inam F (2020) Flexural properties of halloysite nanotubes (HNTS) and carbon nanotubes (CNTS) toughened epoxy composites. Int J Innov Technol Explor Eng 9:1670–1675
18. Abdeen DH, El Hachach M, Koc M et al (2019) A review on the corrosion behaviour of nanocoatings on metallic substrates. Materials (Basel) 12:210–251
19. Hatter CB, Shah J, Anasori B et al (2020) Micromechanical response of two-dimensional transition metal carbonitride (MXene) reinforced epoxy composites. Compos Part B Eng 182:107603–107609
20. Kumar R, Singh RK, Singh DP et al (2017) Laser-assisted synthesis, reduction and micro-patterning of graphene: recent progress and applications. Coord Chem Rev 342:34–79
21. Misra S, Awasthi K, Kumar DR et al (2009) Functionalization effects on the electrical properties of multi-walled carbon nanotube-polyacrylamide composites. J Nanosci Nanotechnol 9:5455–5460

22. Peijs T, Inam F, Wong DWY et al (2010) Multiscale hybrid micro-nanocomposites based on carbon nanotubes and carbon fibers. J Nanomater 2010:1–12
23. Akbari V, Najafi F, Vahabi H et al (2019) Surface chemistry of halloysite nanotubes controls the curability of low filled epoxy nanocomposites. Prog Org Coatings 135:555–564
24. Bourchak M, Nahas MN, Kada B et al (2019) Tensile properties of graphene-based nanocomposites: a comparative study of ultrasonication and microcompounding processing methods. Mech Compos Mater 55:617–626
25. Wei J, Atif R, Vo T et al (2015) Graphene nanoplatelets in epoxy system: dispersion, reaggregation, and mechanical properties of nanocomposites. J Nanomater 2015:1–12
26. Galpaya D, Wang M, George G et al (2014) Preparation of graphene oxide/epoxy nanocomposites with significantly improved mechanical properties. J Appl Phys 116:053518–053527
27. Liao SH, Liu PL, Hsiao MC et al (2012) One-step reduction and functionalization of graphene oxide with phosphorus-based compound to produce flame-retardant epoxy nanocomposite. Ind Eng Chem Res 51:4573–4581
28. Qi B, Lu SR, Xiao XE et al (2014) Enhanced thermal and mechanical properties of epoxy composites by mixing thermotropic liquid crystalline epoxy grafted graphene oxide. Express Polym Lett 8:467–479
29. Inam F, Peijs T (2006) Re-aggregation of carbon nanotubes in two-component epoxy system. J Nanostructured Polym Nanocomposites 2:86–94
30. Lisunova MO, Lebovka NI, Melezhyk OV et al (2006) Stability of the aqueous suspensions of nanotubes in the presence of nonionic surfactant. J Colloid Interface Sci 299:740–746
31. Pourhashem S, Vaezi MR, Rashidi A et al (2017) Exploring corrosion protection properties of solvent based epoxy-graphene oxide nanocomposite coatings on mild steel. Corros Sci 115:78–92
32. Wang X, Xing W, Zhang P et al (2012) Covalent functionalization of graphene with organosilane and its use as a reinforcement in epoxy composites. Compos Sci Technol 72:737–743
33. Yang SY, Lin WN, Huang YL et al (2011) Synergetic effects of graphene platelets and carbon nanotubes on the mechanical and thermal properties of epoxy composites. Carbon N Y 49:793–803
34. Gibson AG, Wan-Jusoh WNB, Kotsikos G (2018) A propane burner test for passive fire protection (PFP) formulations containing added halloysite, carbon nanotubes and graphene. Polym Degrad Stab 148:86–94
35. Jiang Y, Song H, Xu R (2018) Research on the dispersion of carbon nanotubes by ultrasonic oscillation, surfactant and centrifugation respectively and fiscal policies for its industrial development. Ultrason Sonochem 48:30–38
36. Almuhammadi K, Alfano M, Yang Y et al (2014) Analysis of interlaminar fracture toughness and damage mechanisms in composite laminates reinforced with sprayed multi-walled carbon nanotubes. Mater Des 53:921–927
37. Alig I, Pötschke P, Lellinger D et al (2012) Establishment, morphology and properties of carbon nanotube networks in polymer melts. Polymer (Guildf) 53:4–28

Chapter 16
The Degradation of Mechanical Properties Caused by Acetone Chemical Treatment on 3D-Printed PLA-Carbon Fibre Composites

Shakila Ali Nahran, Mohd Shahneel Saharudin, Jaronie Mohd Jani, and Wan Mansor Wan Muhammad

Abstract The paper presents the effect of short-term acetone chemical treatment on PLA-carbon fibre composite mechanical properties, manufactured using the fused deposition modelling (FDM) technique. In this research, PLA-carbon fibre's tensile and dynamic mechanical properties were studied for six different treatment time (0, 60, 100, 120, 180, 220 s). The results suggested that retained acetone significantly reduced the mechanical properties of the PLA-carbon fibre as the treatment time increased. The maximum decrease in tensile strength was observed for the sample treated in acetone for 220 s. The tensile strength decreases about 62% compared to the untreated PLA-carbon fibre composite sample in acetone. SEM images revealed that acetone caused a plasticization effect and caused porosity that acts as a stress concentrator to the composites hence lowering mechanical properties.

Keywords PLA · Carbon fibre · FDM · Acetone · Post-treatment

16.1 Introduction

3D printing, also known as additive manufacturing (AM), refers to processes used to create a three-dimensional object by forming successive layers of material under

S. Ali Nahran (✉) · M. S. Saharudin · J. Mohd Jani
Universiti Kuala Lumpur Malaysia Italy Design Institute, 119, Jalan 7/91, 56100 Cheras, Kuala Lumpur, Malaysia
e-mail: shakila.ali01@s.unikl.edu.my

M. S. Saharudin
e-mail: mshahneel@unikl.edu.my

J. Mohd Jani
e-mail: jaronie@unikl.edu.my

W. M. Wan Muhammad
Mechanical Engineering Section, Universiti Kuala Lumpur Malaysia France Institute, Section 14, Jalan Damai, 43650 Bandar Baru Bangi, Selangor, Malaysia
e-mail: drwmansor@unikl.edu.my

A. Ismail et al. (eds.), *Design in Maritime Engineering*, Advanced Structured Materials 167, https://doi.org/10.1007/978-3-030-89988-2_16

computer control. However, polymer structures generated by additive manufacturing techniques have poor mechanical qualities due to their lack of strength and stiffness, considering pure polymers have low-mechanical properties [1].

The primary aim of incorporating short fibres to the polymer is to increase the strength of the printed object, as pure polymer printed parts have low strength, limiting their usefulness in industrial applications [2]. Carbon fibres significantly improved the strength of the developed model, enhancing both tensile strength and modulus of elasticity [3].

Chemical treatments are commonly used to improve the printed parts' surface quality and enhance the surface fibres for better matrix bonding [2]. Acetone is the most common chemical used to minimize surface roughness. The printed part is wholly immersed into the acetone solution. The surface roughness of the printed part was reduced, and the surface became shinier when it was immersed in the acetone solution, according to Hambali et al. [4]. However, due to the reaction between the solution and the filament, the specimen's tensile strength was lowered by 42.58%.

When acetone dissolves the printed part's surface layer, the dissolved polymer filled in the pores on the surface and the spaces between layers, smoothing up the surface. The treated part's mechanical properties are degraded as a result of this chemical treatment [5].

The purpose of this study was to determine the effect of short-term acetone chemical treatment on the tensile and dynamic properties of PLA-carbon fibre composites. In a previous study by Rasheed et al., acetone was used to disperse graphene in epoxy and was not appropriately removed, which later served as a stress raiser and reduced the mechanical properties of the epoxy/graphene composites formed [6].

16.2 Methodology

In this research, the fused deposition modelling (FDM) technique was used to manufacture the samples for both tensile and dynamic mechanical analysis tests. PLA is a matrix material reinforced by 15% carbon fibre and randomly mixed in PLA pallets to form a filament. In this research, 15% of carbon fibre is filled into 75% of PLA and created as a 1.75 mm diameter filament, with the optimum printing parameters indicated in Table 16.1.

The PLA-carbon fibre samples were produced in two different sizes: ASTM D638 type V for tensile testing and rectangular sample (10 × 35 × 3) mm for dynamic mechanical analysis (DMA). An Instron universal testing machine with a load cell of 30 kN was used with a crosshead speed 1 mm/min. The dynamic mechanical test was run at a heating rate of 3 mm/min from 30 to 12 °C. Figure 16.1 shows the dimension of the tensile testing sample and the DMA sample dimension.

Figure 16.2 shows the 3D printer FDM technology used named Ultimaker S3 3D Printer and the printed samples for tensile test and DMA of 15% carbon fibre reinforced PLA composite using Ultimaker S3.

Table 16.1 Parameter used to produce samples of PLA-carbon fibre (15%) composites

Parameter	Sample Type 1	Sample Type 2
Material	PLA-carbon fibre (15%) composites filament	
Testing machine	Tensile testing	DMA
Standard size	ASTM D638 type V	(10 × 35 × 3) mm
Infill density (%)	100	100
Nozzle used (mm)	0.4	0.4
Layer height (mm)	0.2	0.2
Speed (mm/s)	40	40
Time taken per piece (min)	25	16
Weight per piece (g)	2	1
Total samples (pieces)	30	6

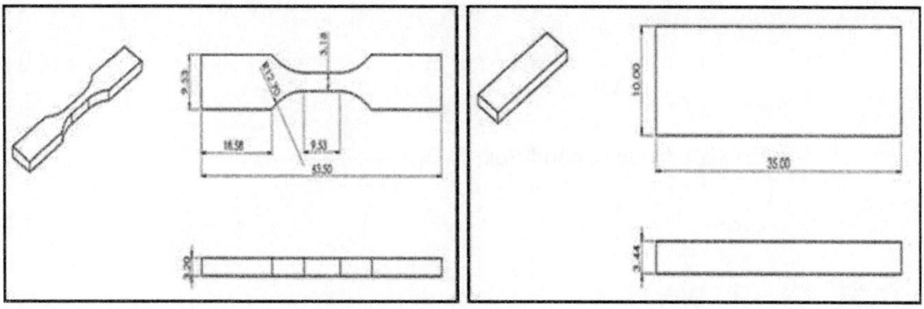

Fig. 16.1 Tensile testing sample dimension ASTM D638 type V (left) and dynamic mechanical analysis (DMA) sample dimension

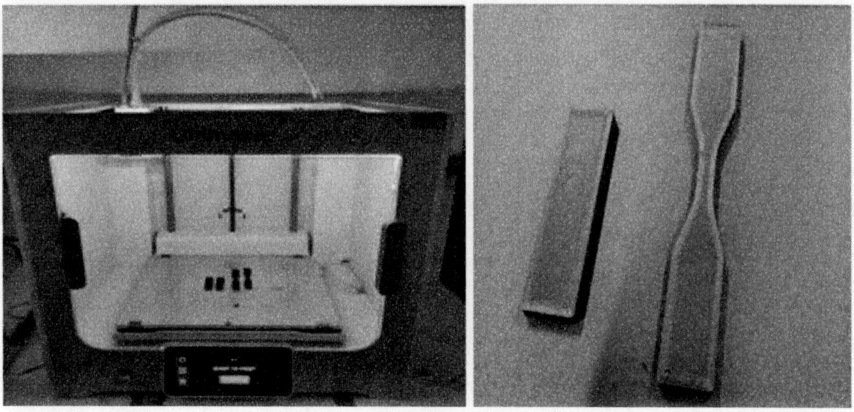

Fig. 16.2 Ultimaker S3 3D printer (left) and 3D-printed PLA-carbon fibre (15%) DMA sample and tensile testing sample using Ultimaker S3

The prepared samples of PLA-carbon fibre were taken into a beaker and fully immersed and soaked into 99% acetone concentration with different soaking time [7] (0, 60, 100, 120, 180, and 220 s). The samples slowly absorb acetone from the surface into the inside layers. The acidic nature of acetone causes molecules to become closer together, arranging them in layers of material and closing gaps, promoting molecular bonding and strengthens the material [7]. The soaked samples were taken and dried in the electrothermal blowing dry box to dry off excess acetone at 120 °C with 120 s of heating time [7]. The prepared 3D printing samples are post-treated and tested to determine their tensile strength longitudinally along the machine axis pushed up to sample breakage. The machine cross heads grips were introduced and used to attach the samples. The machine was connected to a computer interface. The measured force (F) and the sample's minimum section area ($A = 9.54$ mm^2) were used to calculate stress values, and $L_0 = 9.53$ mm is the sample's gauge length [8].

$$\text{Stress}, \sigma = \frac{F}{A} \tag{16.1}$$

$$\text{Young}'\text{sModulus}, E = \frac{FL}{A(\Delta L - L)} \tag{16.2}$$

16.3 Results and Discussion

Figure 16.3 shows stress–strain curves for PLA-carbon fibre samples at different chemical treatment times. Generally, it was observed that the unexposed PLA-carbon fibre sample has the highest tensile stress followed by other samples treated with

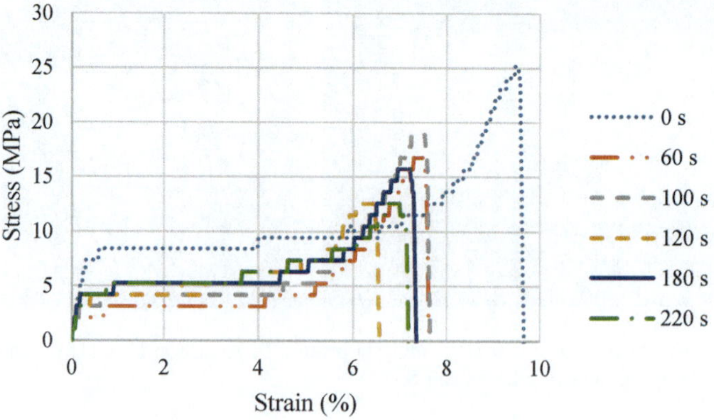

Fig. 16.3 Comparison of stress–strain PLA-carbon fibre (15%) samples with different time of samples soaked into acetone (in seconds)

acetone. However, the chemical treatment caused the tensile stress to drop compared to the untreated sample slightly. This can be associated with the deterioration of samples when exposed to acetone due to liquid penetration into the matrix.

Figure 16.4 shows the ultimate tensile stress, Young's modulus and dynamic mechanical analysis of the PLA-carbon fibre samples. In general, the tensile properties of the samples drop as the soaking time increase. The storage modulus, loss

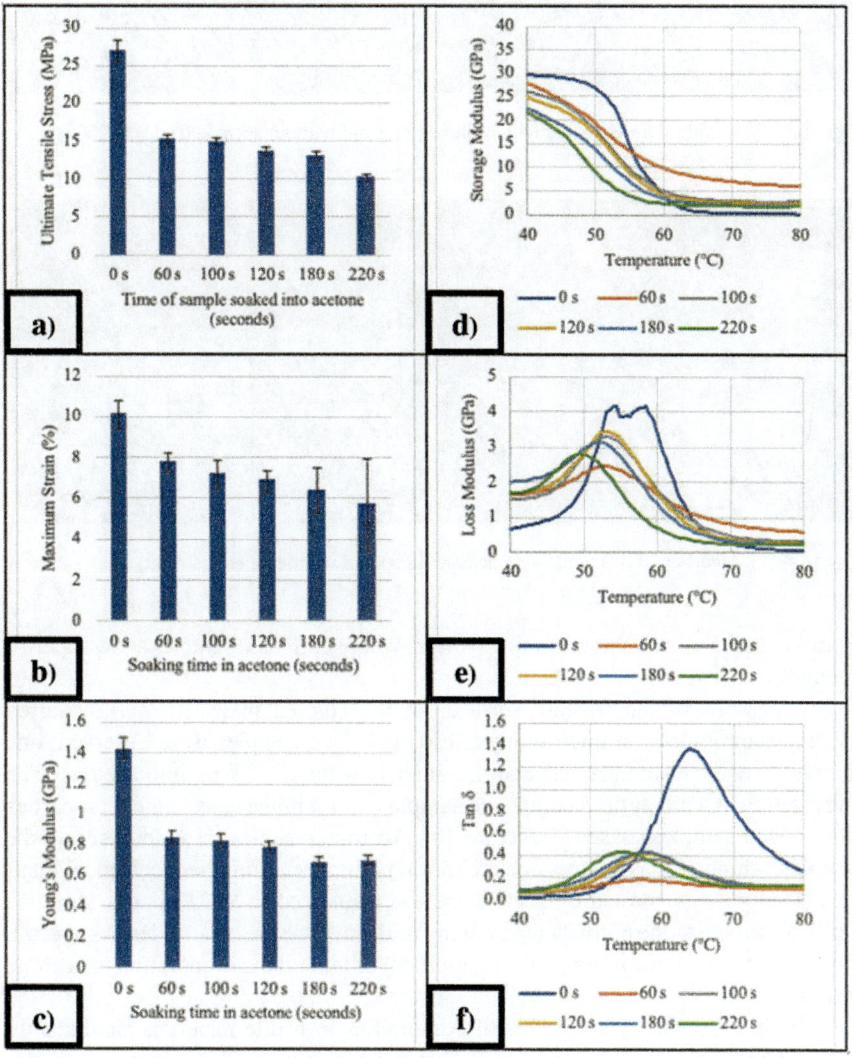

Fig. 16.4 Tensile properties: **a** ultimate tensile stress, **b** maximum strain, **c** Young's modulus, **d** storage modulus, **e** loss modulus, and **f** tan δ

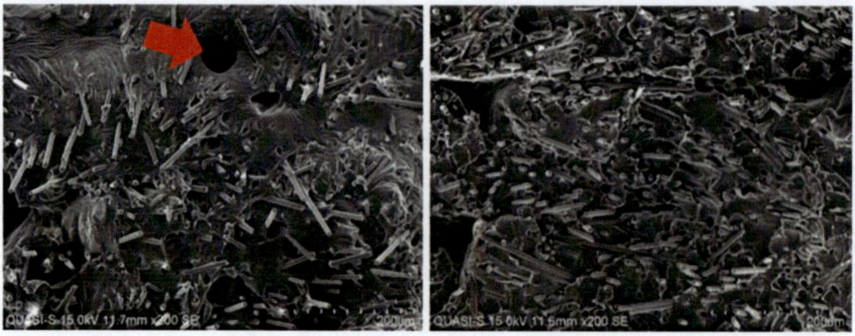

Fig. 16.5 PLA-carbon fibre sample with acetone (left) and without acetone (right) at 200 μm

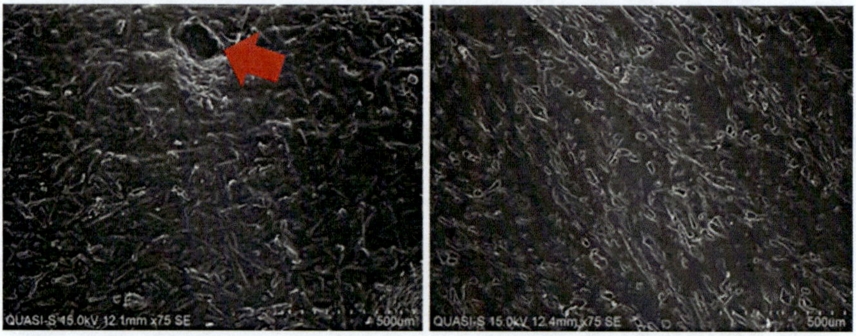

Fig. 16.6 PLA-carbon fibre sample with acetone (left) and without acetone (right)

modulus and tan δ also show a similar pattern, where untreated samples show superior strength.

Micrographs of the fracture surfaces of the studied materials were acquired using scanning electron microscopy (SEM) [9]. Two samples were observed: one sample of with soaked acetone and the other is without soaking into acetone. The untreated PLA-carbon fibre composite sample exhibit higher mechanical properties than treated samples soaked in acetone [6]. Acetone residues create porosity in the samples, which acts as a stress concentration point, impairing the mechanical qualities. The porosity formed by acetone can be visualized in SEM images. Acetone fluids contribute to the porosity by applying uniform pressure on all sides. As a result, this round porosity is caused by the entrapment of air and the evaporation of acetone residues.

Griffith crack theory states that discontinuities in brittle materials significantly reduce their mechanical properties [10]. Due to the fragility of PLA, the stress concentration generated by residual acetone is high, degrading the mechanical properties of PLA nanocomposites drastically.

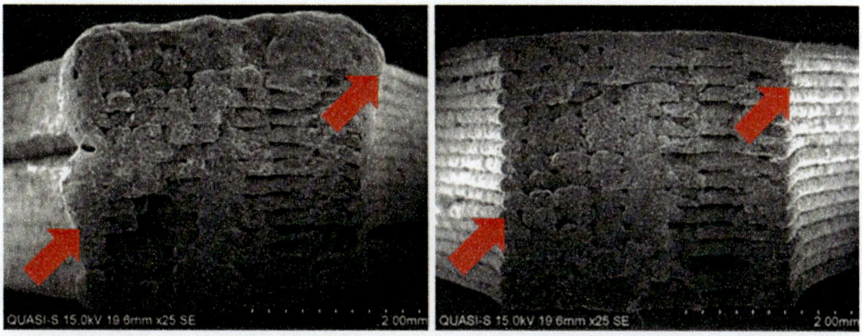

Fig. 16.7 Fractured surface on PLA-carbon fibre treated with acetone (left) and without acetone treatment (right)

As shown in Figs. 16.5 and 16.6, porosity was observed. This porosity could be due to the plasticization effect caused by acetone. Thus, the presence of acetone affected not only the fracture pattern but also the mechanical properties.

Figure 16.7 shows the fractured surface of PLA-carbon fibre composites at low magnification. The current study found that after acetone treatment, the structural integrity of the 3D printed samples were somehow affected. Swelling of the polymer matrix can be seen, especially at the edge of the samples as shown by the arrows, even though carbon fibre is high strength, stress resistance, corrosion resistance and hydrophobic in nature. The PLA matrix seems to be affected by the acetone chemical treatment.

The post-heat treatment of 120 s was not able to completely remove the remaining acetone from the 3D printed samples. As a result, the PLA-carbon fibre composites failed due to matrix dominated failure.

16.4 Conclusion

The following results were drawn from the current post-treatment study work on carbon fibre infused PLA composite filament. The tensile properties were decreased due to the chemical treatment by soaking the PLA-carbon fibre composites in acetone. When treated in acetone, the maximum tensile strength decrease was about 62% and the maximum Young's modulus reduction was about 50%. It can be concluded that PLA-carbon fibre samples treated with acetone need more time in the post-heat treatment, more than 120 s. In the future studies, it is recommended to investigate further the relationship between chemical treatment duration and surface quality of PLA-carbon fibre composites. The liquid barrier property of PLA-carbon fibre exposed to different chemical treatment and time can also be studied to investigate the role of carbon fibre in the tortuosity effect.

Acknowledgements The authors would like to thank Universiti Kuala Lumpur, Malaysia Italy Design Institute for the provision of research facilities, without which the analysis of relevant data was not possible. The authors acknowledge the financial support provided by the Centre of Research and Innovation (CoRI) Universiti Kuala Lumpur (UniKL/CoRI/UER20010).

References

1. Maqsood N, Rimašauskas M (2021) Characterization of carbon fiber reinforced PLA composites manufactured by fused deposition modeling. Compos Part C: Open Access. https://doi.org/10.1016/j.jcomc.2021.100112
2. Wickramasinghe S, Do T, Tran P (2020) FDM-based 3D printing of polymer and associated composite: a review on mechanical properties, defects and treatments. Polymers 12:1–42. https://doi.org/10.3390/polym12071529
3. Saharudin MS, Hajnys J, Kozior T et al (2021) Quality of surface texture and mechanical properties of PLA and PA-based material reinforced with carbon fibers manufactured by FDM and CFF 3D printing technologies. Polymers 13:1671. https://doi.org/10.3390/polym13111671
4. Hambali RH, Cheong KM, Azizan N (2017) Analysis of the influence of chemical treatment to the strength and surface roughness of FDM. In: IOP conf. ser. mater. sci. eng. https://doi.org/10.1088/1757-899X/210/1/012063
5. Galantucci LM, Lavecchia F, Percoco G (2010) Quantitative analysis of a chemical treatment to reduce roughness of parts fabricated using fused deposition modeling. CIRP Ann Manuf Technol 59:247–250. https://doi.org/10.1016/j.cirp.2010.03.074
6. Atif R, Shyha I (2016) Inam F (2016) The degradation of mechanical properties due to stress concentration caused by retained acetone in epoxy nanocomposites. RSC Adv 6:34188–34197. https://doi.org/10.1039/c6ra00739b
7. Guduru KK, Srinivasu G (2020) Effect of post treatment on tensile properties of carbon reinforced PLA composite by 3D printing. Mater Today 33:5403–5407. https://doi.org/10.1016/j.matpr.2020.03.128
8. Wang J, Xie H, Weng Z et al (2016) A novel approach to improve mechanical properties of parts fabricated by fused deposition modeling. Mater Des 105:152–159. https://doi.org/10.1016/j.matdes.2016.05.078
9. Ferreira RTL, Amatte IC, Dutra TA et al (2017) Experimental characterization and micrography of 3D printed PLA and PLA reinforced with short carbon fibers. Compos B Eng 124:88–100. https://doi.org/10.1016/j.compositesb.2017.05.013
10. Griffith AAVI (1921) The phenomena of rupture and flow in solids. Philos Trans R Soc A 221:163–198. https://doi.org/10.1098/rsta.1921.0006

Chapter 17
Tensile and Morphology Analysis of Oil Palm Trunk Specimen Reinforced Epoxy Fabricated via Vacuum-Assisted Resin Transfer Moulding

Wan Nur Fatihah Amirah Nik Wan @ Wan Senik, Anuar Abu Bakar, Suriani Mat Jusoh, Asmalina Mohamed Saat, Zaimi Zainal Mukhtar, Ahmad Fitriadhy, Wan Mohd Norsani Wan Nik, and Mohd Shukry Abdul Majid

Abstract The use of natural plant fibres, especially palm trees, benefits from reinforcement with polymer composites to enhance material properties that can be widely used in various applications. Considering awareness in many areas likely safety, environmental impact, health issues and saving for composite structures, the tensile properties of untreated oil palm trunk (OPT) and oil palm trunk reinforced with

W. N. F. A. Nik Wan @ Wan Senik · A. Abu Bakar (✉) · S. Mat Jusoh · Z. Zainal Mukhtar · A. Fitriadhy · W. M. N. Wan Nik
Faculty of Ocean Engineering Technology and Informatics, Universiti Malaysia Terengganu, Kuala Nerus, 21030 Terengganu, Malaysia
e-mail: anuarbakar@umt.edu.my

S. Mat Jusoh
e-mail: surianimatjusoh@umt.edu.my

Z. Zainal Mukhtar
e-mail: zaimi@unikl.edu.my

A. Fitriadhy
e-mail: a.fitriadhy@umt.edu.my

W. M. N. Wan Nik
e-mail: niksani@umt.edu.my

A. Abu Bakar
Faculty of Maritime Studies, Universiti Malaysia Terengganu, Kuala Nerus, 21030 Terengganu, Malaysia

A. Mohamed Saat · Z. Zainal Mukhtar
Universiti Kuala Lumpur Malaysia Institute of Marine Engineering Technology, Dataran Industri Teknologi Marin, Bandar Teknologi Maritim, Jalan Pantai Remis, 32200 Lumut, Perak, Malaysia
e-mail: asmalina@unikl.edu.my

M. S. Abdul Majid
Faculty of Mechanical Engineering Technology, Universiti Malaysia Perlis, 02600 Arau, Perlis, Malaysia
e-mail: shukry@unimap.edu.my

A. Ismail et al. (eds.), *Design in Maritime Engineering*, Advanced Structured Materials 167, https://doi.org/10.1007/978-3-030-89988-2_17

epoxy (OPTE) composites have been investigated. The untreated OPT and OPTE were varied with angle fibre orientation 0° zone I, 0° zone II, 0° zone III, 30°, 45°, 60° and 90°, respectively. The OPTE composites were fabricated using the vacuum-assisted resin transfer moulding (VARTM) technique. The mechanical properties were analysed by ASTM D3039. The scanning electron microscopy (SEM) analysis results show the different failure modes of tensile specimens. The VARTM technique shows promising tensile strength results when compared to the available data. The results show the increase of tensile strength in OPTE. The 0° of OPTE composite exhibited the highest tensile strength.

Keywords Wood polymer · Biomass waste · Fibre orientation angle · Wood substitute · Natural fibre

17.1 Introduction

Malaysia is well known as one of the world's largest exporters of crude palm oil. As a result, oil palm trees are readily found in Malaysia. The Malaysian Palm Oil Board in 2019 reported that there were 5,900,157 hectares of oil palm plantations in Malaysia [1]. Generally, due to low fruit yield and oil productivity, oil palm trees need to be replaced through replantation after 25 years of lifespan [2].

A major problem in the oil palm cultivation and its related industries is due to the massive amount of biomass waste such as empty fruit bunches (EFB), palm kernel shell (PKS), mesocarp fibre (MF), palm oil mill effluent (POME), oil palm trunks (OPT), oil palm leaves (OPL) and oil palm fronds (OPF) [3]. The oil palm trunks are available during the replanting process [4]. The chopped OPT is usually decomposed naturally or let to be burnt in the plantation area [5]. The OPT can also be developed to turn waste into value-added products.

The utilisation of biomass OPT can be converted into value-added products. In the previous studies, OPT was utilised for the production of new palm-wood material using phenol formaldehyde and by using a high vacuum pump. The results showed better mechanical properties as compared to dried OPT and rubberwood [6]. In the modification of the inner part of OPT with OPS, nanoparticles impregnated with phenol formaldehyde revealed that this might be a good process for the treatment of the inner part of OPT [7]. In a recent study by [8], OPT veneers could be bound without any synthetic adhesive by using the response surface methodology. The results indicated that bending and shear block strength of the panel from optimal manufacturing conditions were 36.07 MPa and 5.20 MPa, respectively with 35.4% of delamination. The impregnation of the OPT and polymer composites were prepared from a combination of the OPT with phenol formaldehyde (PF) and urea formaldehyde (UF) resin in different resin percentages using an impregnation method. The results revealed the dimensional stability of the OPT polymer composites with the highest resin loading being slightly lower when compared to rubberwood [9].

Recently, the vacuum-assisted resin transfer moulding (VARTM) technology has promoted better technologies in fabricating composites. Some research has been conducted using the VARTM technique. In the research conducted by [10], the $CaCO_3$ nanoparticles impregnated kenaf fibres through the vacuum resin infusion process significantly improved the interfacial compatibility between fibres and resin matrix based on the mechanical properties of these composites. The research by [11] showed the bending behaviour of the grafting of KH-560 on jute fabric improved the interfacial compatibility and mechanical properties of laminated composite by using the vacuum resin infusion method. In addition, the research by [12], glass fibres (epoxy)—aluminium hybrid laminated composite by VARTM resulted in displaying excellent mechanical properties. A recent study conducted by [13] involved fibre-reinforced plastic (FRP) confined concrete by using vacuum infusion process that significantly improved the compression strength. The research conducted by [14] solid wood was impregnated by using the vacuum-assisted resin infusion process for manufacturing woody materials and the result showed excellent flexural performances. Therefore, by using this method the mechanical properties of the composite were improved. However, no studies regarding the use of VARTM technology on OPT have been reported.

In this research, oil palm trunks reinforced with epoxy composites were prepared. Different angles with respect to the fibre orientation were tested for tensile properties. The tensile behaviour of untreated OPT and OPTE was compared. The fracture surfaces were observed by using SEM.

17.2 Methodology

17.2.1 Raw Material Preparation

The 25 years old of oil palm trees were obtained from Terengganu, Malaysia, during replanting activities. The sample was taken 2 m from the ground. The bottom of the oil used in this research is as shown in Fig. 17.1. The bottom of the OPT was cut into panels approximately 900 mm (L) × 300 mm (T) × 30 mm (R) as shown in Fig. 17.2.

The epoxy resin used in this research was an epoxy infusion resin (Kinetix R118, ATL Composites, Ltd.) and hardener (Kinetix H120, ATL Composites, Ltd.). Hardener H120 cures at room temperature without post-cure treatment. Epoxy and hardener were prepared with a weight ratio of 4:1.

Fig. 17.1 **a** Oil palm trees. **b** Bottom of the OPT. **c** The cut panel of the OPT before further processing

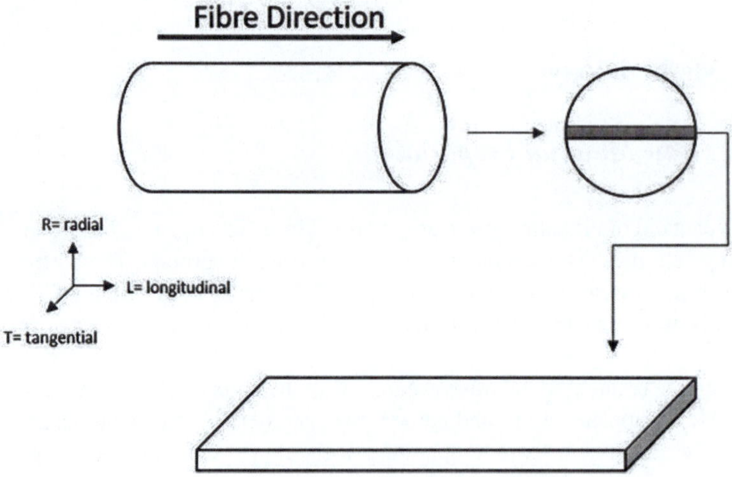

Fig. 17.2 Cutting process of the OPT

17.2.2 Drying Process

The drying process of the OPT was carried out to ensure the OPT reached the bone-dry mass before the fabrication process. The OPT was dried under open-air drying for 14 days before further processing. Before the drying process in the dryer, the OPT panels were cut into panels approximately 900 mm (*L*) × 300 mm (*T*) × 30 mm (*R*). The OPT panels were dried in a dryer at about 40 °C for 24 h until it reached a constant weight.

17.2.3 Fabrication Process

The OPTE was prepared with VARTM as shown in Fig. 17.3. The OPT was set up on a mould. In this process, a high tempered glass table acted as the mould. Before setting up the OPTE on the mould, the mould was wiped carefully in a circular motion with acetone, followed by a wax mould release agent to ensure the mould was free from dust and any particles [15]. The peel ply was laid over the OPT to improve the surface finish. The mesh flow was placed on the top of the peel ply to assist the resin flow during the process.

The inlet and outlet were connected to the laminate specimen. The inlet and outlet fitting pipes were wrapped with a spiral tube covered with a chopped strand mat (CSM) and were connected to the laminate OPT specimen. A vacuum bag was placed over the mould and sealed carefully by tacky tape. The vacuum bag was 20–30% bigger than the actual size of the mould. Pleats were created in the vacuum bag where required.

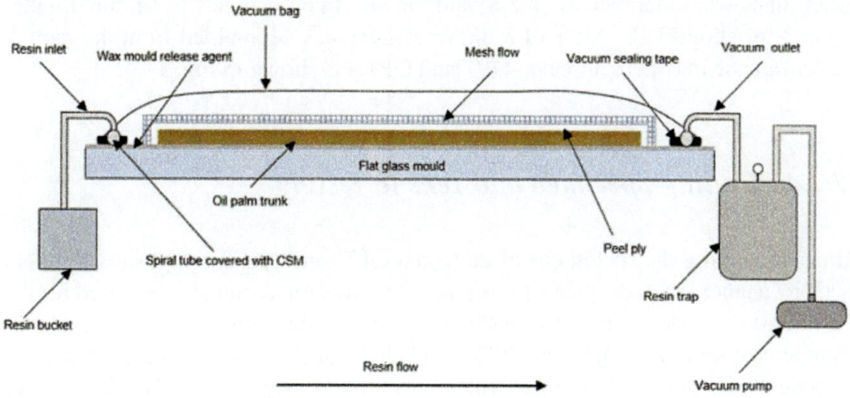

Fig. 17.3 Fabrication process of the OPTE by using the VARTM technique

Fig. 17.4 Difference between untreated OPT and OPTE

Once the set-up was complete, the resin inlet was clamped using a clamp and the resin outlet was connected to the pressure pot of the vacuum pump to set up the drop test and vacuum pressure was applied using a vacuum pump.

After no leaking of air was observed, the pump was switched off, after 20 min, the bag area was rechecked for any leakage. Once the set-up had passed the drop test, the inlet tube was immersed in the resin pot and unclamped to allow the epoxy flow to the system. The epoxy resin flowed parallel to the fibre direction of the OPT. The infusion continued at room temperature for approximately 2 h. The inlet and outlet tubes were clamped after 2 h, and the specimen was left to be cured under vacuum for about 24 h. After 24 h, the specimen was demoulded from the mould. The difference between untreated OPT and OPTE is shown in Fig. 17.4.

17.2.4 Cutting Specimen and Tensile Testing

After fabrication, the specimens of untreated OPT and OPTE composites were cut by using a tenoning and squaring machine into a specific dimension required for the tensile test. In order to investigate the influence of fibre orientation on the tensile behaviour of untreated OPT and OPTE, the OPT panels were cut at the orientation of angle 0°, 30°, 45°, 60° and 90° with respect to the longitudinal fibre direction of OPT as shown in Table 17.1.

Three zones were divided into tensile parallel fibre orientation angle 0° which are zone I (outer zone), zone II (middle zone) and zone III (inner zone). The illustration

Table 17.1 Configuration of tensile experiment of different orientation angle

Fibre orientation condition	Initial	Orientation fibre angle (°)	Zone			Untreated OPT	OPTE composite
			I	II	III		
Parallel to longitudinal fibre orientation angle (0°)	RZI-0	0	/			/	
	RZII-0	0		/		/	
	RZIII-0	0			/	/	
	EZI-0	0	/				/
	EZII-0	0		/			/
	EZIII-0	0			/		/
Angle fibre orientation	RZ-30	30				/	
	EZ-30	30					/
	RZ-45	45				/	
	EZ-45	45					/
	RZ-60	60				/	
	EZ-60	60					/
	RZ-60	90				/	
	EZ-60	90					/

of cutting specimens for different orientation for the tensile behaviour is shown in Fig. 17.5a, b.

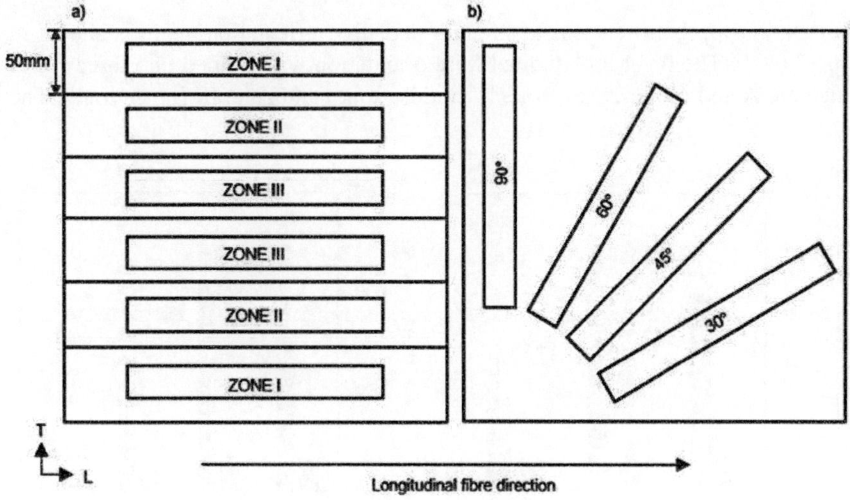

Fig. 17.5 **a** Cutting specimen of zone I (outer), zone II (middle), zone III (inner) of the orientation of angle 0°. **b** The cutting specimen of 30°, 45°, 60° and 90°

The tensile test of untreated OPT and OPTE specimens was performed according to ASTM D3039 using a Shimadzu AGX-50 kN tensile testing machine. The dimension of each flat rectangular specimen for the tensile test was approximately 250 mm (length) × 25 mm (width) × 10 mm (thickness). The gauge length of the specimens was 160 mm. Five specimens of each fibre orientation were tested to obtain the valid results. The crosshead speed was performed in quasi-static condition at 5 mm/min for all test specimens. This was done to reduce the effect of strain rate during the test.

17.2.5 Scanning Electron Microscopy (SEM)

The surface and fracture morphology of the untreated OPT and OPTE were observed by using a SEM (HITACHI model TM 3000). The scanning images were obtained with accelerating voltages of 15 kV.

17.3 Results and Discussion

17.3.1 Tensile Strength of Untreated OPT and OPTE at Different Angles of Longitudinal Fibre Orientation

The relationship between tensile strength and fibre orientation angle is shown in Fig. 17.6a, b. The 0° of longitudinal fibre orientation was divided into three zones which are zone I (outer zone), zone II (middle zone) and zone III (inner zone). The

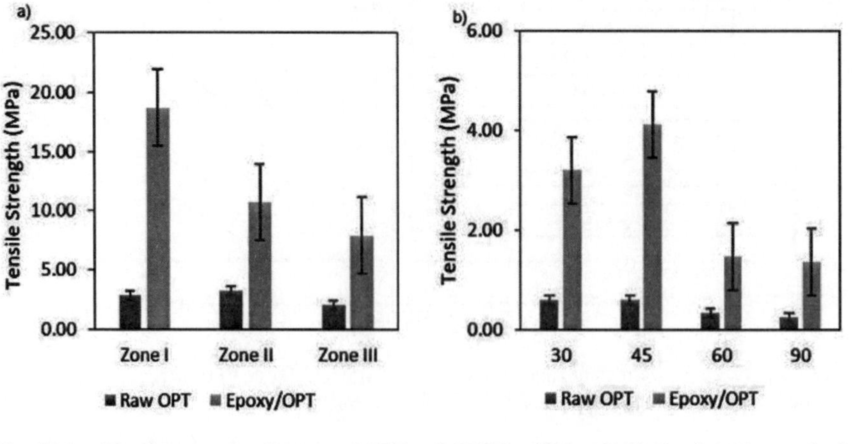

Fig. 17.6 **a** Tensile strength of untreated OPT and OPTE at 0° longitudinal orientation angle. **b** The tensile strength of untreated OPT and OPTE at 30°, 45°, 60° and 90°

average tensile strength of untreated OPT and OPTE at zone I was 2.86 MPa and 18.68 MPa, respectively, corresponding to an 84.68% increase. At zone II, the average tensile strength of untreated OPT and OPTE was 3.25 MPa and 10.70 MPa, respectively, showing 69.63% higher stress compared to untreated OPT at zone II. The average tensile strength at zone III for untreated OPT and OPTE was 1.99 MPa and 7.91 MPa, respectively, corresponding to OPTE being 74.84% higher than untreated OPT at zone III.

Meanwhile, similar results were shown in untreated OPT at zone I and zone II. Therefore, the OPTE at zone I, the outer zone, exhibited the highest tensile strength followed by zone II and zone I. This is due to the previous study by [16, 17] which concluded that the mechanical properties of an OPT in the outer (periphery) zone were better than the inner (central) zone of the OPT.

Average tensile strength of R-30 and E-30 was 0.59 MPa and 3.19 MPa, respectively. Next, the average tensile strength at R-45 and E-45 exhibited 0.59 MPa and 4.12 MPa, respectively. At 60° angle direction, the tensile strength of R-60 and E-60 was 0.33 MPa and 1.47 MPa, respectively. The average tensile strength at R-90 and E-90 was 0.25 and 1.36 MPa.

The tensile strength strongly depends on the fibre orientation angle. The fibre orientation angle 0° zone I exhibited the highest values as compared to other angles.

17.3.2 *Tensile Fracture by Scanning Electron Microscopy (SEM)*

The morphology and fracture surface of untreated OPT and OPTE composites are investigated in this study by using SEM. The number of vascular bundles in three different zones is zone I, zone II and zone III as shown in Fig. 17.7a–c, respectively. Therefore, the increase of tensile strength is due to the higher content of vascular bundles. The vascular bundles contain fibres, phloem and xylem that congest higher

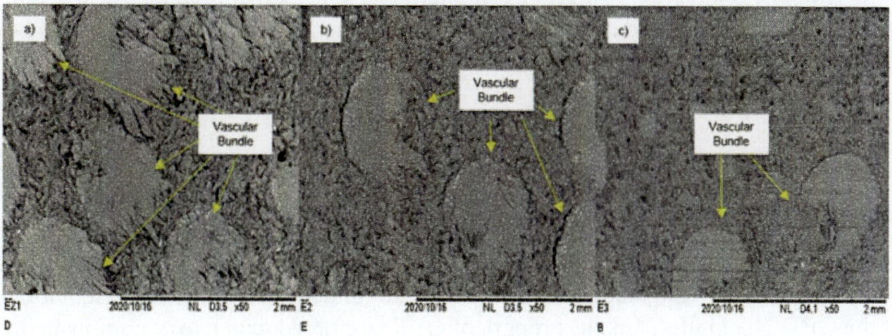

Fig. 17.7 **a** Number of vascular bundles in the outer zone. **b** Number of vascular bundles in the middle zone. **c** Number of vascular bundles in the inner zone

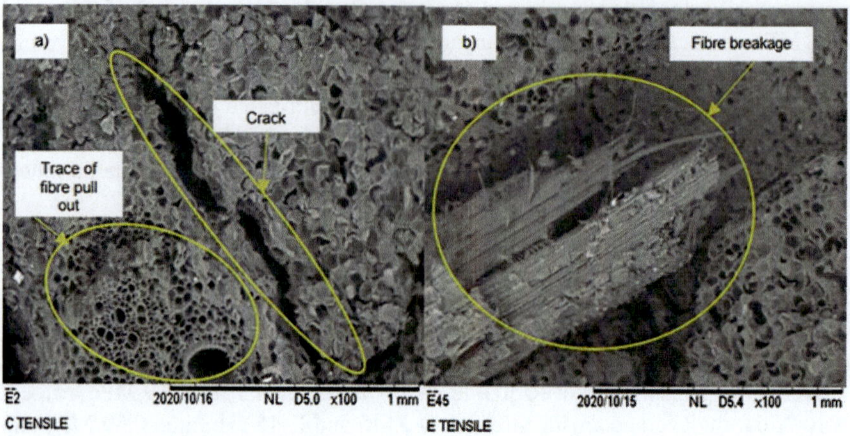

Fig. 17.8 **a** Image SEM revealed traces of fibre pull out and crack propagation in the middle of OPTE composite. **b** Fibre breakage at the 45° cutting fibre orientation angle of OPTE composite

in the outer zone (periphery) and gradually decrease in the inner (central) zone [17]. Therefore, the SEM image of EZI-0 D as shown in Fig. 17.7a revealed that the increased number of vascular bundles, resulted in an increase in the tensile strength.

The analysed SEM image shows traces of fibre pull-out, crack deflection and fibre breakage in the middle zone and 45° cutting fibre orientation angle of OPTE composite, respectively, in Fig. 17.8a, b. This indicates a weak interaction between the OPT and epoxy matrix. The trace of fibre pull-out is clearly shown in the middle zone of the OPTE composite. This shows the weak interfacial bonding between OPT and epoxy matrix. The study report by [18] stated that the fibre pull-out formed due to weak bonding between the sugar palm fibres and the epoxy matrix.

The 30° and 60° cutting fibre orientation angles reveal a misalignment of fibre. It can be observed that there is a presence of misalignment of fibres and entanglement in Fig. 17.9a, b. The misalignments are important in determining the mechanical properties of composites [19]. The weak interfacial adhesion of fibre matrix was due to the misalignment of fibres in natural fibre reinforced composites that can also be the result of the fibre matrix [20].

17.4 Conclusion

The tensile properties and morphological properties of untreated OPT and OPTE composite were investigated. The results are also influenced by the fibre orientation of the OPT. The highest tensile strength of OPT reinforced with epoxy composite in 0° longitudinal fibre angle at zone I (outer zone) is 18.68 MPa. The tensile strength of OPT reinforced with epoxy composite in 0° longitudinal fibre angle shows better

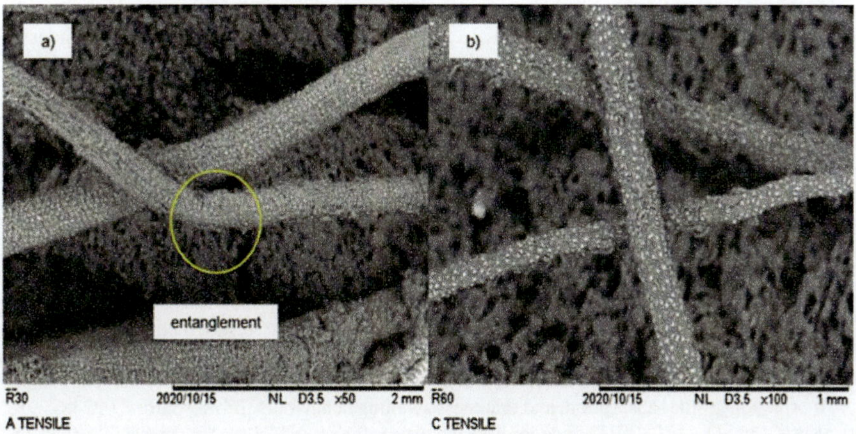

Fig. 17.9 Misalignment of fibre in untreated OPT at the 30° and 60° cutting fibre orientation angle

tensile strength and is comparable with the tensile strength of pure hardwood Meranti Light Red sapwood (*Shorea* spp.) that is commonly used in Malaysia as structural wood [21]. This indicates that the OPTE can be utilised for the fabrication to replace the available wood. The outer zone of OPTE exhibited the highest tensile strength as a result of the fibre content in the outer zone. This was confirmed by the SEM observations on different zones of OPT. Decreased tensile strength of OPT and epoxy composites was due to the low amount of epoxy content penetrating to the OPT during the fabrication process and the low amount of the vascular bundles in the OPT area. Thus, the study shows very promising results that OPTE using VARTM is capable of enhancing tensile strength. This technique can be further studied and improved for commercial application to make use of the abundant available resources.

Acknowledgements This research was supported by the Ministry of Higher Education Malaysia, through the Fundamental Research Grant Scheme (Ref: FRGS/1/2019/TK10/UMT/02/05). The authors would like to thank the Faculty of Ocean Engineering Technology and Informatics, Universiti Malaysia Terengganu and Malaysian Institute of Marine Engineering Technology (UniKL MIMET) for the support of the facility during the course of the research.

References

1. Malaysian Palm Oil Board Economics and Industry Development Division. Oil Palm Planted Area by Category 2019. http://bepi.mpob.gov.my/index.php/en/area/area-2019/oil-palm-planted-area-as-at-dec-2019.html. Accessed 23 Nov 2020
2. Yamada H, Tanaka R, Sulaiman O et al (2010) Old oil palm trunk: a promising source of sugars for bioethanol production. Biomass Bioenergy 34:1608–1613
3. Mushtaq F, Tuan ATA, Mat R et al (2015) Optimization and characterization of bio-oil produced by microwave assisted pyrolysis of oil palm shell waste biomass with microwave absorber. Bioresource Technol 190:442–450

4. Awalludin MF, Sulaiman O, Hashim R et al (2015) An overview of the oil palm industry in Malaysia and its waste utilization through thermochemical conversion, specifically via liquefaction. Renew Sust Energ Rev 50:1469–1484
5. Lim KO, Ahmaddin FH, Malar VS (1997) A note on the conversion of oil-palm trunks to glucose via acid hydrolysis. Bioresource Technol 59:33–35
6. Khalil HPSA, Bhat AH, Jawaid M et al (2010) Agro-wastes: mechanical and physical properties of resin impregnated oil palm trunk core lumber. Polym Compos 1–7
7. Dungani R, Islam MN, Abdul Khalil HPS et al (2014) Modification of the inner part of the oil palm trunk (OPT) with oil palm shell (OPS) nanoparticles and phenol formaldehyde (PF) resin: Physical, mechanical, and thermal properties. BioResources 9:455–471
8. Saari N, Lamaming J, Hashim R et al (2020) Optimization of binderless compressed veneer panel manufacturing process from oil palm trunk using response surface methodology. J Clean Prod 265
9. Abdullah CK, Jawaid M, Abdul KHPS et al (2012) Oil palm trunk polymer composite: morphology, water absorption, and thickness swelling behaviours. BioResources 7:2948–2959
10. Xia C, Shi SQ, Cai L (2015) Vacuum-assisted resin infusion (VARI) and hot pressing for $CaCO_3$ nanoparticle treated kenaf fiber reinforced composites. Compos Part B Eng 78:138–143
11. Wang X, Shi XL, Meng QK et al (2020) Bending behaviors of three grid sandwich structures with wood facing and jute fabrics/epoxy composites cores. Compos Struct 252:112666
12. Vasudevan A, Navin KB, Victor DM et al (2020) Tensile and flexural behaviour of glass fibre reinforced plastic—Aluminium hybrid laminate manufactured by vacuum resin transfer moulding technique (VARTM). Mater Today Proc 37:2132–2140
13. Mukhtar ZZ, Bakar AA, Fitriadhy A et al (2018) Experimental analysis of FRP confined concrete for underwater application. In: Advancement in emerging technologies and engineering applications, pp 65–77
14. Xia C, Wu Y, Qiu Y et al (2019) Processing high-performance woody materials by means of vacuum-assisted resin infusion technology. J Clean Prod 241:118340
15. Salman SD, Sharba MJ, Leman Z et al (2015) Physical, mechanical, and morphological properties of woven Kenaf/polymer composites produced using a vacuum infusion technique. Int J Polym Sci 2015:1–10
16. Erwinsyah (2008) Improvement of oil palm wood properties using bioresin. Technische Universität Dresden
17. Srivaro S, Matan N, Lam F (2018) Property gradients in oil palm trunk (Elaeis guineensis). J Wood Sci 64:709–719
18. Bachtiar DSS et al (2010) Flexural properties of alkaline treated sugar palm fibre. Int J Automot Mech Eng 1:79–90
19. Haameem MJA, Majid MSA, Afendi M et al (2016) Effects of water absorption on Napier grass fibre/polyester composites. Compos Struct 144:138–146
20. Suriani MJ, Ali A, Khalina A et al (2012) Detection of defects in kenaf/epoxy using infrared thermal imaging technique. Procedia Chem 4:172–178
21. Hoque ME, Aminudin MAM, Jawaid M et al (2014) Physical, mechanical, and biodegradable properties of meranti wood polymer composites. Mater Des 64:743–749

Chapter 18
The Analysis of Container Terminal Throughput Using ARIMA and SARIMA

Kasypi Mokhtar, Siti Marsila Mhd Ruslan, Anuar Abu Bakar, Jagan Jeevan, and Mohd Rosni Othman

Abstract Seaport container throughputs are utmost essential indicator for a successful container terminal as it could impact the utilization of resources for terminal operation. The accuracy of throughput forecasting would enable for potential of terminal growth in future. The paper aims to achieve efficient forecasting models by incorporating data throughputs from 2007 to 2015 from the Marine Department of Malaysia. This research focuses on the original ARIMA and the modified model SARIMA for a better model. The forecast results of container throughputs achieved from 2016 to 2018 are then compared with actual figures and then discussed.

Keywords Container terminal · Throughput · ARIMA model · SARIMA model

18.1 Introduction

Over the last decade, containerization and ports have played important roles in international trade [1]. Containerization is an important element of the logistics and security innovations that revolutionized freight handling in the twentieth century. The pattern characteristics of container throughput time series include cycles, seasonality, mutability, and randomity. These traits are determined by the economic structure and market development of the port's hinterland [2].

K. Mokhtar · S. M. Mhd Ruslan (✉) · A. Abu Bakar · J. Jeevan · M. R. Othman
Maritime Operations Research Interest Group, Faculty of Maritime Studies, Universiti Malaysia Terengganu, 21030 Kuala Nerus, Terengganu, Malaysia
e-mail: s.marsila@umt.edu.my

K. Mokhtar
e-mail: kasypi@umt.edu.my

A. Abu Bakar
e-mail: anuarbakar@umt.edu.my

J. Jeevan
e-mail: jagan@umt.edu.my

M. R. Othman
e-mail: rosni@umt.edu.my

A. Ismail et al. (eds.), *Design in Maritime Engineering*, Advanced Structured Materials 167, https://doi.org/10.1007/978-3-030-89988-2_18

Some econometric models have previously been used for container throughput forecasting. Forecasting methods can be derived from qualitative or quantitative, where quantitative can be classified into time series and causal methods [3, 4]. Time series generalize the historical data activities to future and suitable to short-term forecasting. Causal methods assume a relationship between involved variables, for instance, between port throughput and gross domestic product (GDP) and are useful for medium to long-term forecasting [5, 6]. Authors of [7] conducted forecasting methods for short-term prediction of port cargo. Authors of [8, 9] highlighted medium and long of port throughput. Critical reviews mention that a variety of short-term methods have been used to forecast and window analysis of container throughput [10]. Authors of [11] used a modified regression model for short-term forecasting of the volumes of import and export containers in Taiwan. Authors of [12] compared six univariate models for short-term forecasting of container throughput in Taiwan as well. Authors of [13] compared short-term forecasting accuracy of three models, at with genetic programming is the best model. Similarly, Ref. [14] proposed an algorithmic method combining projection pursuit regression and genetic programming for short-term forecasting.

The container throughput time series is usually complex; thus, a single model based on linear assumptions, or a nonlinear dynamic model often cannot obtain satisfactory forecasting performance. An increasing number of researchers have constructed hybrid forecasting models to solve this problem. For example, Ref. [15] proposed three hybrid models based on the least squares support vector regression (LSSVR). In this study, both ARIMA and SARIMA models are applied using the data of container throughput of Malaysia ports between the span of the period of 2007–2015 in order to attain the future of container demand in Malaysia. At the end of the findings, the forecasted value of container throughput for year 2017 and 2018 are compared with the actual value of throughput, to get a clear comparison of both measurements.

The remainder of this paper is organized as follows: The literature review is described in Sect. 18.2. Section 18.3 illustrates the methodology to the forecasting of container throughput by applying ARIMA and SARIMA models. Then, model identification and forecasting are discussed in Sect. 18.4. Section 18.5 presents results and discussion. Finally, Sect. 18.6 concludes the study.

18.2 Maritime Studies

18.2.1 Seaport Operation

The common application methods for maritime studies are either parametric or non-parametric models. Authors of [16], 17 employed regression and neural network methods, respectively, to forecast container growth in Hong Kong. Authors of [15], 18 used genetic programming and modified regression models for container throughput forecasts in Taiwan. Authors of [13] showed six univariate forecasting models for the container throughput at three major seaports in Taiwan. Authors of [19] used a hybrid traditional fuzzy set theory and regression analysis and developed a fuzzy regression model to forecast import and export cargo volumes in Taiwan seaports. Authors of [20] recommended a vector error correction model to forecast, in a long-term, container throughput in Hong Kong seaport. Besides on forecasting, Ref. [21] highlighted on port competition and not port competitiveness, focusing on competitive advantages. Then, it is possible to classify the competition or cooperation models into qualitative and quantitative ones. The qualitative approach reveals port competition, Ref. [22] used questionnaires to identify the criteria of container ship owners of a port. Authors of [23] discussed local and regional competition and cooperation between Hong Kong and South China Ports from administrative and ownership structures. Authors of [24] conducted a study on Copenhagen Malmö Port by referring to [24] and found several advantages of cooperation, such as more effective use of port resources and specialization in which toward port economic of scales by utilizing all resources.

The previous research that is focused on maritime studies discusses on port competition or cooperation; Refs. [25, 26] used market share evaluation, the growth-share matrix, shift-share analysis, and evaluation of the Hirshmann-Herfindahl index (HHI) to analyze the dynamics of container traffic and port concentration within the EU port system. Authors of [27] discussed the port concentration dynamics in Eastern Asia over the period 1975–2005 using a variant of the HHI called the geo-economic concentration index (GECI). Authors of [28] modeled the market share of the North Sea container ports of Rotterdam, Antwerp, Bremen, and Hamburg using the log it model discussed on quality of service. Authors of [29] presented an algorithm based on a multi criteria method called as hierarchical fuzzy process method to identify the competitiveness of container ports in Asia. Apart of that, research on slot capacity analysis for detecting port competition like [30] analyzed developments in container port competition between 10 major ports in East Asia during 1995–2001. Authors of [31] examined competition dynamics between the South-East Asian ports of Lang, Singapore, and Tanjung Pelepas during 1999–2004. Authors of [11] introduced a port market share forecasting model that explicitly modeled port competition by considering origin and destination, as well as cargo type, ship size, maritime access, port capacity and efficiency, and hinterland transport. Authors of [32] examined the competition between the ports of Busan in South Korea and Kobe in Japan using a non-cooperative game model. Authors of [33] highlight qualitative approach for competition and cooperation among Japanese container ports. Authors of [34] highlight a cointegration analysis that has become a popular method for analysing relationships between seaports. Authors of [20] used a structural vector error correction model to conduct a detailed study of competition between the ports of Hong Kong and Singapore. Authors of [30] analyzed the long and short-run competition dynamics between 10 major container ports in East Asia from 1980 to 2001 using the vector autoregressive model (VARM) and Johansen's cointegration test. Authors of [35] analyzed relationships between and among six main Asian ports using a cointegration test and the Granger causality test for short-term relationship. Authors of [36] studied competition–cooperation relationships between four Liaoning ports (China) using the VARM model. They interpreted negative and positive signs of coefficients in regression equations as indications of competition and cooperative competition. VAR-like models have drawbacks, like variables should be stationary in first differences, then decide the number of lags and optionally select interception and trend. As there is no unique VARM from a given data set, the model reveal is statistical and not necessarily realistic. Authors of [37] discussed the expansion of container traffic in the port of Koper at the beginning of this millennium. Authors of [38] compared three forecasting models for quarterly container throughput in the ports of Koper, Trieste, Venice, and Ravenna during 2002–2012. They found the ARIMA model to be superior to the other models for each port in their study. Authors of [37] provided an analysis of NAPs similar to that of [24] for the EU container port system.

Typically, in the literature, parametric or non-parametric forecasting models are adopted for this purpose. Parametric models assume a model structure that can be described by known mathematical expressions, while non-parametric models, on the other hand, do not assume any definite functional form of dependent and independent variables. Despite the fact that many forecasting models have been developed in the literature, most models to date either lack consideration of short-term seasonal variations, measured in terms of monthly container throughput, or they simply focus on a specific seaport or country. Hence, issues such as periodicity, complexity and spatial applicability may not be appropriately addressed. Therefore, there is a need for the development of forecasting model for container terminals in Malaysia by incorporating seasonality with spatial considerations.

a. ARIMA versus SARIMA

ARIMA and SARIMA models are expansions of ARMA lesson in attempts to include more practical elements, in specific, separately, non-stationarity in mean and seasonal behaviors. In practice, numerous financial time series are non-stationary in mean, and they can be modeled only by expelling the non-stationary source of variation [12]. This is typically done by differencing the series. Suppose X_t is non-stationary in mean, and the idea is to build an ARMA model on the series w_t which is defined as the result of the operation of differencing the series d times (in general $d = 1$), $w_t = \Delta^d X_t$.

Hence, ARIMA models (I is defined as integrated) are the ARMA models defined on the dth difference of the original process:

$$\Phi(B)\Delta^d X_t = \theta(B)a_t \tag{18.1}$$

where $\Phi(B)\Delta^d$ is called the generalized autoregressive operator and $\Delta^d X_t$ is a quantity made stationary through the differentiation and can be modeled with an ARMA.

The autoregressive integrated moving average (ARIMA) approach is one method that could be employed for short-term port throughput forecasting. It was found from the literature that this approach tends to have high performance in short-run forecasts

[39–41]. Time series models tend to outperform their counterparts because of the restrictive nature of other econometric models. For instance, they do not incorporate the dynamic structure of time series data and impose improper restrictions on the structural variables. ARIMA models, with the flexibility to incorporate the dynamic structure, have an inherent advantage in short-term forecasting. However, the forecast performance of ARIMA models deteriorates as the forecast span increases, as the model is inefficacious in capturing long-term economic relationships [40]. However, short-term ARIMA forecasts are acceptable.

In many cases, time series have a seasonal component that replicates each s observations. For month-to-month perceptions $s = 12$ (12 in 1 year), or for quarterly observations $s = 4$(4 in 1 year). In order to deal with regularity, ARIMA processes have been modified into SARIMA models [12].

$$\Phi(B)\Delta^d X_t = \theta(B)a_t \tag{18.2}$$

where a_t is such that;

$$s\phi\left(B^s\right)\Delta_s^D a_t = s\Theta\left(B^s\right)a_t \tag{18.3}$$

hence,

$$\Phi(B)s\Phi\left(B^s\right)\Delta_s^D \Delta^d X_t = \theta(B)s\Theta\left(B^s\right)a_t \tag{18.4}$$

and we write $X_t \sim$ ARIMA (p, d, q) × (P, D, Q) s. The idea is that SARIMA are ARIMA (p, d, q) models whose residuals αt are ARIMA (P, D, Q). With ARIMA (P, D, Q) we intend ARIMA models whose operators are defined on B^s and successive powers. Concepts of admissible regions SARIMA are analog to the admissible regions for ARIMA processes, and they are just expressed in terms of B^s powers.

This study explores the use of seasonal autoregressive, integrated, and moving average model (SARIMA) models to forecast container throughput at several major international container ports, while taking into consideration seasonal variations. The SARIMA modeling methodology is described, then a database consisting of yearly container port traffic data from 2007 to 2015, followed by a forecasting model for container terminals in Malaysia.

18.3 Methodology

18.3.1 Undertaking Methodology for ARIMA and SARIMA Model

The data for the study is derived from the container throughputs between the periods of 2007 to 2015, and it is obtained from the Marine Department, Malaysia.

A forecasting technique using ARIMA is applicable for time series in forecasting container throughput. The modified technique SARIMA is used for forecasting container throughput in Malaysia container terminals.

There are several practical phenomena whose data are presented in time series with seasonal characteristic. A seasonal time series is defined as a series with a regular pattern of changes that repeats over S time-periods, i.e., the average values at some particular times within the seasonal intervals are usually significantly different from those at other times. Thus, a seasonal time series is usually a non-stationary series which should be made stationary by using either differencing or logging techniques before ARIMA models are used to do the forecasting for the series.

18.3.1.1 Non-seasonal ARIMA Model

The non-seasonal ARIMA model usually has the form of ARIMA (p, d, and q), where:

- p is the number of lags of the differenced series appearing in the forecasting equation, called autoregressive parameter,
- d is the difference levels to make a time series stationary, called integrated parameter, and
- q is the number of the lags of the forecast errors, called moving average parameter. The "Auto Regressive" term refers to the lags of the differenced series appearing in the forecasting equation and the "Moving Average" term refers to the lags of the forecast errors. This "Integrated" term refers to the difference levels to make a time series stationary.

18.3.1.2 Seasonal ARIMA Model

The variation of a time series is usually affected by several different factors, including seasonality. Seasonality may make several non-stationary time series significantly vary. And, due to the environmental influence, such as periodic trends, the variations induced by seasonal factor sometimes dominate the variations of the original series. A seasonal time series is usually a non-stationary time series that follows some kind of seasonal periodic trend and can be made stationary by seasonal differencing which is defined as a difference between one value and another one with lag that is a multiple of S. Seasonal ARIMA model incorporates both non-seasonal and seasonal factors

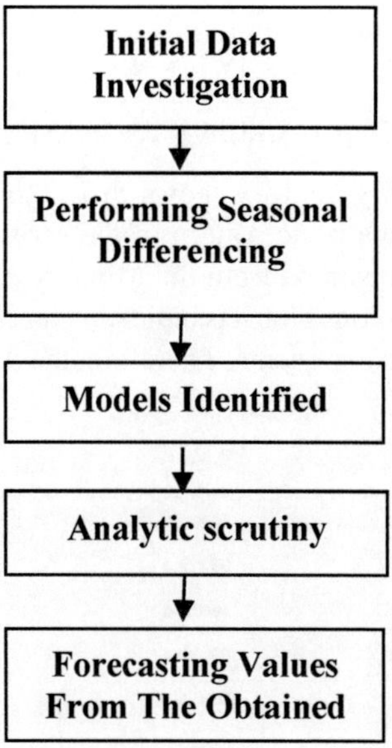

Fig. 18.1 Container throughput forecasting flow

in a multiplicative model with the form of SARIMA (p, d, q) (P, D, Q) S, where: p, d, q are the parameters in non-seasonal ARIMA model that apply to Fig. 18.1 as follows;

- P is the number of seasonal autoregressive order,
- D is the number of seasonal differencing,
- Q is the number of seasonal moving average order, and
- S is the time span of the repeating seasonal pattern

18.4 Model Identification and Forecasting

The method of seasonal ARIMA is commonly applied to time series analysis. The term of ARIMA is in short and stands for the three components which are autoregressive, integrated, and moving average models. The fundamental concept to undertake when we developing models is to understand the characteristics of the series datasets and how it behaves over times. There are some advantages of undertaking this strategy, i.e., to give the freedom to the researcher to select the most appropriate model from all potential models according to the time plot. In our case, we arranged the dataset according to the state which is repeating from 2007 till 2015. The series with seasonal needs the additional differencing to eliminate the seasonal effect. Let z_t be seasonal differenced series,$z_t = y_t - y_{t-15}$ for state data series. If z_t remained non-stationary, then the next step is to perform non-seasonal differencing which is denoted by $w_t = z_t - z_{t-1}$. The specific name for the seasonal model is SARIMA (p, d, q)(P, D, Q)s. Below are the steps of model identification.

Step 1: Initial Data Investigation

Figure 18.2 shows that a simple data investigation was conducted to understand the basic pattern of the series for the Total Throughput Port (TTP). The data was obtained from the Ministry of Transport (MOT) of the maritime section. From the series plot, it is indicating that the series is not stationary with the existence of seasonal components. Therefore, the data are used to build the SARIMA model. Figures 18.3

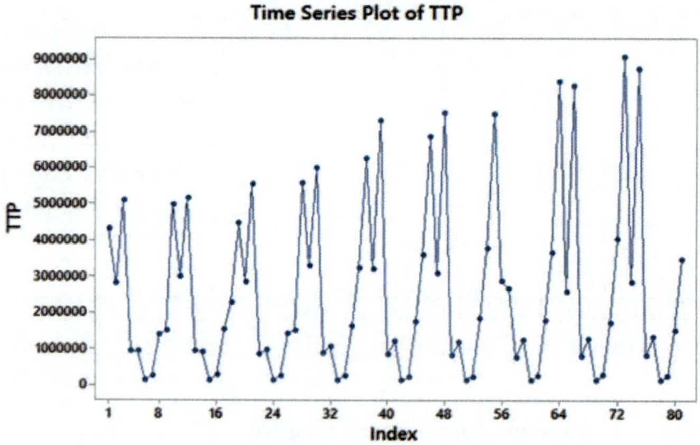

Fig. 18.2 Time plot of throughput

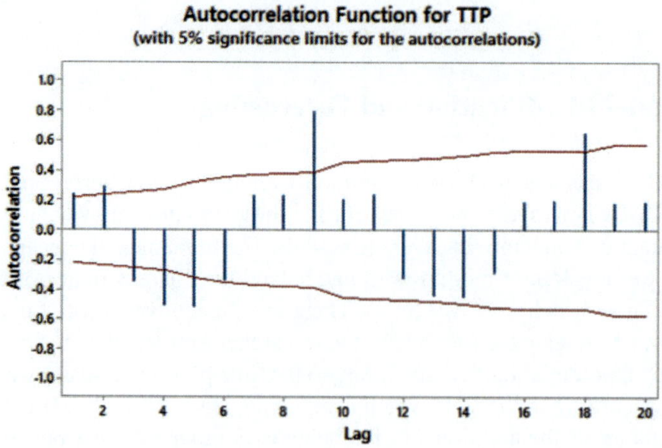

Fig. 18.3 ACF plot original series

and 18.4 show at the lag of 15 that the spike is significant. These characteristics (from Figs. 18.3 and 18.4) show that the seasonal effect is present.

Step 2: Performing Seasonal Differencing

The seasonal difference is given as $z_t = y_t - y_{t-9}$. By observing Figs. 18.3 and 18.4, it show that the original series of the container total throughput port (TTP) could increased at one degree, while in the non stationaries form. Figure 18.5 shows the time series plot of the TTP difference and can be concluded that the series is stationary from the time series plot of TTP Difference. Figures 18.6 and 18.7 show the degree of both non-seasonal and seasonal difference, the series plotted now become stationary. From the stationary series, there are four propose models which are identified in Table 18.1.

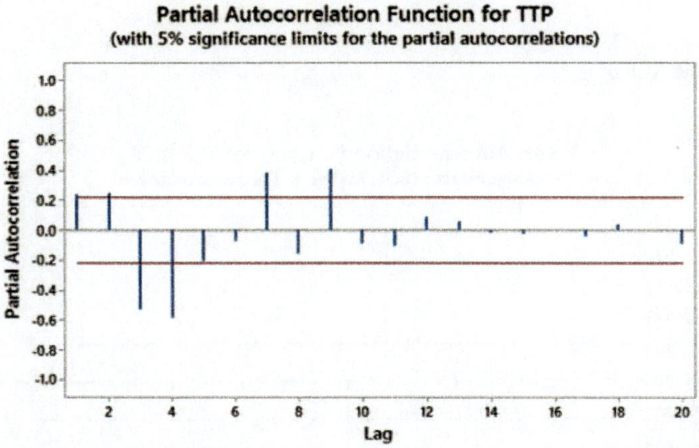

Fig. 18.4 PACF plot original series

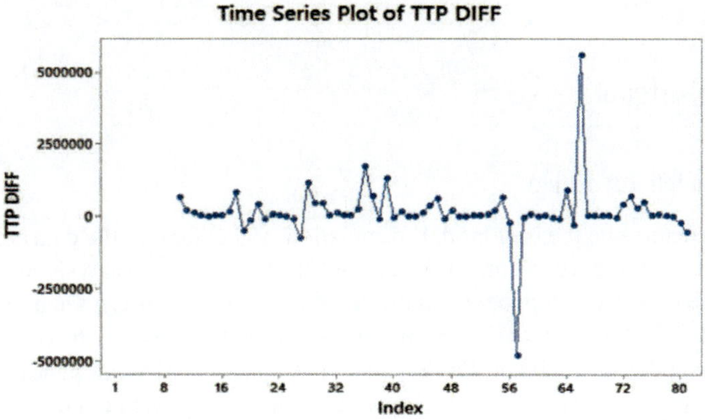

Fig. 18.5 Time plot of series in seasonal difference $z_t = y_t - y_{t-9}$

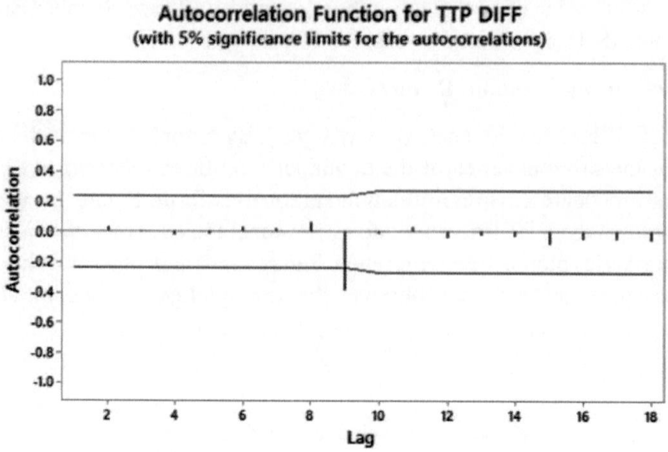

Fig. 18.6 The ACF of z_t

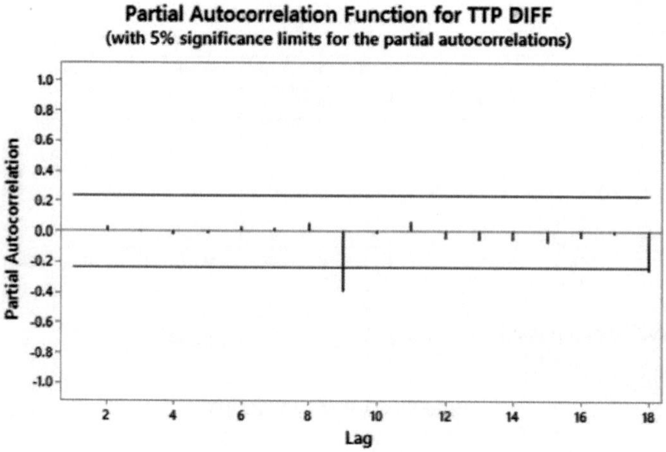

Fig. 18.7 The PACF of z_t

Step 3: Models Identified

In order to determine the best model formulations to be fitted to the data series, one needs to observe for significant spike in Figs. 18.5 and 18.6. Analysis in Fig. 18.1 contains the seasonal component and the general formulation is written as SARIMA (p, d, q)(P, D, Q)$_{15}$. To identify the non-seasonal and seasonal part, one needs to observe the spikes at ACF and PACF of w_t. The spike for MA can be identified by looking ACF plot of w_t and AR by looking at PACFof w_t. While the spike for the seasonal MA and seasonal AR can be obtained by looking at the "irregular" spike for most series. A significant spike is observed at the lag 9 to suggest the seasonal SMA

Table 18.1 Summary of Portmanteau test for each model

	Model			
	SARIMA $(1,0,1)(1,1,1)_9$	SARIMA $(1,0,1)(0,1,1)_9$	SARIMA $(1,0,1)(0,1,0)_9$	SARIMA $(1,1,0)(1,1,0)_9$
Calculated Chi-square (Df)	3.5(8)	2.8(9)	14.0(10)	3.3(9)
P value	0.903	0.972	0.175	0.952
Decision (1% sig. Level)	Do not reject H_0	Do not reject H_0	Reject H_0	Do not reject H_0
Conclusion	The errors are white Noise	The errors are not white noise	The errors are not white noise	The errors are white noise
MSE	735,102, 310,113	737,371, 871,000	949,354, 667,064	824,257, 825,706

(from ACF) and SAR (from PACF). All possible models will be correctly checked for their representative, this to ensure that a well specified model is not missed out.

Table 18.1 depicts that all four proposed models are well specified since the errors are white noise. After considering the concept of parsimony and the size of their respective MSE. Model SARIMA $(1,0,1)(0,1,1)_1$ therefore is being selected as the good model to represent the data. After selection of the model, the next step is to forecast the future value by the gained model.

Step 4: Forecasting Values from the Obtained Model

Figure 18.8 depicts the forecasting result for the model in which the red line represents the forecasting value of the study. The volatile value is significant and consistent for the plot.

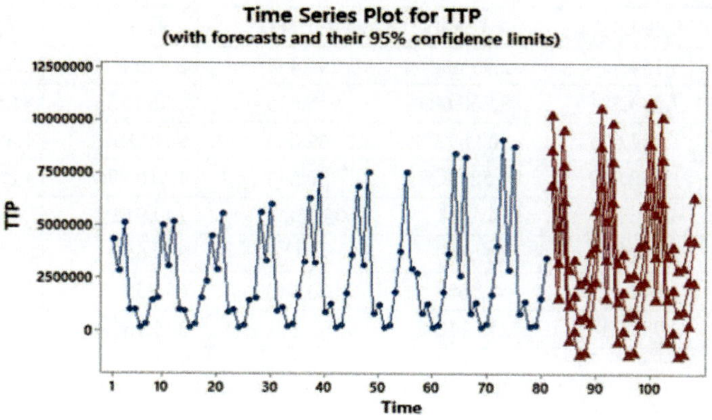

Fig. 18.8 Time series plot with forecasting values

18.5 Results and Discussion

The best model that has been derived from the research is based on throughput data of Malaysia container terminals from 2007 to 2015. The following best model is therefore ARIMA (1,0,1) AR (1) and MA (2), $y_t = \mu + \emptyset_1 y_{t=1} - \theta_1 \varepsilon_{t+1} + \varepsilon_1$. , $y_t - \emptyset_1 y_{t-1} = \mu + \varepsilon_t - \theta_1 \varepsilon_{t+1}$, $(1 - \theta_t \mathrm{B})\varepsilon_t$ and SARIMA (0,1,1)1 SMA(1), $z_t = y_t - y_{t+1}, z_t = \varepsilon_t - \theta_1 \varepsilon_{t+1,} \left(1 - B^1\right) = \left(1 - \theta_1 B^1\right)\varepsilon_t; z_{t=} y_t - y_{t-1}$. Therefore, combining ARIMA and SARIMA with, the model is $(1 - \emptyset_1 B)\left(1 - B^1\right)y_t = (1 - \emptyset_t B)\left(1 - \theta_1 B^1\right)\varepsilon_t$. Table 18.2 depicts the forecasting estimation for 2017, 2018 and 2019 as well as the actual values for throughput for year 2017 and 2018. It is shown from the forecasting results that a significant increase of container throughput for container terminals in Malaysia is seen for all ports. Nonetheless, the actual figures from official data given for the year 2017 and 2018 show a different case. Except for AW, CP and EPP, all other ports recorded a decreased value compared to the forecasted values. This could happen due to several factors, and it is likely contributed by the specification of the period taken for this study. ARIMA and SARIMA models would perform better with longer period rather than 8 years of testing [42]. Other than that such external and internal factors might also lead to the outcome. For example, certain ports were changing their strategic overview, thus leading to a change of regulation by the management level [43]. On the other hand, the lack of demand from importers and exporters in calling to the port could be one of the issues, as well as the projection of competitors that resulted in a stiff competition, which hinder the growth of the container throughput. Apart of that the state interference in determining the pattern of trade could also indirectly affect the direction taken by the port authority in general.

Table 18.2 Estimation of TTP for 2017, 2018 and 2019

Container	Forecast 2017	Actual 2017	Forecast 2018	Actual 2018	Year 2019
AW	8,494,931	9.02 mil	8,637,711	9.5 mil	8,778,038
BN	3,154,949	2.95 mil	3,297,453	2.8 mil	3,437,511
CP	7,773,384	8,260,609	7,915,615	8,960,865	8,055,403
DJ	1,127,675	900,692	1,269,633	941,589	1,409,151
EPP	1,588,859	1,523,828	1,730,543	1,510,376	1,869,793
FK	463,646	147,041	605,058	149,912	744,040
GB	577,192	309,149	718,332	348,665	857,048
HR	1,942,577	55,365	2,083,445	65,333	2,221,894
IS	3,937,214	353,155	4,077,811	374,165	4,215,994

18.6 Conclusion

The forecasting of container throughput is important for operation and management of ports. This study proposes the linear model of ARIMA and SARIMA to predict the outcome of container throughput of Malaysian ports. From that, the predictions of throughput for the forecasting result are mapped out. On the other hand, it is notable to find that some of the actual results show substantial differences between the forecast and actual number of container throughput. This could be happening due to several factors such as the length of sample period, the change of regulation by the port authority in facing the demand, the lack of demand from importers and exporters to the port, state interference in determining the trade, as well as tough competition from the nearby ports that hamper the growth of the related port. For that, this study could be further enhanced with a better representation of data since ARIMA and SARIMA models are better performed with longer period of sample. Other than that, an additional study could be done to investigate the demand coming out from importers and exporters, as well as the extent of the role of state in affecting the direction taken by the port authority in handling up the trade activities.

Acknowledgements The research is conducted under Ministry of Higher Education (MOHE) research grant vote no 59481 with the code FRGS/1/2017/TK08/UMT/02/5 and University of Malaysia Terengganu (UMT) for research facilities.

References

1. Xie G et al (2013) Hybrid approaches based on LSSVR model for container throughput forecasting: a comparative study. Appl Soft Comput 13(5):2232–2241
2. Rashed Y et al (2018) A combined approach to forecast container throughput demand: scenarios for the Hamburg-Le Havre range of ports. Transport Res A-Pol 117(August):127–141
3. Abraham B, Ledolter J (2009) Statistical methods for forecasting. Wiley, New York
4. Chambers JC et al (1971) How to choose the right forecasting technique Harvard business review, Brighton, Massachusetts
5. Carnot N et al (2005) Economic forecasting. Palgrave Macmillan, New York
6. Langen DPW et al (2012) Combining models and commodity chain research for making long-term projections of port throughput: an application to the Hamburg–Le Havre range. Eur J Transp Infrastruct Res 12:310–331
7. Zhang D et al (2013) Incorporation of formal safety assessment and Bayesian network in navigational risk estimation of the Yangtze River. Reliab Eng Syst Safe 118:93–105
8. Hui ECM et al (2004) Forecasting Cargo throughput for the port of Hong Kong: error correction model approach. J Urban Plan Dev 130(4):195–203
9. Van DC et al (2012) A very long term forecast of the port throughput in the Le Havre-Hamburg range up to 2100. Eur J Transp Infrastruct Res 12(1):88–110
10. Kasypi M et al (2016) Window analysis: a container terminal. Adv Sci Let 22(4):2201–2204
11. Zondag B et al (2010) Port competition modeling including maritime, port, and hinterland characteristics. Marit Policy Manag 37:179–194
12. Gerolimetto M (2008) Introduction to time series analysis and forecasting. Wiley, New Jersey
13. Peng WY, Chu CW (2009) A comparison of univariate methods for forecasting container throughput volumes. Math Comp Model Dyn 50(7):1045–1057

14. Huang A et al (2015) Forecasting container throughput with big data using a partially combined framework. In: 2015 international conference on transportation information and safety, pp 641–646
15. Chen SH, Chen JN (2010) Forecasting container throughputs at ports using genetic programming. Expert Syst Appl 37:2054–2058
16. Seabrooke W et al (2003) Forecasting cargo growth and regional role of the port of Hong Kong. Cities 20(1):51–64
17. Lam WHK et al (2004) Forecasts and reliability analysis of port cargo throughput in Hong Kong. J Urban Plan Dev 130(3):133–144
18. Chou CC et al (2008) A modified regression model for forecasting the volumes of Taiwan's import containers. Math Comput Model 47(9):797–807
19. Liang GS, Chou TY (2003) The forecasting of inbound and outbound seaborne cargo volume in Taiwan. J Mar Sci 12:203–218
20. Fung MK (2002) Forecasting Honk Kong's container throughput: an error correction model. J Forecast 21(1):69–80
21. Notteboom T, Yap WY (2012) Port competition and competitiveness. In: The Blackwell companion to maritime economics. Blackwell Publishing Ltd, New Jersey
22. Slack B (1985) Containerization, inter-port competition, and port selection. Marit Policy Manag 12(4):293–303
23. Song DW (2002) Regional container port competition and co-operation: the case of Hong Kong and South China. J Transp Geogr 10:99–110
24. Langen DPW, Nijdam MH (2009) A best practice in cross-border port cooperation: Copenhagen Malmö Port. In: Notteboom T, Ducruet C, Langen PWD (eds) Ports in proximity: competition and coordination among adjacent seaports. Ashgate Pub., Farnham, England; Burlington, VT
25. Notteboom TE (1997) Concentration and load centre development in the European container port system. J Transp Geogr 5:99–115
26. Notteboom TE (2010) Concentration and the formation of multi-port gateway regions in the European container port system: an update. J Transp Geogr 18:567–583
27. Le Y, Ieda H (2010) Evolution dynamics of container port systems with a geo-economic concentration index: a comparison of Japan, China and Korea. Asian Transp Stud 1:46–61
28. Veldman SJ, Buckmann EH (2003) A model on container port competition: an application for the West European container hub-ports. Marit Econ Logist 5:3–22
29. Yeo GT, Song DW (2006) An application of the hierarchical fuzzy process to container port competition: policy and strategic implications. Transportation 33
30. Yap WY, Lam JSL (2006) Competition dynamics between container ports in East Asia. Transp Res A Policy Pract 40:35–51
31. Lam JSL, Yap WY (2008) Competition for transhipment containers by major ports in Southeast Asia: slot capacity analysis. Marit Policy Manag 35:89–101
32. Ishii M et al (2013) A game theoretical analysis of port competition. Transport Res E Log 49:92–106
33. Hoshino H (2010) Competition and collaboration among container ports. Asian J Shipp Logist 26:31–48
34. Juselius K (2006) The cointegrated VAR model: methodology and applications. Oxford University Press, Oxford, New York
35. Chiang CH, Hwang CC (2010) Relationships among major container ports in Asia region. J East Asia Soc Transp Stud 8:2299–2313
36. Ma J, Quian L (2011) A study on Liaoning ports competition and cooperation. Adv Mater Res 181–182:1050–1053
37. Twrdy E, Batista M (2014) Evaluating the competition dynamics of container ports in the North Adriatic. Sci J Marit Res 28:88–93
38. Dragan D et al (2014) A comparison of methods for forecasting the container throughput in North Adriatic Ports. In: IAME 2014 conference, Norfolk VAUSA
39. Litterman RB (1986) Forecasting with Bayesian vector autoregression—five years of experience. J Bus Econ Stat 4(1):25–38

40. Stockton DJ, Glassman JE (1987) An evaluation of the forecast performance of alternative models of inflation. Rev Econ Stat 69(1):108–117
41. Nadal-De SF (2000) Forecasting inflation in chile using state-space and regime switching models. IMF working paper WP/00/162. International Monetary Fund, Washington DC
42. Xie G et al (2017) Data characteristic analysis and model selection for container throughput forecasting within a decomposition-ensemble methodology. Transport Res E-Log 108:160–178
43. Mo L et al (2018) GMDH-based hybrid model for container throughput forecasting: selective combination forecasting in nonlinear subseries. Appl Soft Comput 62:478–490

Chapter 19
Investigation on the Mechanical and Microstructural Characteristics of Diffusional Bonded Gray Cast Iron and Low Carbon Steel

Bakhtiar Ariff Baharudin, Fauzuddin Ayob, Aziz Abdul Rahim, Mazli Mustapha, Azman Ismail, Fauziah Ab Rahman, and Asmawi Ismail

Abstract An investigation has been made of the diffusional bonded couples of gray cast iron and low carbon steel that has been subjected to heat treatment. The objective is to establish the post bond heat treatment's parameters influence on the mechanical and structural properties of the diffusional bonded couples. The tensile strength of the diffusion bonded joints was found to be increased with increased heat treatment's temperature and time. The microstructural examination has also shown that at higher heat treatment temperature and time also had resulted in microvoids and interface lines to be disappeared, and the bond/weld at the interfaces of the diffusional welded couples seemed to be more complete. Correspondently, these resulted in the increased of the tensile strength. The microhardness value at the interface lines of the joints was found to be increased, while the charpy impact strength value was on the opposite way with the increased heat treatment temperature. The microstructural analysis has also shown that much thicker diffusion layers of spherodization zone and carbon rich zone were formed at higher temperature at the interfaces of the

B. A. Baharudin (✉) · F. Ayob · A. Abdul Rahim · A. Ismail · F. Ab Rahman · A. Ismail
Universiti Kuala Lumpur Malaysian Institute of Marine Engineering Technology, 32200 Lumut, Perak, Malaysia
e-mail: bakhtiarab@unikl.edu.my

F. Ayob
e-mail: fauzuddin@unikl.edu.my

A. Abdul Rahim
e-mail: azizar@unikl.edu.my

A. Ismail
e-mail: azman@unikl.edu.my

F. Ab Rahman
e-mail: fauziahabra@unikl.edu.my

A. Ismail
e-mail: asmawiis@unikl.edu.my

M. Mustapha
Universiti Teknologi PETRONAS, 32610 Seri Iskandar, Perak, Malaysia
e-mail: mazli.mustapha@utp.edu.my

A. Ismail et al. (eds.), *Design in Maritime Engineering*, Advanced Structured Materials 167, https://doi.org/10.1007/978-3-030-89988-2_19

joints. These are also correlated with the microhardness and the charpy impact values obtained at higher temperature. These results were in consistent with the principle and theory of diffusion bonding whereby at higher temperatures, more activation energy is available for atoms inter-diffusion to take place, while with longer time, it allows higher volume of diffusion of atoms, and hence it changes the behavior of the joints. Thus, heat treatment's temperature and time were found to have a strong influence on the mechanical and microstructural characteristics of the diffusion bonded gray cast iron and low carbon steel.

Keywords Diffusion bonding · Heat treatment · Mechanical · Microstructure · Diffusion layer

19.1 Introduction

Diffusion welding is a joining process between materials wherein the principal mechanism for joint formation is a solid-state diffusion. Coalescence of the faying surface is accomplished through the application of pressure at elevated temperature. There is no melting and only limited macroscopic deformation, or relative motion of the parts occurs during welding [1]. Diffusion welding is an attractive joining technique for the manufacture of precision apparatus and complex structures [2]. Diffusion welding helps to join difficult-to-weld materials or whenever standard fusion welding methods are unable to be used [3].

The success or failure of the diffusion welding process depends on three variables that require constant watch and careful adjustment. These variables are the welding temperature, the welding pressure (or pressing load), and the holding time (duration of pressure) [1].

A conventional diffusion welding process is usually carried out in a vacuum environment where the mating surfaces are not only protected against further contamination, such as oxidation, but remain clean due to the dissociation, sublimation or dissolution of the oxides present that diffuse into the bulk of the materials [1]. A conventional diffusion welding experiment carried out is using the typical equipment outfitted with the pressurized, heating and vacuum systems.

Various alternative methods were tested in previous studies to perform diffusion welding experiments to seek for an alternative to the conventional method of using the standard machine/equipment known as the hot press used in the preliminary study that had broken down and been rendered unusable [4].

Finally, the equipment and method as described in the previous study are selected for this research. Despite using this method, the results of the tensile test revealed that the joint was weak, however, this method is still successful in producing diffusion welded couples. In the following experiments, the diffusion couples produced through this method are subsequently be placed inside the furnace again without clamping on the fixture for a post bond heat treatment (PBHT). It is expected that with further heat treatment, it would remove voids at the interfaces of the joints and complete the bond/weld by allowing further diffusion process to take place. The objective is to study the effect of heat treatment temperature and time on the diffusion process and the characteristic of the joint. Tensile testing, Charpy impact testing, microhardness and metallographic examination are conducted.

19.2 Methodology

19.2.1 Materials

Cast iron, containing lamellar graphite of grade ASTM A 48 class 35 and equivalent to the ISO 250 and EN-GJL-250, is used with chemical composition and mechanical properties as shown in Tables 19.1 and 19.2, respectively.

Low carbon steel of grade BS 449 grade 250, equivalent to that of ASTM 36 and JIS G 3101 SS 400, is used with chemical composition and mechanical properties as shown in Tables 19.3 and 19.4, respectively.

Table 19.1 Composition of grey cast iron

	Chemical composition (wt. %)				
	C	Si	Mn	P	S
Grey cast Iron (ASTM A48C 35)	3.43	2.21	0.62	0.073	0.069

Table 19.2 Mechanical properties of cast iron

	Tensile strength (MPa)	Elongation (%)	Hardness (HB)
ASTM A48 C 35	250	–	190

Table 19.3 Composition of low carbon steel

	Chemical composition (wt %)				
	C	Si	Mn	P	S
BS 449 grade 250	0.19	0.10	0.46	0.011	0.031

Table 19.4 Mechanical properties of low carbon steel

	Tensile strength (MPa)	Elongation (%)	Hardness (HB)
BS 449 grade 250	408	35	195

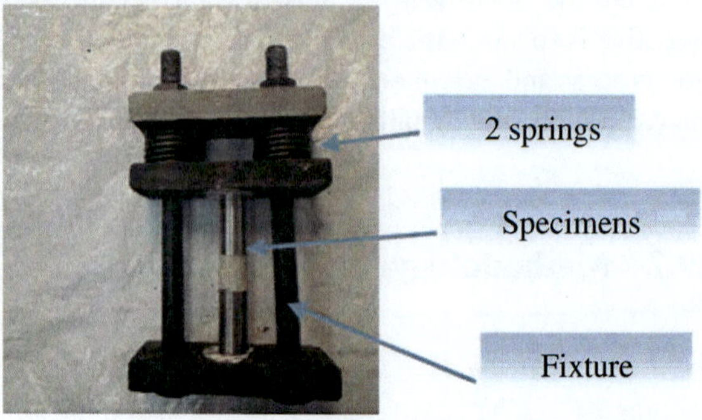

Fig. 19.1 Specimens clamped in fixture

Fig. 19.2 Hydraulic Press with specimens clamped in fixtures

19.2.2 Diffusion Bonding Equipment

Non-standard equipment/unconventional methods were developed to conduct diffusion bonding experiments [4]. In this method, a fixture was developed to clamp the specimen as shown in Fig. 19.1. A press machine was fabricated as in Fig. 19.2 to transfer pressure or load to the specimens. In this way, the specimens were pre-pressed with load before placing them into the heating furnace. Atmospheric furnace, as shown in Fig. 19.3, was used. The heating system is of radiation in nature that consists of a coil resistance furnace as the heat source, with appropriate switches and instrumentations needed to measure and monitor the desired treatment temperature and time.

19.2.3 Specimen Preparation

The specimen materials were cut out on a lathe machine into cylindrical shapes of size ø15 × 50 mm. The specimen ends were joined and polished on 240, 600 and

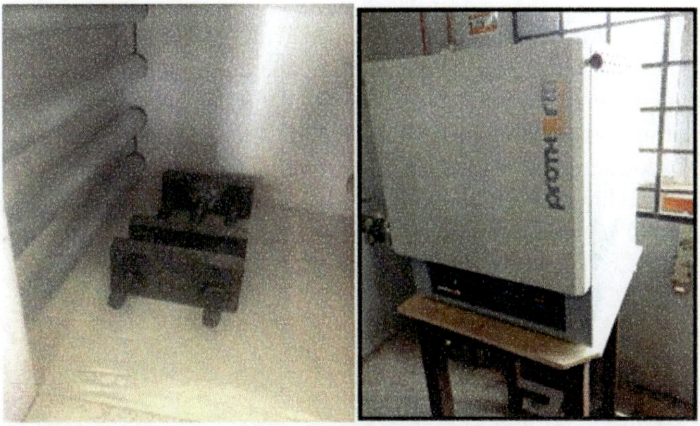

Fig. 19.3 Atmospheric furnace

Table 19.5 Diffusion bonding parameters and conditions

Welding pressure (MPa)	Bonding temp. (°C)	Bonding time (Min)
14	900	60

800 grits abrasive paper and cleaned using alcohol to remove any loose grit, dirt and grease or other contaminants.

19.2.4 Diffusion Bonding Parameters and Conditions

Welding parameters and welding conditions were selected based on best of past experiments and related literature reviews.

The welding process was conducted under atmospheric condition. The welding pressure selected was higher than the one usually carried out under vacuum in the preliminary study. At a higher pressure, it is evident that welding pressure plays an important role in the blocking of oxidation at the weld interface under atmospheric pressure [5].

Experiments to produce diffusion couples were carried out using the welding conditions as shown in Table 19.5.

19.2.5 Diffusion Bonding Procedure

The specimens were mounted, firstly, on fixtures as in Fig. 19.2. Loads or pressures were then transferred to mating surfaces through the fixtures using the press machine as shown in Fig. 19.2. The specimens, together with the fixtures, were placed in the furnace as shown in Fig. 19.3. The heating temperature, duration and heating rate were set through the control panel.

Upon heating and reaching the set temperature, the specimens were subsequently held for the set duration. At the end of the holding time, the heating was stopped automatically to allow the temperature to reduce by itself. The specimens are cooled to less than 1000 °C before removing from the furnace to avoid further oxidation.

19.2.6 Post Bond Heat Treatment

Diffusion couples produced using the above method were subsequently be placed into the furnace again without clamping in the fixture for a post bond heat treatment (PBHT). It is expected that with further heat treatment, voids would be removed at the interfaces of the joints to complete the bond/weld by allowing further diffusion process to take place.

19.2.7 PBHT for Tensile Testing

(a) Effect of PBHT temperature: Experiments were carried out using the PBHT conditions as shown in Table 19.6.
(b) Effect of PBHT time: Experiments were carried out using the PBHT conditions as shown in Table 19.7.

Table 19.6 PBHT parameters and conditions

Temperature (°C)	Time (h)
750	2
850	2
950	2
1000	2

Table 19.7 PBHT parameters and conditions

Temperature (°C)	Time (h)
1000	1
1000	2
1000	4
1000	8

Table 19.8 PBHT parameters and conditions

Temperature (°C)	Time (h)
800	2
900	2
1000	2

Table 19.9 PBHT parameters and conditions

Temperature (°C)	Time (h)
1000	1
1000	2
1000	4

Table 19.10 PBHT parameters and conditions

Temperature (°C)	Time (h)
800	2,4,8
900	2,4,8
1000	2,4,8

19.2.8 PBHT for Charpy Impact Testing

(a) Effect of PBHT temperature: Experiments were carried out using the PBHT conditions as shown in Table 19.8.
(b) Effect of PBHT time: Experiments were carried out using the PBHT conditions as shown in Table 19.9.

19.2.9 PBHT for Metallographic Examination and Microhardness Testing

Experiments were carried out using the PBHT conditions as shown in Table 19.10.

19.2.10 Tensile Testing

Tensile test was conducted on the diffusion couples after PBHT based on the standard DIN 50125, type F test piece, whereby the specimens were in the original form, round bar and un-machined conditions. In this test, the gage length taken was 25 mm, while the rate of extension/crosshead speed is 1 mm per min. An instron universal testing machine was used.

19.2.10.1 Charpy Impact Testing

The Charpy impact test was conducted on the diffusion couples after PBHT based on the standard ASTM E23. ASTM E23-16b as a guideline including the specification of specimen, procedure of testing, safety precaution and data collection. The test specimen is thermally conditioned and positioned on the specimen supports against the anvils; the pendulum is released without vibration and the specimen is impacted by the striker. Information is obtained from the machine and from the broken specimen.

19.2.10.2 Microhardness Testing

The metallographic specimens were also used for hardness testing. In this test, the microhardness tester of the Vickers hardness testing machine was employed with loads of 25 g (gf). The hardness was measured across the bonding interface.

19.2.10.3 Metallographic Examination

After completing the PWHT, the specimens were then prepared for a metallographic examination. They were mechanically polished on succession of 120, 240, 360 and 600, 800 and 1200 grits emery papers before final polishing on metallographic cloths moistened with a diluted solution of diamond particles of 3 and 1 μm sizes successively. The specimens were then etched initial solution (100 ml methanol + 4 ml nitric acid).

Photographs of the prepared metallographic specimens, in the vicinity of diffusion zones, along the bonding interface were then taken by optical microscope.

19.3 Results and Discussion

Diffusion couples were produced arising from the diffusion welding experiments. Figure 19.4 shows one of the diffusion couples still attached to the fixture after removal from the furnace.

Figure 19.5 shows one of the diffusion couples after removal from the fixture. Visual examination and manual testing on the diffusion couples revealed that the joints look good and strong.

(a) Effect of PBHT Temperature

The effect of PBHT temperature on tensile strength of the diffusional bonded joints is shown in Fig. 19.6. The tensile strengths of the joints increased directly with PBHT temperature despite the value being low but slightly increasing.

A typical stress–strain graph of the effect of PBHT's temperature is as shown in Fig. 19.7. Highest tensile strength obtained was 14.71 MPa, that being considered

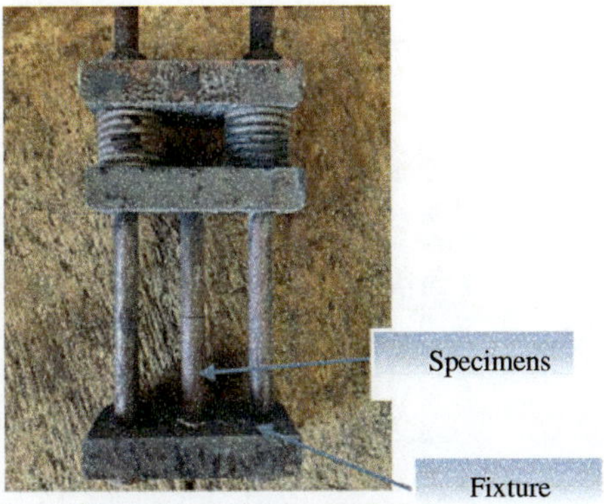

Fig. 19.4 Diffusion couple after removal from furnace

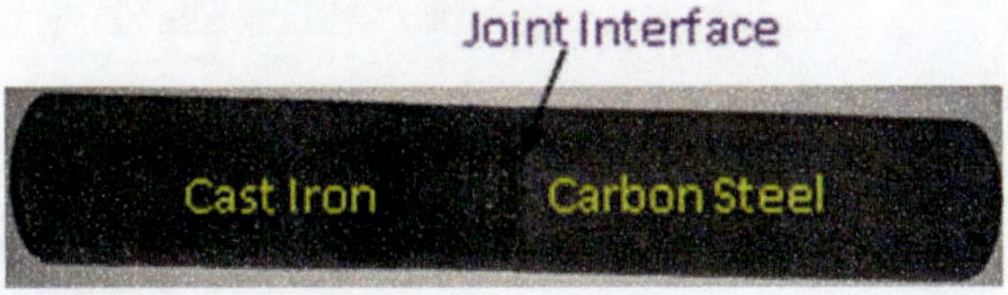

Fig. 19.5 One of successful joint specimens of grey cast iron and low carbon steel (Tw = 900 °C, tw = 60 min, Pw = 13 MPa)

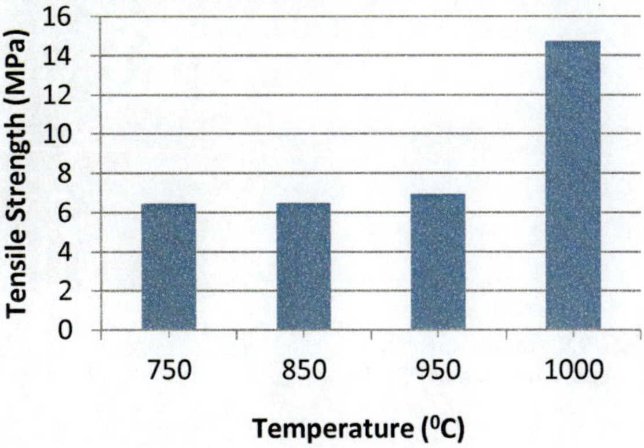

Fig. 19.6 Tensile strength of joints as a function of PBHT temperature

as low, while the tensile strength for gray cast iron, being the lowest material, is 250 MPa. The elongation rate was also very minimal.

A typical fractured surface after a tensile test as function of PBHT's temperature is shown in Fig. 19.8. Bonding or diffusion only occurred at the side of joint interfaces, while no diffusion took place at the center. Effect of PBHT time: The effect of PBHT time on tensile strength of the diffusional bonded joints is shown in Fig. 19.9. The tensile strengths of the joints increased directly with PBHT time despite the value being low but slightly increasing.

A typical stress–strain graph of the effect of PBHT's time is shown in Fig. 19.10. The highest tensile strength obtained was 11.05 MPa and can be considered as low,

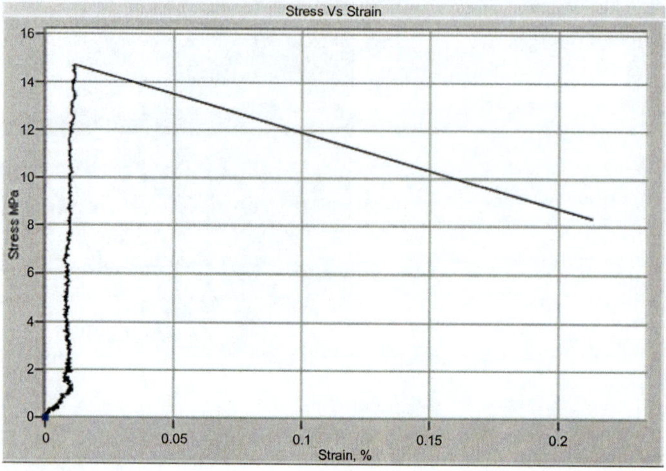

Fig. 19.7 Stress versus Strain graph of effect of temperature

Fig. 19.8 Typical fractures on surfaces of specimen after tensile test

Fig. 19.9 Tensile strength of joints as a function of PBHT time

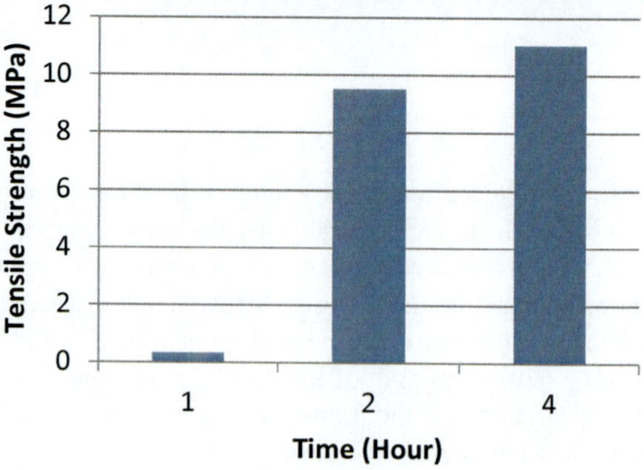

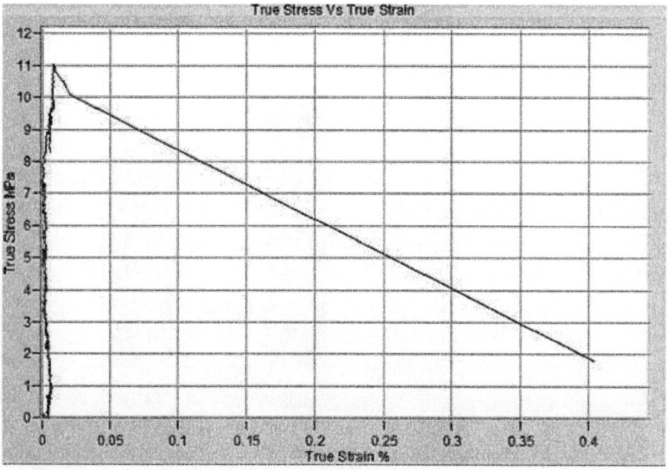

Fig. 19.10 Stress versus Strain graph of effect of time

as tensile strength for the lowest material, gray cast iron, is 250 MPa. The elongation rate was also very minimal.

A typical fractured surface after a tensile test as function of PBHT's time is shown in Fig. 19.11. Bonding or diffusion only occurred at the side of joint interfaces while no diffusion took place at the center.

Effect of PBHT temperature: The effect of PBHT temperature on the Charpy impact value of the diffusional bonded joints is shown in Fig. 19.12. The Charpy impact value of the joints is decreased with PBHT's temperatures and the value is very low as compared to the base metal.

Effect of PBHT time: The effect of PBHT time on the Charpy impact value of the diffusional bonded joints is shown in Fig. 19.13. The Charpy impact value of the joints is also very low and is of reduced value with PBHT times.

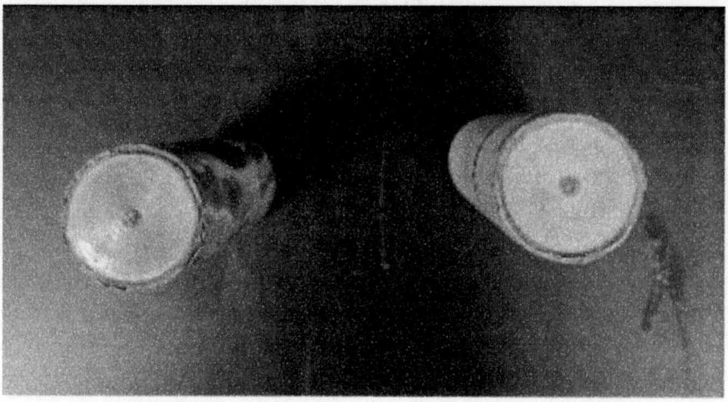

Fig. 19.11 Typical fractures on surfaces of specimen after tensile test

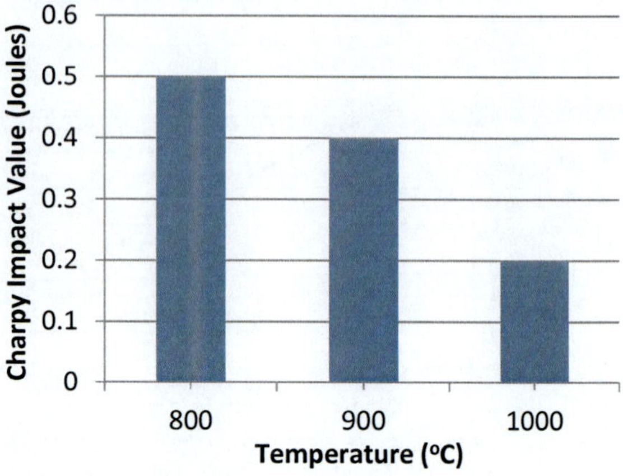

Fig. 19.12 Charpy impact value of joints as a function of PBHT temperature

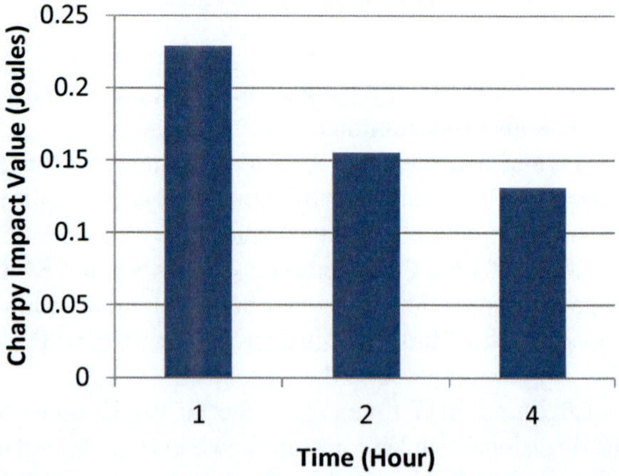

Fig. 19.13 Charpy impact value of joints as a function of PBHT time

The effect of PBHT temperature and time on the microhardness of the diffusional bonded diffusion layers at the interface lines of the joints is as shown in Fig. 19.14. The microhardness of the joints significantly increased directly with PBHT temperature but remained almost constant with the time.

A typical micro-Vickers hardness gradient across the bonding interface of gray cast iron and low-carbon steel diffusion couples after 2, 4 and 8 h at 1000 °C is shown in Fig. 19.15. The hardness of the diffusion layer produced is much higher than the hardness of the respective base metals at all treatment times. The maximum hardness was observed close to the interface on both sides of cast iron and low carbon steel. The maximum hardness was mainly at the interface line as shown in Fig. 19.14.

Figure 19.16 shows the typical optical micrograph of the joint interfaces; of (a) 800 °C (b) 900 °C and (c) 1000 °C for 4 h PBHT before etching. Microvoids and interface lines are very much visible at the interfaces of the specimens treated at temperature of 800 °C, becoming less visible treated at 900 °C and almost disappear

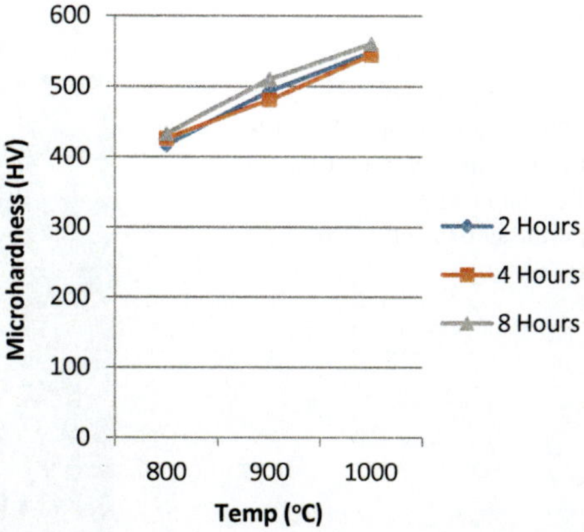

Fig. 19.14 Microhardness of diffusion layer as a function of PBHT temperature at the interface line

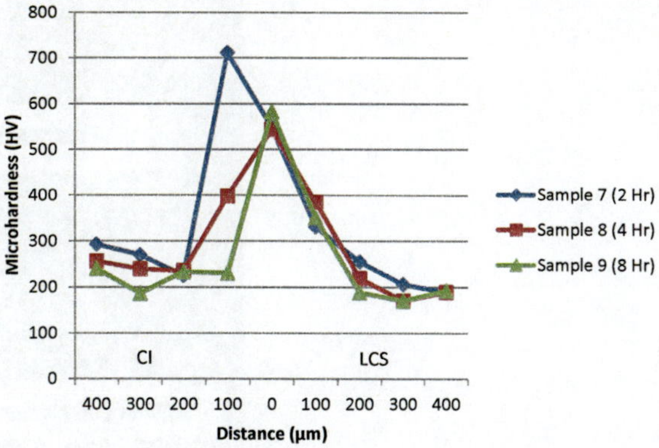

Fig. 19.15 Microhardness profile across the interface of diffusion couples after PBHT (1000 °C) [6]. Reprinted with the permission of IOP Publishing

and the bond/weld also seems to be more complete treated at 1000 °C. Spherodization or dissolution of graphite flakes is observed on the cast iron sides close to the interface of 900 and 1000 °C treated.

Figure 19.17 shows the typical optical micrograph of the joint interfaces; of (a) 800 °C (b) 900 °C and (c) 1000 °C for 4 h PBHT after etching. From Fig. 17a, a diffusion layer of carbon rich zone has formed but spherodization or dissolution of graphite flakes was not observed on the cast iron sides close to the interface. From Fig. 17b, c spherodization or dissolution of graphite flakes was observed on the cast iron sides close to the interface of specimens. As a result of the above, diffusion layers of spherodization zone and carbon rich zone have formed near the interface of cast iron and low-carbon steel side, respectively.

The effect of PBHT temperature and time on the thickness of diffusion layers is shown in Fig. 19.18. A steady increase in thickness of diffusion layers is observed as diffusion temperature and time consecutively increase.

The diffusion coupled of gray cast iron and low carbon steel was successfully produced by the alternative diffusion welding method and equipment described in

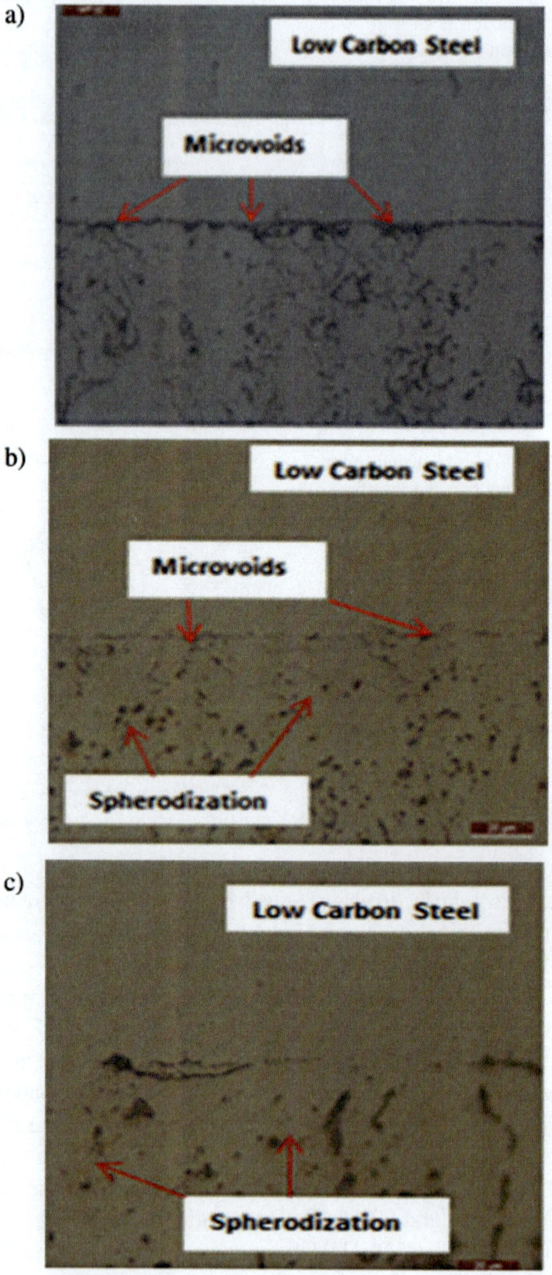

Fig. 19.16 Optical Mocrosopic of the joint interface at **a** 800 °C, **b** 900 °C and **c** 1000 °C for 4 h PBHT before etching [6]. Reprinted with the permission of IOP Publishing

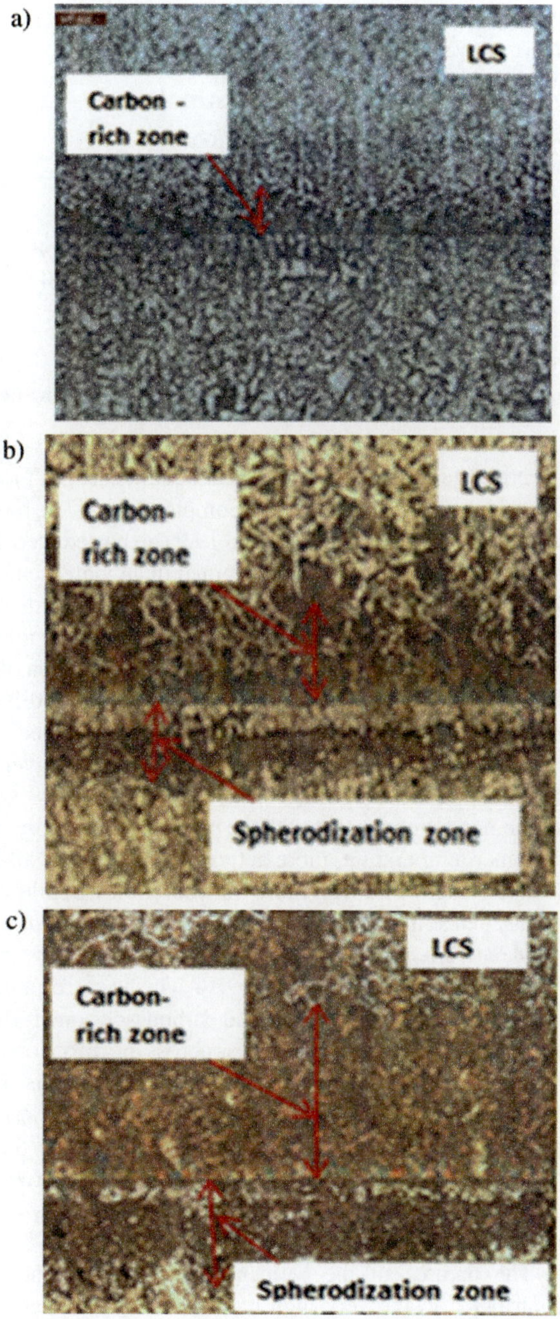

Fig. 19.17 Optical microscopic of the joint's interface at **a** 800 °C, **b** 900 °C and **c** 1000 °C for 4 h PBHT after etching [6]. Reprinted with the permission of IOP Publishing

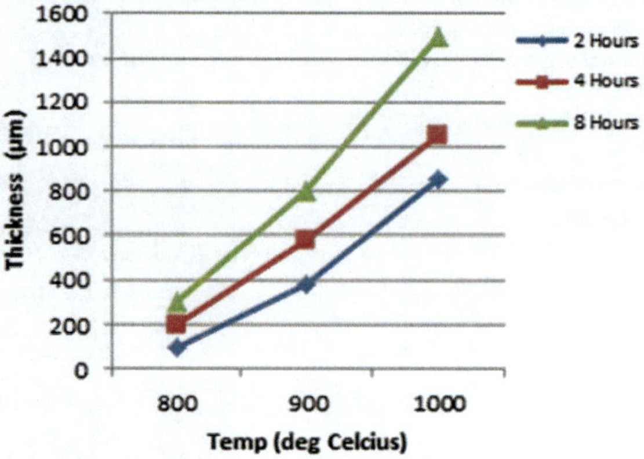

Fig. 19.18 Thickness of diffusion layer [6]. Reprinted with the permission of IOP Publishing

the procedure. Later, post weld heat treatment (PBHT) was conducted to continue the diffusion process to achieve the complete bond. The results obtained from the tensile testing as shown in Figs. 19.6 and 19.9 can be deduced that at higher temperature and longer duration of PBHT had resulted in an increase in tensile strength of the joints. This increase in the strength is correspondent with the reducing and elimination of voids and interface lines observed in the optical microscopic graphs as shown in Fig. 16a–c that complete the bond. This phenomenon of elimination of the interfacial defects (voids) and the original interface at higher diffusion temperatures were also observed in the diffusion bonding of cast iron to medium carbon steel [7].

A steady increase in thickness of diffusion layers is also observed as diffusion temperature and time consecutively increase as shown in Fig. 19.18. All these phenomena is consistent with the principle and theory of diffusion bonding whereby at higher temperature, more activation energy is available for atoms inter-diffusion to take place, while with more time duration will allow more volume of diffusion of atoms, hence improving the bond and the strength of the joints [8].

Meanwhile the hardness of the diffusion layers increased with an increase in temperatures as shown in Fig. 19.14. This is correspondence to the area of diffusion layers of spherodization zone and carbon rich zone formed at higher temperatures as shown at Fig. 17a–c. The hardness across the diffusion layers has it ultimate values at the interface lines and reduced toward the base metals. The above high hardness value can be related to the presence of high amount carbides. According to the researchers [9], increasing hardness value is due to the higher the temperature and the longer times available for carbon to diffuse. This is also correspondently with the decreased of charpy impact values with the increasing times as in Fig. 19.13, which means also increase in brittleness as hardness increases.

The natures of the elements or compositions, the compounds formed that contributed to the very high hardness and brittleness of the diffusion layers especially along the interface lines will be examined in future analysis using scanning electron microscopic (SEM) and the X-Ray diffractometer (XRD). This high hardness values with increased temperature and the formation of carbide at the interface is correspondent with the very low and decreased of charpy impact values at increased temperatures as obtained in Fig. 19.12.

The values, of the tensile and charpy impact strengths as obtained in the results and as discussed above, were found to be low as compared to the base metals. Arising from the fractured surface of the tensile and charpy impact tests specimens it was observed that the bonding or diffusion only occurred at the side of joint interfaces, while there was no diffusion at the center. This effect could be due to the wide gaps or voids still present at the center of the joint interfaces after diffusion bonding prior to the PBHT. The reasons for the incomplete bond of the diffusion bonded couples produced is consistent with the fundamental reason of diffusion welding experiments that were conducted previously but had failed to produce quality joints, namely, due to the welding pressure not being adequately applied, arising from the loss of pressure caused by thermal expansion of the specimens and the fixtures [10]. As weld or bond had still occurred, though minimal, inter-diffusion of atoms across the interfaces was also found to be very minimum. If the pressure is insufficient, some of the voids will be left unfilled, and the strength of the joint will be impaired, to the extent that the bond or joint may not form [8].

Voids at the interface of the joints were not fully eliminated and resulted in complete bond could also be attributed to poor surface preparation [2].

Nevertheless, the subsequent PBHT subjected on the diffusion couples was not fully capable to facilitate further inter-diffusion of atoms across the interfaces to produce complete bond and strong joints.

19.4 Conclusion

The diffusion couples of gray cast iron and low carbon steel were successfully produced by the alternative diffusion welding method and equipment described in the procedure. This allows for the continuance of this research to be done on the diffusion welding and subsequent investigation on the mechanical and metallographic characteristics by conducting PBHT on the diffusion couples. From the tensile, Charpy impact and microhardness test results, a direct correlation between the PBHT's temperature and time and the mechanical strengths was observed. Comparably, the metallographic examinations result also showed a direct correlation with PBHT's parameter. It could be concluded that at higher temperature and longer duration of PBHT would have a strong influence on both the mechanical and the metallographics characteristics of the joints. This is consistent with the principle and theory of diffusion welding whereby at higher temperature, more activation energy is available for atomic inter-diffusion to take place, while with more time duration will allow more volume of diffusion of atoms, hence improving the bond and the strength of the joints.

Acknowledgements Special gratitude to ADTEC Taiping and Universiti Teknologi PETRONAS for facilitating the researchers to conduct mechanical testing and metallographic examinations in their laboratories. Authors would also like to acknowledge other UNIKL MIMET's technicians involved, directly or indirectly, to make this research a success.

References

1. Kazakov NF (1985) An outline of diffusion bonding in vacuum. In: Diffusion bonding of materials. Pergamon Press, Oxford
2. Airu W, Osamu O, Kenji U (2006) Effect of surface asperity on diffusion bonding. Mater Trans 47:179–184
3. Kolarik L et al (2015) Influence of diffusion welding time on homogenous steel joints. Procedia Eng 100:1678–1685
4. Ayob F et al (2020) Unconventional approach in the study of diffusion welding of marine grades grey cast iron to low carbon steel. Mater Today Proc 29:179–174
5. Momono T et al (1990) Diffusion bonding of cast iron to steel under atmospheric pressure. Cast Sci 26:72
6. Ayob F et al (2018) Effect of post heat treatment on the microstructure and microhardness of diffusion coupled gray cast iron and low carbon steel. IOP Conf Ser: Mater Sci Eng. https://doi.org/10.1088/1757-899X/328/1/012001
7. Calvo FA et al (1989) Diffusion bonding of grey cast iron to ARMCO iron and a carbon steel. Mater Sci 24:4152–4159
8. Kazakov NF (1985) Bonding of cast iron and cast iron to steel. In: Diffusion bonding of materials. Pergamon Press, Oxford
9. Abdullah AA, Rawhdan RR (2014) Microstructure characteristics in the interface zone of gray cast iron solid state bonds. Eng Tech J 32:15–22
10. Agilent Technologies (2007) Laser and optics: user's manual. http://www.uzimex.cz/soubory/20080403_laser_optika_cast_1.pdf. Accessed 20 May 2021

Chapter 20
Ergonomic Dynamic Examination Table Innovation Using the Anthropometric Approach and Rational Methods

Aries Abbas, Mohd Razif Idris, Norhisham Seyajah, and Susanto Sudiro

Abstract The examination table is one of the common health facilities. The current examination table condition is a static model which has an overall width of 750 mm, an overall length of 2059 mm, a base height of 650 mm, a weight of 60 kg, and a safe working load (SWL) of 210 kg (2058 N). The thickness of the mat examination table base L 1.900 × W 910 × Th 100–150 (mm). The current use of static examination tables has caused complaints, both from patients and health workers. Health workers stated that it was difficult to help patients to get up from a lying position. The purpose of this research is to innovate the static examination table to become a dynamic table which can be adjusted so that it affects the perception of medical personnel about the quality of care provided to patients. The formulation of the concept of innovation from a static examination table into an ergonomic dynamic using an anthropometric approach and rational methods. Starting from product planning, identifying the needs of patients and health workers, examination table specifications, drafting concepts, concept screening to concept testing, and the use of user anthropometric data as the basis for the dimensions of the examination table. Product innovation uses adjustable hi-lo to meet height-related needs. The use of an adjustable back section to meet the needs related to the process of getting up from a lying position. The innovative dynamic examination table has an overall width of 750 mm, the overall length of 2084 mm, base height 550–870 mm, weight 126 kg, back raise angle 0°–90°, hi-lo adjustment by hi-lo handeland safe working load (SWL) of 210 kg (2058 N).

Keywords Anthropometry · Ergonomics · Product innovation · Examination table

A. Abbas (✉) · M. R. Idris · N. Seyajah
Department Design Engineering Technology, University Kuala Lumpur Malaysia Italy Design Institute, Cheras, Malaysia
e-mail: aries.abbas@s.unikl.edu.my

M. R. Idris
e-mail: mzaiday@unikl.edu.my

N. Seyajah
e-mail: hishams@unikl.edu.my

S. Sudiro
Department of Mechanical Engineering, Pancasila University, Jakarta, Indonesia

A. Ismail et al. (eds.), *Design in Maritime Engineering*, Advanced Structured Materials 167, https://doi.org/10.1007/978-3-030-89988-2_20

20.1 Introduction

An examination table is a common medical device facility that is used to examine patients [1]. The current static examination table has a height that cannot be adjusted to the condition of both medical personnel and patients such as disabled, elderly, and obese patients in the examination table setting in health care [2]. The static examination table function does not yet have settings that can be accessed by the patient and cannot be adjusted so that it has an impact on patient and provider safety [3]. Patients are perceived not only as non-adherent to the examination table and as a barrier to access to equal health care but also as an unsafe and frightening experience, limiting their willingness to go to health care settings. Despite the patient's concerns about adjusting the height of the examination table as a limiting and major safety issue the perception of medical device providers for examination tables on different uses [4]. Patient management has been identified as a significant contributor to musculoskeletal injuries [5]. Musculoskeletal injuries in medical personnel are significant among health workers due to the handling of patients with non-ergonomic static examination tables [6]. Static examination tables that cannot be accessed in placing the patient will cause a high risk of falling and medical personnel at high risk of musculoskeletal injury [7]. These risks can limit the ability of medical personnel to provide adequate health care to patients.

Static examination table: The current examination table condition is a static model which has an overall width of 750 mm, an overall length of 2059 mm, a base height of 650 mm, a weight of 60 kg, and a safe working load (SWL) of 210 kg (2058 N). The use of a static examination table causes complaints/difficulties for patients and health workers. The patient's complaints were as follows: 50 patients (100%) had difficulties reaching the height of the static examination table. This happens because the examination table is too high. Another complaint was the difficulty of getting up from a static examination table stated by 20 patients (43.75%). The difficulty of getting up from a lying position is due to physical limitations, i.e., the difficulty of lifting. Difficulty lifting the upper body when getting up from a lying position [8]. Five health workers stated that it was difficult to help patients to get up from a lying position manually (without assistive devices). With complaints from patients and medical personnel when using the examination bed, it is necessary to redesign a static examination table into a dynamic examination table for ergonomic patients using an anthropometric approach and rational methods [9].

The pickup anthropometric data used is a population sample from anthropometric data. Anthropometric data were taken randomly from 50 people aged 17–35 years. The dynamic examination table design uses an anthropometric approach [10]. The designed dynamic examination table allows the patient to control the bed by himself with a control button. There are three adjustable facilities on this bed, namely: height adjuster, backrest adjuster, and knee fold adjuster [11]. The drivers of the adjustable facility on this bed are the motor (backrest and knee crease) and actuator (height) [6].

The aim of this study was to evaluate the provider of medical devices, namely the examination table which was innovated into a dynamic examination table in an effort to determine the impact of a dynamic examination table that can be adjusted on the safety of patients and medical personnel as health service delivery [12]. Understanding the use of medical equipment can help us to design an ergonomic, dynamic examination table. The dynamic examination table provider opinions will provide useful feedback to continuously improve the quality of medical equipment and ensure that patients always receive the highest standards of care. This approach will help medical personnel to work efficiently to treat patients on any given day, while maintaining the highest standards of patient and medical staff safety [13].

20.2 Methodology

There are two stages in designing a dynamic examination table, namely: the design concept stage and the design stage. The stages of this design concept include (1) identification of consumer needs, (2) determination of product specifications, (3) preparation of product concepts, (4) selection of product concepts and (5) testing of selected product concepts. The next stage is the design stage. This stage is to realize the selected design concept into a dynamic examination table. Anthropometric data is used at this stage. Anthropometric data used in this study came from anthropometric measurements of 50 patients aged 17–35 years. A total of 50 medical personnel from the hospital participated in this study to evaluate their perceptions of the usefulness of the two different examination tables. The medical personnel included doctors, physician assistants, nurse practitioners, nurses, and medical assistants. Thirty-five medical personnel provided feedback on their experience with the static examination table and 20 medical personnel answered a survey based on the dynamic examination table.

20.3 Results and Discussion

Identification of consumer needs is a liaison between customers as the target market with the company's product development. The customers in this study were patients and medical personnel from five hospitals in Jakarta Indonesia. The results of the identification of consumer need in the form of a hierarchy of consumer needs can be seen in Table 20.1.

This research is on medical personnel in connection with the use of examination tables in health examination services for various categories of patients. The literature has identified that health care providers with the highest risk of back pain are those who must assist the patient from the floor to a sitting position on the examination table. Examination table height is the main complaint of patients and health workers. To overcome this complaint, the designed examination table is equipped

Table 20.1 List of questions about using dynamic examination table

How comfortable is the patient sitting on the examination table?	(0 = not at all comfortable, 20 = very comfortable)
How comfortable is the positioning of the examination table?	(0 = not at all comfortable, 20 = very comfortable)
Is the patient comfortable when on the examination table?	(0 = disagree, 20 = agree)
Does the examination table move properly, function properly?	(0 = Never, 20 = Always)
How comfortable are medical personnel when using this dynamic examination table to examine patients?	(0 = not at all comfortable, 20 = very comfortable)
Compared to the static examination table (old product) is it ideal for medical personnel to examine patients?	(0 = not ideal, 20 = ideal)

with an adjustable hi-lo with maximum and minimum height settings referring to the anthropometric dimensions of patients and health workers (Table 20.2).

The strengthened examination table is expected to have a shape and structure that is safe for the patient and comfortable to use [14] besides that it must also:

(a) Able to withstand impact against the support base.
(b) Able to withstand side loads from the mattress support base.
(c) Resistance against the movement of the mattress support base.

This study is limited to the handling and reduction of patient handling activities after the application of dynamic examination tables to health care providers. Furthermore, ergonomically measuring this dynamic examination table to reduce discomfort and body part pain among medical staff providing medical care to patients. In this study, health care providers reported that dynamic examination tables were more convenient and easy to use providing standardized health care when treating different types of patients compared to static examination tables. An ergonomic dynamic examination table can reduce the speed with which medical personnel has to handle patients physically without mechanics. The results of this study indicate that the provider is very well satisfied with the dynamic examination table when the examination table is timely, efficient, and practical for all patients. Furthermore, the height of the examination table can be adjusted according to the height and comfort of medical personnel when observing and examining patients so as to increase patient comfort. This increased comfort can help to reduce the fear or discomfort in patients visiting the clinic for medical treatment.

An up and down movement of the examination table is possible to adjust the position of the mattress to the patient. The mattress is set in the lowest position for ease of patient positioning on the examination table, while for comfort after the patient is lying down. Up and down motion is done by raising the platform mat through vertical motion, while the platform mat remains in a horizontal position.

Table 20.2 List of principles of dynamic Examination table design innovation solutions

List of patient and medical personnel needs	Prerequisite components	Parameters	Data source	Drive components
Examination table is easy to use in the process of ascending and descending the patient	High and low position Examination table and ergonomics according to patient's anthropometry	Examination table settings Max height: Low max: Data from patient an-trometry	Straight height: Waist height: Patient	Pump hydraulic system
Examination table is easy to use in patient examination by medical personnel	Components with adjustable system HI-Lo Examination table that is comfortable and safe to use Ergonomic according to the anthropometry of medical personnel	Examination table settings Max height: low max: Data from medical personnel an-trometry	upright height: waist height: Medical personnel	Pump hydraulic system
The examination table can be moved from a lying position to a sitting position	Components with adjustable system back section Examination table that is comfortable and safe to use with adjustable tilt angle	Minimum tilt angle: 0° max: Referring to SNI	SNI	Pump hydraulic system

The basic principle of the up and down motion is to use a 4 bar mechanism (*fourbar linkage*) which can be seen in Fig. 20.1.

The up and down movement is a vertical motion on the floating link 3 (*floating link*). The link moves vertically without rotation, so the length of link 2 and link 4 must be the same. To generate this movement, it is necessary to place an actuator as a force generator so that the mechanism can function. On link 2 so that link 2 rotates, and the mechanism works, link 3 performs up and down which can be seen in Fig. 20.2.

Fig. 20.1 Fourbar linkage

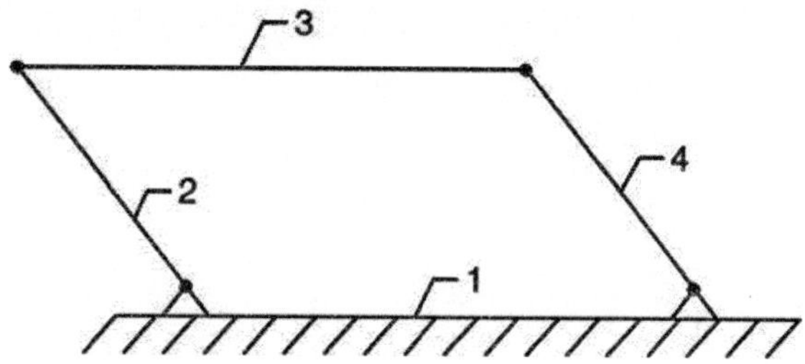

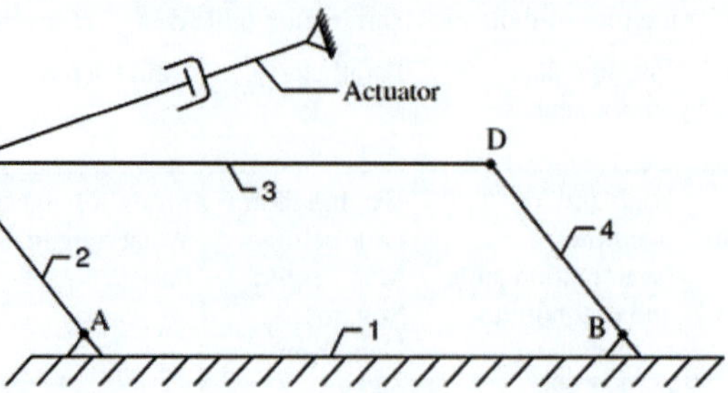

Fig. 20.2 Floating link

Sattic Examination table

This examination table is a static model which has an overall width of 750 mm, an overall length of 2059 mm, a base height of 650 mm, a weight of 60 kg, and a safe working load (SWL) of the 210 kg (2058 N). The thickness of the mat examination table base is L 1.900 × W 910 × Th 100–150 (mm), which can be seen in Fig. 20.3.

Dynamic Examination Table

The innovative dynamic examination table has an overall width of 750 mm, an overall length of 2084 mm, a base height of 550–870 mm, a weight of 126 kg, back raise angle of 0°–90°, hi-lo adjustment by hi-lo handeland safe working load (SWL) of 210 kg (2058 N), which can be seen in Fig. 20.4.

Fig. 20.3 Static examination table

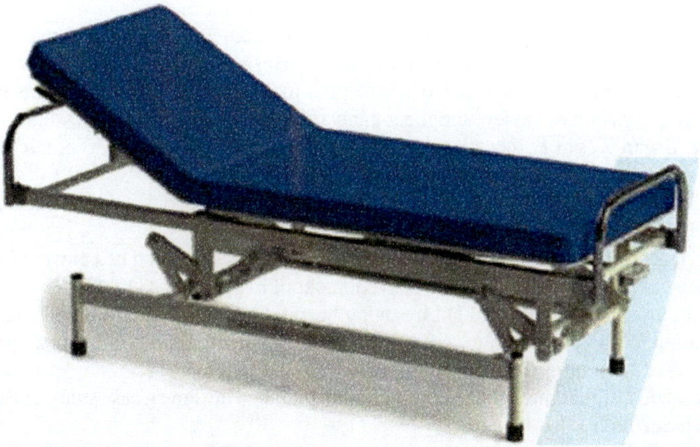

Fig. 20.4 Dynamic examination table

20.4 Conclusion

The designed dynamic examination table has dimensions of an overall width of 750 mm, the overall length of 2084 mm, base height 550–870 mm, weight 126 kg, back raise angle of 0°–90°, hi-lo adjustment by hi-lo handeland safe working load (SWL) of 210 kg (2058 N). This design results in an improved innovation and improved experience of medical personnel (i.e., convenience and usability) in providing health services to various groups of patients, including those with disabilities. The results of interviews with health care providers and patients about the risk outcomes from using the dynamic examination table during the medical examination are ergonomic as measured by the results of patient risks (i.e., risk of falling; risk of injury, worsening of the patient's condition, pain, etc.) and musculoskeletal risks of the medical personnel (e.g., body aches, discomfort or strain, injury, etc.). The dynamic examination table also requires a comprehensive understanding of how applied ergonomics can improve patient's and medical staff safety.

Acknowledgements The authors would like to thank Universiti Kuala Lumpur kampus Cawangan Malaysia Italy Design Institute (UniKL MIDI) for the financial support provided through Grant Scheme.

References

1. Catalano B, Coolidge T (2006) Evaluation and design of a hospital bed to be manufactured and used in China. Huazhong University of Science and Technology Wuhan, China
2. Mehta RK, Horton LM, Agnew MJ, Nussbaum MA (2011) Ergonomics Ergonomic evaluation of hospital bed design features during patient handling tasks. Int J Ind Ergon 41(6):647–652

3. Hopman J, Nillesen M, Both ED, Witte J, Teerenstra S (2015) Mechanical vs manual cleaning of hospital beds: a prospective intervention study. J Hosp Infect 90(2):142–146
4. Carroll R, Metcalfe C, Gunnell D (2014) Hospital management of self-harm patients and risk of repetition: systematic review and meta-analysis. J Affect Disord 168:476–483
5. Liang K, Bryan K, Pin H, Zhecheng Z (2018) Applying Gravity model to predict demand of public hospital beds. Opert Res Health Care 17:65–70
6. Zhou J, Wiggermann N (2017) Ergonomic evaluation of brake pedal and push handle locations on hospital beds. Appl Ergon 60:305–312
7. Daniell N, Merrett S, Paul G (2014) Effectiveness of powered hospital bed movers for reducing physiological strain and back muscle activation. Appl Ergon 45(4):849–856
8. Berge L, Lurås H, Dahl FA (2012) Improving hospital bed utilisation through simulation and optimisation With application to a 40 % increase in patient volume in a Norwegian general hospital. Int J Med Inform 82(2):80–89
9. Islam A (2013) Ergonomics consideration for hospital bed design : a case study in Bangladesh. J Modern Sci Tech 1:30–44
10. Alojado R, Custodio B, Lasala KM, Marigomen PL (2015) Designing an ergonomic chair for pedicurists and manicurists in Quezon City, Philippines. Procedia Manuf 3:1812–1816
11. Christman M et al (2015) Analysis of the influence of hospital bed height on kinematic parameters associated with patient falls during egress. Procedia Manuf 3:280–287
12. Nath T, Kumar A, Lata S (2017) Study of particle dispersion on one bed hospital using computational fluid dynamics. Mater Today Proc 4(9):10074–10079
13. Roemer M (2018) Hospital bed supply and economics of health. Case Stud Public Heal 357–382
14. BSN (2014) Peralatan Elektromedik bagian 2-52: persyaratan khusus keselamatan dasar dan kinerja esensial tempat tidur pasien, SNI IEC 60601-2-52

Chapter 21
The Effect of Corrosion Depth on the Ultimate Strength of an Aging Fixed Offshore Structure

Mohd Hairil Mohd, Nor Adlina Othman, Siti Nur Ain Nazri, Mohd Asamudin A. Rahman, Mohd Azlan Musa, Muhammad Nadzrin Nazri, and Ahmad Fitriadhy

Abstract Due to the high demand for the life extension of offshore structures, it is important to reassess the sustainability of aging offshore structures. Nowadays, many offshore structures in Malaysian waters are now operating beyond their intended design life. Due to the corrosive marine environment, the conditions of the offshore structures deteriorate with time, resulting in corrosion damage, and thus affecting the structural liability and function of the offshore structures. In this study, the four-legged offshore jacket platform was used to determine the effect of the corrosion depth on the ultimate strength of an aging offshore structure. The reserve strength ratio (RSR) was analyzed through pushover analysis by considering the effect of corrosion on global ultimate strength of the structure. The 100-year metocean data condition was applied to two different time-dependent corrosion models. The splash zone was considered as the critical area being corroded among the other areas of the platform. Safety evaluation was considered by the guidance of the PETRONAS Technical Standard guidelines for manned and unmanned conditions. The results from this study show that the platform is safe to operate up to 50 years with average corrosion rate, and it is unlikely unsafe to operate beyond 35 years with a severe corrosion rate.

Keywords Ultimate strength · Reserve strength ratio (RSR) · Pushover · Corrosion · Structural health monitoring

---

M. H. Mohd (✉) · N. A. Othman · S. N. A. Nazri · M. A. A. Rahman · M. A. Musa · M. N. Nazri · A. Fitriadhy
Offshore Engineering Research Interest Group, Maritime Technology Program, Faculty of Ocean Engineering Technology and Informatics, Universiti Malaysia Terengganu, 21300 Kuala Terengganu, Malaysia
e-mail: m.hairil@umt.edu.my

M. A. A. Rahman
e-mail: mohdasamudin@umt.edu.my

M. A. Musa
e-mail: mohdazlan@umt.edu.my

A. Fitriadhy
e-mail: a.fitriadhy@umt.edu.my

A. Ismail et al. (eds.), *Design in Maritime Engineering*, Advanced Structured Materials 167, https://doi.org/10.1007/978-3-030-89988-2_21

21.1 Introduction

Over the past decade, there has been an increase in the number of offshore structures around the world due to the growth in demand and supply for human activities and also industrial activities. Despite the rapidness technologies in constructing the offshore structure, the structural engineer must ensure that the offshore structure is safe to operate over its design life and can withstand several amount of severe load condition such as environmental loads, earthquake loads, ship collisions, and other unforeseen circumstances. One of the compulsory effect needs to always taken into consideration while designing the offshore structure is the effect of corrosion throughout the lifespan of offshore structure. This corrosion effect will cause the overall integrity of the structure to decrease and if not seriously taken into account, collapse and failure of the structure may occur.

There were several researches available in the field of corrosion effect on ships and offshore structures. In particular, the fixed offshore platform is one of the main subjects of research interest. According to the PETRONAS Technical Standard, the offshore structure subjected to extreme environmental conditions must be able to function safely throughout its designated design life of 30 years and more [1]. In the context of Malaysian water, most of the offshore platforms nowadays were operated beyond its design life and most of them already operated over 30 years of age [2]. This particular situation occurred due to several possible reasons such as high cost of removing the existing platform and also the high cost of installing and builds new structure [3]. Same situation was also reported by Shuhud [4], whereby approximately 48% of current operating offshore platforms in Malaysian waters have exceeded their designated design life. In short, although the majority of offshore structures in Malaysian waters were already passing its design life, it is still in service and operating.

Over the years, there were several researches regarding the corrosion effect on ships and offshore structures. The recent study conducted by Bai et al. on the uniform corrosion re-evaluation of offshore jacket structures [5]. Besides, a study was also conducted by Mohd and Paik on the corrosion progress in offshore subsea oil well tubes [6]. They presented an empirical method to estimate time-dependent corrosion damage in offshore oil well tubes based on statistical analysis of corrosion data. A time-dependent ultimate strength performance on corroded FPSOs was conducted by Kim et al. [7]. Other research also was conducted by Melchers [8] on the effect of corrosion was primarily concerned with the structural integrity of steel offshore structures. The prediction of time-dependent corrosion wastage was conducted by Paik et al. [9] on single and double hull tankers, FSOs and FPSOs. Paik et al. also make and investigation on time-dependent risk of aging ship structures [10]. Furthermore, there was a study conducted by Mohd et al. [11] on the time-variant of corrosion wastage model for subsea gas pipelines. On the other hand, the study on the corrosion progress on X70 carbon steel in tropical region was performed by Lim et al. [12].

As stated by Kim et al. [13] in his study, due to corrosive marine environments, environmental effects, and accidental damage, the platform's effectiveness has been steadily decreasing over the years. This situation can lead to structural collapse and intense corrosion, and as consequences, the operator may face hazardous working conditions. In order to avoid these uncertainties, reassessment of the safety of aging fixed offshore platforms is the key factor of extending the design life of the offshore structure. A reassessment of existing offshore structures was conducted by Fayazi and Aghakouchak [3] in the Persian Gulf by considering reliability-based assessment. Reliability reassessment research was also conducted by Bai et al. [14] on the jacket offshore structure.

The effect of corrosion wastage on global ultimate strength is measured by the reserve strength ratio (RSR) through pushover analysis. Pushover analysis is a nonlinear analysis in which the static procedure is used. The vertical load (which functions as a continuous load) is transferred from the deck to the jacket. The lateral load is the amount of the load that would force the structure to exceed its maximum capacity [7]. Through the increasing lateral environmental load, the structure is being "pushed" until the platform fails or collapses.

A study was conducted by Mat Soom et al. [15] is used to determine the RSR value of the jacket platform in Malaysia. They applied a nonlinear analysis, probabilistic model, and pushover analysis for the jacket in order to determine the relation between RSR and exposure time [14]. A study conducted by Kurian et al. applied the approach of a pushover and regression analysis to obtain the RSR of platforms in Malaysian water [16]. On the other hand, there is a research attempted to evaluate the impact of the environmental load factor on the reliability and ultimate strength of existing jacket platforms in Malaysia using another method, the WSD method [17]. Recently, there was a study assessing the reserve strength ratio value through the method of pushover analysis of the corroded jacket platform [13].

There are two different time-dependent corrosion models adopting from Kim et al. [13] (average corrosion model and severe corrosion model) that were used throughout this study. The pushover analysis method was used to determine the ultimate strength of aging offshore structure using the finite element software (SACS).

21.2 Methodology

21.2.1 Offshore Jacket Platform Description

21.2.1.1 Structural Modeling

The platform subjected to this study is named as platform A. It was already in service for several years and consists of a living tetrapod installed in Malaysian waters, and the water depth is 108.40 m. Figure 21.1 shows a three-dimensional representation of platform A. The four-legged platform is a fixed structure which supports a wellhead topside that is also designed to cater a tender assisted drilling rig. On the west face of

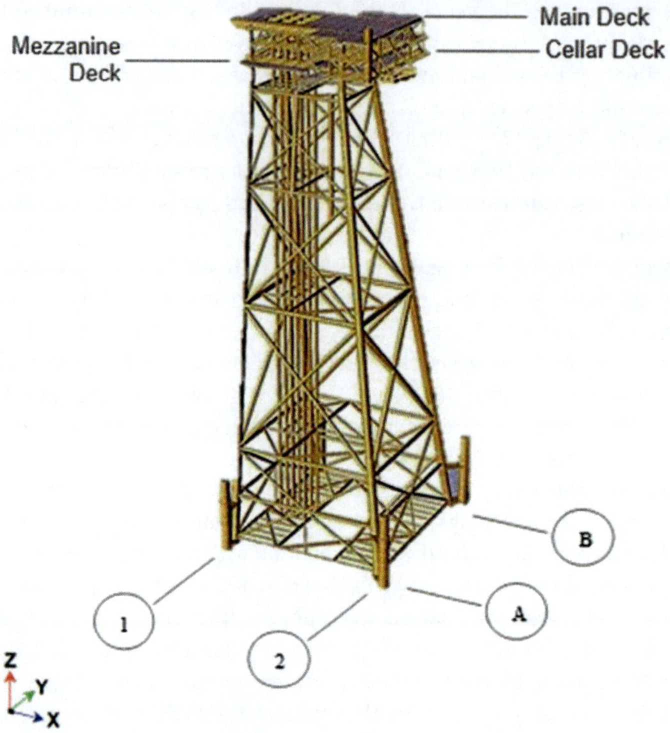

Fig. 21.1 Three-dimensional representation of platform A

the jacket, there is boat landing installed and it is designed to cater seabed subsidence of 4.0 m. The jacket has five framing elevations and is designed to be lift installed. Table 21.1 shows the specification of targeted platform structural design.

Table 21.1 Platform A specification

Feature	Description
Field	Malaysian waters
Design service category	Fixed platform
Design safety category	Manned
Water depth	108.4 m
Platform orientation	135° west of the true north
No. of legs	4
No. of piles	4

Table 21.2 Gravitational load

Dead load	Live load
Modeled structural self-weight	Topside live load
Equipment weight	Mooring load
Piping weight	Main deck live load
Jacket miscellaneous weight	Mezzanine level live load
Topside miscellaneous weight	Cellar deck live load

21.2.1.2 Gravitational Load

Gravity loads consist of structural and facility dead loads, fluid loads, and live loads [18]. Loads being considered in this study are shown in Table 21.2, and they were retained as the design basis of offshore structure.

21.2.1.3 Environmental Modeling

Wind loads, wave loads, current loads, buoyancy loads, ice and mud load as well as seismic loads are the list of environmental loads that are basically needed to be considered [18]. This study will be focused on wind, wave, and current loads. Eight directions of environmental load impact are considered on the targeted aging platform which are 0°, 45.35°, 90°, 134.65°, 180°, 225.35°, 270°, and 314.65° as shown in Fig. 21.2.

Basically, the most important design considerations for an offshore structure during its service life are storm, wind, and storm wave loads [18, 19]. Hence, the metocean data of 100-year return period data was applied to the platform analysis. This condition is the maximum environmental load acting on the offshore structure and is considered as critical condition. This approach was used to calculate the

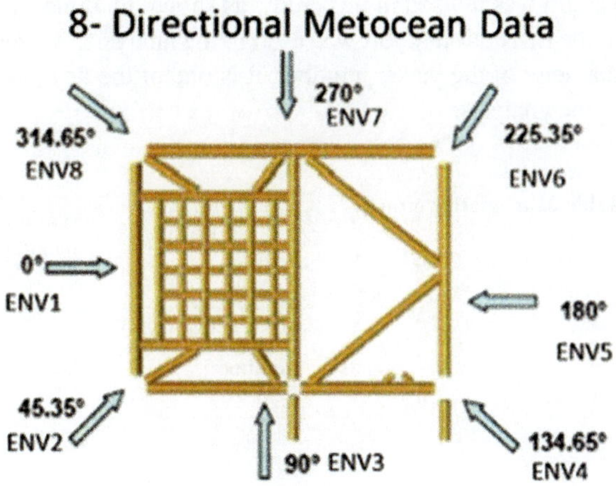

Fig. 21.2 Omni directional metocean data

Table 21.3 Wind load

Location	100-year directional wind	
	X-direction (kN)	Y-direction (kN)
Topside	163.55	190.37
Bridge	13.20	92.24

Table 21.4 Wave load

Parameter	100-year directional wave
H (m)	9.60
T (s)	10.90

Table 21.5 Current load

Depth above mudline	100-year directional current
1.00d	1.88
0.50d	1.49
0.01d	0.87
0.00d	0.40

maximum force and response that is most possible, acting on the targeted jacket platform. Tables 21.3, 21.4, and 21.5 show the 100-year return period metocean data that were applied to the analysis of the structure.

21.2.1.4 Marine Growth

Environmental phenomena such as marine growth can impose additional forces on the offshore structure. Marine growth will increase the member size and surface roughness and thus will increase the wave and current forces [18, 19]. The marine growth pattern was revised in this study, as shown in Table 21.6. The marine growth profile in the inspection report was used in the analysis. As the marine growth increases the diameter of the jacket member, it is one of the key parameter need to be considered in the analysis.

Table 21.6 Marine growth

Elevation (m) (relative to MSL)	Marine growth thickness (mm)	Surface roughness (mm)
At MSL	80	20
1/3 WD from MSL	80	20
Mudline	25	6.25

21.2.1.5 Time-Dependent Corrosion Wastage Model

The corrosion rate acting on the offshore structure comprises three different parts. The three different parts are at the atmospheric zone, splash zone, and immersion zone [13]. By referring to Fig. 21.3, the splash zone is the area where the most critical and fastest rate of corrosion occurred, and as consequence, this particular zone was chosen for the analysis. There were two types of corrosion damage as stated by Bai et al. [5], which are uniform corrosion and localized corrosion. In this study, uniform corrosion was selected to reduce the wall thickness of each platform member [13]. As for localized corrosion, it is normally caused by deterioration in a certain area.

A study by Melchers [8] showed that pitting corrosion pattern was the main reason that causes material loss in offshore and ocean structures. Due to the pitting corrosion, the modeling required more time and effort, as well as uncertainty in corrosion location and nonlinearities [13]. In most cases, uniform corrosion was preferred over pitting corrosion when it comes to numerically modeling corroded structures [13]. Uniform corrosion may give some advantages for the case of whole structural modeling (computational time).

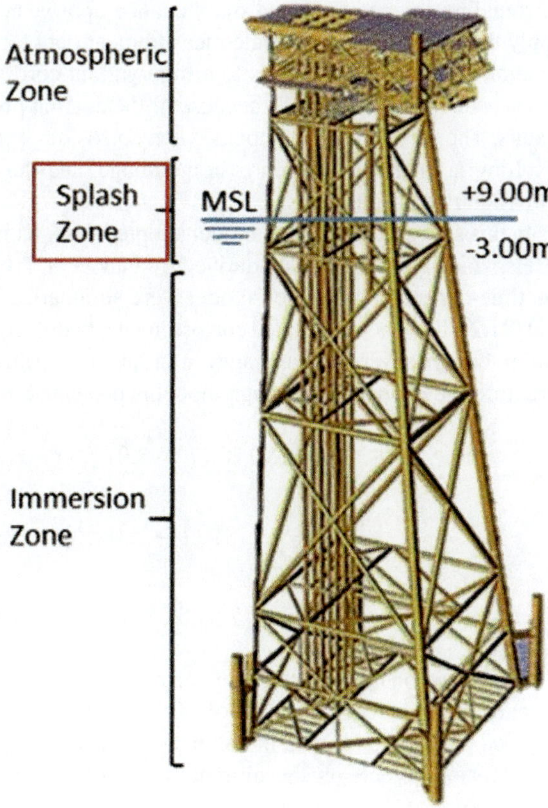

Fig. 21.3 Splash zone area of platform A

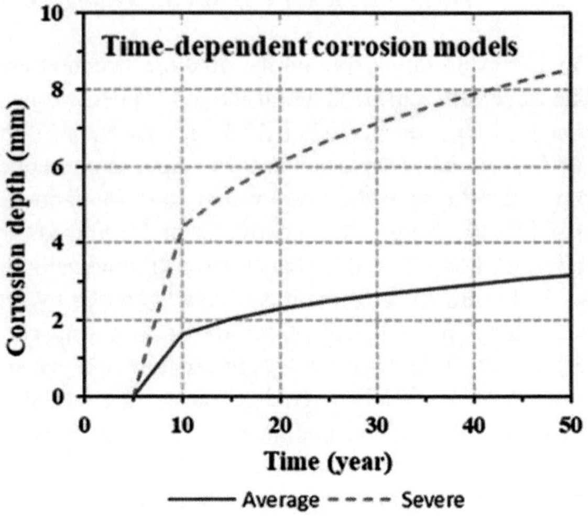

Fig. 21.4 Corrosion depth over time [13]

The corrosion rate may differ based on coating properties and composition, surrounding temperature, and maintenance approach. Basically, it is important to apply the actual and appropriate measurement data for the corrosion rate model. In general, there are three phases to a time-variant corrosion rate model [5]. The first phase is there is no corrosion because of the coating protection; hence, no corrosion occurs. The second phase happens when corrosion protection is no longer effective, resulting in the starting of corrosion damage, and the third phase occurs when the corrosion rate is constant.

In this study, the corrosion model adopted from Kim et al. [13] has two corrosion levels which were previously derived by Paik et al. [10]. The procedures to develop the time-dependent corrosion model were summarized in [9, 13]. Equations (21.1) and (21.2) show the formulated corrosion rate model with different levels of corrosion rate intensity (severe and average) in order to determine the corrosion depth of the structure. The corrosion wastage model is presented in Fig. 21.4.

$$td = 1.0170.(T - 5.0)^{0.3} \text{for average case} \tag{21.1}$$

$$td = 2.7181.(T - 5.0)^{0.3} \text{for severe case} \tag{21.2}$$

where t is the corrosion depth, T is the exposure time in years after the breakdown of coating [13].

The corrosion is assumed to happen after 5 years due to coating protection. Within 5 years' time interval, a period of 50 years of corrosion damage is applied to the targeted platform as shown in Table 21.7. By utilizing pushover analysis, the effect of corrosion defects on the ultimate strength of the aging fixed platform will be examined.

Table 21.7 Corrosion data for numerical simulation

Year	Corrosion depth (mm)	
	Average case	Severe case
5	0.000	0.000
10	1.648	4.405
15	2.029	5.423
20	2.292	6.125
25	2.498	6.677
30	2.671	7.139
35	2.821	7.541
40	2.955	7.897
45	3.076	8.220
50	3.186	8.516

21.2.2 Methodology and Finite Element Analysis

21.2.2.1 Procedure of Analysis

SACS software was used to perform the pushover analysis to the platform A. Nonlinear properties of a structural model and environmental loads are subjected to an incremental lateral load until the structure is unable to resist further loads. This analysis will consist of three main parts.

The first part is for platform modeling and data preparation. The second part is the pushover analysis p using the SACS software. This analysis is used to obtain the platform RSR value and lastly, the identification of structure critical member at the end of the analysis. Figure 21.5 shows the overall flow methodology for this study.

21.2.2.2 Data Preparation

SACS requires three input files when running the pushover analysis, which are the model structure input files, pushover input files, and pile–soil interaction input file (optional). The model input files are the gravitational loads that consist of live load and dead load. For the environmental data, the wind speed, wave height, and current profile are applied in this study. For more accurate result, pile–soil interaction data must also be included.

21.2.2.3 Pushover or Collapse Analysis

Kim et al. stated that static pushover is a regular analysis used to assess the reliability of fixed offshore platforms [13]. Likewise, static pushover analysis is also commonly used nowadays to examine the nonlinear behavior and ultimate capacity

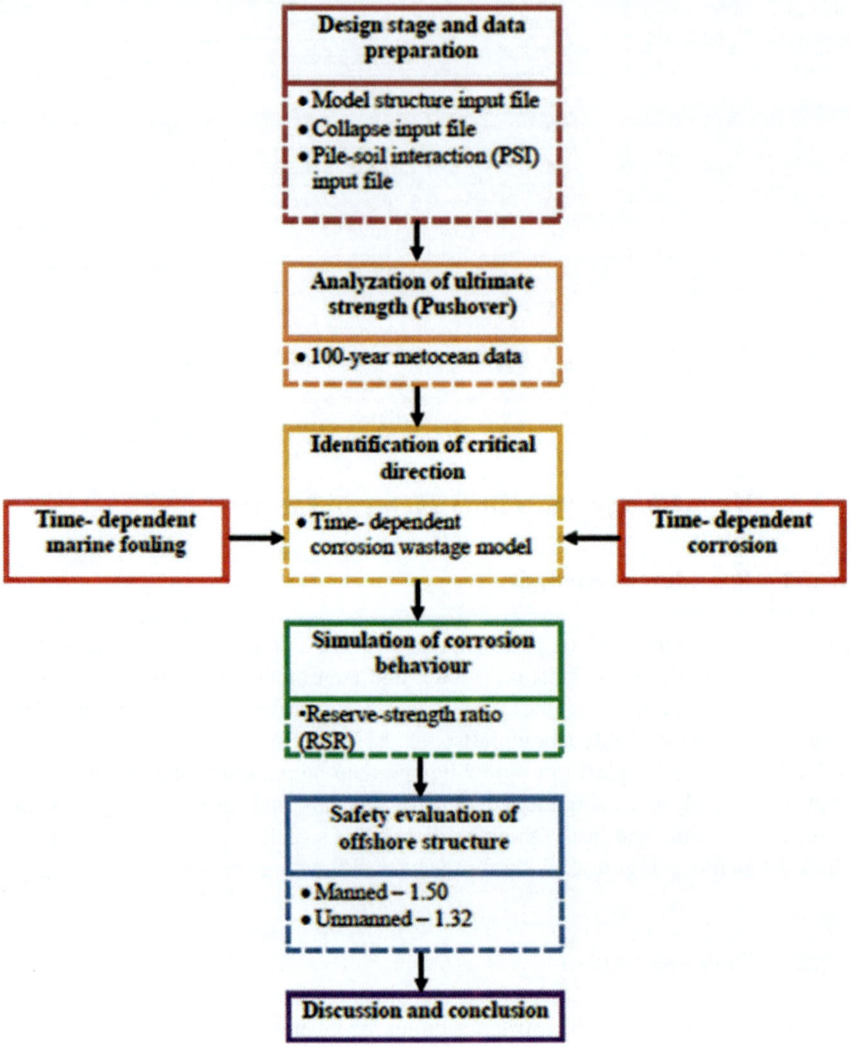

Fig. 21.5 Flowchart of the progress

of the offshore structures in the presence of wind, current, and wave loading [20]. The collapse or static pushover analysis generally includes the combination of extreme lateral storm loading with vertical operational loading acting on the platform structure [13]. Figure 21.6 shows the application of environmental load until the jacket structure collapsed.

Either eight or twelve different directions of environmental loading to determine the ultimate strength of substructure can be used for the analysis. The vertical load is distributed from the offshore platform's deck to the offshore jacket legs, and it

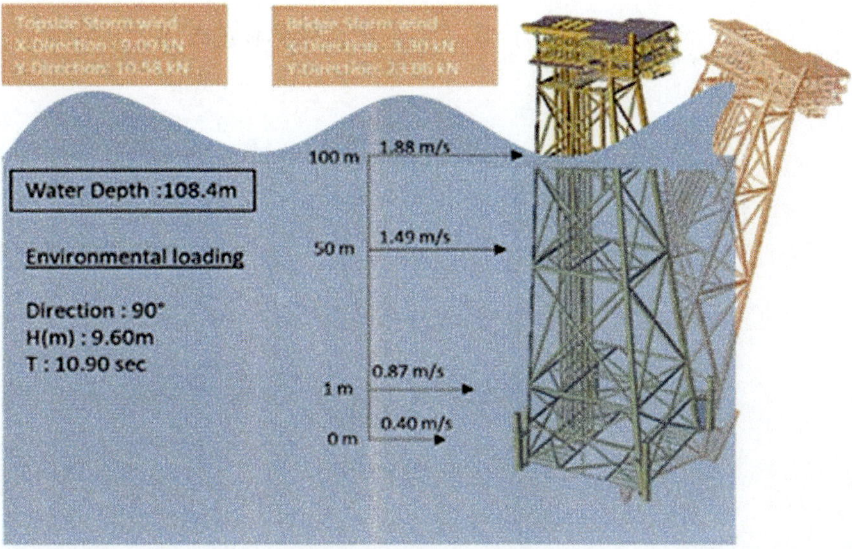

Fig. 21.6 Environmental loading acting on platform A

serves as continuous load [13]. Additionally, the lateral load is the load that requires the amount of force to push the offshore structure to its ultimate capacity [7]. In this study, platform A was subjected to the static pushover analysis and the platform dead and live loads were used in accordance with the design basis.

21.2.2.4 Reserve Strength Ratio (RSR)

The reserve strength ratio (RSR) was calculated using the results of a static pushover analysis. RSR is defined as a ratio of the platform's lateral load (ultimate versus design), which carries the total load at 100 years of design life [21]. It can be used to discover the critical components within the structure and can be implemented to prioritize the inspection and the repair plan. With the increase of lateral environmental load, the structure is being "pushed" until the platform fails or collapses. Water depth, omnidirectional wave height, wind speed, and ocean current velocity are the important factors that can affect the structure when performing the pushover analysis as in Eq. (21.3) [2]:

$$\text{RSR} = \frac{\text{Base Shear Ultimate}}{\text{Base Shear Design}} \tag{21.3}$$

where RSR is the ratio of the platform's base shear design to its ultimate base shear [2].

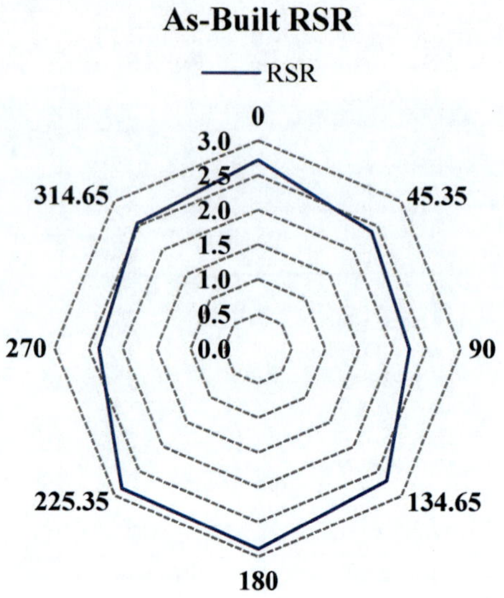

Fig. 21.7 Platform A as-built ultimate strength

21.3 Results and Discussion

21.3.1 As-Built Ultimate Strength

In this study, a method of nonlinear pushover analysis is adopted. The gravitational loads are applied first on the structure, followed by the increment of environmental loads until the aging structure collapses. In pushover analysis, the strength capacity of the structure will be obtained in the expression of RSR value, which is defined as the collapse ratio of the platform's lateral load carrying the total load of design life [21].

Based on the analysis outcome, the minimum RSR was discovered to be at 90° of the fixed structure. Figure 21.7 shows the results of the pushover analysis, which determines the RSR for each direction. As a result, a 90° direction was selected for the application of 100-year metocean loading as an extreme condition, together with thickness loss in the splash zone. Figure 21.7 also demonstrates the RSR value distribution in each of the eight directions.

21.3.2 Impact of Thickness Reduction

Corrosion reduced the thickness of each member at the splash zone. By using Kim's average corrosion model, the RSR value obtained for each year was determined to be almost similar. This is due to the splash zone area of targeted platform that has corrosion allowance of 6 mm for primary members and 3 mm for others. Hence, the reduction in thickness lower than 3 mm was assumed to have no effect on the structure. Instead, it was discovered that the RSR value dropped gradually up to 15 years, remarkably similar to Kim's severe corrosion model's linear trend of the thickness loss.

According to Fig. 21.8, when average corrosion is applied, the lowest RSR occurs at year 50 and is greater than the acceptable value of 1.50. Hence, by not considering the accidental damages or any potential hazards, platform A is safe to operate up to 50. However, when the severe corrosion was applied, the RSR values start to gradually decrease below the allowable value of 1.50. It shows that after around 35 years of the operation, the platform may not safe to operate as shown in Fig. 21.9. It was observed that, at the year 50, the RSR is decreasing to nearly zero and was not able to resist its own gravitational load.

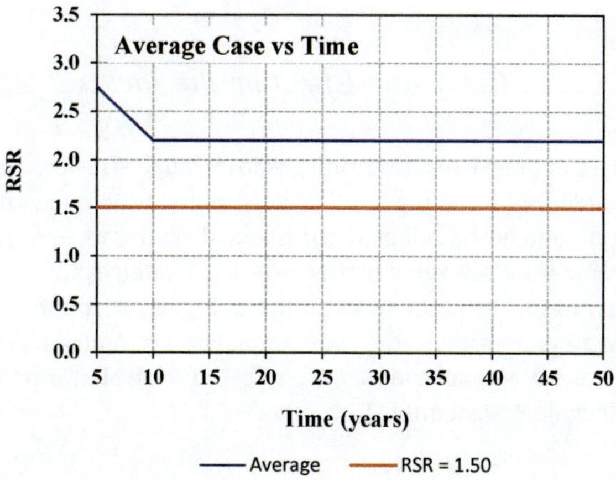

Fig. 21.8 Corrosion at average case

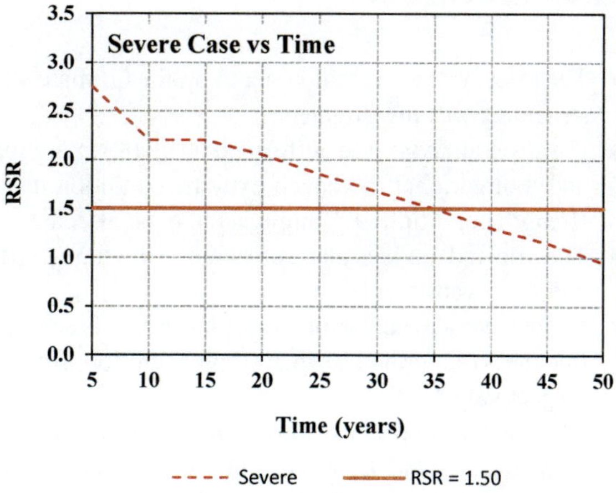

Fig. 21.9 Corrosion at severe case

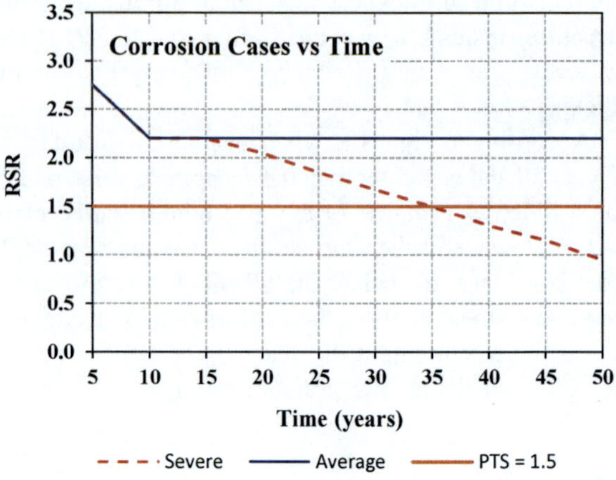

Fig. 21.10 Corrosion at average case and severe case

21.3.3 Corrosion Effect on the Jacket Platform

The obtained results from pushover analysis were compared to the PETRONAS guidelines for aging fixed jackets safety evaluation, where RSR that is greater than 1.50 should be adopted for manned platforms and 1.32 for unmanned platforms [13]. The RSR value of 1.50 was used in this study since platform A was a manned platform. Furthermore, as shown in Fig. 21.10, platform A was safe to operate for up to 50 years with average corrosion, but it is limited to 35 years with severe corrosion. Year 30 was selected as the safe limit for platform A, based on the PETRONAS Technical Standard [1].

21.4 Conclusion

- The effect of the corrosion depth on the ultimate strength of aging fixed offshore structures was investigated.
- Pushover analysis was performed on platform A with maximum gravitational load and 8-directional metocean extreme condition data. When subjected to as-built strength condition, 90° angle is the most affected.
- The affected angle is also subjected to the time-dependent average corrosion case and the severe corrosion case.
- For the average corrosion case, platform A is safe to operate up to 50 years due to thickness reduction less than 6 and 3 mm for primary member and other members respectively
- When exposed to severe corrosion, platform A is no longer safe to operate beyond 35 years. It is also observed that the RSR decrement for severe corrosion was dependent on the trend of the wall thickness reduction.

Acknowledgements This research was funded by the Ministry of Higher Education (MOHE) (Grant No: Matching Grant UTM-UMT-UMP-UTP VOT-53267). The authors are very pleased to acknowledge the help of the anonymous oil and gas company that provided the input files. Special thanks to Universiti Malaysia Terengganu (UMT) for their supports.

References

1. PETRONAS (2012) Design of fixed offshore structures (working stress design). PETRONAS, Kuala Lumpur, Malaysia
2. Mohd MH, Zalaya MA, Latheef M et al (2019) Ultimate bending capacity of aged fixed platform by considering the effect of marine fouling. Lat Am J Solids Struct. https://doi.org/10.1590/1679-78255479
3. Fayazi A, Aghakouchak A (2015) Reliability based assessment of existing fixed offshore platforms located in the Persian Gulf. Int J Marit Technol 4:37–50
4. Shuhud IM (2008) Decommissioning: a Malaysian overview. In: Ascope workshop - regional guidelines for decommissioning and removal of platforms. Denpasar, Indonesia
5. Bai Y, Kim Y, Yan HB et al (2016) Reassessment of the jacket structure due to uniform corrosion damage. Ships Offshore Struct 11:105–112
6. Mohd MH, Paik JK (2013) Investigation of the corrosion progress characteristics of offshore subsea oil well tubes. Corros Sci 67:130–141
7. Kim DK, Liew MS, Youssef SAM et al (2014) Time-dependent ultimate strength performance of corroded FPSOs. Arab J Sci Eng 39:7673–7690
8. Melchers RE (2005) The effect of corrosion on the structural reliability of steel offshore structures. Corros Sci 47:2391–2410
9. Paik JK, Lee JM, Hwang JS, Park Y (2003) A time-dependent corrosion wastage model for the structures of single- and double-hull tankers and FSOs and FPSOs. Mar Technol SNAME News 40:201–217
10. Paik JK, Thayamballi AK, Park YI, Hwang JS (2004) A time-dependent corrosion wastage model for seawater ballast tank structures of ships. Corros Sci 46:471–486
11. Mohd MH, Kim DK, Kim DW, Paik JK (2014) A time-variant corrosion wastage model for subsea gas pipelines. Ships Offshore Struct 9:161–176
12. Lim KS, Yahaya N, Md NN et al (2017) Effects of soil properties on the corrosion progress of X70-carbon steel in tropical region. Ships Offshore Struct 12:991–1003
13. Kim DK, Zalaya MA, Mohd MH et al (2017) Safety assessment of corroded jacket platform considering decommissioning event. Int J Automot Mech Eng 14:4462–4485
14. Bai Y, Yan HB, Cao Y et al (2016) Time-dependent reliability assessment of offshore jacket platforms. Ships Offshore Struct 11:591–602
15. Mat SE, Abu HMK, Mohd ZNI et al (2015) Global ultimate strength assessment (GUSA) for lifetime extension of ageing offshore structures. In: Proceedings of the Twenty-fifth international ocean and polar engineering conference
16. Kurian VJ, Voon MC, Wahab MMA, Liew MS (2014) System reliability assessment of existing jacket platforms in Malaysian waters. Res J Appl Sci Eng Technol 8:2305–2314
17. Cossa NJ, Potty NS, Idrus AB et al (2012) Reliability analysis of jacket platforms in Malaysia-environmental load factors. Res J Appl Sci Eng Technol 4:3544–3551
18. Dehghani A, Aslani F (2019) A review on defects in steel offshore structures and developed strengthening techniques. Structures 20:635–657
19. El-Reedy MA (2012) Offshore structures: design, construction and maintenance, 1st edn. Gulf Professional Publishing, Waltham, USA

20. Abdel RSE (2013) Nonlinear response of fixed jacket offshore platform under structural and wave loads. Coupled Syst Mech 2:111–126
21. Ronalds BF, Trinh S, Cole GK et al (2003) Environmental load distributions: influence of platform configuration. J Constr Steel Res 59:215–231

Chapter 22
An Assessment of Mechanical Properties on Self-Cleaning Concrete Incorporating Rutile Titanium Dioxide

Mohd Syahrul Hisyam Mohd Sani, Fadhluhartini Muftah, and Nazree Ahmad

Abstract Concrete is made of cement, sand, gravel and water to produce an applicable paste, which is continuously hardened over time. There was a lot of problems that occurred in concrete when the strength lost its integrity subjected to load and also a dirty concrete surface due to aggressive air pollution in crowded cities, industrial areas and/or marine structured buildings. Self-cleaning concrete using rutile titanium dioxide (TiO_2) as an additive known as nanomaterial was proposed for this study. The utilisation of photocatalytic materials in concrete was able to solve the problem and also assisted to reduce and degrade air pollution under ultraviolet radiation. The main objective of this study was to determine the mechanical properties of self-cleaning concrete using rutile TiO_2. TiO_2 with percentages of 0.50, 1.00, 1.50, 2.00, 2.50, 3.00 and 3.50% of the total weight of the cement was used. There were two parts in the experimental activity, which included the physical and chemical of the TiO_2 and the mechanical properties of the self-cleaning concrete. From the results, the self-cleaning concrete with 0.50 and 1.00% was reported to have a compressive strength value more than the control mix, approximately 0, 1.16 and 6.0%.

Keywords Self-cleaning concrete · Mechanical properties · Titanium dioxide · Photocatalytic action

M. S. H. Mohd Sani (✉) · F. Muftah
School of Civil Engineering, College of Engineering, Universiti Teknologi MARA (UiTM) Cawangan Pahang, 26400 Bandar Jengka, Pahang, Malaysia
e-mail: msyahrul210@uitm.edu.my

F. Muftah
e-mail: fadhlu@uitm.edu.my

N. Ahmad
Faculty of Applied Sciences, Universiti Teknologi MARA (UiTM) Cawangan Pahang, 26400 Bandar Jengka, Pahang, Malaysia
e-mail: nazreeahmad@uitm.edu.my

A. Ismail et al. (eds.), *Design in Maritime Engineering*, Advanced Structured Materials 167, https://doi.org/10.1007/978-3-030-89988-2_22

22.1 Introduction

Concrete is a building composite material, which is used in construction activities that consists of cement, gravel, sand and water, and added with some admixtures in required proportions for modifying and updating purposes. The additive or additional material used in the concrete paste is depended on the application and solving of the problem's purposes. Concrete is considered throughout the years because it is categorised as the cheapest material in construction, for example, for onshore or offshore buildings, pavements, tunnels, bridges, etc. Nowadays, concrete is studied to produce better strength, durability, workability, flexibility and capability in solving environmental problems. Spiesz et al. [1] reported that concrete was popular due to its mechanical properties, durability, flexibility, architectural options and fairly low price. Therefore, the concrete's ingredients were replaced or added with waste material or chemical substance to obtain a better result and minimising the problems.

The improvement of concrete, especially in the strength, which is depended on the composition and quality of ingredients, is an important way to resist or sustain the load. Recently, the study of improving the strength of concrete was by adding nanoparticles such as titanium dioxide (TiO_2), aluminium oxide (Al_2O_3), and zinc oxide (ZnO). Normally, nanoparticles used in concrete were responsible to fill the gap between the concrete ingredients, to produce the solid condition of the concrete and to promote the strength.

Dikkar et al. [2] reported that TiO_2 had become increasingly popular in recent years and frequently used in food and other consumer goods as an additive. There were four million tons of TiO_2 used in consumer goods, for instance papers, medicine, sunblock cream, toothpaste, paints and inks [3]. Elia et al. [3] reported that TiO_2 was divided into three major forms which were known as anatase, rutile and brookite. TiO_2 was added to the concrete or replaced the cement for exhibiting a self-cleaning property due to photocatalyst actions with the presence of TiO_2. A previous study determined the mechanical properties of self-cleaning concrete by replacing or substituting the cement with TiO_2 [4, 5] and TiO_2 as an additive [6].

Photocatalyst actions were activated by utilising direct UV radiation or direct sunlight and normally could not affect the concrete properties, especially on the strength and durability. Concrete with self-cleaning property is known as self-cleaning concrete or is recognised as photocatalytic concrete [7] or air-purifying concrete [8]. TiO_2 is able to decompose harmful gaseous when the solid body is exposed to sunlight.

The concrete lost its structural integrity when the concrete was observed to have cracked in the shortest time of application and easily broken. Additionally, the concrete was exposed to aggressive environmental problems, especially in crowded cities, developing countries and industrial areas, which was prone to pollutant agents such as transportation, factories and tobacco smoke. Chen et al. [9] mentioned that the air pollution from traffic vehicle's exhaust or vehicle combustion affected the economic development, the environment such as acid rain, ozone depletion and greenhouse effect, and public health such as lung diseases, cardiovascular diseases and respiratory problems. Elia et al. [3] and Odedra et al. [10] stated that the most hazardous air pollutants were recorded such as carbon monoxide, nitrogen oxides (NO_x), lead, sulphur dioxide and volatile organic compounds (VOC). Pollution could contribute a negative impact on human beings, other living organisms and also the concrete. Furthermore, the concrete surface became darker in colour and needed high maintenance cost and cleaning process. Therefore, the study of concrete with additional TiO_2 is to produce a clean concrete surface, reduce the smoke and harmful gases, reduce pollution and improve the strength of the concrete. The pollutants from organic and inorganic sources and particulate matter have the potential to speed up the degradation of concrete in buildings and create significant changes in the material's aesthetic and physical qualities [11].

Several researchers tried to use other materials as replacements or additives, whereby chemical actions were used to produce a better compressive strength. Shchelokova et al. [12] studied the utilisation of an additive on SiO_2-TiO_2 oxides in cement for improving the compressive strength of concrete and parallel with the capability of self-cleaning property. Krishnan et al. [13] studied the effect of compressive strength using the metakaolin-based geopolymer matrix and reported that the result increased with an increasing percentage of TiO_2. Gonzalez-Sanchez et al. [14] used nanosilica, which is known as pozzolanic mineral admixtures and TiO_2 in lime mortar to improve the compressive strength and proposed the self-cleaning property. Satyanarayana and Padmapriya [15] studied the properties of the self-cleaning concrete made by mixing manufactured sand as a sand replacement, palm oil fuel ash as cement replacement and TiO_2. Xu et al. [16] reported that the recycled fine aggregate with a range of 0.16–5 mm and recycled coarse aggregate with a range of 5–25 mm were used in the production of the self-cleaning concrete.

The self-cleaning concrete had created an eco-friendly environment, cleaned the indoor air pollutants inside the concrete and reduced harmful gases from entering the concrete by absorbing ultraviolet radiation [8]. The objective of this study was to determine the mechanical properties of the self-cleaning concrete with the addition of rutile TiO_2.

22.2 Methodology

The study was divided into two parts, which comprised the chemical and physical properties of the TiO_2 and the mechanical properties of the self-cleaning concrete. The physical and chemical properties of TiO_2 were compared with the cement used in the self-cleaning concrete. The cement that was used in the study was the composite Portland cement. The cement and TiO_2 are illustrated in Fig. 22.1. The grade M25 of self-cleaning concrete was prepared with TiO_2 of 0.5, 1.0, 1.5, 2.0, 2.5, 3.0 and 3.5%. A superplasticiser was not used in the study. The specimen of the self-cleaning concrete was with the percentage of TiO_2 on the weight of cement. The proportions of TiO_2 referred to a study by Sorathiya et al. [17] with 0.5, 0.75, 1.0, 1.25 and 1.5% from the weight of cement. The control mix without TiO_2 (0%) was prepared

(a) (b)

Fig. 22.1 Example of the **a** cement and **b** rutile TiO_2

for comparing and verifying the results of the mechanical properties. There were a total of 24 specimens, whereby every percentage of TiO_2 had three specimens. The workability of the self-cleaning concrete was tested and observed for fresh conditions. For hardened conditions, the compressive strength of the cubes (100 × 100 × 100 mm) for seven days, 28 days and 60 days was conducted and determined. The compression machine of 2000 kN capacity determined the compressive strength.

22.3 Results and Discussion

The results and discussion of this study were divided into physical and chemical properties of TiO_2 compared with cement, workability test of fresh self-cleaning concrete and mechanical properties of self-cleaning concrete with different percentages of TiO_2.

22.3.1 Physical Properties

The physical properties of TiO_2 were observed and tested by comparing them with the cement and other previous studies. Table 22.1 tabulates the physical properties of TiO_2 and cement. The result of the physical properties was compared with anatase TiO_2. It showed that the density of rutile TiO_2 was less than cement, but more than anatase TiO_2 liquid-based. Anatase TiO_2 was classified as the most photoactive when compared with rutile, which lowered thermal stability and was inexpensive [18]. Sakthivel et al. [19] reported that the anatase TiO_2 and rutile TiO_2 obtained a tetragonal crystal structure, which was appropriate for photocatalytic action.

Table 22.1 Physical properties of TiO_2 and cement

Physical properties	TiO_2	Cement	TiO_2 [1]
Particles size	1.0–2.0 μm	0.1–250 μm	40–300 nm
Colour and appearance	White powder	Grey powder	Milky liquid
Density (g/cm^3)	3.15	4.23	1.43
Type	Rutile	Composite Portland cement	Anatase

22.3.2 *Chemical Properties*

The chemical properties and surface morphology analysis of cement and TiO_2 were determined using a scanning electron microscope (SEM). Figures 22.2 and 22.3 illustrate the surface morphology of cement and TiO_2 using the SEM images with 500x and 5000x magnification, respectively. From the images, the TiO_2 showed a finer and clearly solid surface, while the cement illustrated consisted of more pores. It was shown that the TiO_2 formed the denser particles, and it was proven that TiO_2 was the finest material and categorised as a nanomaterial group. Therefore, TiO_2 is

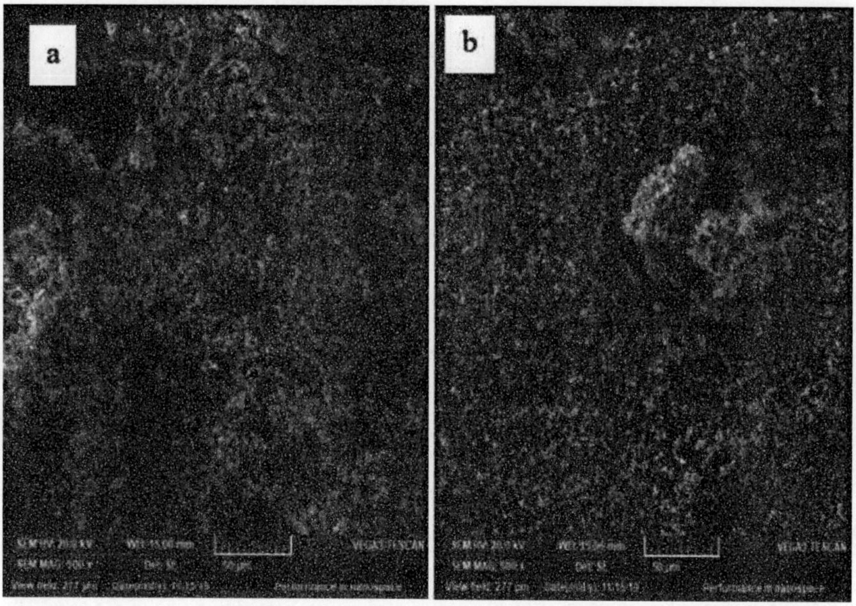

Fig. 22.2 Surface morphology of with 500x magnification of **a** normal cement and **b** rutile TiO_2

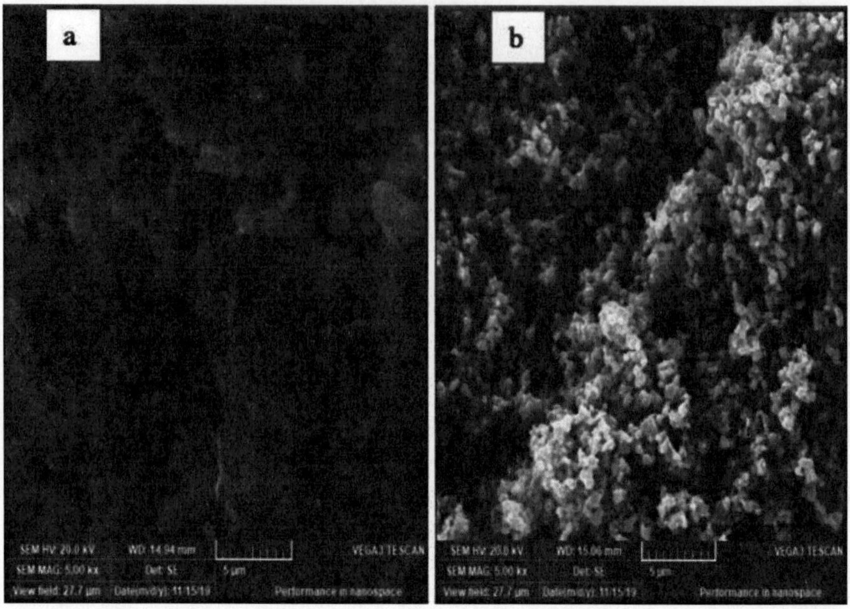

Fig. 22.3 Surface morphology of with 5000x magnification of **a** normal cement and **b** rutile TiO_2

appropriate to be added into the concrete to blend with the cement for producing a compact condition without small air void to produce a stronger concrete.

The energy dispersive X-ray spectroscopy (EDS) technique was used to measure the X-rays emitted from the specimen using an electron beam to determine the elemental composition of the volume. Figure 22.4a and b illustrates the EDS layered electron image and EDS spectrum of cement and TiO_2, respectively. The cement EDS layered electron image was more colourful when compared to TiO_2, due to the content of several elements in it. Besides, the layer of an electron in cement showed to be more compact rather than TiO_2. Table 22.2 illustrates the chemical composition of cement and TiO_2. From Table 22.2, the cement consisted of the highest percentage with the highest peak of calcium (Ca), while titania showed the highest percentage in TiO_2. The percentage difference between rutile TiO_2 with cement and nitrogen-doped TiO_2-SiO_2 for major chemical composition was 1.60% and 10.85%, respectively, for oxygen.

22.3.3 Workability Test

The workability test of fresh concrete was defined by determining the height of the slump. Figure 22.5 represents the result of the height of the slump for different percentages of TiO_2. It shows that the workability of self-cleaning concrete of the

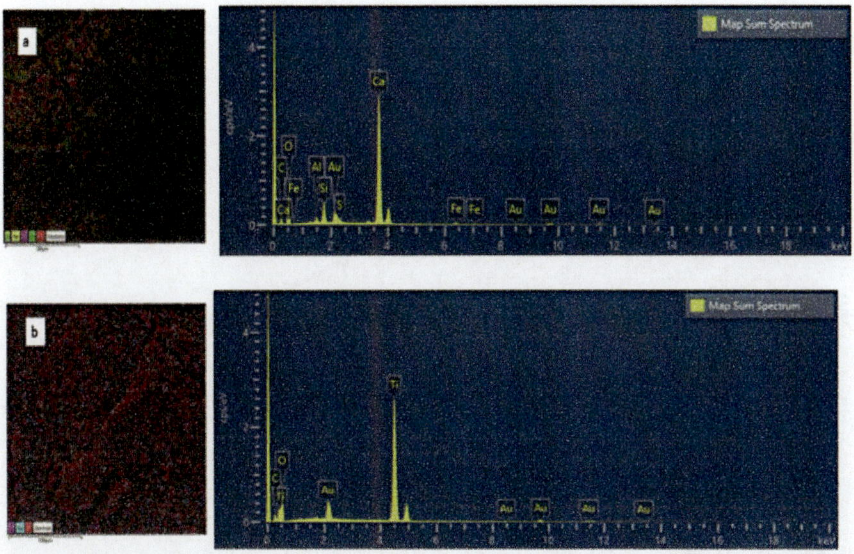

Fig. 22.4 **a** EDS layered electron image and spectrum of cement and **b** the EDS layered electron image and spectrum of rutile TiO_2

Table 22.2 Chemical composition of cement and TiO_2

Chemical composition	Weight of element (%)		
	Cement	TiO_2	Nitrogen-doped TiO_2-SiO_2 [20]
Oxygen (O)	38.60	37.98	42.60
Calcium (Ca)	37.98	–	6.90
Titanium (Ti)	–	38.60	14.42
Aurum (Au)	7.80	3.25	–
Carbon (C)	9.44	9.44	13.10
Silicon (Si)	3.25	–	17.40
Aluminium (Al)	1.04	–	1.30
Iron (Fe)	1.08	–	–
Sulphur (S)	0.81	–	–

control mix (0%) was the same as the specimen with 0.50 and 1.50%. The highest workability amongst the specimens was 1.00% and the lowest workability was 3.50%. The percentage difference between the control mix of 0% with the highest and lowest workability was around 28.57% and 70%, respectively. The slump height was decreased when compared with the percentage of TiO_2 of 1.00% to roughly 28.57% of 1.50%, 50% of 2.00%, 64.3% of 2.50%, 71.4% of 3.00% and 78.6% of 3.50%. Thus, the workability of the concrete by referring to the slump height was decreased when it passed 1.00% of TiO_2. From the results, TiO_2 was reported to have

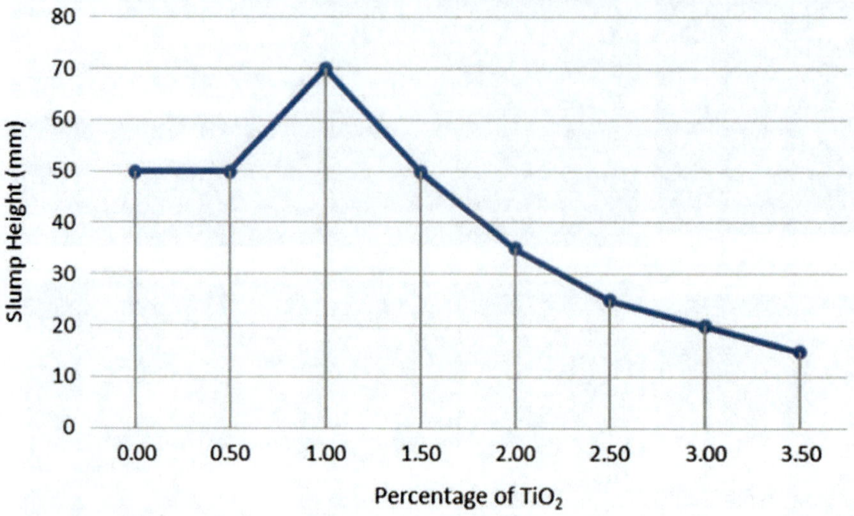

Fig. 22.5 Height of slump with different of the percentage of TiO_2

high water absorption in concrete and should be added with the superplasticiser for maintaining the water content. The result of a study by Sorathiya et al. [17] showed that the slump value was decreased with the increase in the percentage of TiO_2 and the slump value was conservative at 0.5% and 1.0%.

22.3.4 *Compression Strength*

The compressive strength of self-cleaning concrete with different TiO_2 is shown in Table 22.3 and Fig. 22.6. The highest value at 28 days of compressive strength was 28.38 MPa for 1.00% and the lowest value of compressive strength was 20.44 MPa

Table 22.3 Result of compressive strength for 7, 28 and 60 days

Specimen	Compressive strength (MPa)		
	7 days	28 days	60 days
0	18.94	26.68	31.97
0.50%	19.15	26.99	32.90
1.00%	20.12	28.38	35.75
1.50%	17.77	24.68	29.00
2.00%	16.00	22.16	27.80
2.50%	15.67	21.79	24.70
3.00%	14.72	20.44	30.60
3.50%	15.95	22.00	33.80

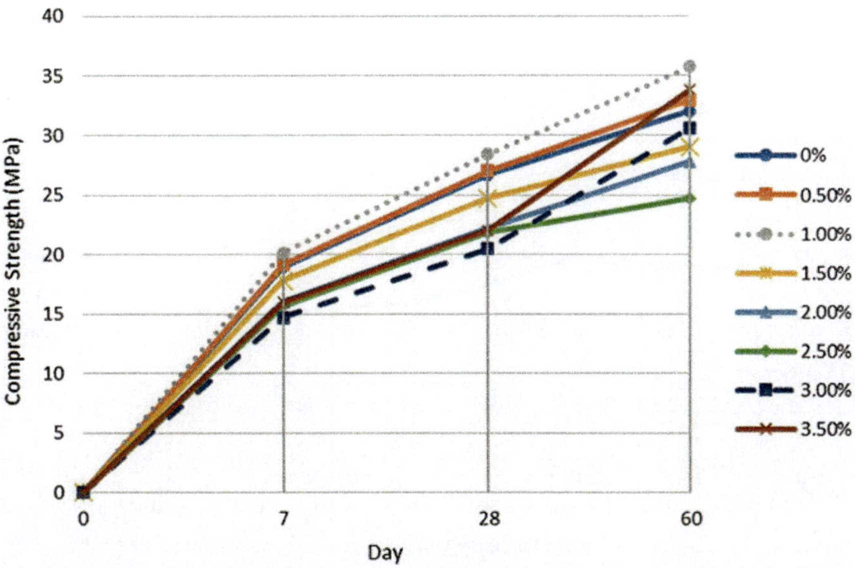

Fig. 22.6 Compressive strength of the self-cleaning concrete

for 3.00%. The percentage difference between the control mix with other specimens were 0.50, 1.00, 1.50, 2.00, 2.50, 3.00 and 3.50% at 28 days, which was reported to obtain 1.17, 6.00, 7.48, 16.93, 18.30, 23.36 and 17.53%, respectively. Two specimens were noted to have a compressive strength value of more than the control mix (0%); a specimen of 0.50 and 1.00% of TiO_2. The compressive strength was increased with the increase in the percentage of TiO_2 from 0.50 to 1.00%, but the compressive strength decreased when the percentage of TiO_2 increased after 1.00% of TiO_2, as shown in Fig. 22.7. Figure 22.7 showed the complete compressive strength with different percentages of TiO_2 for 7, 28 and 60 days. Figure 22.8 illustrated the failure mode of the specimen with 1.0% of TiO_2.

The result of the compressive strength was proven by Dikkar et al. [2] who reported that the strength with 0.5% of TiO_2 was increased, but reduced when the TiO_2 increased to 1.0–1.5%. The reduction of the compressive strength of self-cleaning concrete with 1.0% of TiO_2 from the study of Dikkar et al. [2] was reported to be conservative.

22.4 Conclusion

A series of experimental works was conducted to determine the chemical, physical and mechanical properties of TiO_2 and the self-cleaning concrete with different percentages of TiO_2. From the observation and results, some conclusion could be drawn as follows:

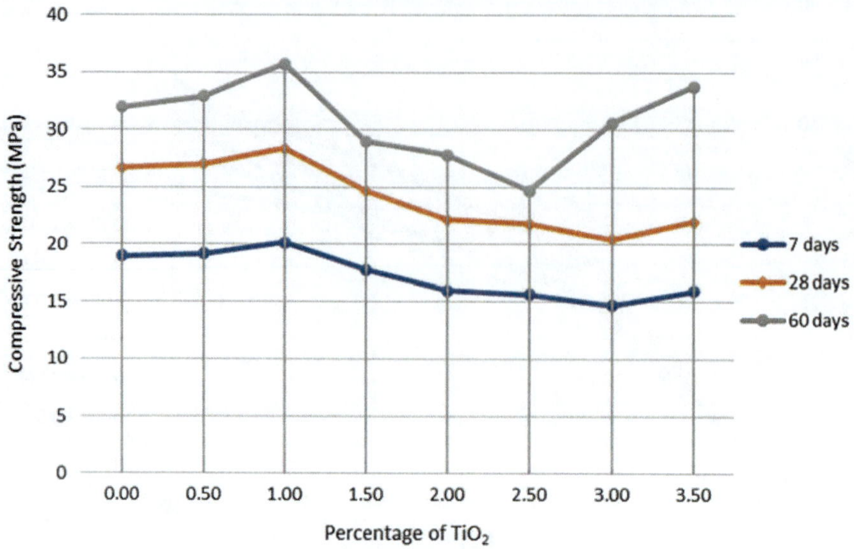

Fig. 22.7 Relationship of the percentage of TiO_2 and compressive strength of self-cleaning concrete

Fig. 22.8 Failure mode of the specimen with 1.0% of TiO_2

1. The workability of specimens was increased from 0% to 1.00% of TiO_2, but the workability of specimens decreased when the percentage of TiO_2 increased, exceeding 1.00%.
2. The compressive strength of self-cleaning concrete increased when the percentage of TiO_2 increased between 0.50 and 1.00%. However, the compressive strength of the self-cleaning concrete decreased with the increase in percentage of TiO_2 from 1.50 to 3.50%.
3. The specimen of self-cleaning concrete with 1.00% of TiO_2 showed the best results amongst the specimens for workability and compressive strength when compared to the control mix.

Further research activities, specifically the experimental, are recommended to obtain the best solution for workability and strength by adding the superplasticiser into the self-cleaning concrete. Additionally, the study of the self-cleaning concrete with an additional percentage of the TiO_2 of up to 10.00% or the study of the self-cleaning concrete with replacement of cement by the TiO_2 should be done to obtain a good relationship.

Acknowledgements The authors gratefully acknowledged the facility's support especially on the laboratory machinery and equipment from the Universiti Teknologi MARA (UiTM), Cawangan Pahang. Special thanks are extended to the lecturers and technicians of Universiti Teknologi MARA (UiTM), Cawangan Pahang for their help during the experimental programmes.

References

1. Spiesz P, Rouvas S, Brouwers HJH (2016) Utilization of waste glass in translucent and photocatalytic concrete. Constr Build Mater. https://doi.org/10.1016/j.conbuildmat.2016.10.063
2. Dikkar H, Kapre V, Diwan A, Sekar SK (2021) Titanium dioxide as a photocatalyst to create self-cleaning concrete. Mater Today Proc. https://doi.org/10.1016/j.matpr.2020.10.948
3. Elia HN, Ghosh A, Akhnoukh AK, Nima ZA (2018) Using nano- and micro-titanium dioxide (TiO_2) in concrete to reduce air pollution. J Nanomed Nanotechnol. https://doi.org/10.4172/2157-7439.1000505
4. Behare HS, Bhosale AN, Kadale JC, Kale SB et al (2021) Investigation of self cleaning concrete by using titanium di-oxide. J Emerg Technol Innov Res 8(6):b548–b554
5. Rajamuniasamy M, Praveen D et al (2018) Experimental study on self cleaning concrete by replacing cement by titanium dioxide. Int J Sci Res Rev 7(11):620–629
6. Vignesh T, Sumathi A, Raja MKS (2018) Study on self-cleaning concrete using nano-liquid TiO_2. Int J Eng Technol. https://doi.org/10.14419/ijet.v7i3.12.16551
7. Shen W, Zhang C, Li Q, Zhang W et al (2015) Preparation of titanium dioxide nano particle modified photocatalytic self-cleaning concrete. J Clean Prod. https://doi.org/10.1016/j.jclepro.2014.10.014
8. Visali C, Priya AK, Dharmaraj R (2021) Utilization of ecofriendly self-cleaning concrete using zinc oxide and polypropylene fibre. Mater Today Proc. https://doi.org/10.1016/j.matpr.2020.06.309
9. Chen C, Tang B, Cao X, Gu F, Huang W (2021) Enhanced photocatalytic decomposition of NO on Portland cement concrete pavement using nano-TiO_2 suspension. Constr Build Mater. https://doi.org/10.1016/j.conbuildmat.2020.122135
10. Odedra RK, Parmer KA, Arora NK (2014) Photocatalytic self-cleaning concrete. Int J Sci Res Dev 1(11):2521–2523
11. Wang Z, Yu Q, Gauvin F, Feng P et al (2020) Nanodispersed TiO_2 hydrosol modified Portland cement paste: the underlying role of hydration on self-cleaning mechanisms. Cement Concrete Res. https://doi.org/10.1016/j.cemconres.2020.106156
12. Shchelokova EA, Tyukavkina VV, Tsyryatyeva AV, Kasikov AG (2021) Synthesis and characterization of SiO_2-TiO_2 nanoparticles and their effect on the strength of self-cleaning cement composites. Constr Build Mater. https://doi.org/10.1016/j.conbuildmat.2021.122769
13. Krishnan U, Sanalkumar A, Yang EH (2021) Self-cleaning performance of nano-TiO_2 modified metakaolin-based geopolymers. Cem Concr Comp. https://doi.org/10.1016/j.cemconcomp.2020.103847
14. Gonzalez-Sanchez JF, Tasci B, Fernandez JM, Navarro-Blasco I, Alvarez JI (2021) Improvement of the depolluting and self-cleaning abilities of air lime mortars with dispersing admixtures. J Clean Prod. https://doi.org/10.1016/j.jclepro.2021.126069
15. Satyanarayana D, Padmapriya R (2021) Performance of photocatalytic concrete blended with m-sand, POFA and titanium dioxide. Mater Today Proc. https://doi.org/10.1016/j.matpr.2020.11.949

16. Xu Y, Chen W, Jin R, Shen J, Smallbone K et al (2018) Experimental investigation of photocatalytic effects of concrete in air purification adopting entire concrete waste reuse model. J Hazard Mater. https://doi.org/10.1016/j.jhazmat.2018.04.030
17. Sorathiya J, Shah S, Kacha S (2017) Effect on addition of nano "titanium dioxide" (TiO_2) on compressive strength of cementitious concrete. Kalpa Publ Civil Eng 1:219–225
18. Kumar J, Srivastava A, Bansal A (2013) Production of self-cleaning cement using modified titanium dioxide. Int J Innov Res Sci Eng Technol 2(7):2688–2693
19. Sakthivel R, Arun KT, Dhanabal M, Aravindan V, Aravindh S (2018) Experimental study of photocatalytic concrete using titanium dioxide. Int J Innov Res Sci Technol 4(11):117–123
20. Koli VB, Mavengere S, Kim JS (2019) An efficient one-pot N doped TiO_2-SiO_2 synthesis and its application for photocatalytic concrete. Appl Surf Sci. https://doi.org/10.1016/j.apsusc.2019.06.123

Chapter 23
Blockchain Interoperability: Connecting Supply Chains Towards Mass Adoption

Bryan Phern Chern Teoh and Bak Aun Teoh

Abstract The rapid development of blockchain technology has gained traction among institutions due to the potential benefits it may have in supply chain management. Many countries and institutions are participating in pilot projects and live projects in hopes that it may one day enhance global supply chains with regards to security, speed, disintermediation, traceability, and cost reduction. As the technology continues to develop, more projects have developed their own blockchain networks with their own network design, governance structure, consensus mechanisms, and protocols, in attempting to address specific issues that are prevalent in existing supply chains. However, the development process is mostly vertical as they work to improve their own blockchain network capabilities. These blockchain silos are unable to communicate with other blockchain networks, which presents a significant problem in the effort towards supply chain mass adoption of blockchain technology. The capability to communicate between blockchain networks is known as blockchain interoperability. Current supply chains have been working towards supply chain integration and collaboration. Supply chain adoption will be highly unlikely if the blockchain interoperability issue is not resolved. This paper will discuss the importance of blockchain interoperability within global supply chains, while identifying the possible use cases of the technology if interoperability solutions are not yet available.

Keywords Blockchain · Blockchain interoperability · Supply chain management

B. P. C. Teoh (✉) · B. A. Teoh
Tunku Abdul Rahman University College, Jalan Genting Klang, 53300 Setapak, Kuala Lumpur, Malaysia
e-mail: teohpc@tarc.edu.my

B. A. Teoh
e-mail: teohba@tarc.edu.my

A. Ismail et al. (eds.), *Design in Maritime Engineering*, Advanced Structured Materials 167, https://doi.org/10.1007/978-3-030-89988-2_23

23.1 Introduction

Entering 2021, blockchain technology and cryptocurrencies have had their fair share of global headlines. Such blockbuster news includes Bitcoin being accepted as legal tender in El Salvador [1], cryptocurrency market value surpassing $2 trillion in April 2021 [2], and various others. Despite being undermined by the magnitude of the global pandemic, blockchain technology has continued to develop in various industries. This includes the financial services industry, logistics and supply chain industry, and many others [3]. The technology promises breakthroughs in many areas due to its inherent security, immutability, programmability, transparency, and decentralization [4]. Despite the many benefits and development in this area, the technology has not experienced any form of mass adoption since the birth of Bitcoin more than a decade ago. There are many possible reasons to this phenomenon. According to Feldman [5], the largest obstacles for blockchain adoption include regulatory uncertainties, lack of trust, inability to scale, intellectual property concerns, and compliance concerns. Among the respondents, a whopping 32% of the respondents stated that the inability to bring separate networks and blockchains together was their largest concern. While a majority of blockchain projects are focused on addressing their own issues such as decentralized finance and supply chain traceability, blockchain interoperability and connectivity has also been a major area of development within the community. However, Sheets [6] stated that this is still one of the largest unanswered questions in the blockchain space despite the rapid development in the recent years. Although there is still much room for development, blockchain interoperability represents an essential, yet missing component before supply chains can handle global trade smoothly [7]. This paper discusses the importance of blockchain technology in the context of various areas within global supply chains. This paper also identifies various areas where blockchain technology can be implemented while interoperability solutions are still under development.

23.2 Literature Review

23.2.1 Blockchain Technology

Fundamentally, blockchain technology is a digital ledger that records transactions permanently between two parties without the need for conventional third parties. Blockchain technology has gained popularity and has potential to transform industries because of its unique characteristics. This includes its security, decentralization, immutability, distribution, consensus, and faster settlement [8]. The digital technology eliminates the need for manual and tedious documentation processes. Every transaction on the blockchain is duplicated and dispersed across geographically distributed nodes, ensuring the safety of the information. Malicious actors will need to simultaneously hack all the nodes to successfully corrupt the network.

Besides, all the parties in the network need to have a consensus before a transaction can be recorded, eliminating the need for a centralized authority. These transactions are time-stamped on the blockchain and are recorded chronologically. Lastly, all these features are backed by cryptography-based security, ensuring a safe environment for storage and transactions [9]. Collectively, these features provide an inherent trust within the system where institutions can develop different applications to conduct business activities [10].

There are currently plenty of live blockchain projects. Many countries are already in later stages of developing their own Central Bank Digital Currencies (CBDC) based on blockchain technology such as China, Sweden, Bahamas, and many others. These CBDCs have the potential to replace cash in the future and can reach the unbanked community throughout the countries [11]. Vadgama and Tasca [12] also observed a high participation in the space within the agricultural sector, groceries sector, and logistics sector, while other sectors include the financial sector and retail sector. Within supply chains, the prominent blockchain projects that have at least 1% of the supply chain market share include Bitcoin, Ethereum, Hyperledger, Oracle Blockchain, Quorum, VeChain, Agnostic, Corda, and Ant Blockchain. There is no shortage of service providers in the space, with many more startups and others attempting to build their own blockchain networks. This creates a fragmented supply chain community where each firm has their own version of a blockchain network which is unable to communicate with other blockchain systems. This is in contrast with the significant efforts to integrate global supply chains over the past decade. Hence, blockchain interoperability among these networks can arguably be one of the most important gaps to fill before supply chain mass adoption can take place.

23.2.2 Blockchain Interoperability

As an emerging technology, blockchain developers have the freedom to design the network as there are little to no standards in this development area. Currently, there are thousands of blockchain projects addressing a myriad of use cases such as smart contracts, payments, data storage, provenance, and many others. All these projects are built in silos, specializing in a specific area, having different levels of scalability, security, decentralization, and governed by a unique consensus algorithm [13]. Due to this heterogeneity, the full potential of these blockchain projects is unlikely to be achieved. Therefore, blockchain interoperability is projected to increase in importance as the individual projects begin to mature and enter the adoption phase. It can be defined as the ability for computer systems to exchange and utilize information and the ability to transfer assets between one another without altering the state of the information or asset [14]. Whereas the World Bank Group [15] included interoperability capabilities with off-chain sources. Ideally, entities in separate blockchains would be able to see, share, and access information from another blockchain while maintaining the decentralized nature of a blockchain. There are various interoperability projects in development now, but the standard functionalities to achieve blockchain interoperability include the ability to integrate with existing systems, conduct transactions on various networks and with other blockchains, and managing transactions between different applications on the same blockchain [16]. In addition, interoperability capabilities need to reduce buyer's remorse after they have selected a blockchain as they can freely transact with users in other blockchain networks [17].

Institutions and supply chains are mainly for profit and are typically cost-conscious. The blockchain implementation process needs to be feasible with regards to the implementation cost, while offering a reasonable sum of benefits post-implementation. Without blockchain interoperability the utilization rate of the blockchain will be limited, while companies might need to adopt another blockchain to complete a separate transaction with other parties, compiling the total cost [18]. For example, the blockchain with the largest sum of money, Bitcoin, is unable to connect to the second-largest blockchain, Ethereum, which has the largest number of smart contracts and decentralized applications to date [19]. To achieve complete interoperability, the governance model between the two blockchains needs to be mutually acceptable, data format needs to be standardized, legal concerns need to be alleviated, consensus mechanisms need to be interoperable, smart contract languages need to be accepted by both blockchains, authorization and authentication methods need to be agreed upon, and proprietary information on a public or private blockchain needs to be protected [14].

23.3 Discussion

23.3.1 The Importance of Blockchain Interoperability in the Supply Chain

As mentioned earlier, there is an increase in the vertical development of blockchain technology where these companies become more specialized in the area they are focusing on White and Skyes [20]. This creates a widening distance between the blockchain projects where each project will only address a limited set of problems. The importance and focus on blockchain interoperability have been gaining traction in the recent years. 84% of corporate executives mentioned that their firms are already involved in blockchain development, which can be in the R&D stage, pilot stage, or already live. In addition, there are at least 9150 active blockchain projects as of 2019 [21]. The rapid development and the sheer number of siloed projects are attempting to exist in a world moving towards supply chain integration. Therefore, blockchain interoperability will play a vital role if mass adoption is to take place within global supply chains. There have been conflicting views on the implementation of interoperable blockchains. Many blockchain projects like Polkadot, Cosmos, and Wanchain are working towards interoperable solutions where others like Williams [7] advocate that there is no economic incentive for multiple companies to work together to ensure an interoperable blockchain and that everyone should use the same blockchain and each run a different node to prevent monopoly. Regardless, this section discusses the importance of the blockchain interoperability in the supply chain context.

There are several cases where blockchain interoperability can prove valuable in the supply chain. The global trade finance industry, a major enabler of global supply chains, is projected to reach $10.9 trillion by 2026 [22]. This represents a large economic incentive from industry players to venture into a technology that can potentially facilitate the whole process more efficiently. Currently, many large players in the market have already set up private blockchains and consortia, and some institutions are also part of multiple networks. This includes We.Trade, Marco Polo, and Voltron [23]. However, these blockchains have different governance and consensus structure, preventing them from communicating with each other, ultimately limiting the potential and use cases for the innovative technology. Parties involved in supply chain trade finance may be involved in transactions with multiple parties using different networks. The possibility of high network switching cost may deter lenders away from borrowers who need cash flow as they limit lenders to transactions only within a certain network. These limits may even deter participants from joining the consortia from the beginning. Blockchain allows smooth and secure execution of smart contracts and information sharing [16]. However, these blockchains need to be interoperable with other blockchains specializing in payments or IoT-based blockchains to provide users with a user-friendly and effortless experience. While eliminating the need for supply chain intermediaries, the networks need to be consistent with the product flow as they move through different locations, being handled by different parties.

Figure 23.1 illustrates typical original equipment manufacturers (OEMs) responsible to manufacture and assemble a variety of products to different buyers [24]. The OEMs have different suppliers, who each have their own set of suppliers and buyers. In addition, the OEM's buyers each have their own supply chain comprising of a set of intermediaries such as warehouses and retailers. All the buyers and sellers can be from different geographical regions and may be large or small industry players. As blockchain technology continues to grow vertically, it is highly likely that some of them may adopt their own blockchain networks, join a consortium, or participate in industry blockchain ecosystems. It will be costly for a company's ERP systems to be compatible with all the different blockchain networks, have multiple operating systems, or to install multiple oracles for each blockchain network. This illustration emphasizes the importance of blockchain interoperability, enabling supply chain entities to operate smoothly while being a part of various supply chains.

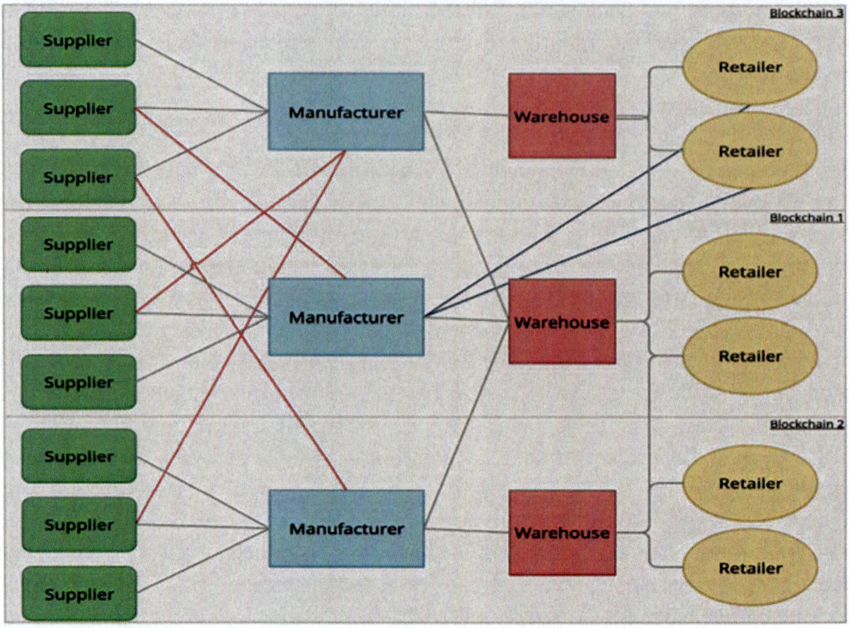

Fig. 23.1 Supply chain complexity illustration

23.3.2 Use Cases for Siloed Blockchain Within Supply Chains

Worst case scenario, if blockchain interoperability does not meet industry standards and is unable to allow blockchains to communicate among each other, institutions can also implement blockchains through industry ecosystems. This is where all the institutions within an industry utilize the same blockchain to conduct supply chain transactions. This enables blockchain implementation without waiting for interoperability solutions to hit the market. For instance, IBM Food Trust is a blockchain ecosystem involving many organizations within the food industry addressing food compliance while Chronicled's Mediledger Project involves many pharmaceutical companies addressing regulatory compliance within the supply chain [25].

IBM Food Trust is built on the Hyperledger Fabric blockchain and has participants from various sectors such as agriculture, food logistics, food manufacturing, groceries, fresh produce, seafood, and restaurants. The common objective throughout the industry is to allow consumers to have access to the full traceability of the products they purchase [26]. This ecosystem has attracted large industry leaders such as Nestle, Walmart, Unilever, Kroger, and many others [27]. For instance, Walmart utilizes the blockchain to trace 25 products from 5 different suppliers, such as milk, vegetables, meat, etc. The implementation reduces the tracing time from 7 days down to 2.2 s [28]. The IBM Food Trust is a private blockchain developed on Hyperledger Fabric specifically for the food industry players to have complete traceability in their food supply chains, which was missing since the beginning of globalization. The private system can speed up tracing time unlike other public blockchains because it operates on a centralized network which significantly increases blockchain scalability. However, some parties may fear the possible monopoly if all food companies were to operate under a single entity such as IBM. Even so, many firms have already signed on as pilot companies because this form of traceability was never available with other centralized non-blockchain applications in the past. The traceability allows food supply chains to ensure the authenticity of their products and prevent food fraud, contain the spread of food contamination, and offers real-time information, ultimately reducing cost and saving lives [29]. The challenges faced by companies in these projects include setting global and industry-wide standards, internal adjustments to adapt to a third-party blockchain network, lack of interoperability with other ecosystems, and determining who controls the governance of the blockchain. These are concerns that need to be addressed before such ecosystems can become an industry-wide standard [25].

In addition, institutions can adopt blockchain technology to enhance specific processes within their supply chains such as data storage, remittance, and product traceability for products involving fewer parties. Blockchain-based data storage may offer users a decentralized alternative to centralized cloud storage such as Google and Amazon. Using the distributed ledger technology, files stored on blockchain-based services like Storj encrypts each file, splits them into pieces, then distributes them to geographically dispersed nodes to ensure a high level of security [30]. This service also reduces data storage costs as they are not required to maintain large, centralized servers. With regards to international remittance services, blockchain-based companies such as Ripple and Stellar enable quick and low-cost remittance by bypassing expensive intermediaries such as conventional remittance companies and financial institutions. The B2B market for cross-border payments is projected to reach $35 trillion in the coming year [31]. Institutions that are involved in global trade can use these services to save cost on transaction fees. Institutions can also acquire the services of blockchain-based product tracking companies like VeChain to ensure product provenance in their supply chain. The turnkey solution does not require much capital investment and can be implemented on a short term or project basis as well [32]. These examples represent ways that blockchain technology can enhance the supply chain, even if interoperability solutions are not yet market-ready.

23.3.3 Possible Solutions

There is a variety of public blockchain projects that are focused on addressing the blockchain interoperability problem such as Polkadot, Cosmos, and various others [16]. Each of these projects has their own unique blockchain designs that addresses the issue in different ways. Kajpust [33] suggested that Cosmos and Polkadot are two of the most promising interoperability projects in the space so far. Cosmos uses a hub and zone model like an airport system. The Cosmos network is the hub while other blockchain networks are zones that connect to the hub. Zones that connect to the hub can interact with the hub and interact with other zones that are connected to the hub. Each zone (individual blockchain networks) can still maintain their own consensus mechanisms. Polkadot uses a similar system but instead of hub and zones, the terms relay chain and parachains are used. Polkadot requires all parachains (individual blockchain networks) to use the same consensus mechanism. These parachains are connected to the relay chain, which is the central connector like the hub used in the Cosmos network [21]. Polkadot has bridges to connect the network to other blockchains such as Bitcoin or Ethereum [13]. Whereas Cosmos has Peg Zones to ensure a blockchain is compatible with the protocols of the Cosmos hub. The main differences between these two projects are the speed, degree of decentralization, security levels, consensus protocols, governance, and developer friendliness [33]. In the future, either both networks can co-exist, one might beat the other, or both networks may be fully interoperable, which might create a large network of interoperable networks that is ripe for supply chain mass adoption. Besides ensuring blockchain networks can communicate with one another, exchanging information and digital assets, companies must also ensure that their networks are able to communicate with off-chain sources to enable smart contract capabilities. This can be done by using blockchain oracles to transmit secure data through hardware oracles (IoT sensors) or software oracles (Web APIs). As mentioned previously, full blockchain interoperability for supply chain mass adoption will require the seamless communication between applications on a blockchain network, between different blockchain networks, and off-chain sources. Currently, Chainlink represents the largest blockchain oracle company that can provide blockchain oracle services to connect blockchains with off-chain sources in a decentralized manner [34].

23.4 Conclusion

Whether it is building an industry ecosystem on Hyperledger Fabric, Cosmos, Polkadot, or other blockchain networks, or using Cosmos and Polkadot to enable blockchain interoperability, the network effect may play an important role in mass adoption. If many players in the same industry have already joined a network, such as in the case of IBM Food Trust, other institutions in the supply chain might be inclined to join the ecosystem as it represents the path of least resistance, or just to avoid being left out. Similarly, if players in the industry were to use Cosmos' or Polkadot's developer kit to develop their own network-compatible blockchain, other players may be inclined to do so. So far, none of the solutions has been able to gain a large market share with regards to institution adoption. This paper discussed and emphasized the importance of blockchain interoperability but does not gravitate towards any of the available solutions. Most of the projects are still in their development stage and can implement new blockchain features soon. For instance, the Polkadot blockchain can implement updates without forks and run on-chain updates [35]. This means that if any blockchain network in the world has found a better solution to interoperability solutions, Polkadot can adopt that idea and implement it on their own chain as an update. Therefore, the sooner a network can get a stable user base, the better the chances of beating the competition. From a supply chain point of view, institutions that are risk averse would wait for the technology to mature before implementation. Whereas larger firms may test the system through pilot projects and may be the first to reap the benefits of blockchain adoption in the supply chain. In conclusion, blockchain interoperability still represents a gap that needs to be addressed before supply chain mass adoption is possible. Future research may study other interoperability solutions that supply chains can implement. In addition, industry-specific interoperability solutions can benefit from more research as well.

References

1. BBC (2021) Bitcoin: El Salvador makes cryptocurrency legal tender. BBC. https://www.bbc.com/news/world-latin-america-57398274. Accessed 1 July 2021
2. Kharpal A (2021) Cryptocurrency market value tops $2 trillion for the first time as Ethereum hits record high. CNBC. https://www.cnbc.com/2021/04/06/cryptocurrency-market-cap-tops-2-trillion-for-the-first-time.html. Accessed 1 July 2021
3. Marr B (2021) The six biggest blockchain trends that everyone should know about in 2021. Forbes. https://www.forbes.com/sites/bernardmarr/2021/03/12/the-six-biggest-blockchain-trends-everyone-should-know-about-in-2021/?sh=2869294d6631. Accessed 1 July 2021
4. Euromoney Learning (2021) What is blockchain? Euromoney Learning. https://www.euromoney.com/learning/blockchain-explained/what-is-blockchain. Accessed 1 July 2021
5. Feldman S (2021) What's blocking blockchain? Statista. https://www.statista.com/chart/17948/worlwide-barriers-to-blockchain-adoption/. Accessed 1 July 2021
6. Sheets R (2021) Interoperability: a big unanswered question in blockchain. University of Arkansas. https://walton.uark.edu/insights/blockchain-interoperability.php. Accessed 1 July 2021
7. Williams N (2021) The false hope of blockchain interoperability. Minespider. https://medium.com/minespider/the-false-hope-of-blockchain-interoperability-258da2af62b. Accessed 2 July 2021
8. Iredale G (2020) 6 key blockchain features you need to know now. 101 Blockchains. https://101blockchains.com/introduction-to-blockchain-features/. Accessed 2 July 2021
9. Deloitte (2017) Key characteristics of the blockchain. Deloitte. https://www2.deloitte.com/content/dam/Deloitte/in/Documents/industries/in-convergence-blockchain-key-characteristics-noexp.pdf. Accessed 2 July 2021
10. Pattison I (2017) 4 characteristics that set blockchain apart. IBM. https://www.ibm.com/blogs/cloud-computing/2017/04/11/characteristics-blockchain/. Accessed 2 July 2021
11. Weisbrodt J, Gross J (2020) CBDS pioneers: which countries are currently testing a retail central bank digital currency? Jonas Gross. https://jonasgross.medium.com/cbdc-pioneers-which-countries-are-currently-testing-a-retail-central-bank-digital-currency-49333be477f4#:~:text=On%2016%20April%202020%2C%20the,Chengdu%2C%20and%20Xiong'an. Accessed 2 July 2021
12. Vadgama N, Tasca P (2021) An analysis of blockchain adoption in supply chains between 2010 and 2020. Frontiers. https://www.frontiersin.org/articles/10.3389/fbloc.2021.610476/full. Accessed 2 July 2021
13. Seba Bank (2020) Blockchain interoperability: towards a connected future. Seba Bank. https://www.seba.swiss/research/blockchain-interoperability-towards-a-connected-future. Accessed 2 July 2021

14. World Economic Forum (2020) Inclusive deployment of blockchain for supply chains: part 6—a framework for blockchain interoperability. World Economic Forum. http://www3.weforum.org/docs/WEF_A_Framework_for_Blockchain_Interoperability_2020.pdf. Accessed 4 July 2021
15. World Bank Group (2020) Blockchain interoperability. World Bank Group. https://documents1.worldbank.org/curated/en/373781615365676101/pdf/Blockchain-Interoperability.pdf. Accessed 4 July 2021
16. De MCRW (2016) Blockchain and interoperability: key to mass adoption. Fin Extra. https://www.finextra.com/blogposting/18972/blockchain-and-interoperability-key-to-mass-adoption. Accessed 4 July 2021
17. Brown RG (2020) The five ingredients of blockchain interoperability. Forbes. https://www.forbes.com/sites/richardgendalbrown/2020/02/13/the-five-ingredients-of-blockchain-interoperability/?sh=37120f4e58a1. Accessed 4 July 2021
18. Ledger Insights (2021) GS1 US: supply chain blockchain has evolved, interoperability isn't optional. Ledger Insights. https://www.ledgerinsights.com/gs1-us-supply-chain-blockchain-has-evolved-interoperability-isnt-optional/. Accessed 4 July 2021
19. Tse S (2021) Blockchain interoperability: why it matters and how to make it happen? Read Write. https://readwrite.com/2021/05/05/blockchain-interoperability-why-it-matters-and-how-to-make-it-happen/. Accessed 4 July 2021
20. White M, Skyes O (2021) Interoperability: is interoperability required? PWC. https://www.pwc.com/m1/en/services/assurance/risk-assurance/digital-and-technology-risk/accelerating-blockchain/overcoming-interoperability-collaboration-challenges.html. Accessed 4 July 2021
21. Akhaldi N (2019) Blockchain interoperability in supply chain. iTransition. https://www.itransition.com/blog/blockchain-interoperability. Accessed 4 July 2021
22. Orbis Research (2021) Global trade finance market anticipating a high boost at a CAGR of 5.4% during 2021–2026. Globe News Wire. https://www.globenewswire.com/news-release/2021/05/26/2236093/0/en/Global-trade-finance-market-anticipating-a-high-boost-at-a-CAGR-of-5-4-during-2021-2026html. Accessed 4 July 2021
23. UnChain (2021) Why is blockchain interoperability so important? UnChain. https://unchain.io/why-is-blockchain-interoperability-so-important/. Accessed 4 July 2021
24. CFI (2021) Original equipment manufacturer. Corporate Finance Institute. https://corporatefinanceinstitute.com/resources/knowledge/other/original-equipment-manufacturer-oem/. Accessed 5 July 2021
25. GS1 (2018) Bridging blockchains: interoperability is essential to the future of data sharing. GS1. https://www.gs1.org/sites/default/files/bridgingblockchains.pdf. Accessed 5 July 2021
26. IBM (2021) IBM Food Trust: a new era for the world's food supply. IBM. https://www.ibm.com/my-en/blockchain/solutions/food-trust. Accessed 5 July 2021
27. Wolfson R (2018) Understanding how IBM and others use blockchain technology to track global food supply chain. Forbes. https://www.forbes.com/sites/rachelwolfson/2018/07/11/understanding-how-ibm-and-others-use-blockchain-technology-to-track-global-food-supply-chain/?sh=2984992f2d1e. Accessed 5 July 2021
28. Hyperledger (2021) Case study: how Walmart brought unprecedented transparency to the food supply chain with Hyperledger fabric. Hyperledger. https://www.hyperledger.org/learn/publications/walmart-case-study. Accessed 6 July 2021
29. Lindley J, Graycar A (2020) Regulating the food supply chain through blockchain. The Regulatory Review. https://www.theregreview.org/2020/12/28/lindley-graycar-regulating-food-supply-chain-blockchain/. Accessed 6 July 2021
30. Storj (2021) Decentralized cloud object storage for developers. Storj. https://www.storj.io/. Accessed 6 July 2021
31. Gupta N (2020) Cross-border payments—Ripple vs Stellar. Akeo Tech. https://medium.com/akeo-tech/cross-border-payments-ripple-vs-stellar-af041523dea1. Accessed 6 July 2021
32. VeChain (2021) What is VeChain. VeChain. https://www.vechain.com/. Accessed 6 July 2021
33. Kajpust D (2018) Blockchain interoperability: Cosmos vs. Polkadot. Dave Kajpust. https://medium.com/@davekaj/blockchain-interoperability-cosmos-vs-polkadot-48097d54d2e2. Accessed 6 July 2021

34. Binance Academy (2020) Blockchain oracles explained. Binance. https://academy.binance.com/en/articles/blockchain-oracles-explained. Accessed 6 July 2021
35. Fransham J (2018) Never fork again. Polkadot Network. https://medium.com/polkadot-network/never-fork-again-438c5e985cd8. Accessed 6 July 2021

Chapter 24
Fire Safety Compliance Amongst Foreign Ships in Malaysian Ports: An Evaluation Using the Flag of Convenience Likelihood Method

Aminuddin Md Arof and Abang Mohammad Syaffiq Idzuan Razak

Abstract As a maritime and trading nation, shipping operation is imperative to the survival of Malaysia's economy. Over a 10 year period, only 13,099 inspections were conducted by the Maritime Operations Division (MOD) of the Marine Department as compared to 607,123 foreign ships that visited Malaysian ports over the same period. By clustering ships using the flag of convenience (FOC) likelihood behaviour approach, this study intends to identify the fire safety performance of visiting ships, whilst at the same time, testing the suitability of the new method to this study. The results generally recorded a total of 6774 inspections with deficiencies (ID) and 2021 inspections with fire safety deficiencies (IFD), which are arguably high compared to the number of inspections performed. Although ships have been clustered into the three FOC likelihood categories, it was found that this does not help to differentiate their performance from other clusters. More than 80% of the ships under the flag states that did not belong to any cluster have recorded inspections with deficiencies, whilst flag states clustered as none/low FOC likelihood behaviour performed poorly against flag states in the other categories. Therefore, some improvements have been suggested to make the new method suitable to be used for other similar studies.

Keywords Flag of convenience · FOC likelihood behaviour · Port state control · Tokyo MOU

24.1 Introduction

Since the volume of international trade continues to increase over the years, applicable rules and regulations on maritime safety are essential to ensure that all ships engaged in transporting seaborne cargo are safe and adequately staffed to prevent

A. Md Arof (✉) · A. M. S. I. Razak
Universiti Kuala Lumpur Malaysian Institute of Marine Engineering Technology, 32200 Lumut, Perak, Malaysia
e-mail: aminuddin@unikl.edu.my

A. M. S. I. Razak
e-mail: abang.razak@s.unikl.edu.my

A. Ismail et al. (eds.), *Design in Maritime Engineering*, Advanced Structured Materials 167, https://doi.org/10.1007/978-3-030-89988-2_24

any risk of accidents. As a maritime country, the maritime sector of Malaysia has been a significant contributor to the country's economic growth, especially in the transportation of more than 90% of the nation's import and export. Maritime transport is also an imperative in connecting the Malay Peninsular with the eastern states of Sabah, Sarawak and Federal Territory of Labuan, supporting the offshore oil and gas industry and cruise tourism. The dependence on the maritime sector for the country's economic growth underscores the need for a dedicated authority to ensure the compliance of domestic and foreign ships to relevant international rules and regulations, whilst visiting her ports. As a result, the government has empowered the Marine Department of Malaysia through its Maritime Operations Division (MOD) to discharge its responsibility in ensuring safety compliance of visiting ships, thereby taking all the necessary measures to prevent any incident that will interrupt the smooth operations of shipping activities and protection of the marine environment. In discharging their duties, the MOD needs to monitor, inspect, and ensure compliance with international safety standards for all merchant ships visiting their ports. The MOD follows regulations enforced under the Tokyo Memorandum of Understanding (MOU) since Malaysia is a member of the Asia–Pacific Port State Control Regime.

Based on the examination of Tokyo MOU annual reports, fire safety measures, which are categorised as Code 07 were the most common detainable defects found onboard merchant ships from 2010 until 2019 [1–10]. Additionally, for four consecutive years between 2016 and 2019, fire protection measures have been the top category of deficiencies found on board merchant ships. According to Transport Statistics Malaysia (2020), 607,123 ships entered Malaysian ports between 2010 and 2019 [11]. In the same period, only 13,099 ships or 2.16% were inspected by the Malaysian authority [10]. Due to the small percentage of ships inspected, there is a high possibility of undetected sub-standard ships visiting Malaysian ports, thus risking the life of other users of the sea and exposing a risk to the marine environment. As fire safety measures continue to become the most prominent detainable deficiencies and are amongst the top categories of deficiencies found onboard ships [9], their compliance with the Tokyo MOU Deficiency Codes concerning fire safety (Code 07) needs to be investigated.

24.2 Flag of Convenience (FOC)

The International Transport Federation (ITF) defines a flag of convenience (FOC) ship as one that flies the flag of a country other than the country of ownership [12]. FOC countries provide a means of avoiding labour regulation in the country of ownership and may become a vehicle for paying low wages and forcing long hours of work and unsafe working conditions on seafarers. Cheap registration fees and low or no taxes are also motivating factors behind a ship owner's decision to "flag out" their ships to the non-traditional flag states. Based on recent statistics, more than 70% of the world's merchant fleet is registered under foreign ship registries led by

Table 24.1 Top 10 flag registration by dead-weight tonnage, 2019

Flag	Number of vessel	Dead-weight tonnage (thousands of tonnes)	Share of world total dead-weight tonnage (%)
(a)	(b)	(c)	(d)
Panama	7860	333,337	17.0
Marshall Islands	3537	245,763	12.0
Liberia	3496	243,129	12.0
Hong Kong (China)	2701	198,747	10.0
Singapore	3433	129,581	7.0
Malta	2172	110,682	6.0
China	5589	91,905	5.0
Bahamas	1401	77,844	4.0
Greece	1308	69,101	3.0
Japan	5017	39,034	2.0

Panama, Marshall Island and Liberia [13]. UNCTAD 2019 report also reveals and illustrates the figure of leading flag registration by dead-weight tonnage for 2019 as per Table 24.1. It demonstrates that the FOC registries such as Panama, Marshall Island, Liberia, Malta and the Bahamas are preferable to the ship owners to register their ships. These FOC countries are amongst the top ten in the world's chosen registries.

It has been argued that FOC countries have been considered as having low safety standards with poor enforcement measures [14]. Since FOC ships have frequently been classified as sub-standard ships, they are described as ships that pose a risk to safety, health and the environment and should therefore be detained [15]. Ships that fly a flag of convenience are prone to accidents due to being manned by a cheaper and less experienced crew. According to Zhang and Drumm, FOC registers are also called open registries that are convenient because they allow ship owners to circumvent more stringent safety and labour laws and more costly tax regimes [16]. The word "open" means that instead of a closed registry that offers ship registration services to ship owners domiciled in the country, the registry is opened to ship owners of any nationality. As sub-standard ships, FOC ship owners tend to compromise on the safety aspect, resulting in the correlation between FOC ships and poor safety records [17]. Arof et al. [18] also conducted a study on compliance with FOC countries' fire safety measures in Malaysian ports. The study found that the level of fire protection provided by ships belonging to FOC countries was still substantially lower than that provided by non-FOC ships. However, it was argued that there was a need for a transparent, repeatable and data-driven approach for prioritisation of all countries based on ship's behaviours associated with FOC rather than a definitive classification whether a particular flag is a FOC or not [19].

Therefore, Ford and Wilcox recently introduced a model-based scoring system that classifies flag states according to how likely they are to act like FOC registries

Table 24.2 Total ships entering Malaysian ports and total inspection conducted

Year	Total ships entering Malaysian port	Total inspections conducted by MPSC	Inspection percentage
S (a)	(b)	(c)	(d)
2019	57,605	1708	2.97
2018	55,754	1891	3.39
2017	55,684	1805	3.24
2016	57,587	1409	2.45
2015	60,862	1275	2.09
2014	61,668	1055	1.71
2013	62,653	1035	1.65
2012	66,761	1064	1.59
2011	64,607	1065	1.64
2010	63,942	792	1.24
Total	607,123	13,099	2.16

especially in analysing illegal, unreported and unregulated (IUU) fishing [19]. The main criteria to identify the FOC likelihood are high ownership ratio, high control of corruption, and low fidelity [19]. This is different from the criteria used by the ITF in defining the FOC that considers factors including those involving number of foreign owned vessels registered; social record of human and trade union rights; the degree of ratification and enforcement of ILO conventions; and safety and environmental record of the flag state [12, 19]. It has been argued that the proposed model has amongst others the advantages of not making prior assumptions about previous or current listing of FOC classification and possible to update as soon as more data becomes available [19].

In the meantime, as Malaysia is one of the signatories of the Tokyo MOU, all the inspections carried out by the MOD will be reported and published in the Asia Pacific Computerised Information System (APCIS) database. Based on Table 24.2, a total of 607,123 ships have visited Malaysian ports from 2010 until 2019 [11]. Compared with preceding statistics, only 13,099 PSC inspections or 2.16% of the total number of ships entering Malaysia were conducted by MOM. Therefore, only 2.16% of vessels have undergone inspections. The total number of ships entering Malaysian ports compared to the number of inspections conducted by MOM from 2010 until 2019 are as per Table 24.2.

24.3 Fire Safety

Fire is considered as one of the biggest threats to safety on board ships [20]. Faulkner also reveals that fire incidents contributed to 60% of marine accidents that lead to the

Table 24.3 Number of inspections and deficiencies between 2010 and 2019

Year	Number of inspections	Inspection with deficiency	Number of deficiency	Number of fire safety deficiency
(a)	(b)	(c)	(d)	(e)
2010	25,762	16,575	90,177	15,998
2011	28,627	18,650	103,549	18,114
2012	30,929	19,250	100,330	20,522
2013	31,018	18,790	95,263	17,539
2014	30,405	19,029	86,560	16,654
2015	31,407	19,142	83,606	15,143
2016	31,678	18,943	81,271	14,960
2017	31,315	18,113	76,108	13,707
2018	31,589	18,091	73,441	13,340
2019	31,372	18,461	73,393	13,178
Total	304,102	185,044	863,698	159,155

loss of ships [21]. Meija et al. [22] analysed deficiencies type under Indian Ocean MOU from January 2002 until December 2009. Fire protection deficiencies are often the first form of deficiencies found and have been the highest rate of deficiencies, with 28.6%. Wu et al. [23] also reveal that fire safety measures were the highest deficiencies found during PSC inspections under Tokyo MOU from 2010 until 2013. Kara and Oksas conducted an analysis based on inspections, detentions and deficiencies rate of all regional PSC MOU performances from 2010 until 2014. Their study proved that the fire safety measure category was the highest type of deficiency recorded in Paris MOU (14.04%), Tokyo MOU (19.19%), Caribbean MOU (16.07%), USCG (21.67%), and Mediterranean MOU (21.89%) [23]. Based on Tokyo MOU annual reports from 2010 until 2019, the fire protection indicator continues to be amongst the top categories of deficiencies reported on board ships over the last ten years. Trend of deficiency for a period from 2010 to 2019 is presented in Table 24.3. In this study, data collection is carried out through the Tokyo MOU online database filtering process for Code 07 (fire-safety-related) deficiencies for the period from January 01, 2010 to December 31, 2019, covering the categories specified as per Table 24.4.

24.4 FOC Behaviour

Since fire has been considered as the main safety threat on board ships, this study intends to observe the likelihood of FOC behaviour as espoused by Ford and Wilcox and fire safety (Code 07) compliances amongst all flag registries that have an inspection recorded by MOD in Malaysian ports between January 01, 2010 and December 31, 2019 [19]. FOC behaviour amongst flag states is determined by observing the

Table 24.4 Tokyo MOU Code 07 descriptions

Code	Descriptions
(a)	(b)
07101	Fire prevention structural integrity
07102	Inert gas system
07103	Division-decks, bulkheads and penetrations
07104	Main vertical zone
07105	Fire doors/Openings in fire-resisting divisions
07106	Fire detection and alarm system
07107	Fire patrol
07108	Ready availability of fire-fighting equipment
07109	Fixed fire extinguishing installation
07110	Fire-fighting equipment and appliances
07111	Personal equipment for fire safety
07112	Emergency escape breathing device and disposition
07113	Fire pumps and its pipes
07114	Remote means of control (opening, pumps, ventilation, etc.) machinery spaces
07115	Fire-dampers
07116	Ventilation
07117	Jacketed high-pressure lines and oil leakage alarm
07118	International shore-connection
07120	Means of escape
07121	Crew alarm
07122	Fire control plan
07123	Operation of fire protection systems
07124	Maintenance of fire protection systems
07125	Evaluation of crew performance (fire drills)
07126	Oil accumulation in engine room
07199	Other (fire safety)

three FOC likelihood indicators that are ownership ratio (OR), control of corruption (CC) and fidelity (FL). OR involves two data sets that were combined to give a measure of ratio of ships nationally flagged to the ships that are nationally owned in dead-weight tonnage (DWT). Meanwhile CC is the estimate of governance performance score that has a possible range of -2.5 (indicating weak governance) to 2.5 (indicating strong governance). Finally, FL is the proportion of time vessels flagged to a nation spend in the home exclusive economic zone (EEZ), territorial sea and archipelagic waters [19]. Overall trends and performance towards compliance with Tokyo MOU Deficiency Codes 07 for the same inspection period were also identified.

24.5 Analysis

Inspections data by MOD on board ships at Malaysian ports between January 01, 2010 and December 31, 2019 were retrieved from Tokyo MOU database. The ownership ratio (OR) for each flag registry that entered Malaysian ports was obtained from the 2019 UNCTAD STAT database. Control of corruption (CC) was obtained from 2019 Worldwide Governance Indicators (WGI) as listed by the World Bank. For fidelity (FL), global AIS data for all MOD 2019 inspected ship flag state vessels from January 01, until December 31, 2019 was used to obtain an average proportion of time (in days) those ships spent in their home exclusive economic zone (EEZ), archipelagic and territorial waters. K mean cluster analysis was conducted to flag registries that were clustered according to their FOC likelihood behaviour based on three indicators that are ownership ratio (OR), control of corruption (CC) and fidelity (FL) as espoused by Ford and Wilcox [19]. Using these indicators, ships were clustered into high, medium and non/low FOC likelihood. Only flag registries with complete data on all three indicators were assessed. There are 84 flag registries that had an inspections history by MOD from 2010 until 2019. Due to incomplete data across all three indicators, only 71 from 84 flag registries were analyzed on their FOC likelihood behaviour. The remaining 13 flag registries were considered as others. Flag registries OR, CC and FL value as per Table 24.5 were clustered thus their FOC likelihood behaviour were obtained as per Table 24.6.

Based on the data at Table 24.5, the mean of OR is 108.72, with a minimum value of 0.03 (United Arab Emirates) and a maximum value of 2500.33 (Antigua and Barbuda). For CC, the mean is 0.31, with a minimum value of −0.05 (Tunisia) and the maximum value of +2.15 (Denmark). For FL, a mean value of 3.52 is obtained with a minimum value of 0.00 (46 out of 71 flag registries with an FL value of 0.00) and a maximum value of 29.57 (Japan). Based on the data at Table 24.5, ships can be clustered into low, medium and high FOC behaviour as per Table 24.6. Overall, MOD conducted 13,099 inspections between 2010 until 2019. Based on Fig. 24.1, cluster 2 has the most significant number of inspections since most of the flag states fall under this cluster with 8868 (68%), followed by cluster 1 with 3388 (26%), others with 517 (4%), and finally cluster 3 with only 326 ships (2%).

Based on Fig. 24.2, cluster 2 has the highest number of inspections with deficiencies (ID) that are 3702 (55%), followed by cluster 1 with 2467 (36%), others with 429 (6%), and cluster 3 with only 176 (3%). However, in terms of percentage of ID over the number of inspections, others have the highest percentage of ID with 82.98%. This is followed by cluster 1 (72.82%), cluster 3 (53.99%) and cluster 2 with 41.75%. The preceding outcomes show that ships in cluster 2 perform even better than other clusters since more than half of their ships inspected did not have any deficiencies. On the other hand, although ships in cluster 1 have been considered as none/low FOC they have high percentage of ID with at least 7 out of 10 ships inspected under their cluster were found to be with deficiencies.

In term of ID on fire safety as presented in Fig. 24.3, cluster 2 has the most significant number of inspections with fire safety deficiencies (IFD) that is 1,088

Table 24.5 Flag Registries OR, CC and FL Value

Flag	OR	CC	FL
(a)	(b)	(c)	(d)
Algeria	0.57	−0.64	0
Argentina	0.42	−0.08	0
Antigua and Barbuda	2500.33	0.28	0
Australia	0.48	1.81	0
Bangladesh	0.91	−0.91	11.2
Bahamas	72.46	1.13	0
Bahrain	2.11	−0.15	29.4
Belize	110.13	−0.14	0
Bolivia	60	−0.63	0
Belgium	0.35	1.51	0
Brunei	22.9	0.8	20.69
Bermuda	0.16	1.24	0
Cayman Island	67.43	0.5	0
Cambodia	56.63	−1.33	0
Cyprus	3.14	0.64	2.95
Chile	0.72	1.01	0
China	0.41	−0.27	28.23
Croatia	0.83	0.13	0
Denmark	0.52	2.15	0.76
Equatorial Guinea	0.11	−1.56	0
Ethiopia	1	−0.49	0
Fiji	9.43	0.38	0
France	0.13	1.32	0
Germany	0.09	1.95	0
Greece	0.2	−0.07	0
Honduras	8	−0.62	0
India	0.7	−0.19	4.89
Indonesia	1.07	−0.25	12.57
Iran	0.24	−0.96	20.82
Italy	0.75	0.24	0
Japan	0.17	1.42	29.57
Jordan	0.43	0.15	0
Kiribati	238	0.34	0
North Korea	1.16	−1.57	0
South Korea	0.17	0.6	9.98

(continued)

Table 24.5 (continued)

Flag	OR	CC	FL
(a)	(b)	(c)	(d)
Liberia	600	−0.85	0
Luxembourg	1.02	2.09	0
Malta	73.78	0.58	0.05
Maldives	0.76	−0.89	0
Marshall Island	342.43	−0.07	0
Myanmar	1.14	−0.59	2.5
Netherland	0.4	2.01	0.99
New Zealand	0.9	2.17	0
Norway	0.36	2.09	1.85
Pakistan	0.94	−0.79	0
Panama	298.35	−0.57	0.51
Philippines	2.57	−0.54	5.93
Portugal	16.1	0.85	1.6
Qatar	0.17	0.72	0
Russia	0.4	−0.85	0
Saudi Arabia	0.7	0.36	11.62
Saint Kitts and Nevis	42.31	0.45	0
Saint Vincent and the Grenadines	239.27	0.77	0
Samoa	0.37	0.64	0
Sierra Leone	312.83	−0.49	0
Singapore	1.05	2.17	3.42
Sri Lanka	1.59	−0.34	6.19
Switzerland	0.05	2.01	0
Syria	0.21	−1.63	0
Thailand	0.83	−0.4	18.97
Tunisia	1.03	−0.05	0
Turkey	0.27	−0.34	0
Tuvalu	1899	0.03	0
United Republic Tanzania	11.38	−0.45	0
United Arab Emirates	0.03	1.15	0
United Kingdom	0.91	1.83	0
United States	0.2	1.32	2.54
Vanuatu	703	−0.16	0
Vietnam	0.88	−0.49	15.17
Hong Kong	1.85	1.68	0

(continued)

Table 24.5 (continued)

Flag	OR	CC	FL
(a)	(b)	(c)	(d)
Taiwan	0.11	1.03	8.37

Table 24.6 Flag registries cluster member and FOC status

Cluster	Status	Cluster Member	Flag
(a)	(b)	(c)	(d)
1	None/Low FOC	10	Bangladesh, Bahrain, Brunei, China, Indonesia, Iran, Japan, Saudi Arabia, Thailand and Vietnam
2	Medium FOC	59	Algeria, Argentina, Australia, **Bahamas**, **Belize**, **Bolivia**, Belgium, **Bermuda**, **Cayman Island**, **Cambodia**, **Cyprus**, Chile, Croatia, Denmark, **Equatorial Guinea**, Ethiopia, Fiji, France, Germany, Greece, **Honduras**, India, Italy, Jordan, Kiribati, **North Korea**, South Korea, **Liberia**, Luxembourg, **Malta**, Maldives, **Marshall Island**, **Myanmar**, Netherland, New Zealand, Norway, Pakistan, **Panama**, Philippines, Portugal, Qatar, Russia, Saint Kitts and Nevis, **Saint Vincent & the Grenadines**, Samoa, Sierra Leone, Singapore, **Sri Lanka**, Switzerland, Syria, Tunisia, Turkey, United Republic Tanzania, United Arab Emirates, United Kingdom, United States, **Vanuatu**, Hong Kong and Taiwan
3	High FOC	2	**Antigua & Barbuda** and Tuvalu
"Others"	Unidentified cluster due to incomplete information	13	**Barbados**, Cook Island, **Comoros**, Dominica, **Faroe Island**, **Gibraltar**, Isle of Man, **Jamaica**, Micronesia, **Mongolia**, Niue, Palau and Togo

Ships categorised as FOC by ITF highlighted in bold

(54%), followed by cluster 1 with 682 (34%), others with 201 (10%), and lastly cluster 3 with only 50 (2%). However, in terms of number of ID in relation to the number of inspections, cluster 2 only experienced 12.27% ID, as compared to cluster 1 with 27.64%, cluster 3 with 15.34% and others with 38.88%, which is the highest compared to other clusters. These findings correlate well with the findings discussed in the preceding paragraph, where cluster 2, which is the medium FOC experienced the lowest percentage of deficiencies and cluster 4 or ships under the

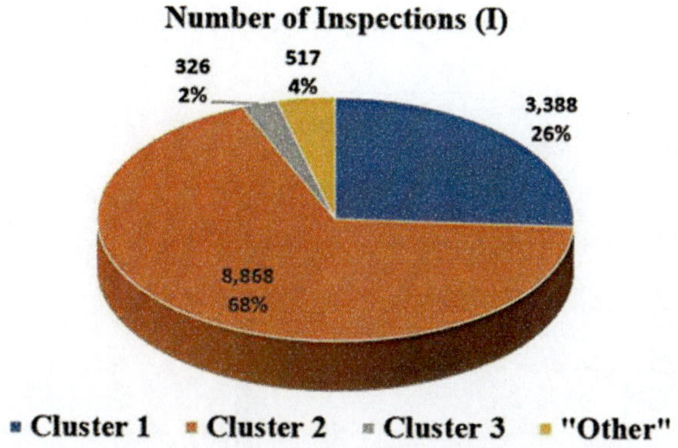

Fig. 24.1 Number of Inspections amongst clusters

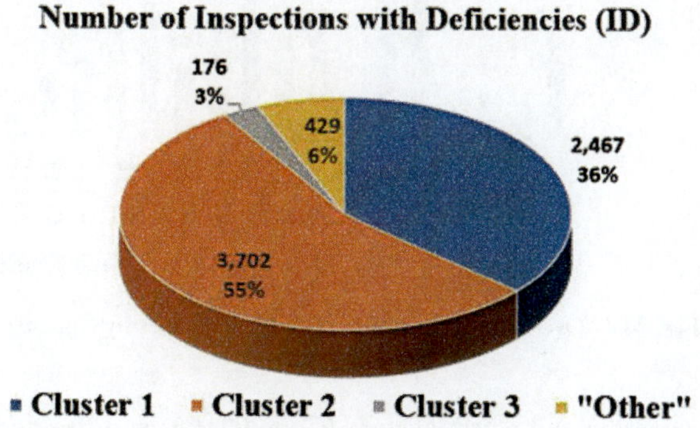

Fig. 24.2 Number of inspections with deficiencies amongst clusters (ID)

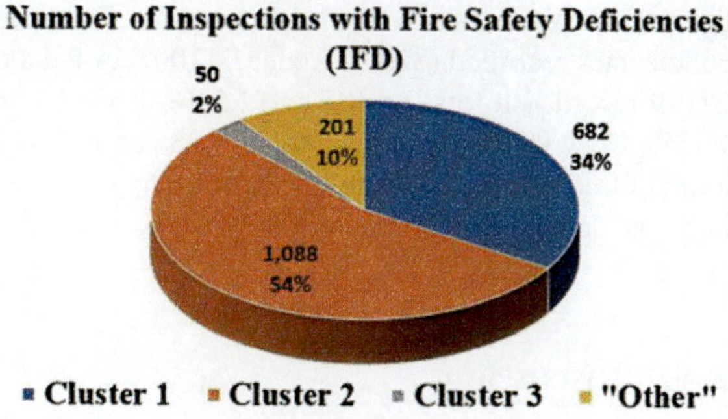

Fig. 24.3 Number of inspections with fire safety deficiencies amongst clusters (IFD)

category of others experienced the highest percentage of deficiencies. Overall trends and performance towards compliance with Tokyo MOU Deficiency Codes 07 for ship inspected by MOD from 2010 until 2019 are summarised as per Fig. 24.4.

A total of 3234 number of fire safety deficiencies in various Tokyo MOU Code 07 Deficiencies (FD) were found during MOD inspections on board ships at Malaysian ports from 2010 until 2019. It can be seen that fire-fighting equipment and appliances (Code 07110), other fire safety (Code 07199), fire control plan (Code 07122), fire

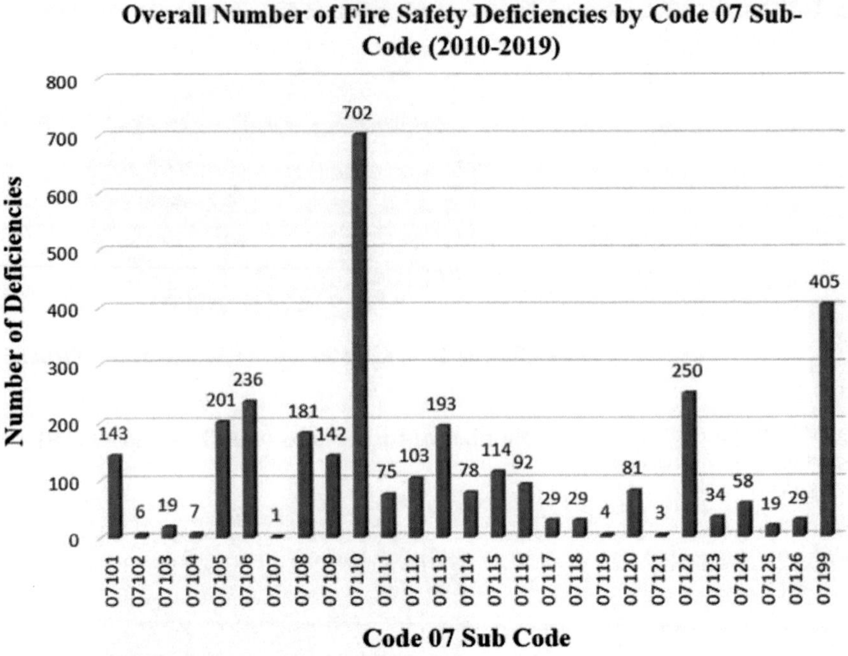

Fig. 24.4 Overall number of fire safety deficiencies by Code 07 sub-code (2010–2019)

detection and alarm system (Code 07106) and fire doors/openings in fire-resisting divisions (Code 07105) become the top five subcategories of fire safety deficiencies discovered onboard ships for the last ten years. From the overall number of fire safety deficiencies recorded (3234), Code 07110 recorded a total of 702 or 21.71%, Code 07199 recorded a total of 405 or 12.52%, Code 07122 recorded a total of 250 or 7.73%, Code 07106 recorded a total of 236 or 7.30%. Lastly, Code 07105 recorded a total of 201 or 6.22%. The lowest subcategories deficiency is Code 07107, with only one deficiency recorded or 0.03%.

24.6 Discussion

The main purpose of this paper was to identify the conformity of foreign ships visiting Malaysian ports between 2010 and 2019 to international fire safety requirements or Code 07 since fire safety has been found as the most significant type of deficiencies on board merchant ships in recent years. The results generally recorded a total of 6774 inspections with deficiencies (ID) and 2021 inspections with fire safety deficiencies (IFD), which are arguably high compared to 13,099 inspections conducted during that period. The ships inspected were clustered using the model-based scoring system introduced by Ford and Wilcox [19]. As discussed earlier, there is a range of advantages to the approach that created a ranked list of all countries based on FOC likelihood indicators. However, due to insufficient data, there are countries that cannot be classified into any of the three categories. Land locked countries will also be at a disadvantage as their ships will not be able to be assessed on the fidelity (FL) component. This is very different from ITF that has identified the countries considered as FOC based on the criteria used.

Based on the list at Table 24.6, six out of 13 countries classified as others (marked in bold) have been classified as FOC by ITF, which can be considered dangerous if these flag states are not classified into any of the categories. This is further strengthened by the findings that placed ships under the unclassified cluster with 82.98% ID compared to the total number of inspections. On the other hand, ships under medium FOC and high FOC clusters performed better than those under none/low FOC cluster although there were 18 out 59 ships and 1 out of 2 ships in the respective clusters that have been classified as FOC by the ITF. Even though all ships in the none/low FOC cluster are also categorised as non-FOC by ITF, their performance in Malaysian ports between 2010 and 2019 in term of PSC inspections with deficiencies or ID had been much worse than those under medium FOC and high FOC clusters. This has further strengthened the argument espoused by Ford and Wilcox [19] that safety performance of ships cannot be solely based on the traditional classification of FOC since some open registries have shown improvement in their safety record especially after the PSC regime was introduced.

However, since there are still weaknesses in identifying ships into the three likelihood of being a FOC categories, it is suggested that the time frame for global AIS data monitoring (for fidelity indicator) to be extended to at least three years. Additionally, one of the criteria used by ITF, i.e. "safety and environmental record of the state" should also be considered as an additional factor as FOC likelihood indicator. Whilst the first suggestion will allow for a better accumulation for a fidelity data especially for ships on time charter that may not have the opportunity to return to their flag states within a particular year, the second will help to reduce the chances of ships with poor track record to be placed into the none/low FOC category.

24.7 Conclusion

There is a range of advantages to the new approach that created a ranked list of all countries based on FOC likelihood indicators. This data-driven approach does not make any claim about the previous or existing listing or classification of FOC. Ford and Wilcox monitored 6 months global AIS data to obtain the fidelity value that was suitable in their study of the IUU fishing vessels. However, this study extended it to a one year monitored time frame to correspond to the seasonal shipping cycle for merchant ships. In spite of the extension, it is suggested that the time frame for global AIS data monitoring (for fidelity indicator) to be extended to three years in order to improve and ensure that every flag registry has an equal chance of being clustered into the three identified FOC likelihood behaviours. Nevertheless, any landlocked country should be excluded from this requirement. Although the number of ships inspected by MOD was arguably low with only 13,099 inspections compared to 607,123 ships that visited Malaysian ports over the ten year period, the number of deficiencies and fire safety deficiencies are arguably high. Although ships have been clustered into the three FOC likelihood categories, it does not help to differentiate their performance from other clusters. More than 80% of the ships under the flag states that did not belong to any cluster have recorded inspections with deficiencies, whilst flag states clustered as none/low FOC likelihood behaviour performed poorly against flag states in the other categories. Therefore, it is also suggested that one of the criteria used by ITF, i.e. "safety and environmental record of the state" should be considered as an additional indicator for the FOC likelihood indicator.

References

1. Tokyo MOU Secretariat (2011) Annual reports on Port state control in the Asia Pacific region, 2010. http://www.tokyo-mou.org/ANN19.pdf. Accessed 8 June 2020
2. Tokyo MOU Secretariat (2012) Annual reports on Port state control in the Asia Pacific region, 2011. http://www.tokyo-mou.org/ANN19.pdf. Accessed 8 June 2020
3. Tokyo MOU Secretariat (2013) Annual reports on Port state control in the Asia Pacific region, 2012. http://www.tokyo-mou.org/ANN19.pdf. Accessed 8 June 2020
4. Tokyo MOU Secretariat (2014) Annual reports on Port state control in the Asia Pacific region, 2013. http://www.tokyo-mou.org/ANN19.pdf. Accessed 8 June 2020
5. Tokyo MOU Secretariat (2015) Annual reports on Port state control in the Asia Pacific region, 2014. http://www.tokyo-mou.org/ANN19.pdf. Accessed 8 June 2020
6. Tokyo MOU Secretariat (2016) Annual reports on Port state control in the Asia Pacific region, 2015. http://www.tokyo-mou.org/ANN19.pdf. Accessed 8 June 2020
7. Tokyo MOU Secretariat (2017) Annual reports on Port state control in the Asia Pacific region, 2016. http://www.tokyo-mou.org/ANN19.pdf. Accessed 8 June 2020
8. Tokyo MOU Secretariat (2018) Annual reports on Port state control in the Asia Pacific region, 2017. http://www.tokyo-mou.org/ANN19.pdf. Accessed 8 June 2020
9. Tokyo MOU Secretariat (2019) Annual reports on Port state control in the Asia Pacific region, 2018. http://www.tokyo-mou.org/ANN19.pdf. Accessed 8 June 2020
10. Tokyo MOU Secretariat (2020) Annual reports on Port state control in the Asia Pacific region, 2019. http://www.tokyo-mou.org/ANN19.pdf. Accessed 8 June 2020
11. Transport Statistics Malaysia (2020) Annual reports on Malaysia Transportation Statistics, 2019. https://www.mot.gov.my/en/Statistik%20Tahunan%20Pengangkutan/Transport%20Statistics%20Malaysia%202020.pdf. Accessed 9 March 2021
12. International Transport Workers Federation (ITF) (2021) Flags of Convenience at www.itfglobal.org/en/sector/seafarers/flags-of-convenience. Accessed 1 June 2021
13. UNCTAD (2019) Investment and new industrial policies. World Investment Report
14. Powell E (2013) Taming the beast: how the international legal regime creates and contains Flags of convenience. Ann Surv Int'l Comp L 19(1):263
15. Cariou P, Mejia MQ, Wolff FC (2009) Evidence on target factors used for port state control inspections. Marine Policy 33(5):847–859
16. Zhang P, Drumm L (2020) The flagging-out strategy: an exam of the impacts on the decreasing German national fleet. Marine Policy 115:103872
17. Alderton T, Winchester N (2002) Flag states and safety: 1997–1999. Marit Policy Manage 29(2):151–162

18. Arof AM, Razak AMSI, Rahmat AK (2019) Compliance on fire safety measures among ships from flags of convenience countries in Malaysian ports. Int J Innov Technol Explor Eng 9(2):4590–4594
19. Ford JH, Wilcox C (2019) Shedding light on the dark side of maritime trade—a new approach for identifying countries as flags of convenience. Mar Policy 99:298–303
20. Azzi C, Pennycott A, Vassalos D (2010) Quant. Risk assessment of shipboard fire by first-principles tools. In: Interflam 2010, the 12th International conference. Nottingham UK
21. Faulkner D (2013) Shipping Saf., a matter of concern, Ingenia, 13
22. Mejia JMQ, Cariou P, Wolff FC (2010) Vessels at risk and the effectiveness of port state control inspections, hal-00470635 at www.hal.archives-ouvertes.fr/hal-00470635. Accessed 9 June 2021
23. Kara EGE, Oksas O (2016) A Comp. Analysis of regional agreements on port state control. Am Sci Res J Eng Technol Sci (ASRJETS) 18(1):259–270
24. Mukherjee PK, Brownrigg M (2013) Farthing on Int' shipping. Springer, Berlin, Heidelberg

Chapter 25
The Retardation Process of Crack Propagation in Unconfined High-Strength Concrete Columns Due to the Introduction of Silica Fume

Ahmad Azmeer Roslee, Johnny Ching Ming Ho, and Dilum Fernando

Abstract High-strength concrete (HSC) has been in high demand as it can offer superior performance, and it is more cost effective over normal strength concrete. Without any addition of admixtures, fillers or supplementary materials, the greatest strength achievable by standard concrete mixes is 80 MPa. Amongst currently available supporting cementitious materials, silica fume offered several benefits. Firstly, its ultrafine particles can fill in the void between the hydrated concrete paste and aggregates. Secondly, as silica fume contains about 97% of silica dioxide, it can prolong the chemical reaction to produce a concrete paste. In other words, the introduction of silica fume in concrete mixes will enhance the structure packing density as the interstitial voids are significantly reduced. Although it is a common practise to have silica fume in concrete mixes, its performance and behaviour have not been accurately analysed. HSC with silica fume is treated as usual HSC or NSC during the analysing process, but studies show that the different is significant (Xiao et al. in J Compos Constr 14:249–259, 2010; Lim and Ozbakkaloglu in Constr Build Mater 63:11–24, 2014). Therefore, a new model that can truly reflect the HSC with silica fume performances needs to be derived and developed. In addition, the HSC column is very brittle, whether it consists silica fume or not. Studies showed that lateral confinement could significantly enhance its axial strength and deformability (Mander et al. in J Struct Eng 114:1804–1826, 1988; Saatcioglu and Razvi in J Struct Eng 125:281–289, 1999; Saatcioglu and Razvi in J Struct Eng 118:1590–1607, 1992). External confinement by steel tubes and fibre reinforced plastic (FRP) tubes provided the best solution as they can act as a formwork hence improving the construction process time. Various studies on these two types of confinement have been carried

A. A. Roslee (✉)
Univerisiti Kuala Lumpur, Malaysian Institute of Marine Engineering Technology, Jalan Pantai Remis, 32200 Lumut, Perak, Malaysia
e-mail: ahmadazmeer@unikl.edu.my

A. A. Roslee · J. C. M. Ho · D. Fernando
The University of Queensland, St Lucia, QLD 4072, Australia
e-mail: johnny.ho@uq.edu.au

D. Fernando
e-mail: dilum.fernando@uq.edu.au

A. Ismail et al. (eds.), *Design in Maritime Engineering*, Advanced Structured Materials 167, https://doi.org/10.1007/978-3-030-89988-2_25

out, but the model proposed only covers the effectiveness of the confinement material and the behaviour of the confined concrete as a whole composite. The composition of concrete mixes is not clearly discussed hence, the effective contribution of silica fume in term of volume, and improved strength in this system are unknown. Therefore, this project will highlight the differences between HSC with and without silica fume concerning axial stress–strain behaviour, lateral-axial strain performance and peak axial stress. The research also will undertake a study on unconfined HSC with silica fume to determine the pre-axial stress peak performance. Experimental investigation on both unconfined and confined HSC with silica fume will be carried out with steel tube and glass fibre reinforced plastic (GFRP) winding tube will be used as a confining material. Then, a more representative model to denote the behaviour of unconfined and confined HSC with silica fume will be presented.

Keywords High-strength concrete · Silica fume · Lateral-axial strain

25.1 Introduction

A final product of cement and water mixture is calcium silicate hydrate (CSH) and calcium hydroxide (CH). CSH, also known as the gel, contributes about 50% of paste volume, whilst CH is a fragile solid that can be easily dissolved with the appearance of excess water. The introduction of silica dioxide (SiO_2) can help producing more CSH by interacting with CH. Silica fume, which is about 1/100 of cement's particles size and consist of around 85–97% SiO_2, definitely the answer in improving concrete structure arrangement. It can fill the interstitial voids between the angular cement particles and other fine powders to increase the wet packing density of concrete or paste [6]. The introduction of silica fume will have three effects which are (1) improving the concrete flowability; (2) improving the strength of concrete, and; (3) increasing the concrete's ductility by demoting the propagation of cracks in concrete [7].

The excess of water in concrete will decrease the particles in concrete paste by acting as lubrication. The added silica fume will free the water by taking up the place in the voids and this water will help to improve the flowability of the cement paste or water demand [8]. Besides, the improved packing density silica fume will contribute to higher concrete strength as the microstructure integrity are strengthened. A denser microstructure will be able to retard the development of concrete cracks by the increased elastic or secant stiffness. It also can increase the difficulty for a crack to propagate. The smaller voids also will limit the crack width during the propagation process. If the lateral confinement is provided, it will only be activated at a later stage or higher axial strain magnitude. Hence, a confined high-strength concrete (HSC) with silica fume will present better ductility and higher deformability than the one without it. Therefore, silica fume is an indispensable material in producing HSC.

Evidently, the behaviour of concrete with silica fume performs differently. Current practise showed the predictions model proposed are based on the unconfined concrete and confining stress. The existing model overestimates the lateral strain value at certain axial strain thus the confining stress and its axial capacity will be inaccurate. A confined concrete subjected to axial compression pressure will shorten and dilate. During the elastic stage, both lateral and axial behaviour can be predicted efficiently using linear elastic behaviour.

However, once the microcrack starts to appear, this relationship will become nonlinear. The lateral strain will begin to increase rapidly along with the development of crack propagation. Therefore, from the start of tensile splitting crack, the lateral strain will only depend on the confining pressure and unconfined concrete strength. As the lateral strain depends on the confining stress, and vice versa, both of them should be evaluated simultaneously [9].

This study will cover the unconfined HSC with silica fume axial stress–strain predictions, lateral-axial strain performance as well as its lateral strain behaviour during the inelastic stage. The proposed model will accommodate the silica fume effect, which is, prolonging the linear elastic phase and delaying the tensile splitting crack formation [10].

25.2 Methodology

The research on both unconfined and confined HSC with silica fume are conducted concurrently. An experimental investigation of confined HSC with silica fume started once there are a clear understanding on (1) relationship between silica fume and unconfined HSC ultimate stress and; (2) unconfined HSC with silica fume axial stress–strain curve.

Eight groups of specimens have been cast with a silica fume content ranging from 0 to 15% of cementitious volumes. The water to cementitious ratio (w/c) range is between 0.25 and 0.5. The first two letters and two numbers following in nomination of the specimens represents silica fume (SF) percentage in the concrete mixtures. The followings two alphabet and numbers represent w/c over 100 (e.g. 30 is equal to 0.3 w/c). As previously mentioned, the water to cementitious material ratio content variation is important in determining its combined effect with silica fume in producing HSC.

The paste volume is defined as the total of water and cementitious material (in this case, cement and silica fume) volume, whilst the aggregate volume is the sum of fine and coarse aggregate volume. The paste volume ratio is the value of paste volume over the sum of the paste and aggregate volume. Else, the aggregate volume ratio is the value of aggregate volume over the total of paste and aggregate volume too. The paste volume effect is the same as w/c on compressive strength [11]. As the study is only interested in defining the silica fume effect, which is a part of the paste, it is understandable to standardise the paste volume ratio as well as the aggregate volume ratio. Table 25.1 shows the composition of the concrete mixtures.

Except for the control specimen, six samples for each group with a diameter of 100 mm and length of 200 mm are prepared for the compression test with three of

Table 25.1 Concrete mixes composition

Specimens	Silica fume content (% volume)	Water to cementitious material ratio content	Paste volume ratio	Aggregate volume ratio
SF15WC30	15	0.30	0.4	0.6
SF10WC30	10	0.30	0.4	0.6
SF5WC30	5	0.30	0.4	0.6
SF15WC50	15	0.50	0.4	0.6
SF10WC50	10	0.50	0.4	0.6
SF5WC50	5	0.50	0.4	0.6
*SF8WC25	8	0.25	0.4	0.6
*SF0WC30	0	0.30	0.4	0.6

* indicates different silica fume content (% volume) and water to cementitious material ratio content

those have strain gauges attached to it as shown in Fig. 25.1. All strain gauges are located at the mid-section of the specimens.

The experimental design is based on several considerations. From the previous study [2], and it is found that the strength starts to have small increment between 8 and 16% of silica fume replacement. It is assumed that the further addition of silica fume may result in a plateau trend of concrete strength. The silica fume will only interact with the calcium hydrate or CH. A further increment of silica fume without any presence of CH will not yield any chemical reaction producing the gel paste or

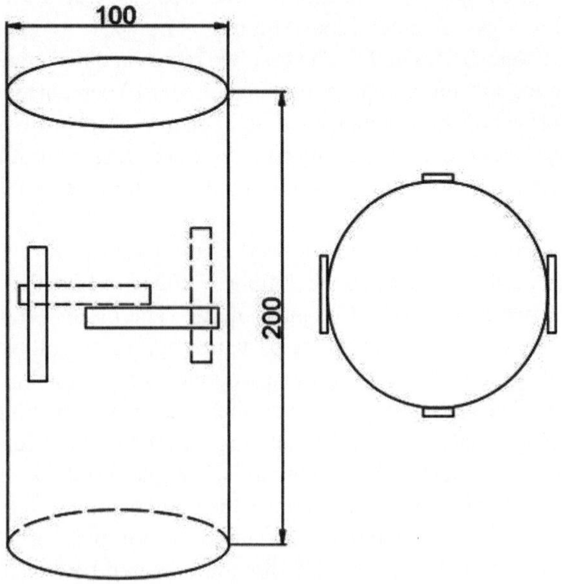

Fig. 25.1 Strain gauge arrangement

CSH. As the primary concern in this research is to understand the effect of silica fume, and not focussing on the concrete strength, the silica fume is limited to a 15% of cement replacement.

On the other hand, it is a common knowledge that the concrete will experience full hydration with w/c of 0.4. However, one research shows that the concrete will undergo complete hydration in the of 0.25–0.75 w/c ratio [12]. Additionally, lower w/c content also shows that a microcrack has difficulty to propagate hence increases the strength of the concrete as opposed to a higher w/c content where a bigger porosity have been observed [13]. Also, another research studying measurement of the w/c ratio and its porosity recognised that the best concrete hydration happened between 0.25–0.45 [14]. Therefore, the research proposed to vary the w/c from 0.25 to 0.4.

For the record, the concrete specimens are cast in July to September 2016 at the University of Queensland, Australia. Adequate superplasticizer (MasterGlenium SKY 8700) is used to achieve a concrete slump spread of 600 mm for workability purpose. The specimens are cured at the ambience condition. All specimens are tested for 28 days compression test. The ultimate stress of each of the specimens and its respective axial strain results are shown in Table 25.2.

In order to develop an actual representative behaviour of confined HSC with silica fume, two models are chosen for theoretical analysis. Xiao et al. [1] provided the most accurate prediction in determining the peak axial stress [15] for unconfined HSC with silica fume. The model which is originally based on the triaxial test [16] are designed for confined normal strength concrete (NSC) and HSC. The proposed model encompassed various experimental data on both actively and passively confined concrete.

On the other hand, Dong et al. [9] provided the only prediction on lateral-axial strain behaviour that incorporates the splitting crack formation. The model distinguishes the linear elastic phase and inelastic phase by determining its turning point in the lateral-axial strain curve. The turning point is where linear relationship between lateral strain and axial strain ended and lateral strain starts to increase exponentially [9]. The splitting crack formation is believed to start from this point.

The following Eqs. (25.1) and (25.2) represent the peak axial stress and its corresponding axial strain, respectively, whilst Eqs. (25.3) and (25.4) describe the axial stress, f_c and axial strain, ε_c relationship [1, 16].

$$\frac{f_{cc}}{f_c'} = 1 + 3.24\left(\frac{\sigma_r}{f_c'}\right)^{0.8} \tag{25.1}$$

$$\frac{\varepsilon_{cc}}{\varepsilon_{co}} = 1 + 17.4\left(\frac{\sigma_r}{f_c'}\right)^{1.06} \tag{25.2}$$

$$f_c = \frac{f_{cc}(\varepsilon_c/\varepsilon_{cc})r}{r - 1 + (\varepsilon_c/\varepsilon_{cc})^r} \tag{25.3}$$

$$r = \frac{E_c}{E_c - (f_{cc}/\varepsilon_{cc})} \tag{25.4}$$

Table 25.2 Peak axial stress and its corresponding axial strain

Specimens	Silica fume (% volume)	w/c ratio	Peak axial stress, f'c (MPa)	Axial strain, εco (× 10^{-6})
SF15WC30-1	15	0.30	71.6	2995
SF15WC30-2	15	0.30	73.3	2923
SF15WC30-3	15	0.30	74.8	3042
SF15WC30-4	15	0.30	69.0	
SF15WC30-5	15	0.30	72.5	
SF15WC30-6	15	0.30	75.4	
Average			72.7	2987
SF10WC30-1	10	0.30	74.4	2544
SF10WC30-2	10	0.30	85.6	2750
SF10WC30-3	10	0.30	84.5	2828
SF10WC30-4	10	0.30	77.8	
SF10WC30-5	10	0.30	90.9	
SF10WC30-6	10	0.30	72.4	
Average			81.0	2708
SF5WC30-1	5	0.30	81.4	3039
SF5WC30-2	5	0.30	81.2	3210
SF5WC30-3	5	0.30	83.4	3056
SF5WC30-4	5	0.30	82.0	
SF5WC30-5	5	0.30	84.1	
SF5WC30-6	5	0.30	75.5	
Average			81.3	3101
SF15WC50-1	15	0.50	46.6	2681
SF15WC50-2	15	0.50	42.4	2414
SF15WC50-3	15	0.50	46.3	2394
SF15WC50-4	15	0.50	49.7	
SF15WC50-5	15	0.50	46.8	
SF15WC50-6	15	0.50	44.6	
Average			46.1	2496
SF10WC50-1	10	0.50	67.2	2421
SF10WC50-2	10	0.50	64.7	2284
SF10WC50-3	10	0.50	65.2	2101
SF10WC50-4	10	0.50	61.7	
SF10WC50-5	10	0.50	60.4	
SF10WC50-6	10	0.50	67.6	
Average			64.5	2269

(continued)

Table 25.2 (continued)

Specimens	Silica fume (% volume)	w/c ratio	Peak axial stress, f'c (MPa)	Axial strain, εco (× 10^{-6})
SF5WC50-1	5	0.50	43.2	2787
SF5WC50-2	5	0.50	45.1	3028
SF5WC50-3	5	0.50	42.1	3361
SF5WC50-4	5	0.50	41.2	
SF5WC50-5	5	0.50	43.0	
SF5WC50-6	5	0.50	45.3	
Average			43.3	3059
SF8WC25-1	8	0.25	54.8	3575
SF8WC25-2	8	0.25	59.8	3156
SF8WC25-3	8	0.25	63.1	2737
SF8WC25-4x	8	0.25	80.9	
SF8WC25-5x	8	0.25	88.2	
SF8WC25-6x	8	0.25	82.6	
Average			59.3	3156

where f_c', f_{cc} and σ_r are the unconfined concrete strength, peak axial stress and the confining stress acting on the concrete, respectively. In addition, ε_{co} and ε_{cc} are the axial strain at peak axial stress of the concrete under the unconfined condition and axial strain at particular confining pressure, respectively. r is a constant parameter to govern the curve [17], and E_c is the elastic modulus of concrete.

25.3 Results and Discussion

HSC column is defined to have a compressive stress of 50 MPa and above. As such, SF15WC50 and SF5WC50 would not be discussed in this section. Figure 25.2 shows the unconfined HSC with silica fume axial stress–strain behaviour.

Both SF10WC30 and SF5WC30 tend to maintain its linear relation between axial stress and axial strain as opposed to SF0WC30 although they possessed a similar peak stress. The similar trend also can be seen in the axial stress and lateral strain curve. It indicates that the specimens with silica fume possess higher load carrying capacity at the same given axial strain. The lateral expansion under axial compression of concrete mixes with silica fume is also lower.

As a result, modification is needed for Eq. (25.4) to adopt this trend in predicting the axial stress–strain for HSC with silica fume is suggested as below.

$$r = \frac{1.26E_c}{E_c - (f_{cc}/\varepsilon_{cc})} \tag{25.5}$$

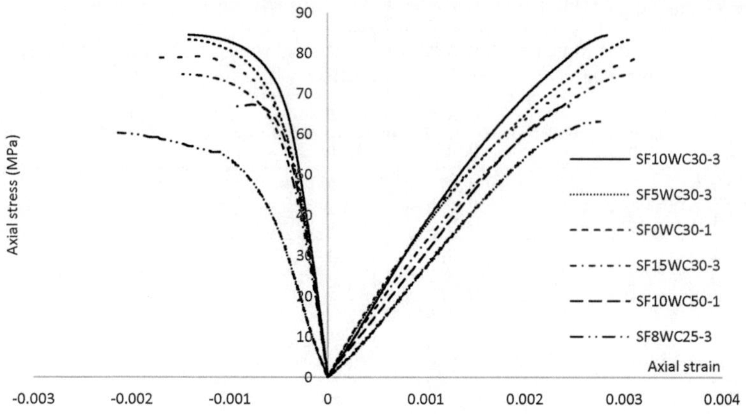

Fig. 25.2 Axial stress–strain of unconfined HSC

As opposed to the existing equation, the alteration made returns higher magnitude showing that the curve proposed, permit longer linear axial stress–strain relationship. The performance of Eqs. (25.3) and (25.5) for unconfined HSC with silica fume are shown in Fig. 25.3.

From the axial stress–strain graph, it is proven that the HSC with silica fume can withstand higher force due to better structural and packing density. The best way to evaluate this event is by determining its ability to withstand tensile splitting crack. In the beginning, the lateral strain increases steadily with axial strain. Once the tensile splitting starts to develop, the lateral strain will experience an exponential rise. The difference of the lateral-axial performance of HSC with and without silica fume can be seen in Table 25.3 and Fig. 25.4.

The addition of 10% silica fume managed to demote the tensile splitting crack of about 54% in term of lateral strain movement. The higher stress experience with the same specimen also proved that HSC with silica fume managed to experience greater deformation before the splitting crack starts to develop.

25.4 Conclusion

The introduction of silica fume in concrete mixtures will not only produce HSC but also behaved differently in terms of axial stress–strain capability and lateral-axial strain relationship. As opposed to current practise, it needs to be treated separately; hence, a more representative model should be developed. On a closer look, silica fume alters the splitting crack development and propagation process. Concrete with silica fume possesses a higher capability to withstand pressure and starts to experience tensile splitting crack at a later stage. On the other hand, it has lowered the peak axial

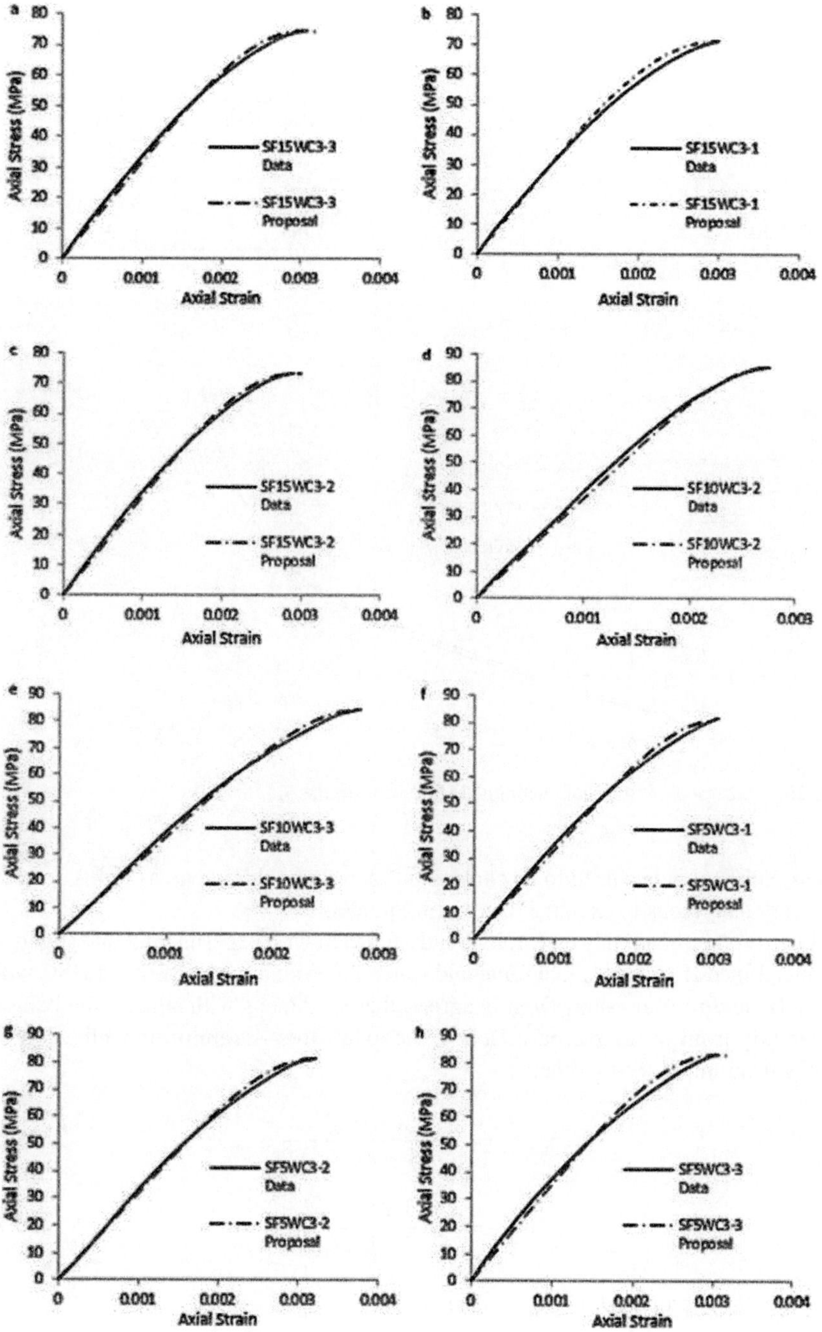

Fig. 25.3 Proposed axial stress–strain performance

Table 25.3 Axial strain and lateral strain onset of splitting crack

Specimen	Axial Stress, f_c' (MPa)	Axial strain onset of splitting crack, ε_{z0} ($\times 10^{-6}$)	Lateral strain onset of splitting crack ($\times 10^{-6}$)	Stress onset of splitting crack (MPa)
SF10WC30-3	84.5	1320	274	49.5
SF5WC30-3	83.4	877	182	33.3
SF0WC30-1	82.8	836	178	32.3

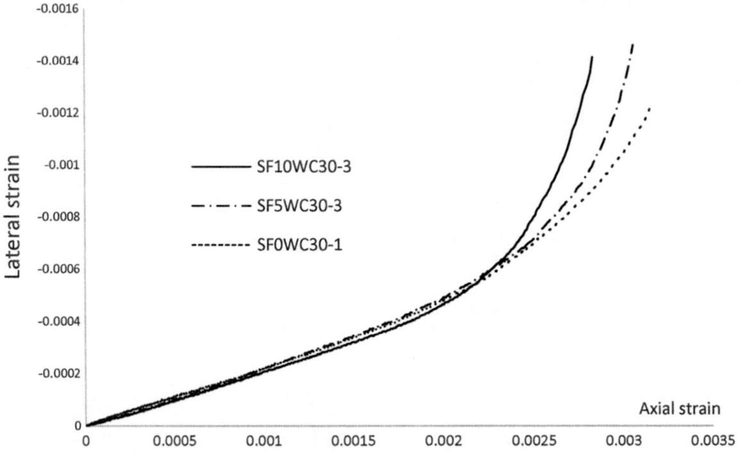

Fig. 25.4 Lateral-axial strain of unconfined HSC with similar f_c'

stress. Further studies need to be carried out to evaluate this event, as if this correct, it is very dangerous to exercise the existing model.

On the other hand, this research already finished with experimental investigation of unconfined HSC with silica fume and statistical evaluation of confined HSC with silica fume. From the analysis, it is agreed that specimens with silica fume behave differently from its counterpart. Hence, the axial stress–strain of unconfined HSC with silica fume can be defined as

$$\text{a} \quad f_c = \frac{f_{cc}(\varepsilon_c/\varepsilon_{cc})r}{r - 1 + (\varepsilon_c/\varepsilon_{cc})^r} \tag{25.6}$$

$$\text{b} \quad r = \frac{1.26E_c}{E_c - (f_{cc}/\varepsilon_{cc})} \tag{25.7}$$

Note: (a) is proposed by Popovics et al. [16]

Acknowledgements I would like to deeply thank Dr Johnny Ho and Dr Dilum Fernando who expertly guided me in conducting the experiment and developing the mathematical model for this research. I also am grateful to be surrounded with a great research group consisting of Dr Shuan

Jiang, Dr Hao Zhou, Dr Ya Ou, Mr Chuang Miao, Ms Ezgi Kaya and Dr Yashar Doroudi. The experiment is also completed with the help of all technical staff at UQ Structural Engineering Laboratory who are very professional and dedicated. Finally, I would like to thank my family who is always being there for me. Thank you.

References

1. Xiao QG, Teng JG, Yu T (2010) Behavior and modeling of confined high-strength concrete. J Compos Constr 14(3):249–259
2. Lim JC, Ozbakkaloglu T (2014) Influence of silica fume on stress–strain behavior of FRP-confined HSC. Constr Build Mater 63:11–24
3. Mander JB, Priestley MJN, Park R (1988) Theoretical stress-strain model for confined concrete. J Struct Eng 114(8):1804–1826
4. Saatcioglu M, Razvi S (1999) Confinement model for high-strength concrete. J Struct Eng 125(3):281–289
5. Saatcioglu M, Razvi SR (1992) Strength and ductility of confined concrete. J Struct Eng 118(6):1590–1607
6. Wong HHC, Kwan AKH (2008) Packing density of cementitious materials: part 1—measurement using a wet packing method. Mater Struct 41(4):689–701
7. Hao H, Zhang C (2019) Demoting splitting crack formation in confined concrete by silica fume. Mech Mater 1651–1656
8. Wong HHC, Kwan AKH (2008) Rheology of cement paste: role of excess water to solid surface area ratio. J Mater Civ Eng 20(2):189–197
9. Dong CX, Kwan AKH, Ho JCM (2015) A constitutive model for predicting the lateral strain of confined concrete. Eng Struct 91:155–166
10. McGarry EJ, Rowe EH (1978) Improving the crack resistance of bulk molding compounds and sheet molding compounds. Polym Eng Sci 18(2):78–86
11. Kolias S, Georgiou S (2005) The effect of paste volume and of water content on the strength and water absorption of concrete. Cem Concr Res 27(2):211–216
12. Wong HS, Matter K, Buenfeld NR (2013) Estimating the original cement content and water–cement ratio of Portland cement concrete and mortar using backscattered electron microscopy. Mag Concr Res 65(11):693–706
13. Prokopski G (1991) Influence of water-cement ratio on micro-cracking of ordinary concrete. J Mater Sci 26(23):6352–6356
14. Wong HS, Buenfeld NR (2009) Determining the water–cement ratio, cement content, water content and degree of hydration of hardened cement paste: method development and validation on paste samples. Cem Concr Res 39(10):957–965
15. Dong CX et al (2015) Effects of confining stiffness and rupture strain on performance of FRP confined concrete. Eng Struct 97:1–14
16. Popovics S (1973) A numerical approach to the complete stress-strain curve of concrete. Cem Concr Res 3(5):583–599
17. Wu YF, Wei Y (2015) General stress-strain model for steel- and FRP-confined concrete. J Compos Constr 19(4):4014069

Chapter 26
Interpretations of Maritime Experts on the Sustainability of Maritime Education: Reducing the Lacuna of Amalgamation Between Maritime Education and Industries

Jagan Jeevan, Mohamand Rosni Othman, Nurul Haqimin Mohd Salleh, Anuar Abu Bakar, Noor Apandi Osnin, Mahendrran Selvaduray, and Noorlee Boonadir

Abstract The assessment of graduate employability, sustainability management, and development in maritime research and education requires an integrated approach to anticipate the sustainability of the subjects offered and the impact on the socio-economic progress. Primarily, this study examines the main topics and themes in the syllabi of maritime studies, disciplines, and related subjects, which may require revision to raise the graduates' quality effectively and recommendations to enhance the research quality in the maritime domain. A qualitative approach has been employed through semi-structured interview sessions to obtain information from 14 participants from various maritime industries to gather their insight and opinion on syllabus enhancement for future graduates and the quality of maritime education. A bibliometric analysis was then conducted on data retrieved from the Scopus database to analyse the research network, current research area, and research collaboration among maritime researchers. These fundamental elements should be developed with continuous improvement as maritime players must stay current and relevant in their profession. The analysis focusses on maritime studies that must be investigated from

J. Jeevan (✉) · M. R. Othman · N. H. M. Salleh · A. A. Bakar · N. A. Osnin · M. Selvaduray
Faculty of Maritime Studies, Universiti Malaysia Terengganu, 21030 Kuala Nerus Kuala, Terengganu, Malaysia
e-mail: jagan@umt.edu.my

N. H. M. Salleh
e-mail: haqimin@umt.edu.my

A. A. Bakar
e-mail: anuar@umt.edu.my

N. A. Osnin
e-mail: apandi@umt.edu.my

N. Boonadir
Universiti Kuala Lumpur Malaysian Institute of Marine Engineering Technology, Bandar Teknologi Maritim, Jalan Pantai Remis, 32200 Lumut, Perak, Malaysia
e-mail: noorlee@unikl.edu.my

A. Ismail et al. (eds.), *Design in Maritime Engineering*, Advanced Structured Materials 167, https://doi.org/10.1007/978-3-030-89988-2_26

the ground up to uplift the actual requirements that may be incorporated into the maritime educational syllabus. Through this survey, marine practitioners and educators were able to assess the consistency and efficiency of the syllabus for educational advancement. To ensure the quality standard in the education system, graduates, educators, and researchers should update their education standards regularly to sustain associated scientific investigations, advanced technology, and practical applications.

Keywords Sustainable development · Employability · Maritime research · And education · Maritime studies

26.1 Introduction

The graduates' first move to the employment sector brings them to the guidance that comprises a range of career processes designed to enable individuals to make informed choices and transitions related to their educational, vocational, and personal development. The higher education community has differing opinions regarding economic conditions and personal characteristics that affect individual graduates' ability to obtain a professional career [1]. Technological developments will radically change the employment patterns in the maritime industry in the forthcoming years. Similarly, skillsets and training required in the immediate, medium-term, and long-term future of the shipping industry will be different than those of today. As mentioned by [2], "skills, understanding, and personal qualities enhance student's capability to be in the job and succeed in their career path for the benefit of workforce, society, and economy". However, it was a fact that the implementation of necessary changes must be included in the higher education system, which plays an essential role in improving working skills among the graduates, particularly with the rise of new digital industrial technologies known as IR 4.0 that started reshaping of the maritime industry's future proactively [3].

This study investigates the elements that determine graduate employability and provides a clear picture of employers' impressions of fresh graduates with soft skills that are more competent than academic level. It is crucial to identify the focus area of subjects offered with appropriately structured content in higher education institutions. The top priority for maritime industry practitioners is to develop the human element by investing in human capital. Human capital theory reinforces the relationship between education and improved productivity and expects university graduates to generate a substantial pool of innovative human capital in future [4]. There are still challenges highlighted by the industry-related competencies, which may affect the capability of the existing competence and professionalism if not overcome. Also, a better collaboration approach between academics and employers in devising comprehensive strategies with new technologies in the maritime industry from industrial and educational perspectives improves job skills [3]. The course curriculum impacts graduates' professional paths, which are also influenced by the graduates' academic interests and long-term goals. This research will analyse the motivation initiator from the feedback of interview session that potentially attracts the graduates to pursue their career in the maritime industry. The following section reviews the different aspects of graduate employability and the relevancy of the academic curriculum incorporated with the industry [5].

26.2 Intertwining Maritime Education and Industries

One significant component to assure quality education is the formulation of policies aimed at a student's development. A student's academic performance may also serve as a measurement of the quality of education, which the institution offers to the students [6]. Technological advancements, the rise of international trade, and supply chain management have caused an increase in operational capacities, contemporary technology, and the recruitment of maritime specialists. Nonetheless, even with technological advancements, equipment operations and core marine activities still require human engagement. Maritime activities comprise a wide variety of services, stimulating the various aspects of human existence. Although the maritime industry is one of the world's economic upholders, a critical issue threatens the industry's long-term viability: talent deficit and a lack of new generations inspired to enter the labour market in the coming decades.

Our global maritime industry was steered towards modernisation in the twentieth century with advanced technologies in the system and practices. The primary goal of maritime education is to gain associated knowledge, skills, and competencies according to the actual techniques and technologies used onboard ships while preparing for future development of global shipping. Shipping is a vital component of the world economy and is involved in almost every supply chain, from the provision of raw materials to the transport of finished products goods to the consumer [7]. These economic activities are built upon a combination of knowledge, skills, and education [8]. There are primary subjects in maritime education covering all of the specific contents required in building a maritime management syllabus, such as logistics, port management, and shipping management. These could help applicants research, practise, and work for a future job.

Safe and efficient shipping is strongly linked to the quality of the maritime education system, especially at the university level and certifying programmes that accredit seafarer's top rank qualifications complying with the Standards of Training, Certification, and Watchkeeping (STCW) Convention [9]. Generally, employability and promotion opportunities are almost guaranteed when individuals receive quality educations, and the industry is adequately managed. The maritime management will review a few quality aspects, such as the efficiency of the teaching process by professional educators and accredited or approved programme curriculum [4].

As the maritime industry's technology advances, the technicality of the shipping increases, and it demands proficient and specialised management skills, procedures, and processes [7]. As such, maritime higher institutions have been working on bettering and incorporating the latest information technology (I.T.) in their education and training to produce competitive seafarers and enable seafarers' employment in shore-based jobs [9]. With digitalisation and an increased level of automation, significant changes occur in the maritime business operation and, consequently, modifying the shipping method. Forthcoming improvements and innovations in biotech, data, materials, and energy in momentum and cognition also influence the shipping industry. Therefore, many modifications are required in the education system to keep graduates up-to-date on organisation, operation procedures and processes, regulations, and technical aspects of the shipping industry [10].

26.3 Methodological Outlines

Two main qualitative approaches have been employed to meet the aim of this study: thematic analysis to analyse the outcome from face-to-face interview sessions and bibliometric analysis to explore the current status of research trends, position of internal and external research networks, and collaborations. First, this study aims to analyse the new scope of the syllabus that needs to be introduced in maritime studies and the main research area that needs to be explored in the maritime disciplines and propose strategies to enhance the quality of graduates in maritime studies. Results are obtained from a questionnaire survey conducted on 14 experts from various maritime industries, including navigation, maritime management, maritime engineering, maritime economic, and maritime transportation. The respondents' opinions on the sustainability of careers related to the blue economy based on their more than ten years of experience are shown in Table 26.1. Most of the surveys involved high-ranking managing teams, contributing to the validity of the outcome to the development of short-term to long-term career planning for maritime graduates.

The respondents in the interview session came from a variety of industries, including shipping lines, seaports, ministries, transport operators, logistics operators, maritime data centres, maritime universities, inland terminals, multimodal operators, manufacturers, and shipping agents, bearing the goal of determining the factors that influence graduate employability and programme quality (see Table 26.1). These samples are derived based on the convenience sampling strategy. Convenience sampling aims to choose the potential participant who is available and willing to be involved in the interview session in the proposed time frame [11]. In addition, this sampling strategy could be carried out by locating potential respondents and selecting them until the sample size proportion is filled [12]. This procedure provides a lot of flexibility to carry out this research during the pandemic.

Bibliometric analysis is then applied to analyse the current trend of research among the lecturers, the network of research collaboration among them, and the keywords that have been generated in the current research agenda. Bibliometric analysis has been widely employed as a qualitative analytical tool [13, 14]. Bibliometric includes descriptive statistics of citation data and network analysis of authors, journals, universities, countries, and keywords based on citations and frequency analysis techniques.

Table 26.1 List of participants in the interview session

Code	Affiliation	Position	Years of experience
R1	Shipping line	Senior Manager	12
R2	Port operator	Chief Operating Officer	15
R3	Port authority	Operation Manager	8
R4	Ministry of transport	Maritime Division	14
R5	Transport operator	Senior Executive	7
R6	Logistics operator	Senior Manager	10
R7	Maritime trade data centre	Senior Data Analyst	6
R8	Maritime university	Senior Lecturer	34
R9	Seaport	Assistant Manager	20
R10	Inland terminal	Engineer	1
R11	Multimodal	Shipping assistant	2
R12	Manufacturer	Executive	4
R13	Logistics operator	Maintenance Manager	6
R14	Shipping agent	Shipping Assistant	2

For this paper, bibliometric data were collected from the Scopus database to maintain the quality of reviewed studies [14].

26.4 Results and Discussion

The interview sessions have been conducted between March until June 2020. The duration of the interview session ranged from 45 to 75 min. In this paper, the sample has been drawn from a significant population from Malaysian maritime agencies and education centres, which can be accessed to garner the required data [15]. This author also emphasises that the reliability and validity of the research outcome depend on the appropriate selection in the sampling frame. Although only three questions were designed for the interview session, the interview sessions with the participants were held for more than one hour to gain more accurate details regarding the proposed questions. Hence, spending prolonged time with the participants to understand the phenomena determines the result's validity [16].

During the interview sessions, the interviews were recorded, and notes were taken to enhance the reliability and organisation of the data so that they are easily retrievable by other researchers for reanalysis [17]. This paper aims to enhance the curriculum of maritime studies by exploring and conducting research on the subject in terms of fundamental content and providing improvement strategies to produce quality, experienced, and competitive graduates in the maritime field. This initiative will also produce future graduates who will be able to serve as maritime specialists in this coastal country in the long run.

26.4.1 Syllabus Enhancement for Future Graduates

Industrial Revolution 4.0 (IR4.0) has enhanced the digitalisation of the entire value chain and the interconnectedness of people, things, and systems through real-time data interchange in maritime education. In line with IR4.0, occupations and jobs were created by quickening demographic, socio-economic, and technological upheavals [10].

Academicians or trainers should take the initiative to perform a maritime skills audit, assessing and evaluating the courses conducted in the syllabus to increase self-expertise, technical coordination, and seafarer capabilities. Referring to Table 26.2, the majority of the feedback demonstrates that the syllabus content of maritime studies is related to the fundamental aspect of the main area in shipping and ports, maritime operation safety, logistic management, and technical competencies regulated in STCW that play a significant role in the maritime industry in the short, medium, and long term. As a result, graduates must be technologically savvy and knowledgeable of the technologies used in the maritime sector to remain competitive and relevant in the market. With rising difficulties in the industry, technical awareness, computing and informatics abilities, and environmental and sustainability concern will be crucial competencies for future seafarer [3].

Generally, the participants' responses indicate the main theme that contributes to enhancing the maritime studies syllabus. These themes include green shipping (*R1*), maritime safety (*R2*), terminal and administration management (*R3*), maritime environment and sustainability as well as maritime digitalisation *(R4)*, marketing and logistics services (*R5*), maritime logistics (*R4, R5, R6*), environmental regulations to the shipowners, ship operator and ship management, and legal principles in shipping (*R7*). Respondent *(R7)* also suggested ocean transportation and shipping business be included in the current syllabus. Respondents also mentioned themes such as human factors, maritime leadership, and entrepreneurship (*R8*), the impact of I.T. in the maritime industry, maritime operation, and ship-to-ship operation (*R9*), harbour management (*R10*), maritime technology (*R11*), maritime archaeology (*R12*), maritime studies through education and experience (*R13*), cargo loading and discharging operation (*R14*). In comparison, the current syllabus of maritime management in the Faculty of Maritime Studies is enforcing five (5) major themes:

Table 26.2 Comparison between new and existing clusters in maritime studies

Participants	Newly proposed cluster in the syllabus of maritime studies and potential area for maritime research	The matches of the responses with current clusters
R1	Green shipping	Maritime policy
R2	Maritime safety	Maritime transportation
R3	Terminal administration and management	Maritime management
R4	*Environment and sustainability* *Maritime digitalisation*	*Not available (*maritime sustainability and maritime digitalisation)*
R5	Marketing and logistic services shipping and logistics marketing	Maritime logistics
R6	Logistics	Maritime logistics
R7	Impact of environmental regulations on the shipowner Legal principles in shipping	Maritime policy
R7	Ship operator and ship management ocean transportation Shipping business	Maritime transportation and management
R8	Human factors Maritime leadership and entrepreneurship	Maritime management
R9	Impact on I.T. in the maritime industry Maritime operation Ship-to-ship operation	Maritime operation
R10	Harbour management	Maritime logistics and management
R11	Maritime technology in operation	Maritime operation
R12	*Maritime archaeology*	*Not available (*maritime archaeology and history)*
R13	*Maritime studies through education and experience*	*Not available (*maritime education and training)*
R14	Cargo loading and discharging operation	Maritime operation

* Newly proposed clusters in maritime education: the Malaysian case

maritime operation, maritime logistics, maritime management, maritime policy, and maritime transportation (see Table 26.2).

The Old Dominion University, Norfolk VA, offered Maritime Management Concentration (Bachelor of Science in Business Administration) as a four-year curriculum. This course consists of four phases comprising freshman, sophomore, junior, and senior. Each phase carries two terms of course completion that gather 120 to 126 credit hours in four years, but each term will only take 13 to 18 credit hours. The students will be taught maritime subjects started in junior to senior phases. The freshman and sophomore syllabi cover writing, calculus, and business courses such as natural science, English, mathematics, introduction of business administration, communication, ethics, and language and culture. Whereas in the junior and senior syllabi, students must take port management, business policy and strategy, upper—division general education, and four elective programmes. All 300–400 level supply chain management and logistics, business analytics, and operation management electives courses are implemented as requirements to complete the four-year programme. In order to enhance their learning and practical abilities, theoretical knowledge, and advanced technology approach, the framework of course organisation must coincide with the student's level of competence. Path-pattern comparison helps us better grasp the uniqueness of strategic maritime management implementation and the educational direction. Strategic maritime management fulfils the requirements to be a mature discipline, either with a proper management framework or with substantial knowledge advancement in maritime studies [18].

Most of the present subjects in the programme align with the responses from the industrial specialists. However, the respondents mentioned a few new themes that are not included in the current syllabus, such as maritime sustainability, maritime digitisation, maritime archaeology, history, and maritime education and training. The establishment of new clusters might improve the percentage of graduate employability (G.E.) affected by the changes in the maritime industry. Surveys show that the programme's student G.E. level was 62% in 2015, 76% in 2016, 82% in 2017, 76% in 2018, and 88% in 2019. The implementation of new clusters comprising maritime sustainability, maritime digitalisation, maritime archaeology, and history, along with maritime education and training, should potentially boost the students' G.E. level.

The inclusion of these new clusters in the curriculum opens up many opportunities for graduates to be prepared for national and international market demands. At one point, the detailed data on Malaysian seafarers' profiles are limited as the percentage of the local seafarers who chose to work onboard foreign-flagged vessels is unclear and confusing among the local shipping industry investors. When the shipping industry expands, it expands the capacity of market demands, transforming Malaysia from a primary producer to a multi-sector economy encompassing services, industries, and agriculture on its way to becoming a high-income nation with the implementation of advanced technology systems. The Review of Maritime Transport 2020, prepared by UNCTAD, highlighted that the container port activity continued to grow in a few countries, including Malaysia, in 2019. In Southeast Asia, the port of Klang, Malaysia, continued to capture more transhipment market share. Malaysia had been a member state of the IMO since 1971 and a council member of the body since 2005. Shipowners and stakeholders were urged to register their ships under the Malaysian flag because the country has ratified 30 international maritime conventions out of the total number of current conventions [19].

26.4.2 Research Trends Among the Academicians in Maritime Studies

Research trends show that maritime studies scholars are currently focussing on five main pillars that are maritime operation, maritime logistics, maritime management, maritime policy, and maritime transportation. However, respondents opined that research on maritime sustainability, maritime digitalisation, maritime archaeology, and history, coupled with maritime education and training, needs to be incorporated into the main programme in maritime research among the academicians. The research at Universiti Malaysia Terengganu (UMT), one of the local higher education institutions, focusses primarily on maritime studies, and this helps students improve their self-education and academic abilities. The courses included in the study programme mainly cover topics such as onshore organisations, quality management, maritime law, and project management.

Nicholas State University in Thibodaux, Louisiana, organised its programme into three phases: university core courses, business core courses, and mandatory courses, all of which accumulate 120 credit hours. The university core courses comprise English composition, university studies, humanities, mathematics, speech, computer literacy, social science elective, and fine arts elective. These act as exposure and prepare the students for the upcoming subjects. Next, the students need to complete the business core courses: accounting, legal environment, business communication, management info system, economics, finance, marketing, management, quantitative analysis, and strategically managing organisation. Students must take these courses to improve their theoretical and calculation skills. These skills will be applied in their mandatory marine studies: admiralty law, economics of shipping, maritime management, human resources management, multinational management, human resources seminar, summer maritime internship, organisation structure and behaviour, marine accident prevention, and approved business elective. All 300–400 level business courses require completing 54 h of mandatory non-remedial coursework by the students, in contrast to 45 semester hours in 300 level courses or above.

The majority of the local institutions, such as Akademi Laut Malaysia (ALAM), UMT, and Sarawak Maritime Academy (SMA), have been providing training to new entrants, specifically deck cadets. The Politeknik Ungku Omar (PUO), Universiti Kuala Lumpur Malaysian Institute of Marine Engineering Technology (UniKL MIMET), ALAM, Ranaco Education & Training Institute (RETI), and SMA offered programmes for engine cadets. Meanwhile, others provide courses for domestic and near coastal voyages and ratings as well. These educational institutions need to adjust the core structure and content of the syllabus based on foreign institutions courses that may be compatible with the prospective labour market. Additionally, the global shipping community and maritime education training centres must focus on operation and advanced technology to produce competitive Malaysian seafarers [19]. While the comparison of the courses manages to describe the regulation and syllabus forte in-depth, which need to be upgraded either theoretical or practical content to ensure it can be applied to industrial practices at the international level with additional support from the collaboration of higher education, research, and industry.

The bibliometric analysis derives main keywords from journal articles published in the Scopus database, and journals that were cited at least once were prioritised. The analysis outcome indicates that the research among the lecturers in the department of maritime studies focusses mainly on seaports, dry ports, competitiveness, hinterland, environment, sustainability, innovation management, supply chain management, container haulage industry, performance, consumer price index, price disparity, cabotage policy, biomass, optimisation, carbon footprint, green supply chain management, and operational performance. The interview responses and the current trend of research area among the lecturers are not aligning well, and the survey participants' feedbacks need to be considered for establishing "real" research which will benefit the maritime industry in this country. This gap shall be reduced by undertaking more research related to maritime sustainability, maritime digitalisation, maritime archaeology and history, and maritime education and training (see Fig. 26.1).

On the other hand, researchers from nautical studies focus on diverse research areas, including port development simulation, navigation safety, inland navigation, marine sustainability, marine tourism, tourism sustainability (see Fig. 26.2). This indicates that the research agenda in nautical studies has a particular similarity with

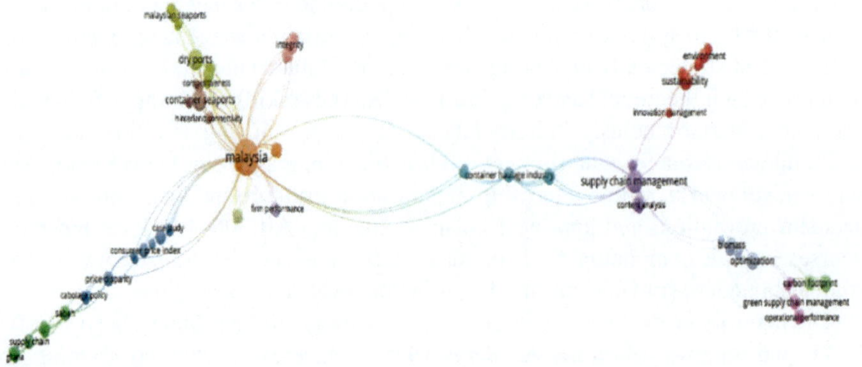

Fig. 26.1 Bibliometric analysis on the main keywords of current maritime research (Based on Scopus database among Maritime Management researchers between 2015 to 2021)

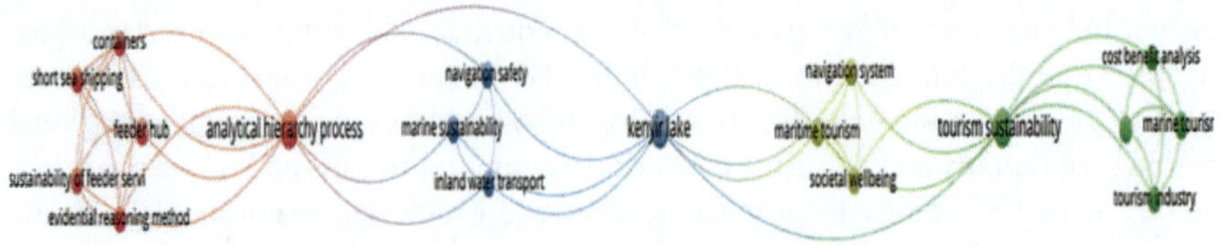

Fig. 26.2 Bibliometric analysis on main keywords of current maritime research (Based on Scopus database among Nautical Science researchers between 2015 and 2021)

the interviewees' responses. For example maritime sustainability has been explored by the researchers in the department. However, some areas have not been specifically explored, i.e. maritime archaeology and history.

In general, the research trend gap emerged due to several justifications, such as the request from their colleagues from different countries to publish journal papers in the given period (see Fig. 26.3). For example the lecturers from this department have significant research collaborations with China, South Korea, Switzerland, Australia, Turkey, the USA, Indonesia, Oman, Czech Republic, Germany, the UK, and Nigeria. Consequently, the researchers from this department need to conform to the requirements, research area, and proposals from their colleagues, although suggestions will be considered. Furthermore, the lecturers' current resources, which they intend to use to complete a journal paper, may not be compatible with the new themes proposed by the participants. They also face limitations in collecting valid data to complete their research papers, and their current research grant may have a different research focus. As a strategy for improvement and producing more trendy research, lecturers from the department of maritime studies can propose to undertake new research, which will carry themes proposed by the experts from maritime industries.

External research collaboration in nautical science, on the other hand, is limited to only a few nations, including Pakistan, Spain, the Czech Republic, and Oman (See Fig. 26.4). This scenario has arisen as a result of the tiny number of researchers currently associated with this department. The department's small number of researchers and the demands of teaching and learning led to a low number of research publications.

Meanwhile, internal research collaborations between academics or research are progressing substantially. This phenomenon is demonstrated by the bibliometric analysis results in Figs. 26.5 and 26.6, revealing that they have formed a research network or collaborated to generate substantial research articles. The results of the interview session with the said researchers, on the other hand, revealed some of the

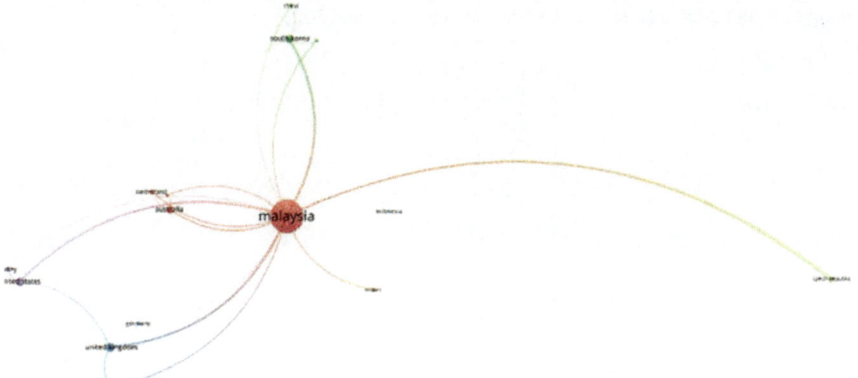

Fig. 26.3 Bibliometric analysis on external research networks (Based on Scopus database among Maritime Management researchers between 2015 and 2021)

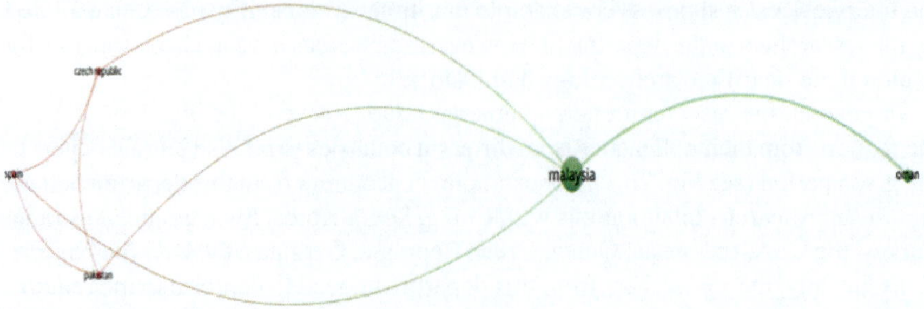

Fig. 26.4 Bibliometric analysis on external research networks (Based on Scopus database among Nautical Science researchers between 2015 and 2021)

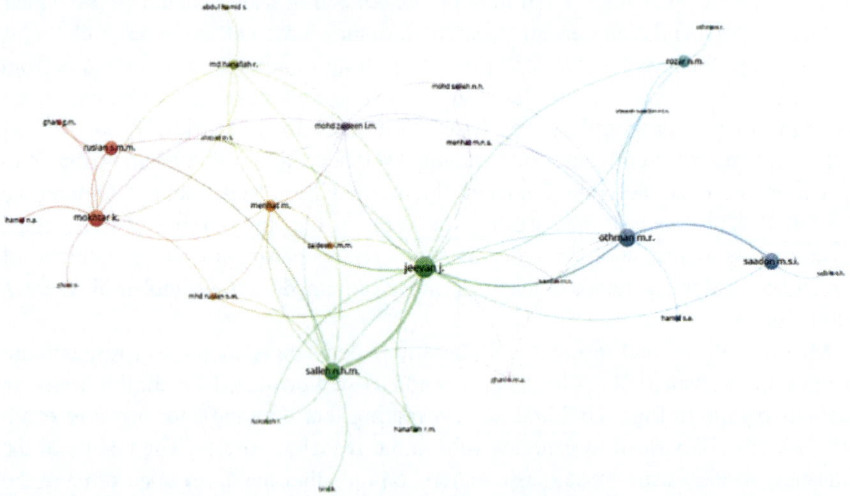

Fig. 26.5 Bibliometric analysis on internal research collaborations (Based on Scopus database among Maritime Management researchers between 2015 and 2021)

Fig. 26.6 Bibliometric analysis on internal research collaborations (Based on Scopus database among Nautical Science researchers between 2015 and 2021)

challenges they face when writing research papers. One of the minor concerns is that they are not informed about the proper method for submitting papers to the appropriate publications. Some are overburdened with administrative responsibilities, which slow them down but do not prevent them from publishing. Additionally, some feel that they are under-exposed to the latest maritime industry trend. Inability to keep up with the rapid pace of research methodology development in maritime-based research, present responsibilities to finish existing research grant projects, a shortage of postgraduate candidates, the focus on the quality of research papers rather than the number of publications produced, limited accessibility for data collection procedure due to global pandemic, and the lack of research fund are some of the issues faced among the academician in this specific department.

Due to several limitations for journal paper publications, a few strategies have been proposed to ensure their research topics aligned with the global trend. For example seeking assistance from mentors or senior lecturers has been suggested to ensure their teaching and research activities that are balanced. Teamwork between senior and junior lecturers achieved through research interest group (RIG) can be another alternative for more exposure in the research agenda. Researchers can take advantage of the current pandemic that forced them to work from home and get used to the new norm to practise good time management, expose themselves to, and learn the requirements for excellent research.

26.4.3 Enhancement of Graduate's Quality in Maritime Studies

All participants and executives in the maritime community must be well-trained and motivated to survive and thrive in the maritime industry. These criteria were assessed from the quality of graduates in maritime studies to the level of employability. The labour market emphasises identifying and developing information, abilities, and traits that help pupils develop operative skills. The higher education institutions are also responsible for the quality of the graduates they produce. As discussed earlier, fundamental maritime education provides applicable knowledge and skills, raises competencies in modern shipping management boards, and prepares the prospective workforce for the global maritime industry. Students need more exposure to the industry to carry out the best practice in technical and shipping management. Furthermore, practical experience in the maritime industry will provide graduates with a thorough understanding of the course content and curriculum objectives. Once the foundation has been laid, efficient communication, operational coding, and strong collaboration with the industry could further improve the graduates' quality. The development of maritime education individually in numerous systematically and successfully education followed and implemented international trends in the maritime studies, as well as progress and the growth of marine technologies. It was mentioned that studying and teaching important subjects in the curriculum are difficult, but it provides students with valuable exposure that allows them to explore a wide range of topics.

Regarding internships and externships, the majority of the respondents provided some views and suggestions (*R1, R10, R12, R13*). They noticed that most students need more exposure to actual practice in real situations to enhance their emotional intelligence and communication strategy. Additionally, the respondents proposed more collaboration with the industries, especially for education and training programmes. They also advised the student to not "hibernate" on the campus but be active outside the campus and exposed themselves to the industries (*R2, R3, R4*). Meanwhile, respondent *(R5)* would like the students to strengthen their foundation in marketing to develop their commercial abilities.

On the other hand, respondent *(R6)* argues that the university should provide a programme that enhances technical skills such as Microsoft Power B.I. for data visualisation, Microsoft Excel, project management, public speaking, data analysis, interpersonal skills, negotiation skills, and conflict resolution. These programmes are essential to preparing the students for challenges in global trade. One of the respondents *(R9)* emphasises the need to apply the 3C model, which incorporates communication, coordination, and collaboration. The respondent *(R7)* has echoed the voice of respondents *(R6, R9, R11, and R14)* by mentioning that the students in this particular programme need opportunities for practical training in maritime-related organisations/companies to understand the complete supply chain of this industry. One of the respondents agrees that the student of this programme needs to pursue their industrial training in maritime-related agencies (*R8*). Table 26.3 summarises the findings on the strategies for enhancing the quality of maritime study graduates.

26.4.4 Synthesise Current Practice in Maritime Studies for Future Direction

In this section, matrix analysis and an analysis of strengths, weaknesses, opportunities, threats have been done to validate the maritime programme's internal and external capacity for further establishment. Regarding internal capacity, 70% of the programme's academicians hold a Ph.D. degree (see Table 26.4), meaning that there is enough capacity for efficient teaching, learning, and supervision of final year projects for undergraduates and research projects for postgraduate and postdoctoral candidates. Besides, undergraduate students have utilised this capacity to produce significant research outcomes in their final year project. These students are also involved in maritime-based international conferences worldwide. In addition, the introduction of internship (6 months) and externship (1 year) provided more opportunities for future graduates to gain real and relevant knowledge and information from the maritime industry. This technique opens up new avenues for increasing G.E. among graduates. Furthermore, Malaysian Logistics And Transport Centre (MALTRAC)'s position as a training institution serving professors, industry professionals, and students provides

Table 26.3 Suggestions to enhance the quality of graduates in maritime studies

Affiliation	Suggestions
R1	• Give more practical exposure
R2	• More industrial connection programmes like training, forum, and projects
R3	• Attend more courses outside the campus
R4	• More exposure to the industry
R5	• Develop a solid foundation in marketing among students because it is significant in developing their commercial abilities
R6	• Provide a programme that enhances technical skills such as Microsoft Power B.I., Microsoft Excel, project management, public speaking, data analysis, interpersonal skills, negotiation skills, and conflict resolution
R7	• To get the opportunity for practical training in maritime-related organisations/companies to better understand the complete supply chain of this industry
R7	• Industry training at the port
R8	• 3C: communication, coordination, and collaboration (training with the industry)
R9	• More exposure to practical studies
R10	• Communication capabilities
R11	• More practical tasks during the semester
R12	• Educate them on good values and how to interact with people in a good way
R13	• More crucial subjects to be introduced and offered in the syllabus • Make sure outdoor activities given can ensure better exposure to the maritime industry
R14	• Easy access to industrial training application with the relevant maritime organisation

additional benefits, mainly through transferring current knowledge and producing new knowledge streams.

The analysis revealed some weaknesses, such as low lecturer to very high student ratio (1:25), English proficiency required for programme entry is too low (Band 1 Malaysian University English Test (MUET), up until 2020), and low programme flexibility due to constraints placed on the selection of elective courses. Besides, it takes three to five years to update the syllabus based on the observation of maritime experts or professional bodies, especially the Chartered Institute of Logistics and Transport (CILT). Numerous methodologies have been developed based on the result of the matrix analysis. Higher education institutions should invite an international lecturer to join the academic team and continue the internship and externship programme locally and globally. MALTRAC, as a dedicated, certified training provider centre for ongoing education, should be fully utilised. Furthermore, international conferences must be organised to encourage professors, undergraduates, and postgraduate students to communicate their research findings using this platform. The institutions also need to enhance their programme's flexibility by revising the credits hours and introducing additional electives courses in the programmes. They have to work

Table 26.4 SWOT analysis on the internal and external domains in maritime studies

SWOT analysis	Items	SWOT metrics
Strength	• Seventy percentage of academician with Ph.D. • Internship at the final semester of the programme • Externship in the final year of the programme • Existence of Malaysian Logistics and Transport Centre (MALTRAC) • High graduate employability rate with high intake Involvement of undergraduate students in international conferences	• Assign international lecturer for the programme • Extensively continue the internship and externship programme locally and internationally • MALTRAC provides additional certified training to the students • Organise international conferences to motivate lecturers, undergraduates, and postgraduate students • Control the student intakes for effective teaching and learning procedures • Mould the local students into successor academicians • Raising MUET requirement from Band 1 to 2 • Enhance the flexibility of the programme by revising the credits hours and introducing additional electives courses in the programmes • Work continuously and closely with maritime-related industries for effective collaborations (between academician and industrial experts) • Ensure continued accreditation from CILT • Establish or enhance related and active MOUs between FPM and industries • Enforce international collaborations through student activities • Expose undergraduates' students in international research agenda • Implement IR4.0 in teaching and learning procedures as well as training programmes through MALTRAC • Prepare students mentally, technologically, and for unexpected scenarios • "Think new" and design its courses to complement similar courses offered by universities around the globe
Weaknesses	• Very low lecturer-student ratio (1:25) • Low English proficiency entry requirement: Band 1 MUET (until 2020) • Low flexibility in the programme • The syllabus needs to be updated due to the dynamism of the maritime sector	
Opportunities	• Getting accredited by the Chartered Institute of Logistics and Transport (CILT) • Developing memorandum of understanding (MOU) with maritime-related entities • International standardised programme (since 2019) • Establishment of Journal of Maritime Studies (Hosted by the faculty) to encourage undergraduate students to publish their final year project	
Threats	• Readiness of the programme to cope with IR4.0 • Economic crisis • Affected by the global pandemic • Competition with other courses/universities	

continuously with maritime-related industries for effective collaborations (between academicians and industrial experts) to maintain and strengthen networking between universities and industries.

From an external viewpoint, this programme is accredited by the Chartered Institute of Logistics and Transport (CILT), which is crucial in ensuring the content of the syllabus in this programme is aligned with international standards and fulfils industrial requirements. Furthermore, this recognition is essential to ensure the credibility of the graduates who meet local and international standards. UMT's maritime programme was recognised as an international programme since 2019, which added more value to this platform [20]. Several memorandums of understanding (MOUs) have also been signed with major marine businesses, including seaports, data-based organisations, logistics organisations, maritime enforcement agencies, and the ministry, to improve the relationship with maritime industries. This is crucial to ensure maritime studies academic excellence and garner full support from adjacent maritime sectors to provide students with authentic and updated information regarding maritime industries. Besides, the Journal of Maritime Studies (JML), hosted by the faculty of maritime studies, has been established to encourage undergraduate students to publish their final year project in collaboration with players in the industry.

Yet, the programme faces considerable challenges such as adapting to IR4.0, economic crisis, implications of the global pandemic, and competition with other universities offering similar courses. Hence, this programme has devised several strategies to reduce the implications of the existing threats, including ensuring continued accreditation from the CILT. Related MOUs can be established with maritime industries, and international collaborations can be strengthened through student activities. Additionally, the programme can provide exposure to undergraduate students in international research agenda, implement IR4.0 in the training programmes and the teaching and learning process through MALTRAC, prepare students mentally, technologically, and for unexpected scenarios. The programme can also always "think new" and design its courses to complement similar courses offered by universities around the globe. These strategies will enhance the programme's sustainability and motivate itself to move forwards as a key platform for students, programme, faculty, and university to sustain itself in the uncertain global scenario.

26.5 Conclusion

The paper explored the themes of the knowledge and abilities that maritime participants gained from their higher education and are extremely satisfied with. The participants expressed their views from many angles, describing essential components in marine studies and continual improvement. The themes are sufficiently descriptive and relevant in a wide range of respondents' education levels and the diversity of seafarers' sea service background and knowledge. Most of the areas relating to advanced technology systems, logistics management, safety aspect, and maritime operation are the important components that required maritime experts to maintain their quality and the efficiency of maritime education. There were currently selected subjects that can be upgraded within the proposed cluster as an alternative for future maritime studies structure, and the syllabus was interrelated with the technical and industrial medium.

The majority of the responses expressed the necessity of practical experience for both lecturers and academicians. The highlighted constraints of the structure of curriculum content in maritime education need to be improved to ensure the sustainability of maritime education in this country. Hence, it requires constant improvement to maintain the quality and nobility of the knowledge. Some institutions could not produce sufficient academicians with expertise in teaching and training, even though a well-functioning human capital market will boost the return on education and, as a result, stimulate academician growth. A competency-based approach to improving employability and sustainability involves holding discussions with the sailors who have advanced to managerial roles. Groups with more work experience are more conscious of the educational interest and are happier with the knowledge gained through formal maritime education. Higher education achievement should align with the noble quality level of individuals who practise appropriate marine industry syllabus.

Future research should focus on how education awareness enhances graduates' ability to explore and learn relevant knowledge, skills, and experience that can help them improve their potential in the workplace through modern technologies. It is advised that the technical capacity of the graduates be assessed and observed, with a focus on the required hands-on skills, the operating system, and the effectiveness of the marine studies' teaching and learning process. To comply with the industrial field's quality education system, graduates must include strategic maritime sustainability, maritime digitalisation, maritime archaeology and history, and maritime education and training, all of which must be on par with a highly competitive global environment. However, to generate high-quality, dependable, and fashionable research, academicians need to have prior expertise with advanced research and collaborate with overseas partners to gain a wider audience. At the same time, the graduates' involvement and constructive collaboration with international maritime players will benefit them in furthering their education and skills with the help of a variety of efficient systems, particularly in maritime studies.

References

1. Mtemeri J (2017) Factors influencing the choice of career pathways among high school students in Midlands Province Zimbabwe. J Soc Sci 33(2):169–178
2. Inge R et al (2020) Unravelling the concept of employability, bringing together research on employability in higher education and the workplace. Stu High Educ 45(12):2588–2603
3. Cicek K et.al. (2019) Future skills requirements analysis in maritime industry, procedia computer science. 158: 270–274

4. Lau Y et al (2021) Maritime undergraduate students: career expectations and choices. J Sustain 13:4297
5. Abelha M et al (2020) Graduate employability and competence development in higher education a fsystematic literature review using prisma. J Sustain 12:5900
6. Taberdo A (2018) The implication of teaching qualities of the instructors on the students' performance. Asia-Pac J Multidiscip Res 6(4):120–125
7. Demirel E (2019) Development of maritime management and maritime economics. Pressacademia. 9(9):242–252
8. Hinrichs S et.al, (2019) Transport 2040: automation, technology, employment—The future of work, world maritime university. World Maritime University,Transport 2040: automation, technology, employment—the future of work. Rep: 58
9. Čampara L et al (2017) Quality of maritime higher education from seafarers' perspective 3(2):137–150
10. Ngcobo LA (2018) Response to technology advancement in maritime education and training: a case study of the South African national maritime institutes. World Marit U Dissert:660
11. Klassen AC et al (2012) Best practices in mixed methods for quality of life research. Qual Life Res 21(3):377–380
12. Robinson O (2014) Sampling in interview-based qualitative research: a theoretical and practical guide. J Qual Res Psychol. 11(1):25–41
13. Rowley J, Slack F (2004) Conducting a literature review. Manag Res News. (ed) 27:31–39
14. Mishra D et al (2016) Big Data and supply chain management: a review and bibliometric analysis. Ann Oper Res 270:313–336
15. Creswell JW (2012) Educational research: Planning, conducting and evaluating quantitative and qualitative research, 4th edn. MA Pearson, Boston
16. Creswell JW (2014) Research design: qualitative, quantitative and mixed methods approaches. 4th edn. Thousand Oaks, CA, Sage
17. Marshall C, Rossman GB (2011) Designing qualitative research, 5th edn. Thousand
18. Wang P, Mileski J (2018) Strategic maritime management as a new emerging field in maritime studies. Marit Bus Rev 3(3):290–313
19. Chuah L et.al. (2021) Profiling Malaysian ship registration and seafarers for streamlining future Malaysian shipping governance. Aus J Marit Ocean Affairs, 1–37
20. Faculty of Maritime Studies (2020) Sustainability report of the maritime programmes. Universiti Malaysia Terengganu, Malaysia

Chapter 27
The Effect of Process Temperature and Holding Time on Weldability in Diffusion Welding of Duplex Stainless Steels and Marine Grade Steels for Oil and Gas Pipes

Bakhtiar Ariff Baharudin, Mazli Mustapha, Mohamad Azmeer Azman, Amirul Naim Shamsuddin, Azman Ismail, Tuan Muhammad Nurkholish Tuan Anuwa, Fauziah Ab Rahman, Fauzuddin Ayob, Darulihsan Abdul Hamid, and Mohd Afendi Rojan

Abstract Diffusion welding is done on specimens of duplex stainless steel and low carbon steel under varied temperature and holding time. The specimens were clamped using jigs and pre-pressed at 10 MPa and heated in a furnace. Impact test were performed on the joined samples. The effect of process temperature and holding time on the impact strength of the diffusion welds were observed. The impact strength

---

B. A. Baharudin (✉) · M. A. Azman · A. N. Shamsuddin · A. Ismail · T. M. N. Tuan Anuwa · F. Ab Rahman · F. Ayob
Universiti Kuala Lumpur Malaysian Institute of Marine Engineering Technology, 32200 Lumut, Perak, Malaysia
e-mail: bakhtiarab@unikl.edu.my

A. Ismail
e-mail: azman@unikl.edu.my

F. Ab Rahman
e-mail: fauziahabra@unikl.edu.my

F. Ayob
e-mail: fauzuddin@unikl.edu.my

M. Mustapha
Universiti Teknologi PETRONAS, 32610 Seri Iskandar, Perak, Malaysia
e-mail: mazli.mustapha@utp.edu.my

D. Abdul Hamid
Kolej Universiti Poly-Tech MARA, Jalan 6/9, Taman Shamelin Perkasa, 56100 Kuala Lumpur, Malaysia
e-mail: darulihsan@kuptm.edu.my

M. A. Rojan
School of Mechatronic Engineering, Universiti Malaysia Perlis, Jalan Wang Ulu, 01000 Arau, Perlis, Malaysia
e-mail: afendirojan@unimap.edu.my

A. Ismail et al. (eds.), *Design in Maritime Engineering*, Advanced Structured Materials 167, https://doi.org/10.1007/978-3-030-89988-2_27

of the joints was examined, and the highest value of 0.9 J is obtained at a temperature of 1050 °C and holding time of 240 min.

Keywords Diffusion welding · Duplex stainless steel · Dissimilar metals

27.1 Introduction

To some extent, advances in fusion/arc welding technology continue to provide problems in joining common materials such as high-tensile steel, stainless steel, aluminum alloys, cast iron, sheet steel, titanium, nickel alloys, and others. With the production of advanced materials such as super alloy metals and composites such as metal matrix composites, the challenges facing the wide range of industries in manufacturing quality joints are even greater. In this research, the relatively new diffusion welding compared to fusion/arc welding is intended to investigate a possible solution to some of the above-mentioned problems. The materials chosen in this study are duplex stainless steel (DSS) and marine grade low carbon steel. DSS can be fused with any grade of steel. The high heat input during fusion welding, however, results in the formation of brittle phase effects, which reduce the properties of corrosion resistance, particularly in the heat affected zone. Diffusion welding can be used to solve fusion-related microstructural problems while producing a joint that is significantly stronger than that created by other non-fusion processes.

DSS 2205 is a well-known stainless steel that is resistant to pitting and crevice corrosion and is utilized in desalination facilities as components for evaporator shells, ventilation systems, and deaerators [1]. It is also used in the manufacture of offshore oil and gas pipelines, offshore concrete structures, offshore umbilicals, ocean mining equipment, ship chemical tankers, fasteners used in marine engineering, bridge building in cold countries, paper, and pulp industries [2]. It is also used for rotors, axles and fans, and acid transport in corrosive conditions [3]. The 2205 DSS is the most inexpensive duplex stainless steel in the world, and it is extremely competitive in terms of pricing, thanks in part to the vast number of worldwide mills and storage businesses. It is generated in almost every type of material, most of which are in stock. This availability made 2205 a viable option for large and small projects and equipment repair and replacement [4]. The driving force for expanding the use of diffusion bonding is the growing development of new and advanced materials, such as metal matrix composites (MMCs), intermetallic and ceramics, where fusion processes are not generally applicable or are of limited use [5]. Diffusion welding is an appealing method for advanced materials, particularly when standard fusion welding procedures impair melt characteristics and those of heating-affected materials regions. The method depends on several factors of variables: time, pressure applied, bonding temperature, environment, product characteristics, roughness, and mating surface contamination degree [6].

Following that, the mechanical properties of bonded specimens must be greater than those of the bulk. Second, the component should only show a single phase microstructure. The third component is the grain growth across the bonding planes, which helps to achieve this. High temperatures are required to meet these conditions, and the diffusion coefficient must be sufficiently high to complete the process within a theoretically reasonable time frame [7].

The aim of this research is to investigate the effects of process time on diffusion welding of dissimilar materials such as A36 and DSS with constant bonding pressure of 10 MPa, variable holding temperatures, and holding time of 240 min.

27.2 Methodology

Further information on the experimental setup is presented.

27.2.1 Equipment for Diffusion Welding

The main equipment used for diffusion bonding is the furnace which provided the heating temperature. The samples are clamped together by using jigs. The jigs are designed and built to hold the specimen in position and fit inside the furnace. A 25 mm thick plate was cut into a rectangular form with a width of 40 mm and a length of 100 mm. Two holes with a diameter of 12 mm were bored into the plate at 15 mm from each end. The jig is secured using bolts and nuts, as well as compression springs, as shown in Fig. 27.1.

27.2.2 Specimen Preparation for Diffusion Welding

Materials used in this study as parent metals were commercial DSS (UNS S31803) and marine grade low carbon steel (ASTM A36). The materials were cut to size for Charpy-V notch experiments. After the specimens had been prepared, inspection has been conducted to make sure the specimens follow the dimension required for the diffusion bonding process. The specimen ends that are to be joined were polished on 240 grit emery paper and cleaned with alcohol to remove any loose grit or dirt and grease or other contaminants. The specimen is then blown dry. Next, a mixture of boron nitride and acetone were evenly coated on to the jig's surfaces and to avoid the specimen from getting attached during the diffusion welding process. Specimen of DSS and ASTM A36 were aligned together in the jig. The jig was pre-pressed at 10 MPa, and the bolts of the jig are tightened.

Fig. 27.1 Specimen placed in jigs

27.2.3 Diffusion Welding Procedure

The specimens were positioned in the furnace. The furnace is programmed to reach the required temperature and holding time. The furnace was heated up at the rate of 36.7 °C/min until it reaches the required holding temperature. The holding temperature was maintained as the required time, for example, for specimen S1, the holding temperature was 800 °C and the holding time was 240 min. When the required holding time was reached, the furnace gradually reduced its temperature. The specimen was left to cool down at furnace temperature until the temperature reaches room temperature. The specimen was then taken out from the furnace and the joint produced was inspected. The diffusion welding parameters are shown in Table 27.1.

27.2.4 Impact Test

In this study, Charpy impact analysis, based on ASTM E23, involves striking a standard notched specimen with a weight pendulum swung from a fixed height. The typical Charpy-V notch specimen is 55 mm long, 10 mm square and has a deep 2 mm notch on one side with a 0.25 mm tip radius.

Table 27.1 Diffusion welding parameters

Welding parameters			
Specimen	Pressure (MPa)	Temperature (°C)	Holding time (min)
1	10	800	240
2	10	850e	240
3	10	900	240
4	10	950	240
5	10	1000	240
6	10	1050	240

The specimen is supported on an anvil by two ends and struck by the pendulum on the opposite side of the notch. The amount of energy consumed is measured in the cracking of the test piece, which gives an indication of the test material's toughness. During the experiment, the pendulum swings around, the height of the swing being a measure of the amount of energy consumed when the sample is broken. The impact test was done using the LS-22 006-CI Charpy Izod Impact Tester.

27.3 Results and Discussion

The results for the impact tests are shown in Fig. 27.2. The highest impact results were 0.9 J, obtained from specimen 6 with parameters of holding temperature at 1050 °C and holding time of 240 min. The lowest impact results obtained were 0.2 J, obtained from specimen 1 with parameters of holding temperature of 800 °C and holding time of 240 min.

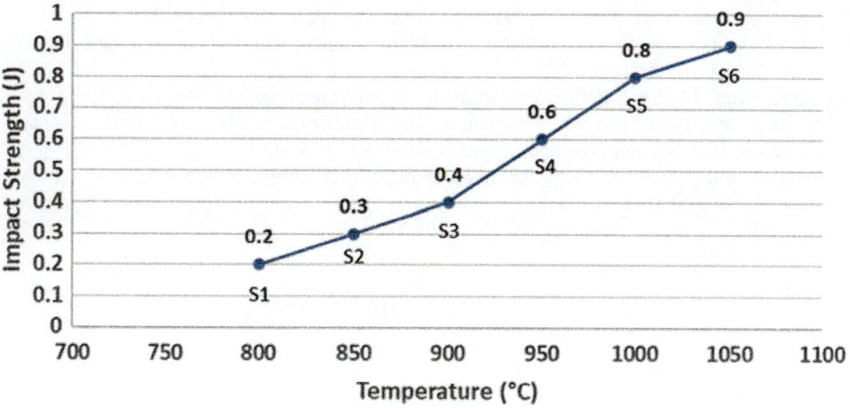

Fig. 27.2 Graph of impact strength versus temperature

27.4 Conclusion

As a conclusion, the experiment of diffusion bonding of dissimilar materials of ASTM A36 and DSS (UNS S31803) mechanical properties values were investigated. The following are the outcomes deduced from the investigation.

Dissimilar metals of ASTM A36 and DSS (UNS S31803) could be joined by the diffusion welding method at temperature range of 800 °C to 1050 °C, with holding time of 240 min and clamped by jigs. However, the impact strength is quite low due to a lack of diffusion.

The highest impact strength is obtained at the parameter of temperature 1050 °C and holding time of 240 min with the value of 0.9 J. The impact strength was observed to be increased as the temperature increases.

Further investigation is needed to obtain the optimum impact strength for the diffusion bonding of DSS (UNS S31803) and ASTM A36.

Acknowledgements The authors are thankful to Ministry of Higher Education (MOHE), Malaysia for the financial support provided via the Fundamental Research Grant Scheme [ref. no. FRGS/1/2020/TK0/UNIKL/03/5]. The authors are also thankful to Universiti Kuala Lumpur, Malaysian Institute of Marine Engineering Technology, Perak, Malaysia for providing necessary facilities and resources to complete this study for publication.

References

1. Jan O, Malin S (2007) Duplex—a new generation of stainless steels for desalination plants. Desalin 205:104–113
2. Vinoth JA, Ajaykumar L, Deepak CR, Aditya KVV (2017) Weldability, machinability and surfacing of commercial duplex stainless steel AISI2205 for marine applications—a recent review. J Adv Res 8:183–199
3. Lagneborg R, Jonson J (1989) Stainless steel in the 90's. Mater Des 10:144–215
4. Schulz Z, Whitcraft P, Wachowiak D (2014) Availability and economics of using duplex stainless steels. Corrosion 4345:1–10
5. Çam G, Koçak M (1998) Progress in joining of advanced materials. Int Mater Rev 43:1–44
6. Violeta T, Mariana L, Lucia L, Georgeta A (2007) Materials bonding by diffusion welding technology. Int Conf Innov Technol Joining Adv Mater 1:134–139
7. Gietzelt T, Toth V, Huell A (2018) Challenges of diffusion bonding of different classes of stainless steels. Adv Eng Mater 20:1–10

Chapter 28
Development of a Floating Solar Platform for River Application

Muhammad Adli Mustapa, Md Salim Kamil, Rohaizad Hafidz Rozali, Mohd Amin Hakim Ramli, and Mohd Idzani Ahmad Jadi

Abstract This paper focuses on designing a floating solar platform for the inland body, consisting of water and/or calm coastal waters mainly near the river mouth and naturally protected channels. In this study, Dinding River in Lumut, Perak has been chosen as a case study location. At present, the use of renewable energy (RE) has been pushed aggressively by policymakers throughout the world to reduce the impact of green gas emission. This paper will focus on the development of a floating solar platform design. The proposed design is developed referring to both basis SPAR type platform and also to the floating instrument platform (FLIP) concept, where the design is able to minimize the vertical movement of the platform/vessel due to vortex-induced vibration (VIV) and vortex-induced motion (VIM). A comparison in the performance of designs is made with the existing design of the floating solar platform. Comparison is made with basic design 1, which consists of an existing design. The vertical movement (heaving) was simulated using a computational fluid Dynamics (CFD) software. Results are presented in the form of heaving for each floating platform design. The numerical results show that at low wave height (0.1 m) the lowest vertical movement is recorded by design 1 with 0.14 m height, while at the highest wave height (0.3 m), the lowest vertical motion is recorded by design 2 with 0.23 m height.

M. A. Mustapa (✉) · M. S. Kamil · R. H. Rozali · M. A. H. Ramli · M. I. A. Jadi
Maritime Engineering Technology Section, Universiti Kuala Lumpur Malaysian Institute of Marine Engineering Technology, 32200 Lumut, Perak, Malaysia
e-mail: adli@unikl.edu.my

M. S. Kamil
e-mail: mdsalim@unikl.edu.my

R. H. Rozali
e-mail: rohaizad@unikl.edu.my

M. A. H. Ramli
e-mail: mohdamin@unik.edu.my

M. I. A. Jadi
e-mail: mohd.idzani@s.unikl.edu.my

A. Ismail et al. (eds.), *Design in Maritime Engineering*, Advanced Structured Materials 167, https://doi.org/10.1007/978-3-030-89988-2_28

Keywords Solar energy · Solar farm · Floating solar platform · Malaysia renewable energy

28.1 Introduction

28.1.1 General

Fossil fuel, as derived in science daily [1], is a hydrocarbon element commonly found as energy in the form of petroleum, coal, and natural gas. This element or material was a by-product consisting of dead plants, plankton and animals and thus renamed as a "fossil" fuel.

In the mid-twentieth century, a new form of power was found and used to supply energy to society and due to increasing in demand has led to the usage of nuclear power. Even though nuclear-generated power is very efficient, it can come at a high cost. The biggest challenges of using nuclear power are the issue of waste management and also a high risk of causing an accident as happened in Chernobyl, Ukraine and Fukushima, Japan. Currently, there are about 250,000 tons of waste [2] generated from nuclear power plants, and this creates a major concern in disposing of the waste safely.

This renewable energy (RE) can be defined as a non-depleting source of energy that can be harvested and used by civilization. This type of energy is available naturally and can be found anywhere on earth. However, it requires a specific method in harvesting each energy type. An example of RE are tidal, wave, solar, wind power and geothermal.

Fossil fuel sources have started to deplete in the time since the revolution of industrialization take place. This caused major greenhouse gaseous emissions, which take place around the world. Thus, the world has begun to look at RE as an alternative to reduce the dependency on fossil fuel usage.

This study focuses on proving the feasibility of deployment of the floating solar field in Dinding River located at Lumut, Perak. The idea of this study is to bring the whole concept of the land solar field to water-surface application. While a floating platform for solar power field use is already in existence, improvement can be done to suit the location selected in this study. The concept of the SPAR platform in oil and gas industry and the floating instrument platform (FLIP) owned by the United States' Office of Naval Research are being idealized for the concept design.

The study on the different floating platform will be referring to the PEC concept as shown in Fig. 28.1. This concept is selected to tackle three important criteria such as preserve, economical and clean.

The definition behind each criterion in the PEC concept is listed as below:

Preserve This research hopes to be able to "preserve" nature by moving a solar field from land to the open sea.

Economical The "Economic" impact of having solar fields offshore will be huge to drive further development of offshore solar fields.

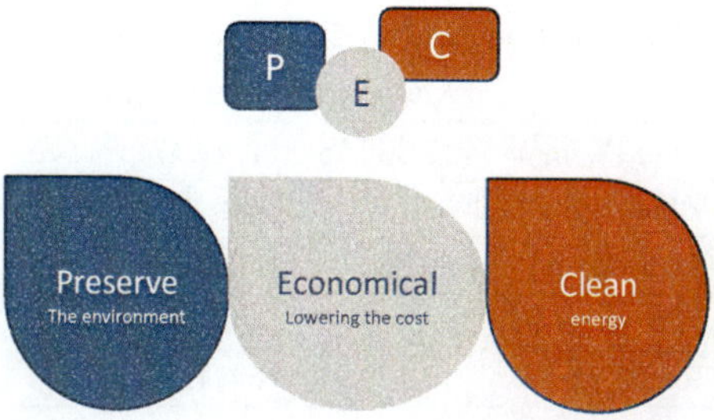

Fig. 28.1 PEC concept

Clean Increasing the share of "Clean" energy in Malaysia's electrical grid is hopefully able to reduce the negative impact of fossil fuel.

There are various problems faced by solar power development. These issues hinder the development of this type of renewable energy for electrical power generation. In Malaysia, the problems faced by renewable energy and in particular, solar power generation are even greater. Being a developing country, these challenges are pronounced, and it has so far pulled the development of solar power technology from advancing at a much higher pace. The authority has the desire to move ahead with the diversification of renewable energy in Malaysia's energy market, but the drive to follow the policies is too slow in research, testing and development. All of these issues have caused the development of solar power technology and also renewable energy technology in Malaysia left behind compared to other countries such as China and Indonesia.

This study will focus on the feasibility of developing a floating solar field in the Manjung/Lumut area specifically in and around Dinding River as well as possible near the shore between the mainland to Pangkor Island. It will also design and develop a said floating platform and perform a software simulation to study the performance of the platform and its effects from natural wave and in addition to manmade such as from ships and boats passing in the area.

At the end of this study, an optimum floating platform that is able to reduce motion caused by wave and current at Dinding River is aspected to design and access the performance of the optimum floating platform using a simulation approach.

28.1.2 Renewable and Solar Energy

> Renewable energy is energy from sources that are naturally replenishing but flow-limited; renewable resources are virtually inexhaustible in duration but limited in the amount of energy that is available per unit of time—EIA [3].

Renewable energy has become the "go-to" energy source by most developed countries to replace the use of fossil fuel to provide sufficient energy for use if humankind from transportation to a day to day routines such as cooking, entertainment and education. In the twentieth century, United States starts using hydropower

Table 28.1 Comparison of power densities based on power source and location in Turkey

	Kastamonu	Istanbul (Marmara)	Istanbul (Blacksea)	Canakkale	Hatay
Wave (W/m^2)	12,338.25	1229.47	12,338.25	1897.38	3815.63
Wind (W/m^2)	640	640	640	770	810
Solar (W/m^2)	592.66	580.72	580.72	505.65	538.12
Current (W/m^2)	0	4464.54	558.06	1883.47	235.43
Total (W/m^2)	13,570.91	6914.73	14,117.03	5056.5	5399.18

and biomass as their main source of renewable energy and later changed to other forms of renewable energy sources such as biofuels, solar, and wind energy.

A study by Sener et al. [4] has found that there are various sources of renewable energy suitable for moderate Mediterranean climate of Turkey. The study focuses on solar, wind, current and wave-based on several locations such as Kastamonu, Istanbul, Canakkale and Hatay. For each location, the most suitable renewable energy differs. As shown in Table 28.1, Hatay may benefit most to use a wave source of renewable energy, while Istanbul should otherwise use solar power as the source of renewable energy.

In Malaysia, more than one hydropower facility has been installed as an alternative to fossil fuel power generation such as the Kenyir Dam which supplies hydropower to Sultan Mahmud Power Station in Terengganu installed in 1988. Hydropower makes up a significant presence in the Malaysian renewable energy by producing 11% of the total energy generation capacity in Malaysia as shown in Fig. 28.2 [5].

As for solar energy, the generation of solar power is still relatively low compared to other types of renewable energy. In the US, solar power generation accommodates only 6% of total renewable energy generated with an approximation of just 1.4% of the total power generated in the country [6]. While worldwide, the top 10 countries with the highest renewable energy generated, solar energy makes up at most 10% of total power generated in Honduras, where else, Denmark (the country with the highest proportion of renewable energy against total power generation) only produces 3% of total power from solar [3].

In Malaysia, the journey of renewable energy started when the administration of the country decided to diversify the energy sources used for power generation in

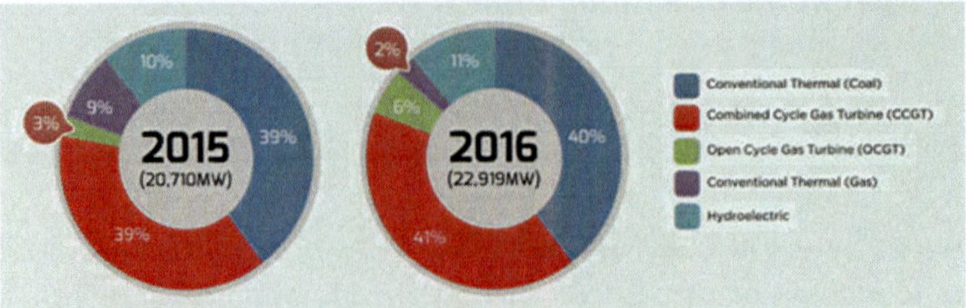

Fig. 28.2 Malaysia's power generation capacity as stated by Suruhanjaya Tenaga

Malaysia. In 1980, with the introduction of the Four Fuel Diversification Strategy [7], it is the nation's goal to shift the balance of fossil fuel usage with renewable energy. According to Sharvini et al. [5], out of the four countries investigated in 2014, China, Indonesia and Japan, Malaysia has the lowest usage of renewable energy.

By 2020, renewable energy accounts for only 9% of the total energy generated in Malaysia with only 1% generated from solar power [7]. And it is projected that solar power generation in Malaysia will remain stagnant until 2025 when energy sourced from renewable sources will increase slightly to 12% of the total power generated [7].

28.1.3 Renewable Energy Policy and Government's Action in Malaysia

Malaysia started its journey toward renewable energy application and to have dedicated policies in 1979 with the introduction of the National Energy Policy which focuses the use of renewable energy, and it is divided into 3 main topics [8]:

- Supply—where the renewable energy supply being discussed to be provided in a cost-effective way, secure and continuous as well as sufficient.
- Utilization—promote the use of renewable energy to generate power in an effective and safe manner.
- Environmental—ensuring that renewable energy is approached with a detrimental effect on the environment.

From the introduction of the Four Fuel Diversification Strategy in 1980, and Five Fuel Diversification Strategy in 1999 [8], Malaysia has set the target of the use of renewable energy to generate power at 5% by 2005. Although the intention was there, this target was still considered small with only 1% of the total energy generated in Malaysia came from renewable energy in 2005. In 2020, it is projected that renewable energy will accounts for 9% of total power generated [5].

According to the Sustainable Energy Development Authority (SEDA) [5], the National Renewable Energy Policy has set out a target of 20% renewable energy mix in Malaysia's power generation system by 2025. But, this target may not be achievable based on the outlook set out by Suruhanjaya Tenaga [5] which estimates only 13% of the renewable energy mix in 2025 as shown in Fig. 28.3.

The latest policy set out by the government of Malaysia was in 2011 with the introduction of the Renewable Energy Act 2011. The act comprises of 9 parts focusing on several issues such as connection, purchase and distribution of renewable energy, enforcement and savings and transitional.

The Malaysian government is committed to renewable energy technology when they pledged to COP-15 (15th Conference of the Parties under the United Nations

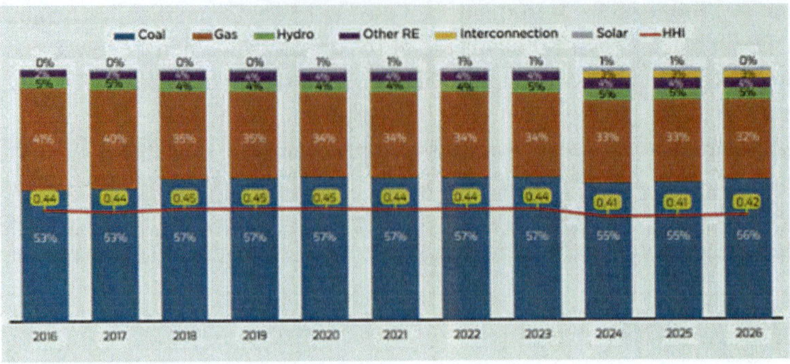

Fig. 28.3 Power generation mix from 2016 to 2026

Framework Convention on Climate Change (UNFCCC)). In this framework, all countries that have pledged to COP-15 commitment agrees in principle to reduce carbon intensity by 40% [7].

The Malaysian government has taken an initiative in the sense of the Malaysian Building Integrated Photovoltaic Project (MBIPV) to promote the installation of solar panels on building rooftops. A Photovoltaic System Monitoring Center (PVSMC) was set up in Universiti Teknologi Mara to monitor the progress of the Building Integrated Photovoltaic Project (BIPV) for a period of 5 years from 2006 to 2011. SURIA-1000 is another government-backed project that aspires to generate interest in BIPV in the residential and commercial sector.

According to a Solar Magazine article [7], Tenaga Nasional Berhad (TNB) as a private entity has also played a significant role in the policy making of renewable energy in Malaysia. This is because TNB is the only player for direct to consumer distribution of electricity in Peninsular Malaysia and Sabah. In addition, the lack of foreign participation in the local power sector is also becoming a key factor in slow-growing renewable energy development in Malaysia.

28.1.4 Potential of Solar Energy in Malaysia

Jamalludin et al. [9] performed a study on the viability of floating solar specifically in Malaysia. The six areas studied are in Langkawi, Kota Bharu, Kuala Terengganu, Kota Kinabalu, Labuan and Miri. He found that Labuan is able to generate the most power annually at 2592 MWh and in the Peninsular, the best location is in Kota Bharu at 2426 MWh per year.

By combining all generated power from these six locations will approximately produce 14,530 MWh per annum [9]. But it is suggested that further studies be performed to find a conclusion on various issues that may arise such as financial sustainability, structural reliability and the most important for the safety.

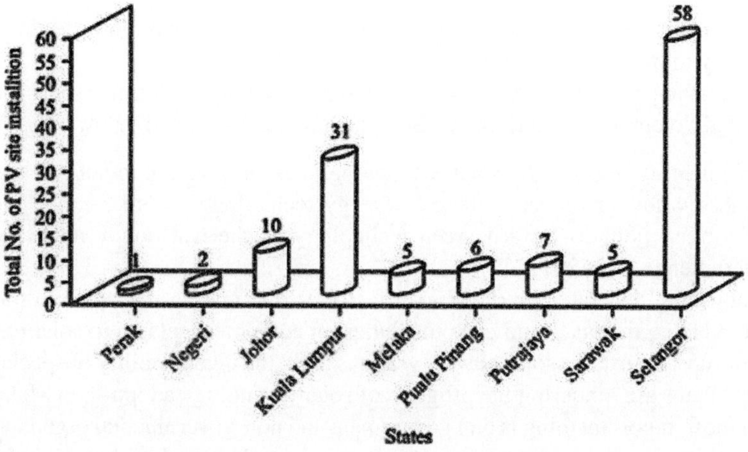

Fig. 28.4 Photovoltaic installation sites by states

Abdullah et al. [8] stated in their research that the irradiation in Malaysia is maximum during the North-East monsoon season in the period from November to March, while it is reduced during the South-West Monsoon which brings the wind from Australia toward Malaysia in the months between May to September. That being said, Abdullah et al. [8] still claimed that the potential for solar energy deployment is still high in Malaysia because it is sunny all year round. Which makes the use of large scale solar (LSS), a great option for deployment in Malaysia.

On average, Malaysia has a mean solar irradiance of 1643 kWh/m^2 with irradiance in Ipoh and Telok Intan scores at 1739 kWh/m^2 and 1824 kWh/m^{2,} respectively [9]. Aziz et al. [10] claimed that solar power in Malaysia is mainly being used for pumping systems and domestic water heating systems.

As shown in Fig. 28.4 [10], the number of sites for the installation of a photovoltaic solar field in Malaysia is still low. In Perak, with high-annual solar irradiance, as of June 2011, PVSMC has identified only one suitable site for installation. This is in contrast with the high number of installation sites known in Kuala Lumpur and Selangor.

With the policymakers being friendly with the idea of renewable energy, coupled with Malaysia being a suitable location for large scale solar projects, the potential of renewable energy in Malaysia is considered promising.

28.1.5 Potential of Solar Energy in Malaysia

In any new technology, they are bound to be challenges faced for it to become accepted and mainstream. In the case of solar technology, several challenges prevent it from becoming adopted in a more fast-paced manner. Some of the challenges are superficial and others are more real and have to be dealt with in order for solar technology to become mainstream.

The challenges include the following based on a study, "Current Solar Energy Policy and Potential in Malaysia" by Solangi et al. [11] as stated below:

- Implementing supportive solar energy policies and giving practical support to those who implement renewable energy technology. There are not enough and vibrant policies toward green technology as the last policy made by the government was in 2011.
- Establishing dedicated credit or loan facilities that make solar power attractive and reducing the taxes and customs duties on equipment related to solar energy. As the cost to install a solar power system is high, the unavailability of specialized credit loans are hindering the progress of solar technology adoption in Malaysia.
- The most important thing is that government and non-governmental organizations (NGO) can do is creating awareness among the public such as benefits of solar energy legal requirements, financial aspects and environmental advantages. This stem from a lack of support to promote awareness on green technology including solar.
- Government can also provide details of information regarding the implementation of the solar technology and build technical capacity. Similar to item no. 3 where support is lacking from the authority.
- It is essential to reduce the subsidies for fossil fuel as a prerequisite for technological development of solar technology and develop a market for solar energy with attractive prices for users as well as suppliers. The use of subsidized fossil fuel is a hindrance to the general public to switch and adopt solar power and green technology in general.
- Although Malaysia set a clear and ambitious target of 5% for the use of renewable energy use out of total electricity production in the Eighth Malaysian Plan, it does not materialize due to poor implementation.
- Independent power producers must get access to the national power grid and Tenaga Nasional Berhad (TNB) should give preference to renewable energy projects. Low participation from external parties (local and foreign) leads to stagnation of the development of renewable energy including solar in Malaysia.

28.1.6 Floating Solar Field and Its Advantages

In the article entitled "Potential of floating solar technology in Malaysia," Jamalludin et al. [9] discussed the viability of having particular renewable energy be employed in the Malaysian region. The researcher initially discussed a floating structure in general which is called "Floating Offshore Solar Field" or "Solar Island" and it then can be further sub-categorized under very large floating structures (VLFS), or otherwise can also be called as very large floating platforms (VLFP).

It also provides the example of current existing very large floating structures, including the Marina Bay Floating Stage in Singapore and also Floating Islands in

Han River in Seoul, Korea. Jamalludin et al. [9] also claimed that the study of the feasibility of a floating solar field to generate energy has never been explored in Malaysia, thus their study is to show with evidence the viability and effectiveness of such projects to be deployed in Malaysia.

The same research further discussed solar technology in general such as the latest development in solar panels. The name used by the researchers is a very large floating solar structure (VLFSS). The researcher initially touched briefly on the concept design, material to use, such as flotation foam and configuration needed for the system. Mooring system was also discussed to ensure that the floating platform stays in place.

Jamalludin et al. listed the advantages of a floating solar system, and these are mentioned below:

- Abundant natural energy sources—Malaysia is a tropical country with year-round sunny days. With average irradiance at around 1900 kWh/m^2, it is a great opportunity to be utilized.
- The oceans provide ample zone to energy demand—with the uninterrupted area and with no deforestation needed, any large enough body of water is able to provide much-required space for a solar field.
- Low operating and maintenance costs—there is no cost associate with land acquisition when a solar field is located on the water. At the same time, the maintenance cost to maintain such a facility is low.
- Solar panels support by the buoyancy characteristic and allowing a simple structure—for a floating structure, as long as weight is distributed correctly, a simple structure without the need for an underground base, is sufficient. Thus, it will indirectly reduce the cost of installation.
- Air and seawater cooling promote higher efficiencies—surrounding uninterrupted air circulation and water beneath the platform, the photovoltaic system can be kept cool and later increase the efficiency of the solar energy generation system.
- Cost-effective to manufacture and environmentally friendly—simple structure and no land acquisition needed leads to low capital for floating solar field installation. And this will also be environmentally friendly as no deforestation needed to make way for a solar field which can be counterproductive.

Sener et al. [4] in their study, "The Renewable Energy Potential of Turkish Coasts and A Concept Design of a Near Shore Sea Platform" has made a comparison between an offshore and nearshore floating platform for use of renewable energy. In Table 28.2, it can be summarized that the selection of whether to use offshore or onshore platform may depend on the needs and financial capability of the project owner.

This decision-making process is critical to ensure the success of any renewable energy projects as it will affect the installation and operating cost of these platforms in the long run. Thus, a project owner has to evaluate based on their needs before proceeding to commit to such projects.

Sener et al. [4] also studied various types of platforms that are suitable in harnessing renewable energy. They also envisioned that the platform may be able to

Table 28.2 Comparison of platform type

	Offshore	Nearshore
Size	Large	Small
Capital cost	High	Low
Power capacity	High	Low
Installation	Hard	Easy
Depth (m)	>60	20–25
Distance from shore	Away	Near

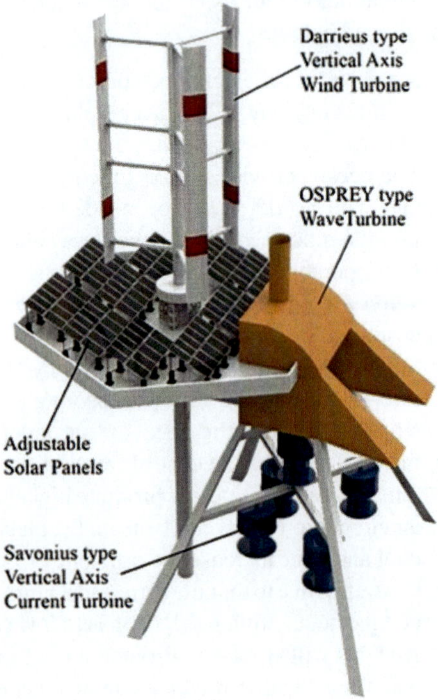

Fig. 28.5 A concept design of multi-sourced renewable energy platform

act as a multi-purpose platform to harness a combination of two or more renewable energy sources as shown in Figs. 28.5 and 28.6.

28.1.7 Concept Design

There are already several designs of floating platforms used in the renewable energy industry particularly for solar power harnessing. Each design has its advantages and disadvantages. Some may be suitable for calm inland waters, while others may be suitable to be deployed in rougher seas, near or offshore areas.

Fig. 28.6 Modules being arranged to create larger floating platform assemblies

This study will focus on two designs. One is currently being deployed in Singapore, and the other has received US patent approval and also have been approved for use by Det Norse Veritas (DNV).

28.1.7.1 Singapore Floating Solar Platform

As being reported by Channel News Asia [12], a floating solar platform field will be deployed that will be able to produce 50 megawatt-peak (MWp) of clean power as shown in Fig. 28.7. This solar field, in theory, will be able to reduce Singapore's carbon emission by 28,000 tons. In comparison with, 28,000 tons of carbon emission is equivalent to the emission of about 6000 cars on the road.

Fig. 28.7 Singapore floating solar platform

Another two smaller projects are also planned, assumedly would be operational as early as 1st half of 2020. Being in place by 2021, these floating solar systems will be able to produce around 57 MWp which can power about 15,500 homes.

28.1.7.2 Hexagonal-Shape Solar Field

Smadja et al. [13] in their submission to the United States Patent Application Publication has submitted a design of a floating solar platform that uses the concept of buoys in a hexagonal shape. This assembly can be attached to create one large solar field.

This design as reported by the Institute of Marine Engineering, Science and Technology (IMAREST) in their website [14] has received approval from DNV for use offshore. The approval from DNV proved that the design conforms to the standard and accepted practices in the sense of design, usage and safety.

28.1.8 Research Gap

With the research on current available studies and publication, it is understood that there are several gaps available that can be improved.

- Current solar field developments in Malaysia lead to deforestation of land and counterproductive to the concept of green technology.
- High cost for land acquisition. With the need for land acquisition for the deployment of the inland solar field, there will be a need for a huge amount of budget.
- High cost of deploying due to the need for a base to hold solar structure on land.
- Lagging of solar power generation development in Malaysia. Even though there is valid interest in solar power generation in Malaysia, the policy making and actual development is lagging even when comparing among similar developing countries such as Indonesia and China.
- Available floating solar platform design is too huge and may pose challenges in deploying which may also increase the cost for installation [13].
- The platform used in Singapore may not be suitable for use in busy waters with traffic that causes waves such as in the channel.
- Not suitable to be deployed on nearshore waters with constant waves such as waters near shore or river estuary such as Dinding River, Lumut, Perak.

28.2 Methodology

28.2.1 Introduction

The methodology helps to determine actions to be taken and steps to be included to achieve a good result for this study. It may also include pre-planning, planning and post-study steps. In the case of this particular study, various computational fluid dynamics (CFD) softwares can be used to access each performance of the developed design. This software allows the simulation to be performed accordingly and data to be extracted for analysis. Once data is available, other tools such as Microsoft's Excel can be used for data compilation. Tables and graphs can be generated to assist the analysis.

All in all, having a clear and proper methodology will ensure that the researcher has a clear path to follow as not to astray from achieving the goal of the study.

28.2.2 Comparison

A decision has been made to compare a result for the proposed design with an existing design:

- Existing design:
  - Smadja et al. [13] Offshore Floating Solar Platform Design.

For comparison, several steps are needed to verify the simulation result. Firstly, all data extracted from the simulation process are compiled. Excel tables are used to allow comparison to be more apparent. Then, proceed by analyzing the data and later identifying differences between each recorded data.

By selecting the needed data, a graph can be plotted. Based on these graphs, the data can be further study and analyze. Using graphs, trends of the data can be seen, and it can then be compared.

Lastly, existing books and journal related to platform designs and fundamentals of vortex-induced motion (VIM) and vortex-induced vibration (VIV) will be used for guidance to verify the result of this study.

28.3 Results and Discussion

28.3.1 Introduction

There are two designs created for this study. All these designs have been simplified and optimized to further improve simulation efficiency. The next stage will consist of

Table 28.3 Example of wave condition for 3-day period in October

Day	Day 1			Day 2			Day 3		
Time of day	Afternoon	Night	Morning	Afternoon	Night	Morning	Afternoon	Night	Morning
Wave height (m)	0.1	0.1	0.2	0.2	0.2	0.1	0.1	0	0.1
Period (s)	3	4	4	4	2	2	4	4	1
Wind (km/h)	10	5	15	5	15	10	10	10	5
Direction	W	ENE	ENE	WSW	SSE	ESE	SSW	E	SE

a process of conducting a grid dependent study (GDS), where an optimum number of the mesh size for every design needs to be identified. This helps to reduce the error on the final result and to reduce simulation running time.

28.3.2 Dinding River Wave Condition Data

The input data for Dinding River has been extracted from Jabatan Meteorologi Malaysia. The data was observed for October to see the variance of the data. This is to ensure that the variance is not drastically changed that it will impact the result of the simulation. A three-day data is collected from the seven, and the result is shown as follows.

As shown in Table 28.3, the wave condition in the Lumut area is observed not too strong throughout the month, and the wave height has a range from 0.5 to 0.1 m. Below are the factors in deciding the wave amplitude:

- Dinding River mouth is presumed to be less affected by the open water wave.
- The month of October is chosen as monsoon starts to strike and becoming best-case scenario for this study.

Thus, for this study based on the above factors, the wave amplitudes used in the CFD software were set at 0.05 m, 0.10 m and 0.15 m.

28.3.3 Design 1

There are two designs considered for this study. Design 1 is an existing design and design 2 is a proposed design, based on a concept of SPAR platform and FLIP vessel. The two existing designs have been explained previously, and the schematic design are shown in Figs. 28.8 and 28.9 which were created using CAD software.

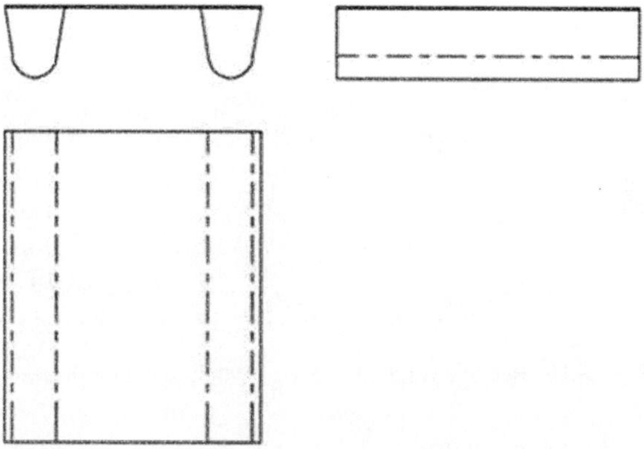

Fig. 28.8 Design 1

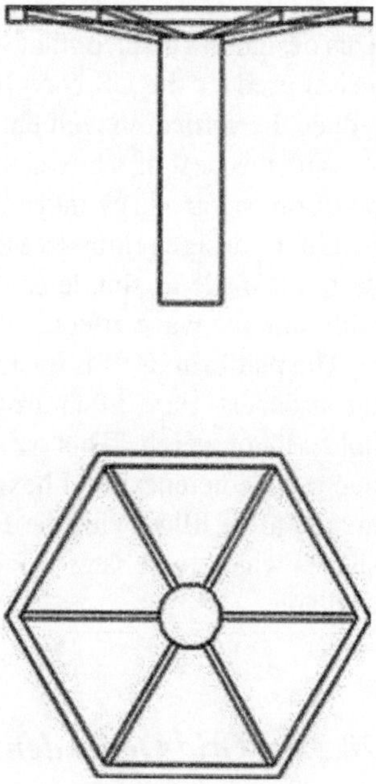

Fig. 28.9 Design 2

Design 1 was replicated but not with an exact dimension as no blueprint of the design was found. Thus, it is based purely on designer creativity to replicate it as nearest as possible to the original design used by Smadja et al. [13].

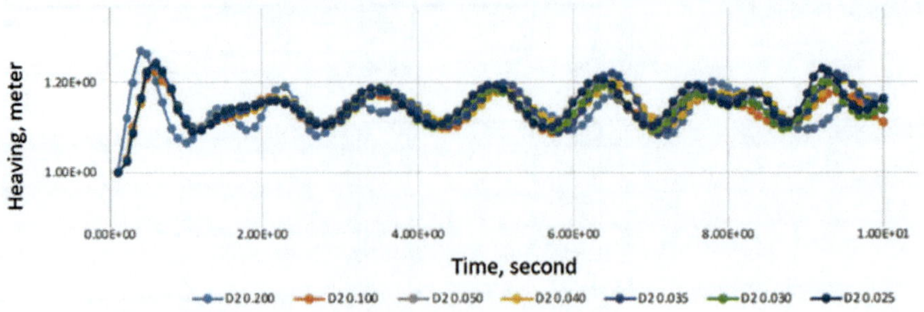

Fig. 28.10 Heaving results for design 1 at different mesh size

28.3.4 *Design 2*

This design is a concept that was taken from the use of the SPAR platform and FLIP vessel used by the US Navy Research team. The idea is that using this design can reduce the vertical movement of the platform.

With this design, a hexagon shape platform is tried to be incorporated as the based platform on top of a cylinder shape floating device. The cylinder shape of the floating part of the design reminisces and borrowing a lot from the SPAR concept. This initial design is made as simple as possible to check its movement characteristic in water with minimal wave effect.

The platform itself is designed to be in hexagon shape where it is well known as the strongest shape in nature and architecture. It can hold its shape, and it also can hold a lot of weight. This particular shape can commonly be found in nature as well due to its efficiency. In a hexagonal grid, each line is as short as it can be if a large area is to be filled with the fewest number of hexagons. It is also one of the only shapes which tessellates perfectly, while also being the most space-efficient shape available.

28.3.5 *Grid Dependent Study (GDS)*

Using a computational fluid dynamic (CFD) software, simulations were performed on the designs. Several initial tries have been made to identify the best setting for the objective set. These tries were very time consuming as various changes need to be made on the weight distribution to achieve a perfect setting.

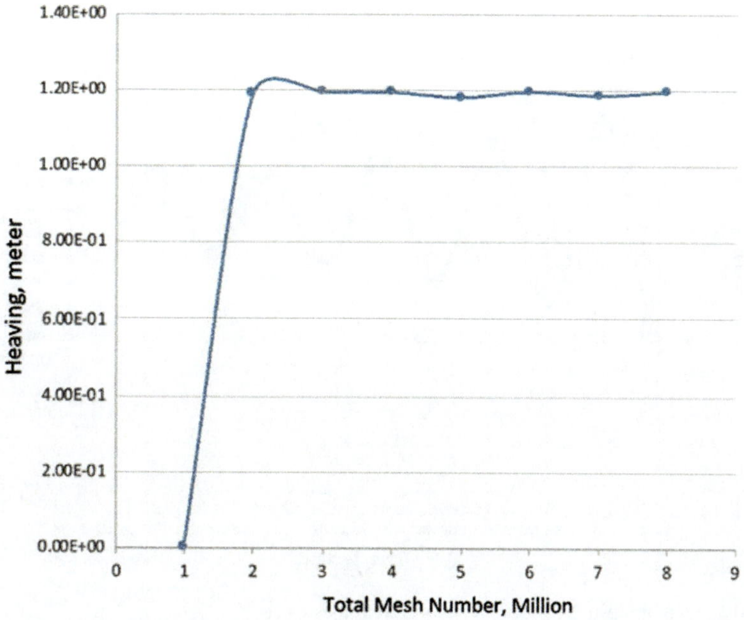

Fig. 28.11 Mesh optimization selection—design 1

28.3.6 Grid Dependent Study (GDS) for Design 1

As per Fig. 28.10, it can be seen that design 1 vertical movement (based on Z-coordinate probe) were not consistent as the wave beginning to fully develop at the starts of the simulation, it became quite stable from 3 to 7 s and then again becoming uneven toward the end of the simulation. Thus, it was determined that the point to be used for the next step would be at peak number four.

The point at peak number four to number seven were recorded for every meshes condition. Another graph was made to determine the most optimum mesh that can be used for the actual simulation as shown in Fig. 28.11.

From Fig. 28.11, it can be seen that the result starts to show a consistent trend which starts at the total mesh number of 107 million. Thus, mesh size 0.05 mm was selected for the next step to align it with design 2 which will be discussed next.

28.3.7 Grid Dependent Study (GDS) for Design 2

Figure 28.12 shows design 2 heaving motion (based on Z-coordinate probe). The result shows a similar pattern but at different period. With the larges mesh size (mesh size of 0.20 and 0.10 mm), the period is shorter for both. But the rest of the meshes started to be in tune with each other. This behavior is expected as the larger the mesh

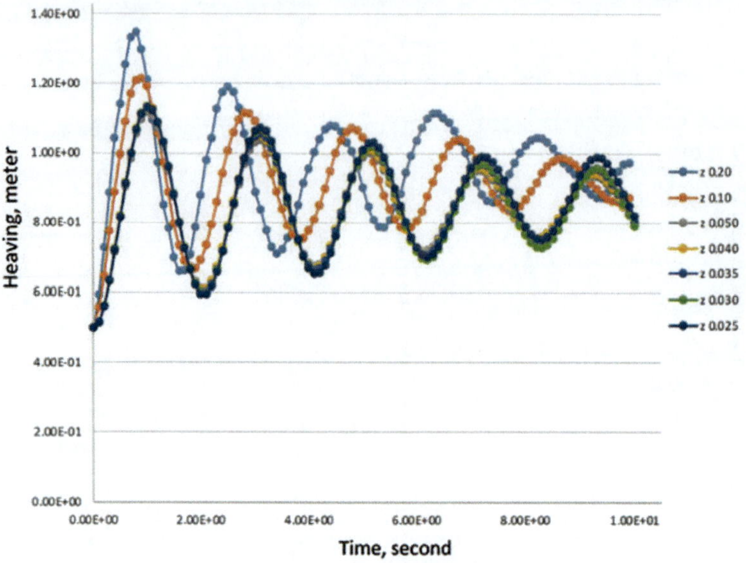

Fig. 28.12 Heaving results for design 2 at different mesh size

size used in a CFD software simulation will tend to show an accurate result. This is because, with a larger mesh size, the design will be read as having larger pixel by the software. Thus, it was determined that the point to be used for the next step would be at peak number 4.

The results from peak number four were recorded and are used to plot in finding the most optimum mesh number for design 2. The optimum mesh size is then selected as shown in Fig. 28.13.

As can be seen in Fig. 28.13, the mesh size only started to show a consistent pattern at a size of 0.05 mm, which is approximately equal to 2.9 million of the total mesh number. Thus, it was decided that 0.05 mm is the optimized mesh size for the simulation. If the larger mesh size was selected and used, then the data may not be as accurate, while by using the smallest mesh size, it will be redundant, as the time taken to run the simulation will be longer and unnecessary as mesh size 0.05 mm is already able to give an accurate result.

28.3.8 Heaving Motion Results

Simulations were made with similar parameters for both design 1 and design 2. They were simulated in 3 waves condition which was determined as being present in the condition of the calm water at Dinding River where there is no incoming wave coming from the open sea. The only wave from tide and water traffic from fishing boats are present.

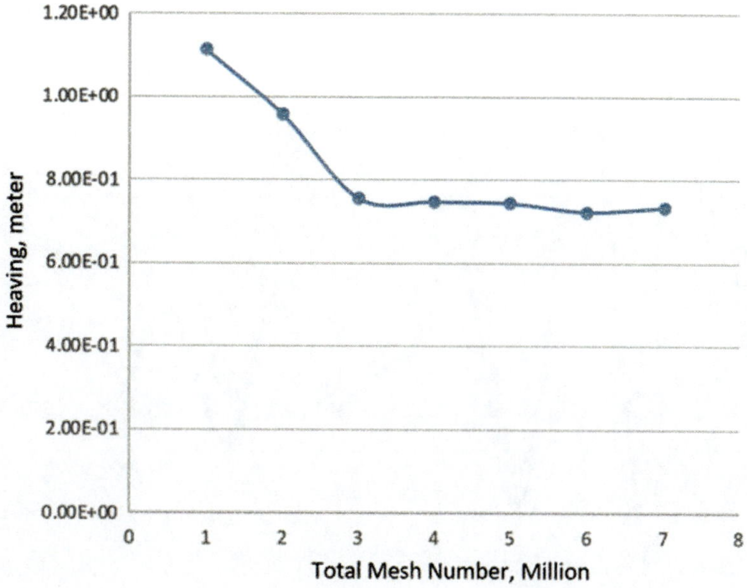

Fig. 28.13 Mesh optimization selection—design 2

Based on Fig. 28.14, it shows the simulation result for design 1. Based on the analyzes made, the platform moved vertically (heaving) and produce a vertical movement 5 times. This is different from design 2 as shown in Fig. 28.15, where it produces less vertical movement (heaving) at only 4 times within the same period. Design 2 then become more stabilized in the next 5 s toward the end of the simulation. Design 1, also based on Fig. 28.14, is becoming not stable which starts from 7 to 10 s.

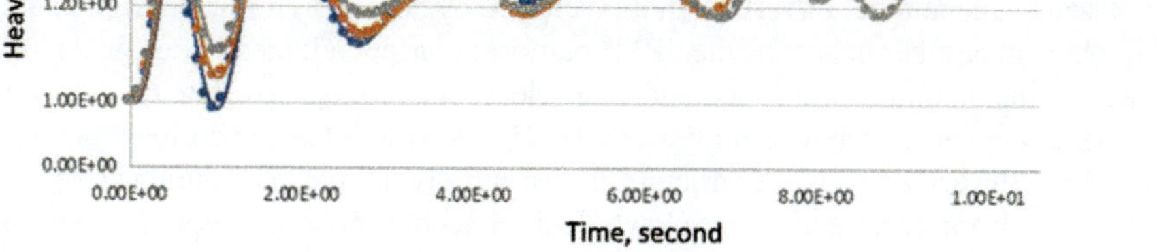

Fig. 28.14 Structure heaving motion for design 1 at different wave amplitude

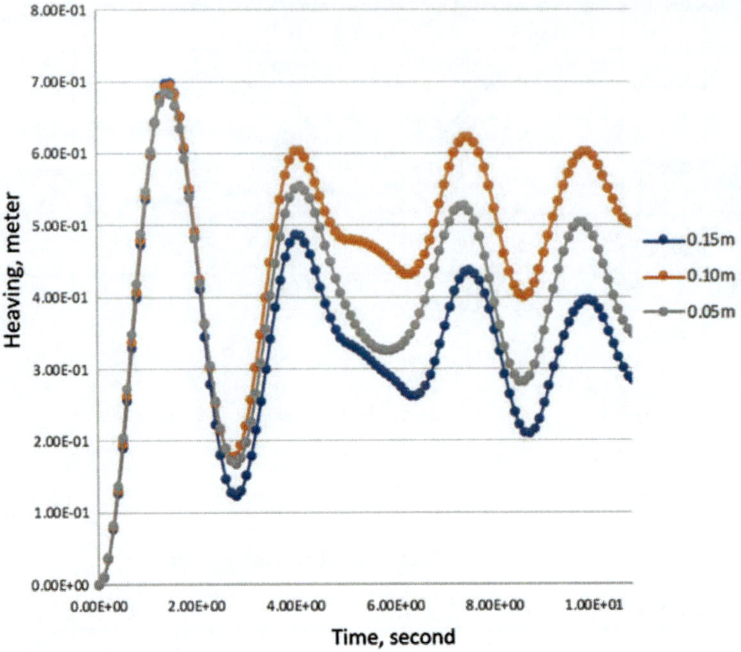

Fig. 28.15 Structure heaving motion for design 2 at different wave amplitude

Table 28.4 Vertical movement difference (m)

Design	Time (s)	1/7–10	2/7–10
Wave amplitude (m)	0.15	0.34	0.23
	0.1	0.21	0.22
	0.05	0.14	0.23

From Table 28.4, considering the output results were taken from 7 to 10 s, it can be seen that at a wave amplitude of 0.15 m design 1 produced a higher vertical motion (Z-coordinate) of 0.34 m compared to design 2 which recorded only 0.23 m. However, the results are vice versa at a wave amplitude of 0.05 and 0.1 m where the vertical motion are recorded higher at design 2 with 0.22 and 0.23 m, respectively.

Based on the number of vertical motion created by both design 1 and design 2, the difference in number occurs as the SPAR platform concept adopted in design 2 tend to absorb the incoming wave well and later release the motion slowly to counter the resonance created by the incoming wave. This helps to stabilize the entire structure. From here, the behaviors of the structure are not entirely riding the incoming wave but rather absorb and release the force slowly. This concept is good in keeping the entire platform float with minimal heaving motion and may reduce unwanted movement on the platform. However, consideration needs to be made as the SPAR platform concept reduces certain deep of avoiding the grounding phenomena.

From these results, design 1 is much suitable for river application as it produces low vertical motion at low wave amplitude, while having great clearance to the river floor compared to design 2. However, design 2 is much more suitable for open sea application as the incoming wave is greatly absorbed well in keeping good stability to the entire structure. This is also supported by the performance in keeping the vertical motion low as the incoming wave amplitude become higher.

28.4 Conclusion

It is concluded that Dinding River can be safely used to harvest energy from the sun as the area has a good geological location consisting of sufficient depth, width and even smaller waves. This is supported by a high-annual solar irradiance received.

This study has proven that the newly created design 2 has an advantage over the existing design as it is not riding the wave but it helps to absorb and release the force slowly in keeping the entire platform structure stable. In addition, the vertical motion behavior is more constant and predictable. However, Dinding River is still prone to waves due to the tides current and water traffic, and this is critical to avoid the platform being damaged from the movement due to waves which in turn can lead to damage on the more expensive solar panels.

To summarize this study, it is hoped that further research and development of floating solar panel can be continued. By doing so, it can lead to the actual development of renewable energy in Malaysia, specifically for solar energy. There are several obvious advantages for human civilization as well as nature if this can be achieved and as an aspiring developing country, Malaysia should envision itself to be the front runner in this renewable energy industry.

With the world is moving even faster to abandon fossil fuel as the main source of energy generation and power consumption, Malaysia should not be left behind. It may not be possible for a developing country like Malaysia to abandon fossil fuel just yet, but efforts, knowledge and resources should be put in place to ensure that the country will be ready when the time comes.

Acknowledgements The authors would like to thank the Maritime Engineering Technology (MET) section, University Kuala Lumpur (UniKL) for the information they have provided and all staff of UniKL for their precious assistance.

References

1. Elnugoumi MG, Ahmed ZA, Almsafir MK (2012) Current status and challenges of solar energy in Malaysia. Int J Adv Sci Eng Inf Technol 2(4):330–337
2. Nuclear Energy Institute (2018) What is nuclear energy? https://www.nei.org/fundamentals/what-is-nuclear-energy. Accessed 4 May 2020

3. US Energy Information Administration (2010) What is renewable energy? https://www.eia.gov/energyexplained/renewable-sources/. Accessed 8 May 2020
4. Şener B, Awtac S (2017) The renewable energy potential of Turkish coasts and a concept design of a Near Shore Sea platform. J Therm Eng 3(3):1211–1220
5. Sharvini SR, Noor ZZ, Chong CS, Stringer LC (2018) Energy consumption trends and their linkages with renewable energy policies in East and Southeast Asian countries: challenges and opportunities. Sustain Environ Res 28(6):257–266
6. Collu M, Borg M (2016) Design of floating offshore wind turbines. Offshore Wind Farms (OWF) 359–385
7. Energy Commission (2017) Peninsular Malaysia electric supply outlook. https://www.st.gov.my/en/contents/publications/outlook/Peninsular%20Malaysia%20Electricity%20Supply%20Outlook%202017.pdf. Accessed 1 Apr 2020
8. Abdullah WSW, Osman M, Kadir MZAA, Verayiah R (2019) The potential and status of renewable energy development in Malaysia. Energies 12(12):2437
9. Jamalludin MAS, Sukki FM et al (2019) Potential of floating solar technology in Malaysia. Int J Power Electron Drive Syst 10(3):1638–1644
10. Aziz PDA, Wahid SSA, Arief YZ, Aziz NA (2016) Evaluation of solar energy potential in Malaysia. Trends Bioinform 9(2):35–43
11. Solangi KH, Islam MR, Saidur R, Rahim NA, Fayaz H (2011) A review on global solar energy policy. Renew Sustain Energy Rev 15(4):2149–2163
12. Channel News Asia (2019) PUB to deploy Singapore's first large-scale floating solar panel system. https://www.channelnewsasia.com/news/singapore/pub-to-deploy-singapore-s-first-large-scale-floating-solar-panel-11601030. Accessed 4 May 2020
13. Smadja L, Smadja P I (2016) Floating solar panel systems. Resource document. United States Patent. https://patentimages.storage.googleapis.com/78/4d/43/df54e64f1afbc6/US10141885.pdf. Accessed 2 July 2021.
14. Institute of Marine Engineering, Science & Technology (2020) DNV GL approves innovative floating solar energy design. https://www.imarest.org/themarineprofessional/item/5484-dnv-gl-approves-design-of-innovative-floating-solar-energy-system. Accessed 25 Apr 2020

Chapter 29
A Review on the Potential Applications of the Ketapang Tree in Different Areas (*Terminalia Catappa*)

Pungkas Prayitno, Mohd Zaid Abu Yazid, Norhisham Seyajah, and Susanto Sudiro

Abstract Ketapang with the scientific name *Terminalia catappa* Linn is known as a shade tree with nutritious fruits, and it has many benefits including medicinal benefits, water treatments for fish farming, natural dyes for textiles, and reinforcing filler in composites. This comprehensive review of the potency of Ketapang, which has chemical constituents such as phenols, flavonoids, carotenoids, and tannins. Many investigations focus on the ability of this plant for traditional medicine that can treat hepatitis, dermatitis and has anti-cancer, antioxidant and anticlastogenic properties. In addition, this plant has antidiabetic, antioxidant, antimicrobial, anti-inflammatory, hepatoprotective, and has anti-cancer properties. Young Ketapang leaf juice is used as a skin medicine in the form of an ointment to relieve scabies, leprosy, and is also used internally for stomach aches and headaches. Partly for water treatment for fisheries, namely Ketapang leaf extract has a positive effect on the quality of culture water for the survival of ornamental fish. In textiles, tannins that produce yellow color can be used as natural dyes textiles. Ketapang fruit has an oil extract in the seeds that can be used as vegetable oil to make biodiesel for industry and has good physicochemical properties.

Keywords Terminalia catappa · Medicine · Water treatment · Biodiesel · Textile

P. Prayitno (✉) · M. Z. Abu Yazid · N. Seyajah
Department Design Engineering Technology, University Kuala Lumpur Malaysia Italy Design Institute, Cheras, Malaysia
e-mail: pungkas.prayitno@s.unikl.edu.my

M. Z. Abu Yazid
e-mail: mzaiday@unikl.edu.my

N. Seyajah
e-mail: hishams@unikl.edu.my

S. Sudiro
Department of Mechanical Engineering, Pancasila University, Jakarta, Indonesia

A. Ismail et al. (eds.), *Design in Maritime Engineering*, Advanced Structured Materials 167, https://doi.org/10.1007/978-3-030-89988-2_29

Table 29.1 Distribution of Ketapang plants

Asia	Afghanistan, Bangladesh, Brunei Darussalam, Cambodia, China, Indonesia, Malaysia, Myanmar, Pakistan, Philippines, Singapore, Sri Lanka, Taiwan, Thailand Vietnam [12]
Africa	Botswana, Burundi, Cameroon, Madagascar, Malawi, Nigeria, Réunion Rwanda, Senegal, Seychelles, Somalia, Sudan, Tanzania, Uganda, Zimbabwe [13]
North America	Bermuda, Mexico, USA [14]
Central America and Caribbean	Anguilla, Bahamas, Barbados, Belize, Costa Rica, Curaçao, Domiica, Dominican Republic, Guatemala, Haiti, Honduras, Jamaica, Martinique, Nicaragua [15]
South America	Argentina, Bolivia, Brazil, Mato Grosso Mato Grosso do Sul Minas Gerais Para Paraiba, Parana Pernambuco, Piaui, Rio de Janeiro, Colombia, Ecuador, Galapagos Islands, French Guiana, Guyana, Paraguay, Peru [16]
Oceania	American Samoa, Australia, Australian Northern Territory, Queensland, Cook Islands, Fiji, French Polynesia, Guam, Kiribati, Marshall Islands, Papua New Guinea, Samoa, Solomon Islands, Tokelau, Tonga, Wallis and Futuna Islands [17]

29.1 Introduction

Ketapang is a beach plant, and this plant is often used as shade in roadside gardens [1], house yards, and grows wild in coastal areas [2]. Ketapang is easy to grow and survives in tropical environments [3]. Ketapang is quite familiar and easy to find in Indonesia [4]. Ketapang with the scientific name *Terminalia catappa* Linn [5] is known as a shade tree with nutritious fruits [6] and has many benefits including medicinal benefits [7], water treatments for fish farming [8], natural dyes for textiles [9] and reinforcing filler in composites [10]. The distribution of Ketapang plants [11] in various countries is shown in Table 29.1.

Ketapang has a characteristic leaf widened, crowned and always sheds its leaves every day [18]. Ketapang trees have various local names [1], in various countries, as can be seen in Table 29.2.

29.2 Description of the Ketapang Tree

Ketapang has easily recognizable characters and characteristics [28]. Morphology wise, the Ketapang tree is a woody plant with many branches and shady leaves [29]. This is the main reason why this tree is often used as a shade tree as can be seen in Fig. 29.1. The parts of it are as follows.

Table 29.2 Local name of Ketapang plants

Australia	*Kotamba* [2]
Brazil	*amendoeira-da-India* [19]
Brunei Darussalam	*Telisai* [19]
Cambodia	*Kapang, barang* [20, 19]
Colombia	*Kotamba* [2]
Cuba	*almendro de la India* [2]
Fiji	*tavali; tivi* [11]
Germany	*Katappenbaum* [11]
Haiti	*amadier tropical* [11]
India	*Adamaram, badambo, white bombwe* [2]
Indonesia	*Ketapang* [21]
Kiribati	*Te kunikun, te ntarine* [20]
Laos	*Hou kouang, hu kwang* [2]
Malaysia	*Jelawai Ketapang, Ketapang, Lingkak* [22]
Myanmar	*Badan* [20]
Netherlands	*Amandel boom, Wilde amandel* [1, 23]
Papua New Guinea	*Jara Almond, reddish-brown Terminalia* [24]
Peru	*Castana* [11]
Philippines	*Dalinsi, kalumpit, talisai* [25]
Sri Lanka	*Kottamba* [26]
Thailand	*Dat mue, hukwang, taa-pang* [27]
Vietnam	*Bang bien* [11]

29.2.1 Roots

This shade plant is from the group of dicotyl plants [30] or in two pieces so that the root system is a taproot [31]. The taproot type of Ketapang is a branched [32] taproot because there is one cone-shaped root [33] that grows straight down and has many root branches that grow sideways as support [34].

29.2.2 Tree Trunk

Ketapang has woody stems and is able to grow to a height of about 35 m [35]. The texture of the trunk is rough because there are grooves or sulcatus on the surface of the bark. The plot will look very clear if you look at it longitudinally [36]. The trunk of the Ketapang tree is round [33] or teres and grows upright [3]. Even so, the stems

Fig. 29.1 Ketapang tree

of this plant are usually difficult to identify, because their size is not much different from the branches [37].

Therefore, Ketapang branching is also called sympodial branching [32]. It is not uncommon for the branch size to be larger [23] than the main stem. This is because the process of growth and development of stems usually stops faster [14], while the branches continue to grow. The Ketapang tree branching system grows horizontally [31] and forms right angles to the trunk [36].

29.2.3 Leaves

Ketapang leaves are included in the incomplete leaf group [38] because there are only two constituent elements, namely the petiole and the leaf blade [39]. While a complete leaf must have three parts, namely the leaf midrib (vagina) [16], petiole (petiolus), and also the leaf blade (lamina) [40] (Fig. 29.2).

The leaf stalk of the Ketapang tree is the same as the leaves in general [26], is cylindrical with the base side widened, and tends to be flat [41]. Meanwhile, the leaves are shaped like an inverted egg or like a heart [42]. The texture of the upper surface is slightly slippery, while the lower surface has fine hair [23].

Fig. 29.2 **a** Seeds, **b** Fruits and **c** Leaves

The Ketapang leaf repetition system is pinnate because it has one large leaf bone as its parent [42]. Leaf bones are at the base of the leaf. In addition, there are branches that appear from the center of the leaf toward the outer edge of the leaf [43]. If one touches the leaves of the Ketapang tree, they also feel soft and thin. The leaves are oval with rounded edges and tapering at the stalk, the color [44] of the leaves is green and if they are old they will turn red when they fall [25].

29.2.4 Flowers

Ketapang tree flowers are small and resemble a bell [45]. They are about 4–8 mm in size with white, cream, to yellow colors [46]. Ketapang flowers do not have a crown but there are 5 petals for each flower. Ketapang flower growing points gather at the ends of twigs 8–25 cm long [47].

29.2.5 Fruit

Ketapang trees also produce fruits that looks like almonds [44]. Therefore, this tree is also referred to as tropical almond [20]. Ketapang fruits are measuring between 4 and 5.5 cm and green when young [3], then turn brownish-red when cooked [28]. This fruit has a seed inside which is protected by a slippery rind [48].

29.2.6 Seeds

Inside the fruit of the Ketapang tree, there are seeds wrapped in fiber [33]. These seeds are divided into two parts, namely the seed coat and the umbilical cord [22]. The seed coat consists of two layers [16], namely the testa or the outermost layer [49] of the skin and the tegmen of the innermost layer of the skin. The outer layer serves as a protection, because it has a hard texture like wood [39].

Table 29.3 Classification of Ketapang plants [20]

Kingdom	Plantae
Division	Magnoliophyta
Classis	Magnoliopsida
Ordo	Myrtales
Familia	Combretaceae
Genus	Terminalia
Species	*Terminalia Catappa L*

29.3 Classification of Ketapang Plants

Ketapang plants belong to the Combretaceae family, according to Backer [50]. It is explained that the Ketapang plant has the following classification, as can be seen in Table 29.3.

29.4 Potential of Ketapang

29.4.1 Medicine

Development of quality standards for Ketapang leaves standardization is an important measure for quality, purity, and sample identification [51]. Some of the simplest and cheapest methods are quantitative analytic microscopy with macromorphology and cytomorphology. Ketapang leaves have good medicinal value in traditional medicine systems [47]. But the pharmacognostic and phytochemical standardization of leaves was not validated until recently [16].

Traditional medicine treat hepatitis, dermatitis and has anti-cancer, antioxidant, and anticlastogenic properties [39]. The juice of the young Ketapang leaves is used as an ointment skin preparation for scabies, leprosy and is also used internally for stomach aches and headaches. The aqueous and ethanol extract of the leaves were reported to have hepatoprotective activity [5].

In addition, antioxidant, anti-diarrhea, and anti-inflammatory properties are reported [25]. Ethanol extract of Ketapang leaves bark as a treatment for inflammatory bowel disease, inflammation in oxidative stress, and immune dysfunction [48].

However, Ketapang has pharmacological properties that the neuro-modulation effects of Ketapang on chronic light stress have rarely been explored. Potential antidepressant effects such as the hydro alcohol (Ketapang) leaf extract using the chronic mild stress model for 7 weeks. It has been demonstrated that Ketapang supplementation can suppress stress-induced stress by adjusting the neurotransmitter monoamine, levels of acetylcholine esterase in improving oxidative stress, so that Ketapang can be used as a drug against depressive disorders [38].

While Ketapang in the Caribbean is registered as a vegetable, this medicinal plant that treats ulcers [44]. Ketapang is known as a treatment for digestive disorders and inflammatory conditions. The effect of prevention and healing of Ketapang leaf infusion on gastric lesions due to ischemia and reperfusion injuries and characterize its mechanism of action on the gastric of rats [52].

To provide a potential scientific basis for traditional use with testing, the activity of Ketapang against quorum sensing (QS) bacteria [53] in the form of Ketapang leaves extract as an antiseptic. Ketapang leaves have tannin-rich components that are able to inhibit certain phenotypic expressions in the test pathway used.

The extract of the leaves and bark of the Ketapang tree with methanol as the observed inhibition of violacein pigment as opposed to complete clearance of bacterial growth was detected through inhibition as an antiseptic [46].

The use of Ketapang leaf tannins from India for the biosorption of dicloxacillin from pharmaceutical wastewater, Ketapang leaf adsorbent is very good at removing dicloxacillin from pharmaceutical wastewater through biosorption [12].

29.4.2 Water Treatment

Ketapang leaves as an antibacterial agent have their multifaceted applications in water treatment, coating antibacterial, and as a colorimetric mercury sensor [54]. The effect of giving Ketapang leaf extract on the quality of culture water for the survival of ornamental fish, the phytochemical content of Ketapang leaf extract on the phytochemical test [42] saponins, triterpenoids, quinones, phenolics, tannins, and flavonoids were detected [55].

The use of Ketapang leaf biomass for water treatment of pharmaceutical waste [22] on the removal of dicloxacillin from pharmaceutical wastewater by was reported in [12]. Ketapang leaf biomass is considered to be suitable to remove dicloxacillin from pharmaceutical wastewater [56].

Temperature and dissolved oxygen (DO) were not influenced by the concentration of Ketapang leaf extract. The lowest power of hydrogen (pH 5.05) was found in the fish medium immersed in 625 ppm Ketapang leaf extract. Adding Ketapang leaf extract above 375 ppm resulted in higher survival [57]. Ketapang dried green leaves were used to prevent streptococcal infection in tilapia (Oreochromis niloticus), Ketapang leaf powder was added to feed (in a concentration of 5 and 10%) the Fish for 15 days before challenging the fish with the disease, and the feeding was kept up to 21 days after challenge. The mortality and cumulative of disease were significantly reduced. Growth rate and body weight increased for the specific fish [58].

29.4.3 Textile

Ketapang leaves serve as a natural polymer reservoir for applications in areas such as textiles, paints, and papermaking [22]. Natural dyes in textiles make use of Ketapang leaves, where these leaves contain tannins which produce a yellow color [9]. One of the alternatives to synthetic dyes are natural dyes [54] which have potential because they are environmentally friendly, and possess technical and economic advantages. Ketapang leaves are also widely available, especially in Indonesia. The cloth was tested for color fastness using the detergent washing method and the scrubbing method. The test results show the fabric's fastness [21].

29.4.4 Ketapang Seed Oil

In regards to the utilization of Ketapang seeds (KS) as food, no study has focused on their use as a food supplement although these edible seeds [59] could serve as, a new source of dietary fat [60]. KS have a high oil content (600 g/kg) and have an optimal balance of fatty acids shown in the dietary guidelines in order to make it a better fat compared to soybean oil [61], vegetable oil, which are popular in the world, making it suitable as a new vegetable oil. The chemical composition and content of KS make it suitable for the possible use for humans and/or animals [62].

KS are a good source of minerals. Based on an analysis [63], KS contain 4.13% water, 23.78% crude protein, 51.80% fat, 4.94% crude fiber, 16.02% carbohydrates, 548.78 kcal calorific value, and 4.27% ash, potassium (9280 $\pm$ 0.14 mg/100 g), calcium (827.20 $\pm$ 2.18 mg/100 g), sodium (27.89 $\pm$ 0.42 mg/100 g), and magnesium (798.6 $\pm$ 0.32 mg/100 g). Other than that KS contain high levels of unsaturated fatty acids, especially oleic (31.48%) and linoleic (28.93%) [27]. The dominant saturated acids are palmitic (up to 35.96%) and stearic (up to 4.13%) [41].

Ketapang oil can be classified in the oleic-linoleic acid group [64]. The oil extract exhibits good physicochemical properties and can be useful as vegetable oil and for industrial applications [65] (Table 29.4).

In regards to the characterization of the Ketapang fruit as a potential alternative source of biodiesel [69], the oil obtained from seeds yields of 49% (mass%) with a fatty acid composition similar to conventional oil [70]. KS oil is transesterified using methanol as a catalyst to form biodiesel as a fuel for diesel engines [71]. KS oil methyl ester fuel properties can be used as a separate fuel or as a mixture [68].

29.4.5 Reinforcing Filler in Composites

Ketapang leaf extract as a reducing agent is used as a mixture of copper nanoparticle composite films (CuNPs) in the cellulose matrix. Cellulose composites were used to

Table 29.4 Industrial applications of Ketapang plants

Name of material	Function	Industrial applications	Description
Leaves	Natural dye	Textile	Natural dyes from extracting Ketapang leaves by using the maceration method. That resulted in colors were light yellow, brownish yellow, and gray near black [21]
	Biosorption	Biosorbent adsorb palladium, and platinum	Biosorption of platinum and palladium ions on Ketapang leaves from an aqueous solution [26]
	Material composites	Construction applications	Mechanical studies of the tensile and flexural natural fibers and hybrid composites were made using Ketapang leaf fiber and kenaf fiber with epoxy resin using the hand layup method [66]
	Material composites	Shoe inshole antibacterial	The blending of thermoplastic and elastomer reinforced with Ketapang leaves fiber can apply to the antibacterial properties shoe insole [67]
	Water treatment	Aquaculture industry	Dried leaves of Ketapang to apply in the aquaculture industry the alternative treatment of fungi, bacteria, and fish ectoparasite. In place of chemicals and antibiotics in treating fish pathogens [56]

(continued)

compare antibacterial properties and activity with silver nanoparticles and antibiotics [72]. The composite film has sufficient tensile strength, thus can replace polymer packing materials such as polyethylene. Furthermore, cellulose/CuNP composite films have good antibacterial activity against E. coli [45].

Ketapang leaf fibers as reinforcement for polypropylene composites [73], fibers obtained from fallen Ketapang leaves are mixed into a composite polypropylene as

Table 29.4 (continued)

Name of material	Function	Industrial applications	Description
Fruit and leaves	Anticancer, antioxidant, anti-HIV reverse transcriptase, anti-inflammatory, antidiabetic effects, and hepatoprotective activities	Medicine	The juice of the young Ketapang leaves is used in the preparation of ointments for leprosy, scabies and is also used internally for colic and headaches, Pharmacological studies Ketapang leaf and fruit extracts for anticancer, antioxidant, anti-HIV reverse transcriptase, anti-inflammatory, antidiabetic effects, and hepatoprotective activities [91]
Fruit and seeds	New dietary source	Dietary oil	The Ketapang seeds contained high levels of protein, oil, and essential minerals [51]
	Material fuel	Biodiesel	The results obtained in this study regarding the fuel properties of Ketapang oil methyl esters, we conclude that this material can be used as a separate fuel (B100) or as a mixture [68]
	Material oil	Especially as a transformer fluid	Ketapang seeds potential industrial applications, especially as a transformer fluid solvent extraction method was used for oil extraction, with n-hexane as the extraction solvent [61]

(continued)

matrix [74] and has the clear potential to replace neat polypropylene in common applications such as packaging [75].

The development of recyclable natural fiber composites from Ketapang fruit is discussed in Ref. [76]. This fiber is more environmentally friendly because it can decompose naturally, and the price is low [77]. Characteristics of Ketapang fruit fiber composite board with polyvinyl acetate matrix are discussed in Ref. [78].

Table 29.4 (continued)

Name of material	Function	Industrial applications	Description
	Animal feed	Animal feed production	Ketapang fruit contains the main nutrients carbohydrates, oils, and metal ions which are needed for the biochemical activity of vertebrate organs. a good source of alternative local raw materials or supplements for animal feed formulations [63]

29.5 Engineering Applications

Research that has been done by researchers on Ketapang tree products is intended for engineering applications [79]. The raw material utilization of Ketapang leaves in the production of carbon supercapacitor electrodes is discussed in Ref. [80], where the supercapacitor is an energy storage device [81]. The supercapacitor consists of two electrodes, a separator, an electrolyte, and two current collectors [82]. The carbon material from Ketapang leaves is used as electrode material in the form of activated carbon [83]. The optimum specific capacitance of carbon electrodes from Ketapang leaf is obtained at a carbonization temperature of 700 °C as high as 54 F g^{-1} [80]. At that temperature, the best supercapacitor electrochemical properties were found [84].

Other researchers examined the biocomposite use for insole application at which the filler is made with a percentage of Ketapang leaves fibers, namely 15%, with optimum tensile strength and modulus of strength [85] which increases with increasing fiber load, besides proving that Ketapang leaves fibers have antibacterial properties [86]. With high fiber content, the impact strength is reduced. An antibacterial test showed inhibitory activity against bacteria on the shoe sole [67].

Other researchers investigated Ketapang leaves as a composite material for construction applications [66]. The hybrid composite was made using the Ketapang fruit and kenaf fibers using the hand layup method. Mechanical studies, water absorption, and biodegradation were carried out [87].

Studies show that natural fibers and hybrid composites degrade more and lose more than 50% of their weight within 60 days compared to epoxy resin [88]. The study showed that tensile and bending increased by 31% and 53% compared to Ketapang/kenaf/Ketapang which showed an increase of 16%, testing was carried out by water absorption (>10%) [84], compared to other materials because there are more cracks on the surface of the Ketapang fruit fibers and a biodegradation study was carried out.

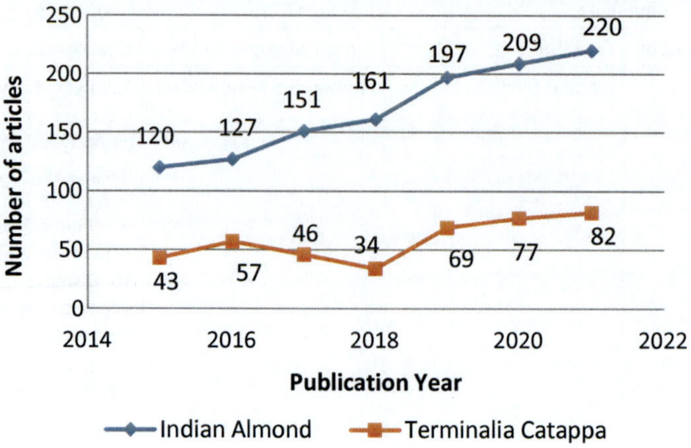

Fig. 29.3 Publication trend chart

29.6 Conclusion

The Ketapang tree (*T. Catappa*) is a species of the Combretaceae family. Based on literature studies, it is shown that various parts (leaves, fruit, and bark) have functions that can be used in industrial fields such as medicine, textile, water treatment, oil ingredients fuel, and as a composite reinforcing filler. From the review, in this regard, further studies need to be carried out to explore whether the Ketapang tree can still be developed as composite material industrial applications for future generations. The trend of research on Ketapang always rises from year to year as shown in Fig. 29.3.

Acknowledgements The authors would like to thank Universiti Kuala Lumpur kampus Cawangan Malaysia Italy Design Institute (UniKL MIDI) for the financial support provided through Grant Scheme

References

1. Thomson LAJ, Evans B (2006) Terminalia catappa (tropical almond). Perm Agric Resour 1–20
2. Terminalia catappa, Indian almond-wood, bastard almond, Andaman badam http://apps.worldagroforestry.org/treedb2/AFTPDFS/Terminalia_catappa. Accessed 24 May 2021
3. Somta C, Winotai A, Ooi PAC (2010) Fruit flies reared from Terminalia catappa in Thailand. J Asia Pac Entomol 13(1):27–30
4. Marjenah (2017) Morphological characteristic and physical environment of Terminalia catappa in east Kalimantan, Indonesia. Asian J For 1:33–39
5. Agrawal S (2009) Brief review on medicinal potential of Terminalia catappa. J Herb Med Toxicol 3:13–17
6. Bironke AA et al (2008) Short-term toxicological evaluation of Terminalia catappa, Pentaclethra macrophylla and Calophyllum inophyllum seed oils in rats. Food Chem 106:458–465

7. Janporn S, Ho C, Chavasit V, Pan M (2014) Physicochemical properties of Terminalia catappa seed oil as a novel dietary lipid source. J Food Drug Anal 23(2):201–209
8. Nozaleda BM (2019) Treatment Terminalia catappa (talisay) leaves as coagulant for preliminary surface water treatment. https://doi.org/10.12692/ijb/14.5.324-329
9. Masilungan-Manuel JT (2016) High molecular weight carrier-stabilized spray drying of aqueous talisay (Terminalia catappa) leaf extracts as natural dye for textiles. https://scholar.google.com.tw/citations?view_op=view_citation&hl=en&user=gEK3KCEAAAAJ&citation_for_view=gEK3KCEAAAAJ:YsMSGLbcyi4C
10. Muthulakshmi L et al (2016) Preparation and properties of cellulose nanocomposite films with in situ generated copper nanoparticles using Terminalia catappa leaf extract. Int J Biol Macromol 16:1–27
11. Rojas-Sandoval J (2017) Terminalia catappa (Singapore almond). Invasive species compendium. CABI, Wallingford, UK. https://doi.org/10.1079/ISC.53143.20203483198
12. Sunsandee N et al (2020) Biosorption of dicloxacillin from pharmaceutical waste water using tannin from Indian almond leaf: kinetic and equilibrium studies. Biotechnol Rep 27:e00488
13. Nithaniyal S, Parani M (2016) Evaluation of chloroplast and nuclear DNA barcodes for species identification in Terminalia L. Biochem Syst Ecol 68:223–229
14. Begoude D et al (2010) Botryosphaeriaceae associated with Terminalia catappa in Cameroon, South Africa and Madagascar. Mycol Prog 9:101–123
15. Bryan MN (2017) Terminalia catappa (Talisay) leaves for preliminary surface water treatment: an eco-friendly approach. Nat Prod Chem Res 5:249. https://doi.org/10.4172/2329-6836.1000249
16. Kadam PV et al (2011) Development of Quality Standards of Terminalia catappa Leaves. Pharmacognosy J 3:19–24
17. Andrea CCB, Laise DHC (2016) Mixobiota corticícola em terminalia catappa l. (combretaceae). Sitientibus Serie Ciências Biológicas 7:154–160
18. Mau J, Ko P, Chyau C (2003) Aroma characterization and antioxidant activity of supercritical carbon dioxide extracts from Terminalia catappa leaves. Food Res Int 36:97–104
19. Segaran GRDV et al (2019) Phytochemical profiles, in vitro antioxidant, anti-inflammatory and antibacterial activities of Terminalia catappa. Int J Pharm Sci Rev Res 55(2):51–59
20. Thomson B, Evans L (2019) Terminalia catappa, tropical almond, The IUCN red list of threatened species. https://doi.org/10.2305/IUCN.UK.2019-3.RLTS.T61989853A61989855.en
21. Chafidz FRA (2019) Extraction of natural dye from ketapang leaf (Terminalia catappa) for coloring textile materials. IOP Conf Sci Mater Ser. https://doi.org/10.1088/1757-899X/543/1/012074
22. Sharma R, Rana V (2017) Effect of carboxymethylation on rheological and drug release characteristics of Terminalia catappa gum. Carbohydr Polym 175:728–738
23. Chimezie E (2015) Epidermal structures and stomatal ontogeny in Terminalia catappa L. (Combretaceae). Int J Bot. 11:1–9. ISSN 1811-9700
24. Mohale D, Dewani AP, Researcher I, Chandewar AV (2009) Brief review on medicinal potential of Terminalia catappa. J Herb Med Toxicol 3:7–11
25. Ahmad MS (2014) Terminalia catappa, an anticlastogenic agent against MMS induced genotoxicity in the human lymphocyte culture and in bone marrow cells of Albino mice, Egypt. J Med Hum Genet 15(3):227–233
26. Ramakul P et al (2012) Biosorption of palladium (II) and platinum (IV) from aqueous solution using tannin from Indian almond (Terminalia catappa L). leaf biomass: kinetic and equilibrium studies. Chem Eng J 193–194:102–111
27. Nzikou JM et al (2009) Composition and nutritional properties of seeds and oil from Terminalia catappa L. Adv J Food Sci Technol 1:71–76
28. Lasekan O, Abbas K (2011) Investigation of the roasting conditions with minimal acrylamide generation in tropical almond (Terminalia catappa) nuts by response surface methodology. Food Chem 125(2):713–718
29. Berg W (2021) Indian almond leaves. http://www.indianalmondleaves.com. Accessed 20 June 2021

30. Ekop AS, Eddy NO (2005) Comparative studies of the level of toxicant in the seeds of Terminalia catappa (Indian almond) and Coula edulis (African walnut). J Chem Class 2:74–76
31. Begoude BAD (2010) Characterization of Botryosphaeriaceae and Cryphonectriaceae associated with Terminalia SPP. in Africa. University of Pretoria
32. Santos OV et al (2020) CO_2 supercritical fluid extraction of pulp and nut oils from Terminalia catappa fruits: thermogravimetric behavior, spectroscopic and fatty acid profiles. Food Res Int 139:109814. ISSN 0963-9969
33. Kaneria MJ et al (2018) Nontargeted metabolomics approach to determine metabolites profile and antioxidant study of tropical almond (Terminalia catappa L.) fruit peels using GC-QTOF-MS and LC-QTOF-MS. J Pharm Biomed Anal 160:415–427
34. Azad S, Alam J, Mollick AS, Khan NI (2017) Rooting of cuttings of the wild Indian almond tree (Sterculia foetida) enhanced by the application of indole-3-butyric acid (IBA) under leafy and non-leafy conditions. Rhizosphere 5:8–15
35. Baratelli TG et al (2012) Phytochemical and allelopathic studies of Terminalia catappa L. (Combretaceae). Biochem Syst Ecol 41:119–125
36. Pierantoni M, Brumfeld V, Addadi L, Weiner S (2019) A 3D study of the relationship between leaf vein structure and mechanical function. Acta Biomater 88:111–119
37. Divya N et al (2018) Phytotherapeutic efficacy of the medicinal plant Terminalia catappa L. Saudi J Biol Sci 26:985–988
38. Chandrasekhar Y et al (2020) Antidepressant like effects of hydrolysable tannins of Terminalia catappa leaf extract via modulation of hippocampal plasticity and regulation of monoamine neurotransmitters subjected to chronic mild stress (CMS). Biomed Pharmacother 86(2017):414–425
39. Toghueo RMK et al (2017) Enzymatic activity of endophytic fungi from the medicinal plants Terminalia catappa, Terminalia mantaly and Cananga odorata, South African. J Bot 109:146–153
40. Yang S et al (2010) Antimetastatic effects of Terminalia catappa L. on oral cancer via a down-regulation of metastasis-associated proteases. Food Chem Toxicol 48(4):1052–1058
41. Durowaye S et al (2018) Synthesis and characterization of hybrid polypropylene matrix composites reinforced with carbonized Terminalia catappa shell particles and Turritela communis shell particles. J Taibah Univ Sci 12(1):79–86
42. Katiki LM et al (2017) Veterinary parasitology Terminalia catappa: Chemical composition, in vitro and in vivo effects on Haemonchus contortus. J Vetpar 246:118–123. https://doi.org/10.1016/j.vetpar.2017.09.006
43. Marjenah M (2017) Morphological characteristic and physical environment of Terminalia catappa in East Kalimantan, Indonesia. Asianjfor 1:33–39. https://doi.org/10.13057/asianjfor/r010105
44. Silva LP et al (2015) Terminalia catappa L.: a medicinal plant from the Caribbean pharmacopeia with anti-Helicobacter pylori and antiulcer action in experimental rodent models. J Ethnopharmacol 159:285–295
45. Muthulakshmi L, Varada A, Saravanan G (2019) Preparation of cellulose/copper nanoparticles bionanocomposite films using a bioflocculant polymer as reducing agent for antibacterial and anticorrosion applications. Compos Part B 175:107177
46. Taganna JC et al (2011) Tannin-rich fraction from Terminalia catappa inhibits quorum sensing (QS) in Chromobacterium violaceum and the QS-controlled biofilm maturation and Las A staphylolytic activity in Pseudomonas aeruginosa. J Ethnopharmacol 134(3):865–871
47. Ohara R et al (2020) Terminalia catappa L. infusion accelerates the healing process of gastric ischemia-reperfusion injury in rats. J Ethnopharmacol 256:112793
48. Algieri F et al (2016) Antiinflammatory and immunomodulatory activity of an ethanolic extract from the stem bark of Terminalia catappa L. (Combretaceae): in vitro and in vivo evidences. J Ethnopharmacol 192:309–319
49. Mahendra P, Kheta R, Harish A, Gupta K, Shekhawat NS (2012) Micropropagation of mature Terminalia catappa (Indian almond), a medicinally important forest tree. J For Res 17(2):202–207. https://doi.org/10.1007/s10310-011-0295-0

50. Backer CA, Brink RC, Bakhuizen VD (1963) Flora of Java (Spermatophytes only)/C.A. Backer, R.C. Bakhuizen Van Den Brink. N.V.P. Noordhoff, Netherlands
51. Weerawatanakorn M et al (2015) Terminalia catappa Linn seeds as a new food source. Songklanakarin J Sci Technol 37(5):507–514
52. Muhammad NO, Oloyede OB (2010) Effects of Aspergillus niger-fermented Terminalia catappa seed meal-based diet on selected enzymes of some tissues of broiler chicks. Food Chem Toxicol 48(5):1250–1254
53. Lampi M, Kanninen M, Luukkanen O (2018) Terminalia laxiflora and Terminalia brownii contain a broad spectrum of antimycobacterial compounds including ellagitannins, ellagic acid derivatives, triterpenes, fatty acids and fatty alcohols Enass. J Ethnopharmacol 227:82–96
54. Stable H, Nanoparticles S (2017) Highly stable silver nanoparticles synthesized using Terminalia catappa leaves as antibacterial agent and colorimetric mercury sensor. Mater Lett 207:66–71
55. Krishnaveni M et al (2015) A Preliminary Study on Terminalia catappa L. leaf, stem. Res J Pharm Technol 8(12)
56. Chitmanat C et al (2005) Antiparasitic, antibacterial, and antifungal activities derived from a Terminalia catappa solution against some Tilapia (Oreochromis Niloticus) pathogens. Acta Hortic 678:179–182. https://doi.org/10.17660/ActaHortic.2005.678.25
57. Nugroho RA et al (2016) The effects of Terminalia catappa L. leaves extract on the water quality properties, survival and blood profile of ornamental fish (Betta sp) cultured. Biosaintifika 8:241–248
58. Lusiastuti M et al (2017) Dry green leaves of Indian almond (Terminalia Catappa) to prevent streptococcal infection in juveniles of the Nile tilapia (Oreochromis niloticus). Bull Eur Ass Fish Pathol 37(3):119
59. Omeje EO et al (2008) Kinetics of autoxidation of an oil extract from Terminalia catappa. Indian J Pharm Sci 70(2):260–262
60. Goswami RK et al (2020) Growth and digestive enzyme activities of rohu labeo rohita fed diets containing macrophytes and almond oil-cake. Anim Feed Sci Technol 263:114456
61. Agu CM et al (2020) Modeling and optimization of Terminalia catappa L. kernel oil extraction using response surface methodology and artificial neural network. Artif Intell Agric 4:1–11
62. Oliveira JTA et al (2000) Composition and nutritional properties of seeds from Pachira aquatica Aubl, Sterculia striata St Hil et Naud and Terminalia catappa Linn. Food Chem 70:185–191
63. Nwosu F, Dosumu O (2015) The potential of Terminalia catappa (Almond) and Hyphaene thebaica (Dum palm) fruits as raw materials for livestock feed. Afr J Biotechnol 7(24):4576–4580
64. Ladele B et al (2016) Chemical composition and nutritional properties of Terminalia catappa L. oil and kernels from Benin. C R Chim 19:876–883
65. Iha OK et al (2014) Potential application of Terminalia catappa L. and Carapa guianensis oils for biofuel production: physical-chemical properties of neat vegetable oils, their methyl-esters and bio-oils (hydrocarbons). Ind Crop Prod 52:95–98
66. Nampoothiri EN et al (2020) Experimental investigation on mechanical and biodegradation properties of Indian almond–kenaf fiber-reinforced hybrid composites for construction applications. J Nat Fibers. https://doi.org/10.1080/15440478.2020.1739592
67. Ahmad AA, Baharum A (2018) Biocomposite of epoxidized natural rubber/poly (Laticacid)/Catappa leaves as shoe insole. J Polym Sci Technol 3(2):20–28
68. Santos ICF et al (2008) Studies of Terminalia catappa L. oil: characterization and biodiesel production. Bioresour Technol 99:6545–6549
69. Sani L, Abechi S (2019) Production and characterization of biodiesel-diesel blends from Terminalia Catappa seed oil. FJS 2:214–220
70. Nzikou JM et al (2009) Composition and nutritional properties of seeds and oil from Terminalia catappa L. Adv J Food Sci Technol 1:72–77
71. Silva JCMD et al (2020) Thermal and oxidative stabilities of binary blends of esters from soybean oil and non-edible oils (Aleurites moluccanus, Terminalia catappa, and Scheelea phalerata). Fuel 262:116644

72. Volova TG et al (2018) Antibacterial properties of fi lms of cellulose composites with silver nanoparticles and antibiotics. Polym Test 65:54–68
73. Ninomiya K et al (2017) Lignocellulose nanofibers prepared by ionic liquid pretreatment and subsequent mechanical nanofibrillation of bagasse powder: application to esterified bagasse/polypropylene composites. Carbohydr Polym 182:8–14
74. Al-oqla FM et al (2015) Natural fiber reinforced conductive polymer composites as functional materials: a review. Synth Met 206:42–54
75. Souza RCHS et al (2020) Steam-exploded fibers of almond tree leaves as reinforcement of novel recycled polypropylene composites. J Mater Res Technol 9(5):11791–11800
76. Jurd D, Pole M (2017) Miocene 'fin-winged' fruits and Pliocene drift fruits—the first record of Combretaceae (Terminalia) from New Zealand. Geobios 50(5–6):423–429
77. Dahlan I, Doyan A (2018) Mechanical test characteristics of Terminalia catappa fruit fiber composite material measurement of tensile strength. Proc ICRIEMS Publ by Fac Math Nat Sci Yogyakarta State Univ 40–47.
78. Dahlan I, Doyan A (2018) Tensile test of Terminalia catappa fruit fiber composite material. IOSR J Appl Phys 10(3):63–67
79. Bharath KN, Basavarajappa S (2016) Applications of biocomposite materials based on natural fibers from renewable resources: a review. Sci Eng Compos Mater 23(2):123–133
80. Series C (2018) The physical and electrochemical properties of activated carbon electrode made from Terminalia catappa leaf (TCL) for supercapacitor cell application. IOP Conf Ser: J Phys: Conf Ser 1120:012007. https://doi.org/10.1088/1742-6596/1120/1/012007
81. Das D, Bordoloi U, Muigai HH, Kalita P (2020) A novel form stable PCM based bio composite material for solar thermal energy storage applications. J Energy Storage 30:101403
82. Rajendhiran R et al (2020) Terminalia catappa and carissa carandas assisted synthesis of TiO_2 nanoparticles—a green synthesis approach. Mater Today Proc 45(Part 2):2232–2238
83. Mohan D et al (2011) Development of magnetic activated carbon from almond shells for trinitrophenol removal from water. Chem Eng J 172(2–3):1111–1125
84. Narayanan EN, Bensam RJ, Thanigaivelan R, Karuppasamy R (2020) Experimental investigation on mechanical and biodegradation properties of Indian almond-kenaf fiber-reinforced hybrid composites for construction applications. J Nat Fibers. https://doi.org/10.1080/15440478.2020.1739592
85. Asano A (2017) Polymer blends and composites. Library of Congress Cataloging-in-Publication Data, USA. ISBN 978-1-118-11889-4
86. Sivaranjana P et al (2018) Cellulose nanocomposite films with in situ generated silver nanoparticles using Cassia alata leaf extract as a reducing agent. Int J Biol Macromol 99:223–232
87. Jeyapragash R, Srinivasan V, Sathiyamurthy S (2019) Mechanical properties of natural fiber/particulate reinforced epoxy composites—a review of the literature. Mater Today Proc 22(Part 3):1223–1227
88. Shankarganesh PSP et al (2019) Investigation of tensile, flexural and impact properties of Neem-Indian almond hybrid fiber based epoxy composites. IOP Publ 6:1–10

Chapter 30
A Review on Contactless Power Transfer Using Matrix Converter Topology for Battery Charging Application

Mohd Zaifulrizal Zainol, Wardiah Mohd Dahalan, Mohd Rohaimi Mohd Dahalan, and Mohd Fakhizan Romlie

Abstract DC-AC high-frequency inverter is the most preferred topology for primary side converter for contactless power transfer system because of their simplicity over other topologies. However, two stages energy conversion make the system more expensive, bulky and losses during the power conversion. In addition, the electric vehicle (EV) battery charging system requires a constant current characteristic, which can be relied on the current source converter. Therefore, there are possibilities for alternative topologies and controllers to improve the efficiency and the performance of contactless power transfer. This study reviews the contactless power transfer matrix converter topology for EV charging application.

Keywords Contactless power transfer · Electric vehicle · Battery charging · Current source converter · Matrix converter

30.1 Introduction

The utilization of one-third world petrolatum energy and 40% of ozone depleting substance in 2016 is caused by transportation [1, 2]. In order to decrease the CO2 emission, the enactment is being acquainted in a few nations to decrease the emission from vehicles. The utilization of natural resources has made a shortage in the future; subsequently, the world may confront the effect of the reliance on these depleting natural energies especially petroleum. Electric power has turned out to be a suitable option for transportation fuel, so the electric vehicle (EV) has cleared the best way to

M. Z. Zainol (✉) · W. Mohd Dahalan · M. R. Mohd Dahalan
Universiti Kuala Lumpur Malaysian Institute of Marine Engineering, Technology, Jalan Pantai Remis, 32200 Lumut, Perak, Malaysia
e-mail: ymzaifulrizal@unikl.edu.my

W. Mohd Dahalan
e-mail: wardiah@unikl.edu.my

M. Z. Zainol · M. F. Romlie
Department of Electrical and Electronic Engineering, Universiti Teknologi PETRONAS, 32610 Seri Iskandar, Perak, Malaysia
e-mail: fakhizan.romlie@utp.edu.my

A. Ismail et al. (eds.), *Design in Maritime Engineering*, Advanced Structured Materials 167, https://doi.org/10.1007/978-3-030-89988-2_30

reduce the reliance on oil energy. The EVs are exclusively driven by electric power and create zero emissions of greenhouse gasses. This makes significant changes to decrease the air pollution, along these lines making the populated surroundings cleaner and satisfying for all livings around. With the advancement of contactless charging of EVs, people do not need to bother with any wires to charge them securely. With contactless charging, the people just need to stop the EV in the plugless charging pad. Through the improvement in innovation, the element charging of electric vehicles has been presented, which empowers the charging of moving vehicles on the roadway. The charging framework can work dependably even in the most unfavorable climate.

30.2 Contactless Power Transfer

In the most recent years, the research on contactless power transfer systems (CPTs) encountered a major development. CPTs can be very much related to wirelessly power small wearable or portable electronics, to contactless charging systems for electric vehicles, to factory automation systems, to medical and healthcare. Many papers, benchmarks, books, and reports about ICPT configuration have been distributed, with particular applications, diverse topologies, and compensation networks resonance frequency and soft switching, system efficiency and power. It has been known that signals and power can be transferred electromagnetically. The term contactless power transfer, in general, can be used to describe the power transfer between two objects that are physically unconnected. The contactless power transfer systems have many advantages compared to conventional power transfer due to the elimination of conductive electrical contacts. Inductive power transfer systems are used to transfer power efficiently over a reasonably large air gap 15 cm using a loose magnetics coupling. Table 30.1 shows the research on ICPT for electric vehicle charging system since 1994, which is started by the University of Auckland, New Zealand. The work on ICPT is intensively discovered by the researchers around the world as summarized in Table 30.1 in terms of their power rating, frequency, air gap, and efficiency.

The fundamental principles of inductively coupled power transfer (ICPT) are almost the same with transformers and induction motors. Generally, an ICPT system comprises two independent electrical systems with mutual coupling. Since the two windings of the transformer are physically separated by an air gap, this air gap brings to a low coupling efficient and high leakage inductance which result in a poor power

Table 30.1 Research on IPT for electric vehicle application

Year	Institute	Power rating. (kW)	Frequency (kHz)	Air gap (mm)	Efficiency
1994 [3, 4]	Univ. Auckland, New Zealand	0.6	10.2	–	–
1997, [5]	Grenoble Institute of Technology, France	3	22.95	6–8	89
2000 [6]	Univ. Auckland, New Zealand	17	12.9	~ 50	–
2004[7, 8]	Institut f. Automation u. Kommunikation, Germany	1	100	30 ~ 300	90
2005[9, 10]	Univ. Auckland, New Zealand	30	20	45	–
2008[11]	Univ. Auckland, New Zealand	0.39	10	–	–
2009[12–14]	Univ. Auckland, New Zealand	2	5 ~ 50	50 ~ 80	85
2010 [15]	Saitama University, JAPAN	1.5	20	70	95
2010 [16]	Univ. Auckland, New Zealand	0.5	38.4	–	89
2010[17–21]	KAIST, Korea	3 ~ 25	20	10 ~ 250	71 ~ 87
2011 [22]	University of Tokyo, Japan	1	–	300	88
2011 [9, 23–26]	Oak Ridge National Laboratory, USA	2 ~ 4; 6 ~ 7	22, 48 ~ 81	125 ~ 254	82 ~ 92,95
2011 [27]	Univ. Auckland, New Zealand	7	20	125	-
2012 [28]	Utah State University, UT	5	20	246	90
2012 [29]	Kyoto Universiti, Japan	30	20	140	92
2013 [30]	KAIST, Korea	180	20	100	85
2014 [31]	Korea Railroad Research Institute	300–600	60	30	83
2016 [32, 33]	Florida International University, USA	0.267	12.25	200	88.2

conversion [34, 35]. Technology of CPT is enabled by power electronics which the power conversion from a low-frequency system such as a DC or 50/60 Hz mains power supply to a much higher frequency system with a frequency of about 10–100 kHz. The researcher uses IGBTs as their semiconductor switching device, and the result achieved is that the system is able to operate at higher switching frequencies and increased efficiency. Operating frequency is limited below 100 kHz because of switching losses [36–38].

Another problem occurred when the magnetizing current is quite large, and this results in high circulating energy loss. To solve this issue, various resonant capacitor compensation networks have been proposed. In order to minimize converter switching losses and to compensate for the high leakage inductance of the transformer, the compensation capacitor is used in ICPT applications. As shown in Fig. 30.1, there are four typical compensation topologies which consist of series-series (SS) compensation, series-parallel (SP) compensation, parallel-series (PS) compensation, and parallel-parallel (PP) compensation.

DC-AC inverters are broadly utilized for a range of applications of IPT systems. Various advancements of inverter have been developed and are based on typical current-fed and voltage-fed topologies, respectively. Every topology and control scheme have its own advantages and disadvantages. Soft switching is well-known for achieving high-power efficiencies and low EMI, but these types of power converter have transient process which is very complex and difficult to analyze. Theoretically, an AC-AC converter can likewise be utilized to convert an AC power source such as the mains supply to another AC power source, where the output frequency can be set

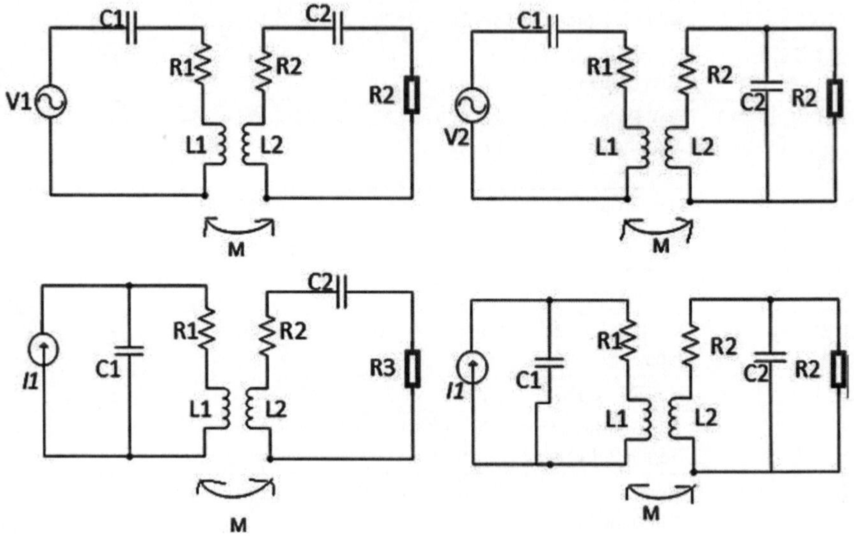

Fig. 30.1 Compensation Topologies

subjectively. High-frequency conversion cannot be achieved by conventional phase-controlled AC-AC choppers, PWM controlled AC choppers, and cycloconverters.

In order to encounter this drawback, matrix converter using multiple bidirectional switches and local resonance offers a possibility to generate a high-frequency power from a low-frequency power by its matrix conversion structure. A matrix converter will provide direct AC-AC conversion without a DC link. However, synchronizing the operation of the switches between the instantaneous input and output can be extremely troublesome, and the output waveform quality is typically poor because the complicated switching combinations required to shape the waveforms. The difficulties involved in the switching synchronization and the magnitude regulation of the high-frequency output make such matrix converters difficult to use in viable IPT applications.

Even though DC-AC inverters and AC-AC power converters are available for high-frequency applications, they have a lot of drawbacks for this type such as difficulties in implementation, poor output waveforms, or high switching losses. It is still necessary to conduct the research and development of new types of high-frequency converter for ICPT applications.

30.3 Topologies of Converter for Contactless Power Transfer

In a contactless power transfer, the purpose of the power converter is to create a high-frequency current on the primary side. Resonant topology is applied to increase the switching frequency and to enhance the efficiency [3–10]. A wide range of control strategies was proposed to control the transferred power. The control scheme could be classified as primary side control [11], secondary side control [12], and double side control [13] depending on where the control scheme is connected.

Power converters are regularly used to produce a high-frequency current on the primary side of a contactless power transfer system. Basically, there are two ways to obtain high-frequency current using linear amplifiers or switch mode power converter. Even though the advancement of various types of linear electronics amplifiers had been improved, the characteristic of high-power loss in linear region constraint a linear amplifier to only can be used by low power application. Comparing to linear amplifier, power converter by fully on/off is preferred as it is efficient [14]. Therefore, as the power efficiency is one of the major concerns, switch mode control and switch mode power converters are widely used to generate high-frequency track currents for medium to large contactless power transfers applications [15].

Various switch mode power converters are as of now being used to create high-frequency system for IPT frameworks. The high-frequency power converter in the primary part can be either a DC or directly from AC mains [16]. In this manner, the switch mode control converter for an IPT framework can be ordinarily grouped further into two classifications as indicated by the kind of their primary part sources. They are DC-AC inverters and direct AC-AC converters.

At present, most of the power converters used two stage AC-DC-AC converters compared to direct AC-AC converters. To get a consistent DC control source, a front-end AC-DC rectifier is required. Many methods are applied to the AC-DC conversion stage, such as synchronous rectifiers, filters, power factor correction circuits to meet the requirements of international standards and to improve the efficiency. Once a steady DC power is acquired, a second-stage DC-AC inverter is required for the high-frequency track current generation. Many research projects have been done in the region of DC-AC inverters for IPT frameworks. A few topologies and switching control techniques have been produced.

AC-AC converter which is also known as matrix converter is converting frequency AC supply to another frequency supply without involving an intermediate DC link capacitor. Matrix converter has the potential to be applied in ICPT systems. By using a matrix converter, the low-frequency AC grid can be converted to high frequency without the need of multistage converter and DC link capacitor.

30.4 Matrix Converter Topology for Electric Vehicle Battery Charging

A plug-in electric vehicle (PEV) has two different charging approaches which are conductive charging and inductively coupled power transfer (ICPT) charging. Conductive charging application has a wire connection between the power supply to power electronic interface for charging. Otherwise, ICPT charging or contact-less charging does not use a wired connection between the power supply to power electronic interface for charging.

Matrix Converter for Conductive Charging Application.

A comparison of matrix converters for conductive charging application with respects to the parameters including topology, frequency, soft switching, capacity, and compensation configuration is shown in Table 30.2.

The topology of matrix converter has been first investigated in 2010 [17] to replace other topologies for conductive charging application. The aim of this replacement is due to other topologies lack of bidirectional power flow characteristics. A novel MC topology for plug-in EV with bidirectional power flow capability and controllable power factor has been proposed in this paper. Referring to [18], the author is using the PWM switching signal which is generated by comparing the duty ratios of each switch against a triangle wave as carrier signal. As shown in Fig. 30.2, only six switches are used at source side. At one period, only two switches can be closed in the top row and bottom row, respectively. Therefore, there are nine states of switching as the matrix converter (MC) is connected to three-phase AC and the output is connected

Table 30.2 Comparison of matrix converter for conductive charging application reported in literatures

No.	Switching/control technique	Topology	f_s(kHZ)	Soft switching	Capacity (KW)	Compensation configuration	Power quality	Strong points	LIMITATIONS	YEAR	REF
1	PWM (comparing duty ratios against a carrier)	Single-phase/3X 1 matrix converter	10	No	1–5	Transformer is considered ideal (no magnetizing inductance) Ratio 1:1	Not reported. Just mentioned to eliminate the high-order harmonics in the future research	• A novel MC topology for plug-in EV with controllable power factor	• Just made analysis on power flow from grid to vehicle(charging mode)	2010	[39]
2	SPWM (comparing unipolar SPWM against high-frequency square wave)	3X 1 matrix converter	6	No	45	Ideal transformer Ratio 8:15	THDi < 5% PF ~ unity	• Proposed for bidirectional flow using MC-based topology • Provide current regulation capability for battery charging/discharging	• Does not allow PF control maintaining at unity	2012	[40]
3	Phase shift modulation (voltage control)	Single-phase matrix converter	10	No	1.5	Leakage inductance lumped together and referred o secondary circuit	Simulation THD = 14.6% Efficiency = 78% Hardware THD = 4.4% Efficiency = 80.4%	• Modulation phase shift and DC bus current • Output voltage control • Open-loop power factor correction	• Low efficiency	2013	[41]

(continued)

Table 30.2 (continued)

No.	Switching/control technique	Topology	f_s(kHZ)	Soft switching	Capacity (KW)	Compensation configuration	Power quality	Strong points	LIMITATIONS	YEAR	REF
4	SPWM (comparing unipolar SPWM against high-frequency square wave)	3X 1 matrix converter	15	ZVS	30(simulation) 0.5(hardware)	Primary: LC	$THD_i < 3\%$ PF ~ unity	• Achieve soft switching (ZVS) • Had resonant compensation tank to achieve soft switching • Fast charging with in both G2V and V2G		2015	[42]
5	Proposed new switching scheme named as switched rectifier inverter	3X 1 matrix converter	40	Yes, ZCS in selective switches	8	Leakage inductance = 5μH Ratio 1:3	THD 4.09% P.F. 0.997	• Less switching transition • Reduced current and voltage slew rates		2015	[43]
6	PWM (dual H-bridge modulation-DHBM)	Single-phase matrix converter	5	Yes. ZCS	1.5	Primary: LC	Not reported	• ZCS achieved in switches of MC		2013	[44]
7	Space vector modulation (SVM)	3X 1 matrix converter	40		0.5	Primary: LC	THD 3.06%	• Soft commutation of high-frequency AC current • Minimization of duty cycle loss		2017	[45]
8	Space vector modulation (SVM) + phase shift modulation (PSM)	3X 1 matrix converter	20		10	Ideal transformer	THD (G2V) 2.58%	• Simultaneously control the PF in grid and current in battery		2018	[46]

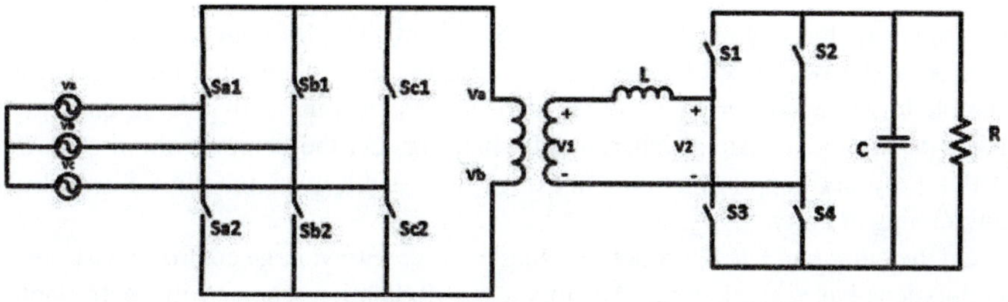

Fig. 30.2 MC with three-phase system

to a two-terminal high-frequency transformer. However, the work is only analyzed on power flow from grid to vehicle (charging mode).

In 2012 [19], the author proposed a new three-phase converter topology based on a 3 × 1 MC for plug-in EV battery charging as shown in Fig. 30.3. The proposed topology, which named as MC-PWM rectifier, is developed to operate like a dual active bridge (DAB). Figure 30.4 shows the switching scheme for this proposed

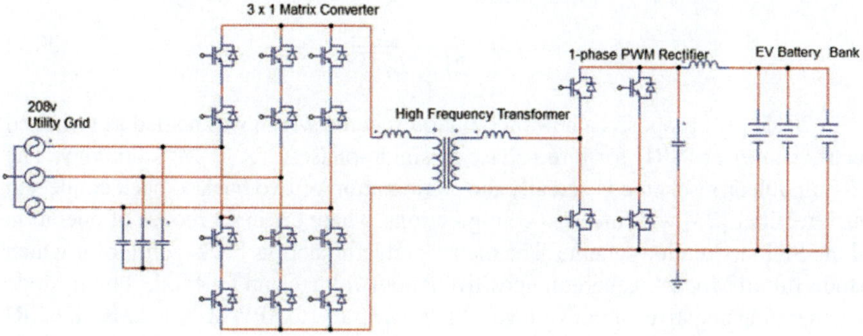

Fig. 30.3 3 × 1 matrix converter

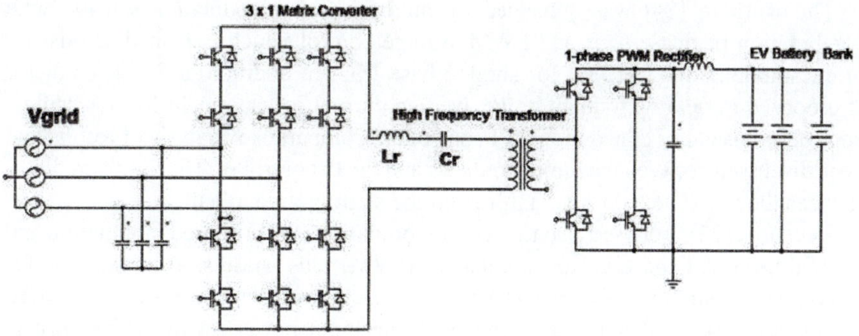

Fig. 30.4 Three-phase MC with series resonant

topology, which is a product of a unipolar SPWM with high-frequency square wave. This proposed MC-based topology not only provides bidirectional flow, which is suitable for vehicle-to-grid (V2G), but also, at the same time, provides current regulation capability for battery charging/discharging. On the contrary, the drawback of this proposed topology is the control of the system does not allow PF control maintaining at unity.

On the other hand, [20] proposed a single phase with voltage control (phase shift modulation)-based single-phase MC topology. Modulation scheme with controllable phase shift δ and the average DC bus current was derived. PI controller is designed based on the transfer function that has been developed. The work has proven of power factor correction (PFC) for the application of EV battery charging. Nevertheless, obviously, the drawbacks in power quality as the proposed topology with phase shift modulation is observed. It has less efficiency and high THD.

Improving the previous work in [21] had a resonant compensation tank to achieve soft switching with fast charging system in both G2V and V2G. In order to achieve zero voltage switching (ZVS), an L_rC_r resonant tank is connected across the output of MC as shown in Fig. 30.4. For modulation of the MC, the square wave frequency, fsq, is set above the resonant frequency of the LrCr tank which is given by

$$f_r = \frac{1}{2\Pi\sqrt{L_r C_r}} \tag{30.1}$$

In 2015, [22] proposed a new modulation scheme which was named as switched rectifier inverter (SRI) for three-phase to single-phase (3 × 1) MC topology. The SRI modulation scheme is actually the combination of two back to back connected buck rectifier [23] with inverted configurations. There are eight modes of operation of the SRI modulation scheme. The matrix converter acts as buck rectifier for which Mode 1 until Mode 4, generating positive output voltage and for Mode 4 until Mode 8, generating negative output voltage. On the contrary, to SPWM-based MC, the SRI modulation scheme used is symmetrical and fewer number of switching quadrant and transition. Hence, it increased the performance of power quality as it reduced the current and voltage slew rates.

The work in [24] was enhanced by modifying the modulation scheme with combination of phase shift and PWM voltage control which is named as dual H-bridge modulation (DHBM) for single-phase MC. In addition, a Smith predictor is proposed to deal with delay in the average current measurement for power flow controller instead of conventional PI controller. Comparative analysis of results had been discussed between the Smith predictor and the PI controller. The Smith predictor reduces the rise time to around a third and the settling time to half.

Recently, [25] proposed a space vector modulation (SVM)-based modulation and [9] had proposed space vector modulation (SVM) plus phase shift modulation for three-phase to single-phase (3 × 1) MC. As a result, both these works had improved power conversion efficiency and enhanced the input power quality. As shown in Fig. 30.5, two additional switches had been added to make extra one leg acted as

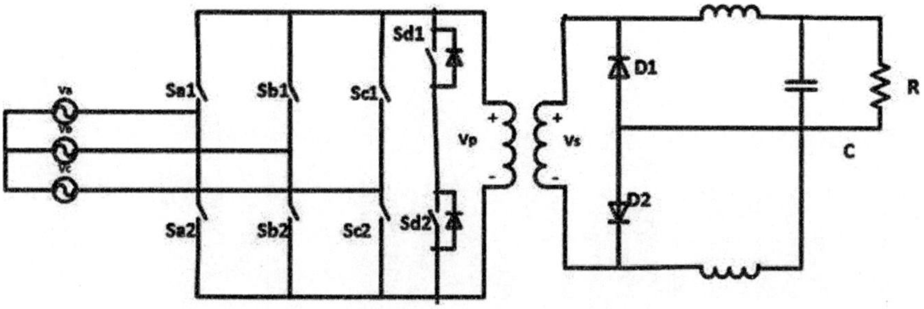

Fig. 30.5 Space vector modulation (SVM)-based modulation with additional switches

shorting when these switches operated at the same time. This topology is successfully converting the three-phase AC voltages to bipolar single-phase high-frequency AC voltage with good input power quality. In addition, this topology reduced the control complexity as only one-gate drive signal for one matrix switch. However, the compensation tank LrCr is built by adding a series capacitor in the primary side. The significant advantage of the proposed modulation scheme [26] is the ability to control the power factor in the grid and current to the battery simultaneously.

Matrix Converter for Inductive Charging Application

A comparison of matrix converters for inductive charging application with respect to the parameters including topology, frequency, soft switching, capacity, and compensation configuration is shown in Table 30.3. The matrix converter topologies had been firstly applied for inductive power transfer in [27]. In this paper, conditions for bidirectional switch commutation are derived from the commutation of the full-bridge series-resonant converter. The comparison had been made between the full-bridge series-resonant converter with the three-phase to single-phase matrix converter. The author highlighted the minimization of commutation steps which is independent from the load current sign as the main advantages of the derived strategy. This work had embarked the possibility of matrix converter to replace two-stage power conversion for AC-DC and then DC-AC. Nevertheless, the hardware implementation had developed just for series-resonant load which is not specifically for ICPT.

A single-stage three-phase AC-AC matrix converter as shown in Fig. 30.7 is proposed for ICPT systems. The three-phase matrix converter consists of six reverse-blocking switches and one regular switch, which is in parallel with resonant tank. A series combination of an IGBT or a MOSFET with a diode had been done to create a reverse-blocking switch. The author employed a control strategy based on energy-injection and free oscillation [28] of the resonant circuit. The switching frequency is varied until 50%, and duty cycle is fixed. Three different control modes consist of resonant current regulation control mode, voltage regulation control mode, and power regulation control mode which are all based on zero current switching (ZCS) operation. The proposed converter has minimized the number of switches with only seven switches with the use of reverse-blocking switches. Moreover, the proposed

Table 30.3 Comparison of matrix converter for IPT (inductive charging) application reported in literatures

No.	Switching/control technique	Topology	f_s(kHZ)	Soft switching	Capacity	Compensation configuration	Power quality	Strong points	Limitations	Year	Ref.
1	Fixed f_s, square wave control	3X 1 matrix converter	5	$f_s > f_o$ -ZVS $f_s <$ f_o-ZCS	–	–	–	• The possibility of using switching techniques • Two-step commutation switching	• Had just validated by simulation • The hardware implementa-tion had developed just for series-resonant load	2007	[47]
2	Fixed $f_{s,}$ phase shift modulation	3X 1 matrix converter	50	Not achieved	300 W	Primary: LC series Secondary: LC series	–	• Detailed analysis on power loss of switching devices and diodes • SiC properties had been used for switching devices	• The feasibility of the proposed system ignores the analysis on power quality	2014	[48]
3	Variable $f_{s,}$ fixed duty cycle	3X 1 matrix converter	12.3	Yes, ZCS of all devices	267 W	Primary: LC series Secondary: LC parallel	THD 14.3% P.F 0.59–0.67	• The resonant current had been regulated based on the energy-injection and free oscillation	• Unidirectional IPT system	2016	[49]

(continued)

Table 30.3 (continued)

No.	Switching/control technique	Topology	f_s(kHZ)	Soft switching	Capacity	Compensation configuration	Power quality	Strong points	Limitations	Year	Ref.
4	Fixed $f_{s,}$ and variable duty cycle modulation	Single-phase matrix converter	20	Not Reported	1 KW	Primary: LCCL Secondary: LCCL	THD 9.4–10.6%	• Had grid interface for V2G and G2V applications	• Difficult to maintain quality current of grid as the source current cannot be controlled	2017	[50]

converter has soft switching operation and elimination of the bulk reactive element of electrolytic capacitor as DC link. However, the efficiency of the converter is low which is 88.2%, and it is just applicable for a unidirectional system when connecting to the grid (Fig. 30.6).

In [29], a single-phase matrix converter for IPT system for V2G and G2V applications had been proposed. A four-step current commutation algorithm had been

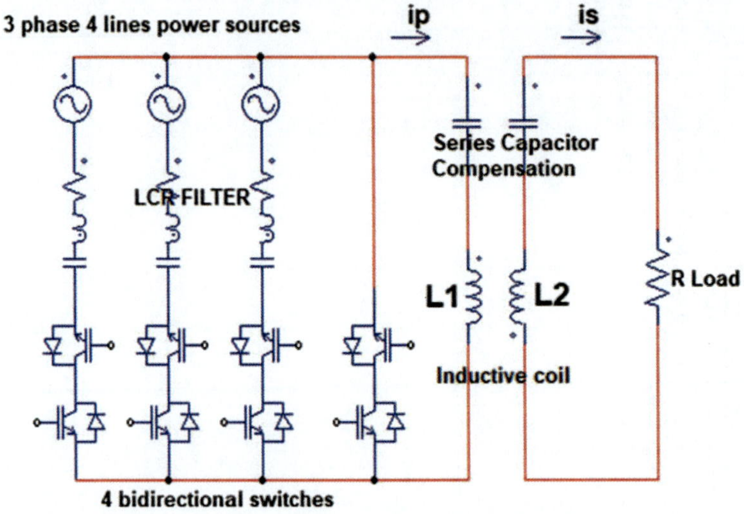

Fig. 30.6 Matrix converter-based IPT system with using SiC devices

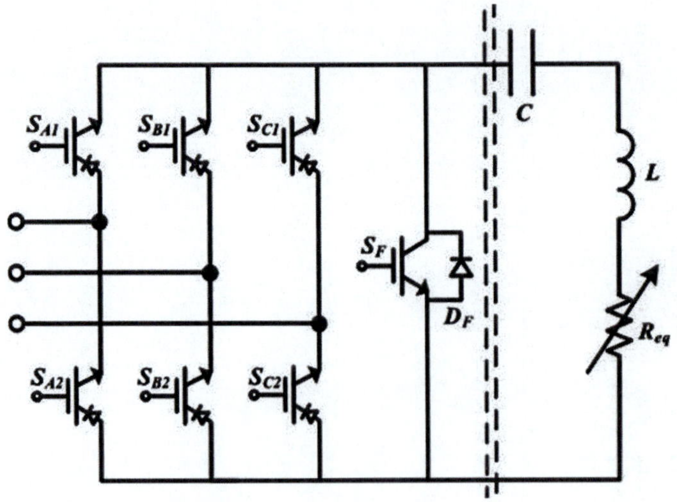

Fig. 30.7 Three-phase matrix converter with one additional switch

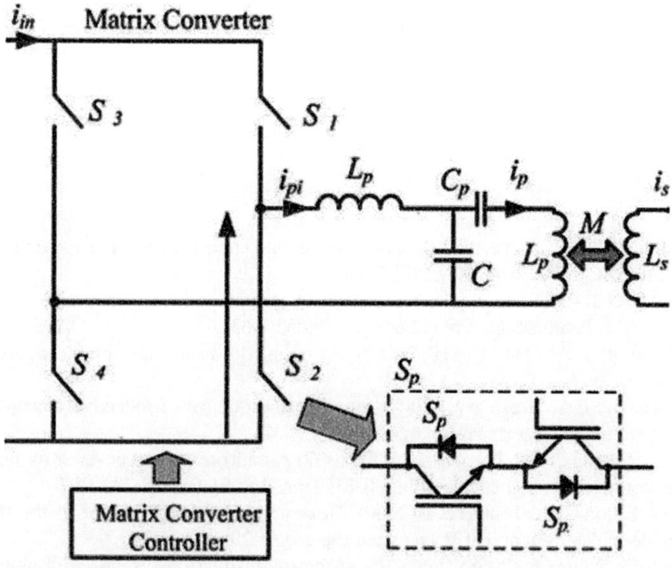

Fig. 30.8 Matrix converter grid-connected system

applied to maintain that the operated frequency is equal to resonant frequency. As shown in Fig. 30.8, a single-phase matrix converter with a LCCL resonant tank configuration consists of four bidirectional switches, which conduct current and block voltages in both directions. The variable duty cycle modulation is used as the control and commutation strategy for this converter. The converter is grid interfaced, but it is quite difficult to maintain the quality of grid current source, and the current cannot be controlled.

30.5 Conclusions

Based on the review, the AC-AC matrix converter could be the best alternative to obtain high-frequency power from AC main supply without the double or three stage of power conversion. In spite of current source converter inherits advantages over voltage source converter, most of the current researchers focused on voltage source (voltage-fed converter) for ICPT due to the fact that the current source topology (current-fed converter) has main drawbacks such as efficiency degradation and limited dynamic performance caused by the large DC link inductor. By improving the performance and efficiency, the current source converter characteristic can be employed for ICPT of electric vehicles. A matrix converter promises an improved efficiency and reduced cost compared to the conventional converter for ICPT system. However, the source current has high numbers of harmonics as source current cannot actively be controlled. Nevertheless, ICPT system with current source is the best choice to achieve zero voltage switching (ZVS) during turn ON and turn OFF of converter switches.

References

1. Tie SF, Tan CW (2013) A review of energy sources and energy management system in electric vehicles. Renew. Sust. Energ. Rev. 20:82–102
2. Tripathi AK et al (2009) A review on prospects of essential oils as biopesticide in insect-pest management. J. Pharmacogn. Phytotherapy 1(5):052–063
3. Green W, Boys J T (1994) 10 kHz inductively coupled power transfer-concept and control. ICPESA: 694–699
4. Boys JT, Covic GA, Green AW (2000) Stability and control of inductively coupled power transfer systems. IET Electr Power Appl 147(1):37–43
5. Laouamer R, Brunello M, Ferrieux JP et al (1997) A multi-resonant converter for non-contact charging with electromagnetic coupling. IEEE Trans Ind Electron 2:792–797
6. Covic GA, Elliott G, Stielau OH et al (2000) The design of a contact-less energy transfer system for a people mover system. IEEE Trans Ind Electron 1:79–84
7. Kurschner D, Rathge C (2008) Contactless energy transmission systems with improved coil positioning flexibility for high power applications. IEEE Trans Power Electron pp 4326–4332
8. Chwei-Sen W, Stielau OH, Covic GA (2005) Design considerations for a contactless electric vehicle battery charger. IEEE Trans Ind Electron 52(5):1308–1314
9. Chwei-Sen W, Covic GA, Stielau OH (2004) Power transfer capability and bifurcation phenomena of loosely coupled inductive power transfer systems. IEEE Trans Ind Electron 51(1):148–157
10. Covic GA, Boys JT et al (2008) Self tuning pick-ups for inductive power transfer. IEEE Trans Power Electron 3489–3494
11. Budhia M, Covic GA, Boys JT (2011) Design and optimization of circular magnetic structures for lumped inductive power transfer systems. IEEE Trans Power Electron 26(11):3096–3108
12. Budhia M, Covic GA, Boys JT (2009) Design and optimisation of magnetic structures for lumped Inductive Power Transfer systems. IEEE Energy Convers Congr Expo ECCE. 2081–2088
13. Nagatsuka Y, Ehara N, Kaneko Y et al (2010) Compact contactless power transfer system for electric vehicles. IEEE Int Conf Power Electron Intell Control Energy Syst ECCE pp 807–813
14. Wu HH, Boys JT, Covic GA (2010) An ac processing pickup for ipt systems. IEEE Trans Power Electron 25(5):1275–1284
15. Huh J, Lee SW, Lee WY et al (2011) Narrow-width inductive power transfer system for online electrical vehicles. IEEE Trans Power Electron 26(12):3666–3679
16. Huh J, Lee S, Park C et al (2010) "High performance inductive power transfer system with narrow rail width for On-Line Electric Vehicles, ECCE pp 647–651
17. Lee S, Huh J, Park C, Choi NS et al (2010) "On-Line Electric Vehicle using inductive power transfer system. IEEE Energy Convers Congr Expo: 1598–
18. Choi SY, Gu BW, Jeong SY, Rim CT (2015) Advances in wireless power transfer systems for roadway-powered electric vehicles. IEEE J Em Sel Top P 3(1):18–36
19. Ning P, Miller JM, Onar OC, White CP (2013) A compact wireless charging system for electric vehicles. ECCE: 3629–3634
20. Wu HH, Gilchrist A, Sealy K et al (2011) A review on inductive charging for electric vehicles. IEEE Energy Convers Congr Expo: 143–147
21. Chinthavali MS, Onar OC, Miller JM, Tang L (2013) Single-phase active boost rectifier with power factor correction for wireless power transfer applications. IEEE Energy Convers Congr Expo: 3258–3265

22. Budhia M, Covic GA, Boys JT, Huang CY (2011) Development and evaluation of single sided flux couplers for contactless electric vehicle charging. IEEE Energy Convers Congr Expo: 614–621
23. Wu HH, Gilchrist A, Sealy KD, Bronson D (2012) A high efficiency 5 kW inductive charger for evs using dual side control. IEEE T Ind Inform. 8(3):585–595
24. Kim JH, Lee BS (2015) Development of 1-MW inductive power transfer system for a high-speed train. IEEE Trans Ind Electron 62(10):6242–6250
25. Moghaddami M, Sarwat A (2017) Self-tuning variable frequency controller for inductive electric vehicle charging with multiple power levels. IEEE Trans Transp Electrification 3(2):488–495
26. Kissin MLG, Huang CY, Covic GA, Boys JT (2009) Detection of the tuned point of a fixed-frequency lcl resonant power supply. IEEE Trans Ind Electron 24(4):1140–1143
27. Hsu JUW, Hu AP, Swain A (2009) A Wireless power pickup based on directional tuning control of magnetic amplifier. IEEE Trans Industr Electron 56(7):2771–2781
28. Boys JT, Chen CI, Covic GA (2005) Controlling inrush currents in inductively coupled power systems. Power Eng Int 2:1046–1051
29. Krishnamoorthy HS, Garg P, Enjeti PN (2012) A matrix converter-based topology for high power electric vehicle battery charging and V2G application. IEEE Ind Electron Mag: 2866–2871
30. Kwon OH, Yang JY, Kwon DS (2013) An exercise motivational system for walking rehabilitation using the Smart Walker. In: 2013 IEEE RO-MAN. IEEE, pp 376–377
31. Kim JS, Rie DH (2014) A study of comparative of evacuation time by platform type according to the propagation speed of smoke in subway platform fire. J Korean Tunn Under Space Assoc 19(4):577–588
32. Posner BZ (2016) Investigating the reliability and validity of the Leadership. Practices Inventory®. Adm Sci 6(4):17
33. MacKenzie RA, Foulk PB, Klump JV, Weckerly K, Purbospito J, Murdiyarso D, ... Nam VN (2016) Sedimentation and belowground carbon accumulation rates in mangrove forests that differ in diversity and land use: a tale of two mangroves. Wetlands Ecol Manage 24(2):245–261
34. Sergeant P, Bossche AVD (2008) Inductive coupler for contactless power transmission. IET Electr Power Appl 2(1):1–7
35. Valtchev S, Borges B et al (2009) Resonant contactless energy transfer with improved efficiency. IEEE Trans Power Electron 24(3):685–699
36. Kazmierkowski MP, Moradewicz AJ (2012) Unplugged but connected: review of contactless energy transfer systems. IEEE Trans Ind Electron 6(4):47–55
37. Foote, Onar OC (2017) A review of high-power wireless power transfer. Electrif. Conf. Expo (ITEC), pp 234–240
38. Jiang KT, Chau CL, Lee CHT (2017) An overview of resonant circuits for wireless power transfer. Energies 10(7):894
39. Weise ND, Mohapatra KK, Mohan N (2010, July). Universal utility interface for plug-in hybrid electric vehicles with vehicle-to-grid functionality. In: IEEE PES General Meeting. IEEE, pp 1–8
40. Yan Z, Zhang K, Li J, Wu W (2012) A novel absolute value logic SPWM control strategy based on de-recoupling idea for high frequency link matrix rectifier. IEEE Trans Industr Inf 9(2):1188–1198
41. Shi X, Wang Z, Tolbert LM, Wang F (2013, September). A comparison of phase disposition and phase shift PWM strategies for modular multilevel converters. In: 2013 IEEE Energy conversion congress and exposition. IEEE, pp 4089–4096
42. Namin, A, Chaidee E, Sriprom T, Bencha P (2018, June) Performance of inductive wireless powertransfer between using pure sine wave and square wave inverters. In: 2018 IEEE Transportation electrification conference and expo. Asia-Pacific (ITEC Asia-Pacific). IEEE, pp 1–5
43. Xia Y, Roy J, Ayyanar R (2017) March). A GaN based doubly grounded, reduced capacitance transformerless split phase photovoltaic inverter with active power decoupling. In: 2017 IEEE Applied power electronics conference and exposition (APEC). IEEE, pp 2983–2988

44. Shi B, Zhou B, Hu D (2017) A novel dual-output indirect matrix converter for more electric aircraft. In: IECON 2017–43rd Annual conference of the IEEE industrial electronics society. In: IECON 2017–43rd Annual conference of the IEEE industrial electronics society. IEEE, pp 4083–4087
45. Guo X (2017) Three-phase CH7 inverter with a new space vector modulation to reduce leakage current for transformerless photovoltaic systems. IEEE J Emerg Sel Top Power Electron 5(2):708–712
46. da Silva MM, Nicolini A, Jucuni J, Pinheiro H (2018, June). Optimized model predictive control space vector modulation for three level flying capacitor converters. In: 2018 IEEE 19th Workshop on control and modeling for power electronics (COMPEL). IEEE, pp 1–7
47. Sandberg H, Lanzon A, Anderson BD (2007) Model approximation using magnitude and phase criteria: Implications for model reduction and system identification. Int J Robust and Nonlinear Control: IFAC-Affiliated J 17(5–6):435–461
48. Frey S, Sadlo F, Ma KL, Ertl T (2014) Interactive progressive visualization with space-time error control. IEEE Trans Visual Comput Graphics 20(12):2397–2406
49. Luo A, Xu Q, Ma F, Chen Y (2016) Overview of power quality analysis and control technology for the smart grid. J Mod Power Sys Clean Energy 4(1):1–9
50. Comin D, Gazarini L, Zanoni JN, Milani H, de Oliveira RMW (2017) Vitamin E improves learning performance and changes the expression of nitric oxide-producing neurons in the brains of diabetic rats. Behav Brain Res 210(1):38–45

Chapter 31
Fatigue and Drowsiness Detection System Using Artificial Intelligence Technique for Car Drivers

Mohd Azlan Abu, Izzat Danial Ishak, Hafiz Basarudin, Aizat Faiz Ramli, and Mohd Ibrahim Shapiai

Abstract Road traffic accident in Malaysia is a heavy concern in these days. Among the top factors of traffic accidents, the fatigue and drowsiness of drivers often times contributed to the increasing number of cases and fatality rate of accidents. This research aims to develop a computer vision system to detect such fatigue and drowsiness of the drivers and wake them up from the split-second nap. The implementation of this research is to develop a drowsiness detection system implemented in a compact development board to assist drivers to awaken from microsleep during driving on fatigue due to long driving hours and various other reasons. This research used a Raspberry Pi 4 along with the official Raspberry Pi camera module V2 and an active buzzer module as waking mechanism for the system. The development used and experimented on the Haar cascade classifier and Histogram of Oriented Gradient + linear Support Vector Machine in the effort of determining the best suitable model to be used for drowsiness detection in terms of speed and accuracy. Both models were run and tested to work properly. The implementation of the Haar cascade classifier produced the best performance in terms of speed and response time to detect drowsiness. On the other hand, the HOG + SVM had better accuracy when compared to the Haar cascade classifier even in low illumination. Having said that, the response time is significantly slower than Haar model which caused a problem regarding the

M. A. Abu (✉) · M. I. Shapiai
Malaysia-Japan International Institute of Technology, Universiti Teknologi Malaysia, Jalan Sultan Yahya Petra, 54100 Kuala Lumpur, Malaysia
e-mail: mohdazlan.abu@utm.my

M. I. Shapiai
e-mail: md_ibrahim83@utm.my

I. D. Ishak · H. Basarudin · A. F. Ramli
Universiti Kuala Lumpur British Malaysian Institute, 53100 Gombak, Batu 8, Jalan Sg Pusu, Selangor, Malaysia
e-mail: izzat.ishak07@s.unikl.edu.my

H. Basarudin
e-mail: hafizb@unikl.edu.my

A. F. Ramli
e-mail: aizatfaiz@unikl.edu.my

A. Ismail et al. (eds.), *Design in Maritime Engineering*, Advanced Structured Materials 167, https://doi.org/10.1007/978-3-030-89988-2_31

reaction time of drivers to react on time. To conclude, the Haar cascaded classifier is decided as the most appropriate model to be applied for the development of a drowsiness detection system.

Keywords Computer vision · Applied AI · Artificial intelligence · Eye aspect ratio · Embedded system

31.1 Introduction

Road traffic accident is a heavy concern all year long, ranging from different factors of traffic crashes that are often the cause of fatality, particularly in Malaysia. Regrettably, one of the top factors of traffic accidents includes the fatigue and drowsiness of drivers especially during long hours of drive as well as occupational driving scenario which is easily solvable by closing the eyes and take a rest from driving for a moment of time. The director the general of Health Ministry Datuk Dr. Noor Hisham Abdullah also stressed that drivers are not being on the road with the feeling of fatigue and tiredness as to avoid road accidents [1]. According to the Malaysian Institute of Road Safety Research (MIROS) in their annual statistical report [2], driver fatigue is one of the most critical factors to be concerned to reduce fatality due to traffic accidents as they cause more fatalities per case which is on par with the risky driving and speeding cases. The report also states that fatigue cases are recorded as the most prominent factors of crash occurrence which is 8.6% of total cases. The trend tends to be similar throughout the year-long duration.

The highlight of the proposed research is to monitor the eye movements of the driver's eyes whereas the subject is recognized as drowsy when the eyes are shut after even a short time. The way to achieve this is by implementing the facial landmarks to extract eye features from the face and afterwards calculating the eye aspect ratio between the height and width of the eyes. By referring to the paper "Real-Time Eye Blink Detection using Facial Landmarks" by Tereza Soukupova and Jan Cech [3], the concept is represented in Fig. 31.1 where the landmark points are determined at the corner, top and bottom for the height and width of the eye. According to the paper, the eye landmarks are detected for every frame while calculating the eye aspect ratio (EAR) to determine the state of the eyes using the formula given in Eq. (31.1), where

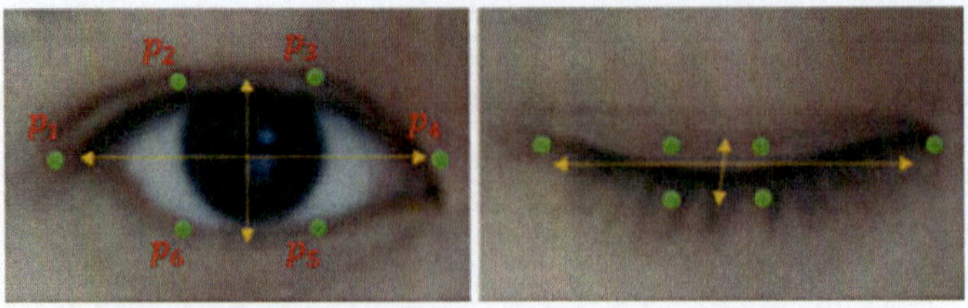

Fig. 31.1 Facial landmarks pinpoint the eye features

"P" are the 2D landmark locations [3].

$$\text{EAR} = \frac{\|P_2 - P_6\| + \|P_3 - P_5\|}{2P_1 - P_4} \tag{31.1}$$

Equation (31.1) represents the average of eye aspect ratio on each eye because of the simultaneous blinking of both. When in open state, the EAR is typically constant due to the eye's minimal movements. While in closed state, the ratio becomes close to zero in which the values can be used to program for indicating shut eyes. Onto the subject of varying eye size, this however does not affect the reliance for this concept as the aspect ratio of open eyes of different individuals has the unnoticeable small variance to compute. With the use of this EAR concept, the system can be programmed to signal for alert from the driver's drowsiness through the person's shut eyes for more than at least a second.

31.2 Literature Review

This segment highlights the studies of components and elements considered as essential in constructing from a basic face detection system to an advanced and efficient one, utilizing recent technologies. The section also comprises of latest and recent studies involving the contributions, challenges and varying development of face detection systems. Several journals, articles and reports are referenced for the purpose of this research.

31.2.1 Large-Scale Human Detection Using HOG

Detecting humans and other living things is a challenging task to execute because of the wide variations of outward forms and postures they take on in images and videos. To add to that, difficult lightings and cluttered surrounding settings further complicate the task of detecting humans even on a smaller scale. A robust and efficient detection algorithm is needed to overcome the difficulties of filtering human forms from difficult backgrounds and illuminations.

The implementation of HOG for a large-scale human detection is experimented in the journal by Dalal and Triggs [8] using the pedestrian database as dataset, which consists of 509 training and 200 test images of pedestrians in city scenes. Among the sample images used, the human subjects tend to be upright, with addition to limited obstructions and a wide range of variations in appearances, lighting, backgrounds, postures and apparels in a crowded city spot. Basically, the HOG-based detectors greatly outperform that of other wavelet-based detectors especially the Haar-like wavelets that comes second after all other HOG-based detectors. The performances of final rectangular (R-HOG) and circular (C-HOG) detectors that utilize the SVM algorithm that possess similar low miss rates of detection contribute to the near-perfect accuracy of human detection using the HOG with linear SVM algorithm.

This paper proved that using the HOG gives very good results for person detection, reducing false positive rates by more than an order of magnitude relative to the best Haar-based detector. As stated by the authors [8], "although our current linear SVM detector is reasonably efficient by processing a 320 × 240 scale-space image (4000 detection windows) in less than a second there is still room for optimization and to further speed up detections it would be useful to develop a coarse-to-fine or rejection-chain style detector based on HOG descriptors".

31.2.2 Face Detection and Face Recognition on Raspberry Pi

The combination used of both Haar-like features and HOG + SVM is precisely demonstrated in the paper by Deshmukh [10] where the Raspberry Pi 3 is used to create a security system that is trained to identify unknown individuals using the HOG + SVM algorithm in Python language.

The first and foremost step is the face detection using the Haar cascade classifier built in the OpenCV libraries to create a database of positive and negative images. From there, the HOG handles the normalization process of the images to result in better invariance towards changes in lightings, shadowing and black points. This helps for a better and effective training by the SVM algorithm. The SVM classifier then trains data to recognize the designated face from many faces and background objects.

The author stated in the journal [10] that the use of Raspberry Pi for a face recognition system can make for a lighter and lower power consumption of a security device, so it makes for a more convenient system than a PC-based face recognition. Due to the open-source nature, it is easier to program a software development on Linux because of wide availability of source codes and references in Python language. The analysis made by the authors revealed that the proposed system shows promising performance in face recognition while being able to be used for face detection even from poor quality images.

In contrast to the above discussion, the study by Gupta and team [11] makes use of the eigenface approach by using the principal component analysis (PCA) algorithm for face recognition process. This algorithm utilizes a mathematical procedure that transforms several possibly correlated variables into variables. The authors state that the eigenface approach used can aid in reducing the size of the database required for recognition. This conventional eigenface approach is implemented into the Raspberry Pi's ARM Cortex for the purpose of face recognition using face recognition modules in Python programming.

31.3 Methodology

This section gives the insight and overview of details of the proposed research methodology to achieve the purpose of developing an effective drowsiness detection system,

The discussions and procedures included in this chapter are important to be achieved. The aim and objectives of the research in developing the drowsiness detection system will be discussed.

31.3.1 Block Diagram

The drowsiness detection system essentially comprises of the Raspberry Pi 4 board itself as the device, connected with the official Raspberry Pi camera module for eye scanning and a display to monitor the perspective of the camera for debugging, through its I/O ports. The eye features will be extracted upon detection of the facial landmarks, particularly the eye regions. The process is seen as a block diagram in Fig. 31.2 which contains the input, process and output of the system.

Input: Live feed obtained from the camera module will be taken as the input to be sent to the central unit.

Process: Left and right-hand eyes from the video stream will be detected. The dlib library from OpenCV will be utilized to deliver the information for extracting only the eye region from the facial landmarks. The camera continues to monitor the eye aspect ratio to be calculated for indications of shut eyes in a set period in seconds. The video stream of the camera module will be shown on the display in real-time mainly for monitoring and debugging purposes. Green outlines will appear around both eyes for eye aspect ratio calculations and drowsiness alerts will be indicated by a red sentence "drowsiness alert" on the screen.

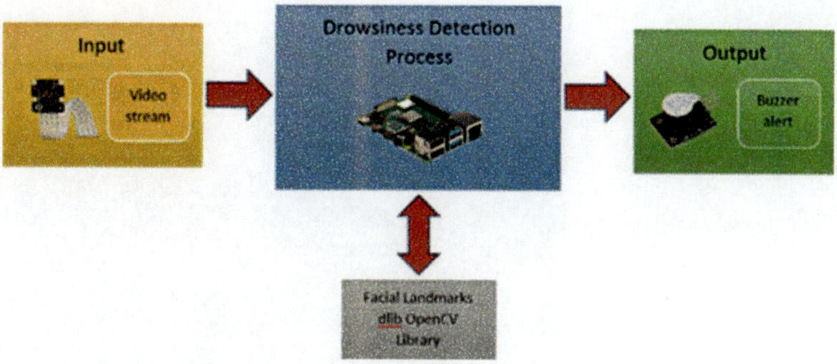

Fig. 31.2 Drowsiness detection system block diagram

Output: An active buzzer module will produce beep sounds to alert the driver upon drowsiness. The high frequency beep sounds serve the purpose of waking a person from the drowsiness.

The drowsiness detection will start with detecting any visible eyes from the camera video stream. Then, the drowsiness detection process will be executed by extracting the eye region from the facial landmarks referencing from the dlib OpenCV library. The system will then monitor the eye aspect ratio with range of 0.3 and below, in which once the ratio approximately below 0.3 is detected for a set time, the system will recognize as the driver being drowsy. Thus, a signal will be sent to the buzzer to ring the driver from drowsiness. Otherwise, the detection system will continue to monitor the eye ratios from the camera's perspective.

31.4 Result and Discussion

This section presents and discusses the result obtained from the program actual test run using a subject's eyes to detect drowsiness in real time. The figures presented in this chapter are the screenshots of the actual live video stream on different conditions.

31.4.1 *Haar Cascade Classifier*

Tests are run using the Haar cascade classifier to gauge the algorithm's performance involving the framerate per second (fps), response time (ms) and different conditions during the test run. The test was first running on normal condition to determine the performance on the fundamental level. Figure 31.3 shows the screenshot of real-time video feed of the subject with outlines around the eyes' shape.

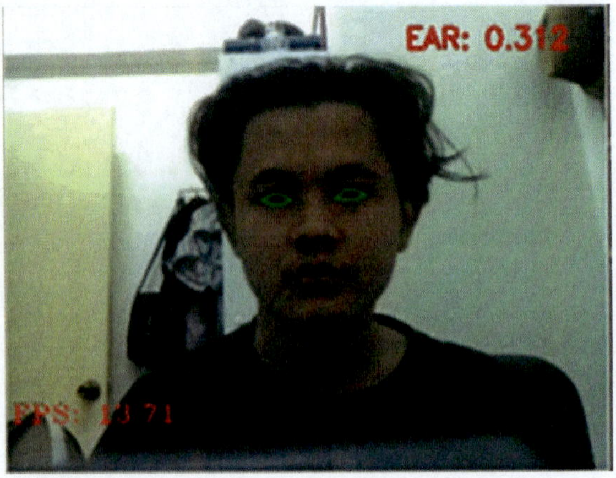

Fig. 31.3 Video feed on normal condition

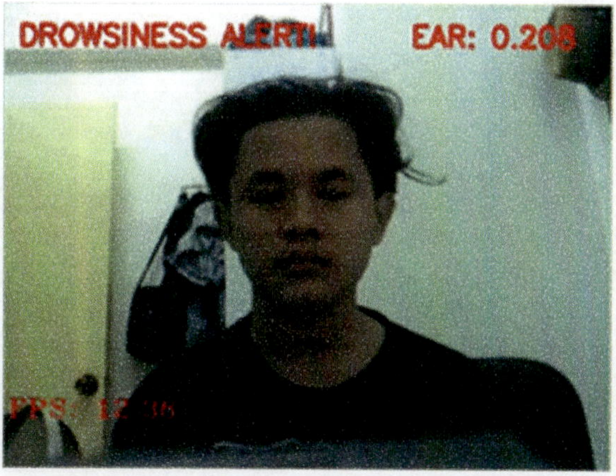

Fig. 31.4 Drowsiness alert

Figure 31.4 shows both screenshot of real-time feed without outlines of the shut eyes to test the detection process of the Haar algorithm. When the eyes are shut, the EAR rate has dropped from the threshold of 0.3 to an average of 0.21 within 2 s which is supposed to be the indication of the subject's drowsiness. The drowsiness alert indication then popped up on top left as well as the beeping sound of the buzzer to wake the subject. The fps rate is still maintained on the average of 13.5 which demonstrated a lesser lag in using the Haar cascade classifier in monitoring the eye features.

The findings showed favourable results using the method of eye blink detection by implementing the facial landmarks detector from the works of Soukupova and Cech [3]. However, instead of using HOG + SVM, this method used the Haar cascade classifier to compensate for the lesser processing power of the compact Raspberry Pi 4.

31.4.2 Hog + svm

In Fig. 31.5, the implementation of HOG + SVM on the Raspberry Pi 4 is successful. However, the recorded fps is substantially lower which is two times lower than that of applying the Haar cascade classifier. Although the drowsiness detecting succeeded as shown in Fig. 31.6 even in poor illumination, the response time to detect is significantly slower in contrast with the Haar cascade classifier, which is undesired.

In Table 31.1, it is shown that the average fps of implementing the Haar cascade classifier is two times higher than that of HOG + SVM. The response time for the Haar detector to detect drowsiness is also significantly faster at 1.12 s than HOG + SVM at 6.95 s. This makes the Haar cascade classifier substantially faster in processing detection on the Raspberry Pi 4 as well as putting less burden to the processor.

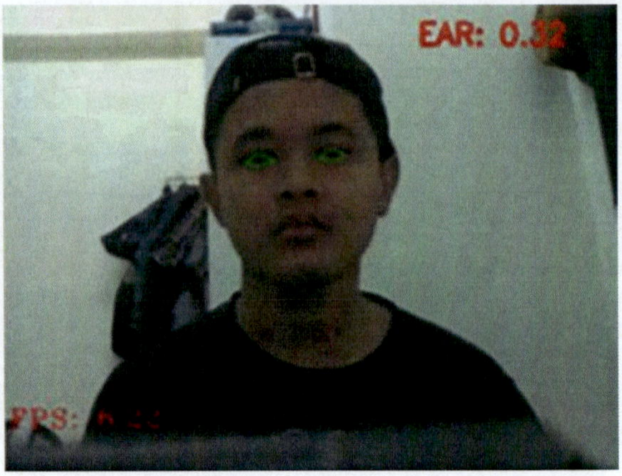

Fig. 31.5 HOG + SVM normal condition

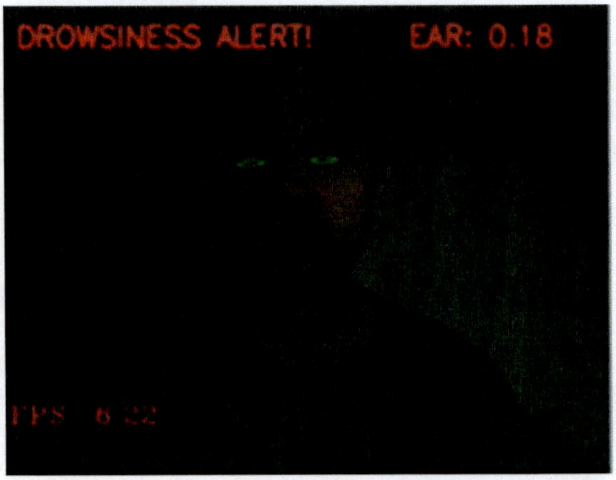

Fig. 31.6 HOG + SVM detects in low illumination

Table 31.1 Comparison of Haar and HOG + SVM

	Haar cascade classifier	HOG + SVM
Average FPS	13.5	6.2
Response time (s)	1.12	6.95

Overall, the HOG + SVM has undeniably better accuracy than the Haar detector especially in the findings of detecting in dark and low illumination condition. But the response time posed a problem in terms of the response of the subject to wake from microsleep during driving. The subject might not have the time to react by the time the HOG + SVM detects drowsiness. Therefore, the Haar cascade classifier seemed to be the best fit specifically in this case of implementing on the small and compact Raspberry Pi 4.

31.5 Conclusion

Choosing the appropriate AI model is of utmost importance when developing a safety feature that intends to keep people from harm, not vice versa. A low performance drowsiness detection system could not possibly satisfy the minimum safety standards and therefore beats the purpose of the system. From the results, the HOG with SVM and Haar cascade classifier are tested to measure and gauge the performance of the models upon running the program in Raspberry Pi 4. It is observed that the Haar cascade classifier favours more in performance and detection speed in contrast with the HOG with SVM that puts more burden on the Raspberry Pi processing power and consequently reducing the detection speed of the system. Given the surrounding setting of the driver seat in the vehicle interior where the driver is positioned, it is decided that detection accuracy is less emphasized due to lesser faces and eyes are to be detected in a driver's seat, in which the use of the Haar cascade classifier is sufficient enough for the task. The detection accuracy upon implementing this model is fairly compensated with its detection speed offered. For future recommendation, the current camera module used in the research which is the official Raspberry Pi camera module lacks the ability to see clearly in dark and poor-light time of the day. The possible solution to overcome this is by swapping out the current camera with the Raspberry Pi NoIR camera module. This module not only gives everything the regular camera module offers, but also has the night vision capability. The NoIR that stands for "No Infrared" simply means it does not employ an infrared filter. This means that images and videos taken by daylight will look distorted on the colours, but in turn gives the ability to see in the dark with infrared lighting. The only concern is how well the detection performs with distorted colours and dim vision. This should be the task for future works progress.

The research sacrifices a slight bit of accuracy in return for the speed of detection by using the Haar cascade classifier model instead of the more accurate HOG with SVM. This is due to the demanding processing power from the HOG with SVM to process the detection with better accuracy. The current performance could potentially bring about undesired false positives and false negatives upon the detection process. Further research and tests should be made to find the best possible model to be implemented into the system that favours both speed and accuracy attributes.

Acknowledgements Author would like to acknowledge Malaysia-Japan International Institute of Technology, Universiti Teknologi Malaysia, Jalan Sultan Yahya Petra, 54100 Kuala Lumpur for the funding provided.

References

1. Bernama (2017) "Don't drive if you are fatigued, sleepy—health DG," http://english.astroawani.com/malaysia-news/dont-drive-if-you-are-fatigued-sleepy-health-dg-144224. Accessed 14 Jun 2020

2. Zainal AANS, Mohd FSA, Lamin F, Abdul MAR (2012) MIROS crash investigation and reconstruction annual statistical report 2007–2010
3. Cech J, Soukupova T (2016) "Real-time eye blink detection using facial landmarks," Cent Mach Perception, Dep Cybern Fac Electr Eng Czech Tech Univ Prague, pp 1–8. https://doi.org/10.1017/CBO9781107415324.004
4. Farley P (2019) "Face detection and attributes concepts—azure cognitive services | microsoft docs,". https://docs.microsoft.com/en-us/azure/cognitive-services/face/concepts/face-detection#face-landmarks . Accessed 16 Jun 2020
5. Dwivedi D (2018) "Face detection for beginners—towards data science," https://towardsdatascience.com/face-detection-for-beginners-e58e8f21aad9 . Accessed 15 Jun 2020
6. Vanderplas J (2016) "Python data science handbook," https://jakevdp.github.io/PythonDataScienceHandbook/ . Accessed 15 Jun 2020
7. Berger W (2020) "Deep learning haar cascade explained." http://www.willberger.org/cascade-haar-explained/. Accessed 15 Jun 2020
8. Dalal N, Triggs B (2018) "Histogram of oriented gradients for human detection," Proc 2018 5th Int Conf Bus Ind Res Smart Technol Next Gener Information, Eng Bus Soc Sci ICBIR 2018. https://doi.org/10.1109/ICBIR.2018.8391187
9. Frigerio A, Hadlock TA, Murray EH, Heaton JT (2014) Infrared-based blink-detecting glasses for facial pacing: toward a bionic blink. JAMA Facial Plast Surg 16(3):211–218. https://doi.org/10.1001/jamafacial.2014.1
10. Deshmukh SV et al (2017) Face detection and face recognition using raspberry pi. Ijarcce 6(4):70–73. https://doi.org/10.17148/ijarcce.2017.6414
11. Gupta I, Patil V, Kadam C, Dumbre S (2017) "Face detection and recognition using raspberry pi," WIECON-ECE 2016–2016 IEEE Int WIE Conf Electr Comput Eng, pp 83–86. https://doi.org/10.1109/WIECON-ECE.2016.8009092
12. Kwolek B (2014) "Face detection using convolutional neural networks and gabor filters," Lect. Notes Comput Sci 3696. https://doi.org/10.1007/11550822
13. Kamencay P, Benco M, Mizdos T, Radil R (2017) "A new method for face recognition using convolutional neural network," Adv Electr Electron Eng 15(4): 663–672. https://doi.org/10.15598/aeee.v15i4.2389
14. Reddy B, Kim Y H, Yun S, Seo C, Jang J (2017) "Real-time driver drowsiness detection for embedded system using model compression of deep neural networks." In IEEE computer society conference on computer vision and pattern recognition workshops. https://doi.org/10.1109/CVPRW.2017.59
15. Zhang K, Zhang Z, Li Z, Qiao Y (2016) "Joint face detection and alignment using multi-task cascaded convolutional networks," IEEE Signal Process Lett. https://doi.org/10.1109/LSP.2016.2603342

Chapter 32
Development of an International Framework for Private Maritime Security Companies in Malaysia

Ahmad Faizal Ahmad Fuad, Aimie Qamarina Anwar, Mohd Sharifuddin Ahmad, Mohd Hafizi Said, and Amir Syawal Khamis

Abstract MOHA and the Malaysian NSC developed a standard operation procedure (SOP) for private maritime security companies (PMSC) in Malaysia for local security companies to follow. However, the SOP they developed was based on local regulations relating to the land-based Private Security Company and Firearms Act. Consequently, these regulations are not compatible with local and international maritime law and practices. This happened because no framework for PMSC had been established to guide the development of a standard operation procedure that is compatible with existing national and international maritime law and practices. A framework is defined as 'A basic structure underlying a system or a concept'. The objective of this study is to create a conceptual framework for PMSC in Malaysia so that the resulting SOP would be fully compatible with national and international maritime law and practices. The research methodology adopted in this study consists of several activities based on the Delphi method. Surveys were conducted and safety experts were consulted and invited to express their views, either through interviews or workshops. It is concluded that this new framework was developed based on relevant agencies and expert consensus and should be implemented in Malaysia in order to produce a consistent, effective and secure action plan which addresses the main role of PMSC.

Keywords Maritime security · Private maritime security · Piracy · Conceptual framework

---

A. F. Ahmad Fuad (✉) · A. Q. Anwar · M. H. Said
Nautical Science and Maritime Transportation Program, Universiti Malaysia Terengganu, 21030 Kuala Terengganu, Terengganu, Malaysia

M. H. Said
e-mail: hafizi@umt.edu.my

M. S. Ahmad
Maritime Management Program, Universiti Malaysia Terengganu, 21300 Kuala Terengganu, Terengganu, Malaysia
e-mail: mohdsharifuddin@umt.edu.my

A. S. Khamis
Akademi Laut Malaysia Batu 30, Kampung Tanjung Dahan, 78200 Kuala Sungai Baru, Melaka, Malaysia

A. Ismail et al. (eds.), *Design in Maritime Engineering*, Advanced Structured Materials 167, https://doi.org/10.1007/978-3-030-89988-2_32

32.1 Introduction

Private maritime security companies (PMSC) were established in Malaysia to address the problem of piracy in the high seas around Malaysia such as Malacca straits and Sulu Celebes sea [1]. The need for PMSC is due to the increase in piracy in the area around Malaysia and also the title 'war risk area' at Malacca straits declared by the Lloyd Joint War Committee (JWC) on 13 September 2005 [2]. The establishment of PMSC indirectly needs to formulate policies and standard operation procedure (SOPs) for operations carried out effective and follow the regulation. Current SOP of PMSC in Malaysia are shown in Table 32.1.

Table 32.1 shows the current PMSCs SOP in Malaysia consisting of Attachment A, B, C, D, E, F, and G. Attachment A explains about maritime security control services' handling methods, which contains maritime safety control services application requirements, handling and use of firearms, allowed areas to operate, types of maritime security control services allowed, travel, operational and accident handling, escort boat or ship specifications, communication equipment, insurance coverage, uniform clothing, report, training and closing.

Attachment B concerns on SOP of handling and use of weapons which is contains the application of firearms, firearms licence, authorised firearm, firearm storage, number or ratio firearm, firearm movement and monitoring. For attachment C, D, E, F and G that focus on SOP for services provided by security services such as on board, escort, port, offshore installation, and island. All of these services provided have the same main point that they need to consider but different main requirements, area of responsible, surveillance and security control and accident handling.

32.2 Problem Statement

The existing SOP has weaknesses where there are issues arising such as the role of the flag state is not explained its responsibilities while each ship must follow the laws of the flag state when sailing. Besides, the area of operation allowed to operate is also limited to allowed in Malaysian maritime waters only, while according to International Maritime Organization (IMO), PMSC is allowed to operate on board ships wherever the direction of navigation [3]. In addition, the report needs to be explained more clearly, for example, when there is a threat in the port, it needs to be reported to Royal Malaysia Police (RMP), when in Malaysian waters it needs to be reported to Malaysia Maritime Enforcement Agency (MMEA), and at high sea it needs to be reported to Royal Malaysian Navy (RMN). Reports are sent only to the main body of security. All these issue lead research to study the development of local framework and international framework for PMSC in Malaysia. By referring to international conventions, international guidelines, and local regulation can assist the study in developing a standard and effective international framework for PMSC in Malaysia.

Table 32.1 Shows the current standard operating procedures for a private security company in Malaysia as follows:

Attachment	Main heading	Supporting sub-headings
Attachment A	MARITIME SECURITY CONTROL SERVICES HANDLING METHODS	1. Maritime Safety Control Services Application Requirements 2. Handling and Use of Firearms 3. Allowed Areas to Operate 4. Types of Maritime Security Control Services Allowed 5. Travel, Operational and Accident Handling 6. Escort Boat/Ship Specifications 7. Communication Equipment 8. Insurance coverage 9. Uniform clothing 10. Report 11. Training 12. Closing
Attachment B	STANDARD OPERATING PROCEDURE (SOP) HANDLING AND USE OF WEAPONS	1. Introduction 2. Application firearm licence 3. Authorised firearms 4. Firearm storage 5. Number or ratio firearm 6. Firearms movement 7. Monitoring
Attachment C	STANDARD OPERATING PROCEDURE (SOP) ON-BOARD MARITIME SAFETY CONTROL (ON BOARD)	1. Main requirement 2. Area responsible 3. Surveillance and security control 4. Accident handling
Attachment D	STANDARD OPERATING PROCEDURE (SOP) MARITIME SECURITY CONTROL USING ESCORT BOATS	1. Main requirement 2. Area responsible 3. Surveillance and security control 4. Accident handling
Attachment E	STANDARD OPERATING PROCEDURE (SOP) OF MARITIME SAFETY CONTROL IN PORT AREA	1. Main requirement 2. Area responsible 3. Surveillance and security control 4. Accident Handling
Attachment F	STANDARD OPERATING PROCEDURE (SOP) MARITIME SAFETY CONTROL ON OFFSHORE STRUCTURE	1. Main requirement 2. Area responsible 3. Surveillance and security control 4. Accident handling
Attachment G	STANDARD OPERATING PROCEDURE (SOP) MARITIME SAFETY CONTROL IN ISLAND	1. Main requirement 2. Area responsible 3. Surveillance and security control 4. Accident handling

32.3 Methodology

The overall research activity is shown in Fig. 32.1. The detail of each step is explained in the following paragraphs.

Step one was to identify factors from the existing SOP, which was obtained from Ministry of Home Affair (MOHA) and the National Security Council (NSC). This was done by interviewing the officers-in-charge of PMSC, at MOHA and the NSC. Based on the information gathered from these interviews, the factors were identified.

Step two was to develop the preliminary PMSC framework for Malaysia. The factors identified from SOP, MOHA and NSC consisting of services and company detail about handling and monitoring were used to develop the preliminary PMSC framework. Other important input to the framework came from national and international regulations, international guidelines, the type of services provided and agencies involved. However, this framework cannot be accepted before a validation process and at this stage was considered the first draft.

Step three was to develop the framework by using the Delphi method round one. Prior to the survey, questions were developed based on identified factors. The standard response questionnaire was the 6-point Likert scale. The Likert scale produced the ordinal data used to measure the participant opinions or perceptions, which are related to a single 'latent' variable [4]. The experts selected for this survey were relevant officers from MOHA, the NSC, MMEA, MARDEP, PETRONAS (HSE) and PMSC. According to Belton (2019), the number of experts for a survey should be at least five

Fig. 32.1 Flow chart of overall research activity

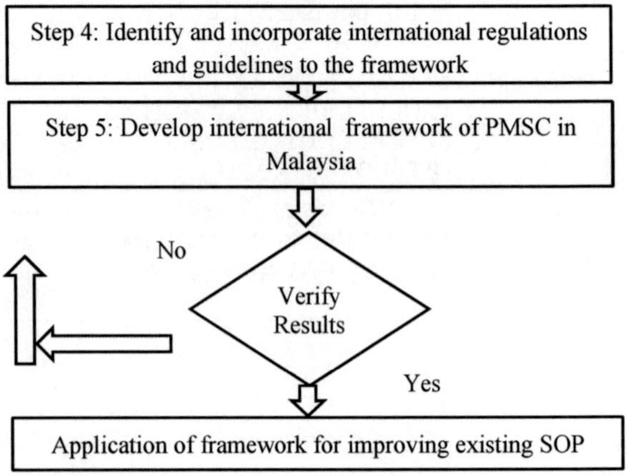

persons [5]. The Delphi expert criteria requires an expert to be at the top of his/her field of technical knowledge, long working experience and the knowledge to judge [6]. The study does not stop here, a framework that follows international standards needs to be built. Thus, the study needs to find and read the laws related to PMSC practices.

Step four was to identify and incorporate international regulation and guidelines to the framework, the regulation are listed in Table 32.2. Table 32.2 shows the most relevant international regulations and guidelines for practices towards private military and security companies based the on United Nation Conference on Trade and Development (UNCTAD) part 2 title of International legal framework and multilateral cooperation to combat piracy on annex 1 [7]. After the development of the framework was done, the study proceeded to identify and incorporate international regulations and guidelines to develop the international framework of PMSC in Malaysia.

32.4 Results

The results are shown in Fig. 32.2 and Table 32.3 which represent step four and five to develop the international framework of PMSC of Malaysia.

Step four was to develop the international framework of PMSC in Malaysia based on the international regulations in Fig. 32.2. The framework was developed based on services, company, regulation and agencies. Under services, only have on board and escort. Under company, it is considered management and operation. For the regulation, we have national and international input, national gain from the interview expert session while international regulations are from websites and UNCTAD documented. This international framework is also verified by expert using a meeting session.

Step five was the application of the framework of PMSC to improve the existing SOP shown in Table 32.3.

32.5 Conclusion

The development of the conceptual framework for PMSC operation in Malaysia has followed the proper methods suggested in the available reputable literature and engagement with experts. Furthermore, the framework has been validated by related government agencies and industry. The purpose of the developed framework is to serve as a guide for reviewing the existing SOP and developing a new SOP which would be compatible with existing national and international regulations, guidelines and practices. The developed framework has fulfilled the study objectives and addressed the problem statement.

Table 32.2 International regulation and guideline towards PMSC [7], [8]

Source and Year	Literature review
IMO 2005	Convention on Facilitation of International Maritime Traffic (FAL), 1965 incorporated 2005 amendment
FDFA 2008	The Montreux Document on Pertinent International Legal Obligations and Good Practices for States Related to Operations of Private Military and Security Companies During Armed Conflict
Montreux Document 2008	On pertinent legal obligations and good practices for states related to operations of private military and security companies during armed conflict. The Montreux document is the result of an international process launched by the government of Switzerland and the ICRC. It is an intergovernmental document intended to promote respect for international humanitarian law and human rights law whenever private military and security companies are present in armed conflicts
ICOC-PSSP 2010	The International Code of Conduct for Private Security Service Providers. This Code builds upon the Montreux document and is the result of an active collaboration of members of the private security industry along with the Swiss Department of Foreign Affairs, the Geneva Centre for the Democratic Control of Armed Forces (DCAF) and the Geneva Academy of International Humanitarian Law and Human Rights (ADH) to support the development of a Global Code of Conduct that lays down international industry norms and standards for the provision of private security services
IMO 2011	**MSC-FAL.1/Circ.2.** Questionnaire on information on port and coastal state requirements related to privately contracted armed security personnel on board ships
IMO 2012	**MSC.1/Circ.1405/Rev.2** Revised Interim Guidance to Ship owners, Ship Operators and Shipmasters on the use of Privately Contracted Armed Security Personnel on Board Ships in the High-Risk Area (25 May 2012)
IMO 2012	**MSC.1/Circ.1406/Rev.3** Revised Interim Recommendations for Flag States regarding the use of Privately Contracted Armed Security Personnel on board ships in the High-Risk Area (25 May 2012)
IMO 2012	**MSC.1/Circ.1443** Interim Guidance to Private Maritime Security Companies Providing Privately Contracted Armed Security Personnel on Board Ships in the High-Risk Area (25 May 2012)
IMO 2012	**MSC.1/Circ.1408/Rev.1** Revised Interim Recommendations for Port and Coastal States Regarding the Use of Privately Contracted Armed Security Personnel On Board Ships in the High Risk Area
GAIHLHR 2013	The **International Code of Conduct** (ICOC) for Private Security Service Providers
IMO 2015	MSC.1Circ./1406-Rev.3 Revised Interim Recommendations for Flag States regarding the use of Privately Contracted Armed Security Personnel on board ships in the High-Risk Area
ISO 2015	ISO 28007–1:2015 Guidelines for Private Maritime Security Companies (PMSC) providing privately contracted armed security personnel (PCASP) on board ships

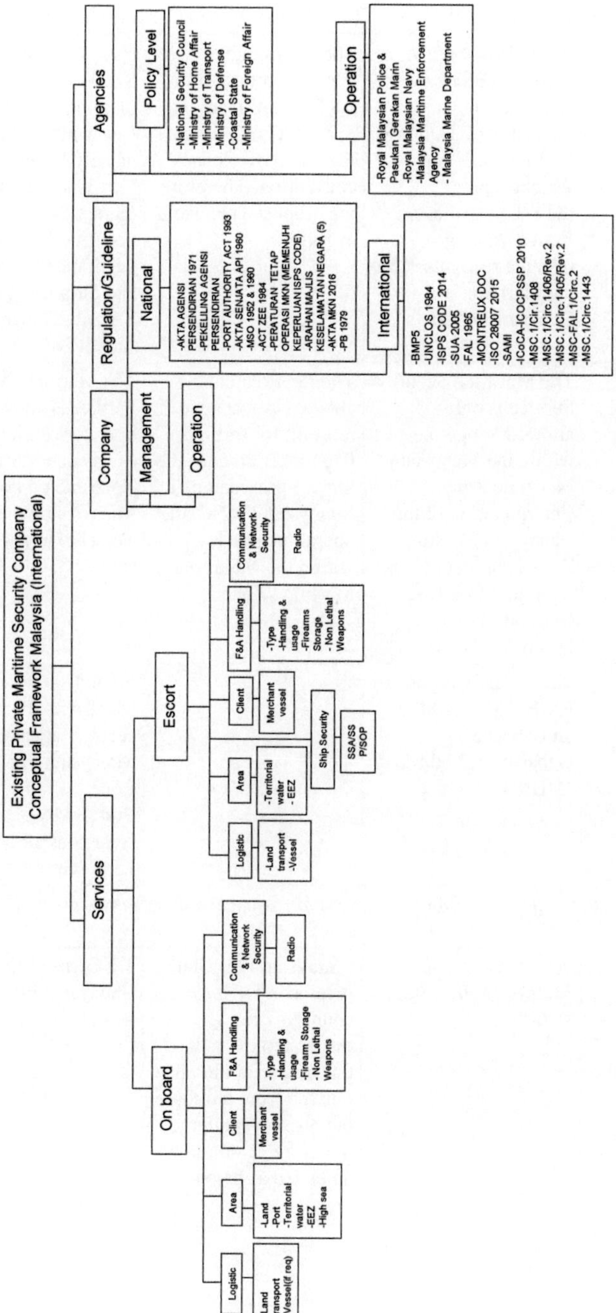

Fig. 32.2 International Framework of PMSC in Malaysia

Table 32.3 Application of framework for improving existing SOP

Attachment A: METHODS OH HANDLING THE MARITIME SECURITY SERVICE			
Item 2: Recruitment of Maritime Security Guards	Has a Basic Safety Training (BST) certificate and a Standards of Training, Certifications and Watchkeeping (STWC) certificate from a training institute recognised by the Marine Department Malaysia	The requirement above is not accurate. BST is certificate under STCW and not regulated by other convention. Therefore, the proposed change is as follows:	Have the required STCW certificates required by Marine Department Malaysia, namely Basic Training Certificate (BT) (compulsory), Ship Security Awareness Certificate (optional), and Ship Designated Security Duties (DSD) (optional)
Item 4: Areas Allowed to Operate	The Maritime Security Service is only allowed to operate within the Malaysian Maritime Zone (interpretation of the Malaysian Maritime Zone is in accordance with the MMEA Act 663 and Map of Malaysia 1979)	The requirement above is correct and relevant for foreign flag vessel that use the service provided by a Malaysian PMSC, the operation area is within the Malaysian Maritime Zone	The Maritime Security Service is allowed to operate within the Malaysian Maritime Zone and on board Malaysian Flagged Vessel to any waters it sails
Item 3: Security monitoring and control	Submit daily reports to RMP, RMN, MMEA and Marine Department Malaysia (MDM)		Should submit only to MMEA as the main security agency in Malaysian Maritime Zone Port: RMN Near coastal: MMEA High sea: RMN
Attachment D: Standard Operating Procedure (Sop) Maritime Security Service Using Escort Boats			
Item 1: Key Conditions	1.1 Owns a vessel of Malaysian Flag State status;	Malaysian Flag State is referring to a country. The ownership of the vessel should not only be from purchase but also by charter; therefore, it should be amended as below:	1.1 Owns or operate a Malaysian Flagged vessel

(continued)

Table 32.3 (continued)

Attachment A: METHODS OH HANDLING THE MARITIME SECURITY SERVICE			
	1.2 Have appropriate and adequate communication equipment	Should be changed as follows:	Carry the GMDSS Communication Equipment according to the vessel's operational Sea Area under GMDSS, namely A1, A2, A3 and A4
	1.4 Have adequate safety equipment	Should be changed as follows:	Carry the safety equipment as stipulated in SOLAS 74 as amended and MSO 52/60
Item 2: Area of Responsibility	2.1 The area of responsibility for maritime security service providers using escort boats is to escorted ships	Should be changed as follows:	The area of responsibilities for the escort boat is an area within 500 m of the escorted vessel and to escort the vessel in Malaysian Maritime Zone, at high seas and in other country's EEZ

Acknowledgements The research team would like to acknowledge Ministry of Higher Education of Malaysia support our financial project FRGS/1/2019/SS02/UMT/02/1 and also the good cooperation received from MOHA, MMEA, the NSC and PETRONAS (HHSE) during the development of the conceptual framework.

References

1. Osnin NA (2006) Private maritime security company (PMSC) in the Strait of Malacca: options for Malaysia. WMU J Marit Aff 5(2):195–206. https://doi.org/10.1007/BF03195104
2. Arifin B, Damanik NA (2019) "The implementation of indonesia's counter piracy strategies through multilateral cooperation in the malacca strait (2004–2009),"UPH J. Int. Relat: 5–20
3. IMO (2012) "Interim guidance to private maritime security companies providing privately contracted armed security personnel on board ships in the high risk area—MSC.1/Circ.1443," vol 44
4. Joshi A, Kale S, Chandel S, Pal D (2015) Likert scale: explored and explained. Br J Appl Sci Technol 7(4):396–403. https://doi.org/10.9734/bjast/2015/14975
5. Belton I (2019) "Improving the practical application of the Delphi method in group-based judgment: A six-step prescription for a well-founded and defensible process," Technol Forecast Soc Change 147(April):72–82. https://doi.org/10.1016/j.techfore.2019.07.002
6. Skinner R, Nelson RR, Chin WW, Land L (2015) The Delphi method research strategy in studies of information systems. Commun Assoc Inf Syst 37:31–63. https://doi.org/10.17705/1cais.03702

7. UNCTAD (2014) "Part II: an overview of the international legal"
8. International Committee of the Red Cross (2008) "The montreaux document,": 11–27

Chapter 33
Development of Floating Buoy Technology Using a Modular Method

Rohaizad Hafidz Rozali, Mohd Yuzri Mohd Yusop, Wardiah Mohd Dahalan, Noorazlina Mohamid Salih, Siti Noor Kamariah Yaakob, and Aminatul Hawa Yahaya

Abstract The previous researchers had developed an ocean device called a buoy. The buoy acts as a flotation device with its purpose to distinguish ocean characteristics and weather characteristics. The buoy is bright or fluorescent in color generally. The buoy's different color indicates another buoy function because it is easier for any vessel to detect a buoy in the middle of the ocean. As the newest technology was introduced, buoy's application became wide, and its purpose was upgraded depending on the researchers. In this paper, we describe the development of a floating buoy and the system integration for it. The construction process uses the modular method, and fiberglass is used as the primary material to install electrical, electronic, and wireless network communication systems. The final weight captured for the buoy is 56.5 kg, and the designs consist of an anemometer, solar panel, camera, antenna, microcontroller, telemetry, solar charge controller, and battery. In conclusion, the

R. H. Rozali (✉) · M. Y. Mohd Yusop
Maritime Engineering Technology Section, Universiti Kuala Lumpur, Malaysian Institute of Marine Engineering Technology, 32200 Lumut, Perak, Malaysia
e-mail: rohaizad@unikl.edu.my

M. Y. Mohd Yusop
e-mail: myuzri@unikl.edu.my

W. Mohd Dahalan · N. Mohamid Salih
Marine and Electrical Technology Section, Universiti Kuala Lumpur, Malaysian Institute of Marine Engineering Technology, 32200 Lumut, Perak, Malaysia
e-mail: wardiah@unikl.edu.my

N. Mohamid Salih
e-mail: noorazlinams@unikl.edu.my

S. N. K. Yaakob
Teknoputra Section, Universiti Kuala Lumpur, Malaysian Institute of Marine Engineering Technology, 32200 Lumut, Perak, Malaysia
e-mail: sitikamariah@unikl.edu.my

A. H. Yahaya
Economic Section, Universiti Kuala Lumpur, Business School, Jalan Pesiaran Gurney, Wilayah Persekutuan Kuala Lumpur, Kampung Datuk Keramat, 54000 Kuala Lumpur, Malaysia
e-mail: aminatulhawa@unikl.edu.my

A. Ismail et al. (eds.), *Design in Maritime Engineering*, Advanced Structured Materials 167, https://doi.org/10.1007/978-3-030-89988-2_33

complete floating buoy was successfully developed and is ready to be tested and performed a sea-trial for the buoy to serve its purposes.

Keywords Buoy technology · Modular concepts · Fiberglass · Wireless Technology

33.1 Introduction

Over the years, several types and styles of marine buoys have been designed to serve varied tasks. The research focuses on the design that brings out stability, which is challenging because of the geometric shapes of the product and the intricacy of the phenomena. Generally, the conventional material of a buoy is steel [1]. A steel buoy has advantages such as high strength and easy production, employing cutting and welding techniques to create a shape. However, it might lead to maritime pollution due to steel corrosion and maintenance cost due to its heavy weight. To boost the potential of buoy development, a novel design buoy that incorporates a modular concept was shown [2]. The advantages of the modular design are that the entire weight was minimized. The weight of the existing buoy makes it slightly better for an operation to install or uninstall the buoy, making it a friendlier user, and do not need much expenses during the process. This paper discusses the development of buoy utilizing the modular method. In addition, the system used for modular buoy will be suggested.

33.2 Literature Review

33.2.1 Malaysian Tsunami Buoy

The buoy was designed by review of the previous design and which are still actively employed. One of the buoy designs in Malaysia, named the Wavescan, has been deployed in the Malaysian open sea. The role of this buoy is for a tsunami detection system. The preferred design for this project is the buoy designed by a firm named Furgo [3]. The Fugro company has designed a buoy that used a base material to manage the Earth's ecology by employing polyethene. The notion of wave scan buoy is very nice, but it lacks modularity. It cannot be called modular with a total weight of 924 kg and net buoyancy of 3000 N [4].

33.2.2 Concept Design

Traditionally, the creation of the new design is based on brainstorming ideas among the designers. The client will specify buoy features according to their requirements. A design spiral concept is a technique that can be employed to develop a new product [5]. The process starts with an idea design, rough design, and detailed design. The key to a successful approach is the resolution of all critical design trade-offs. The new buoy concept design has previously done by the concept that prior researchers focused on the buoy idea design. The buoy design applies the diamond form, which means the buoy will not glide on the water. The V-shaped bottom will make it more stable, the same as the ship hull design [6]. The concept design is shown in Figs. 33.1 and 33.2.

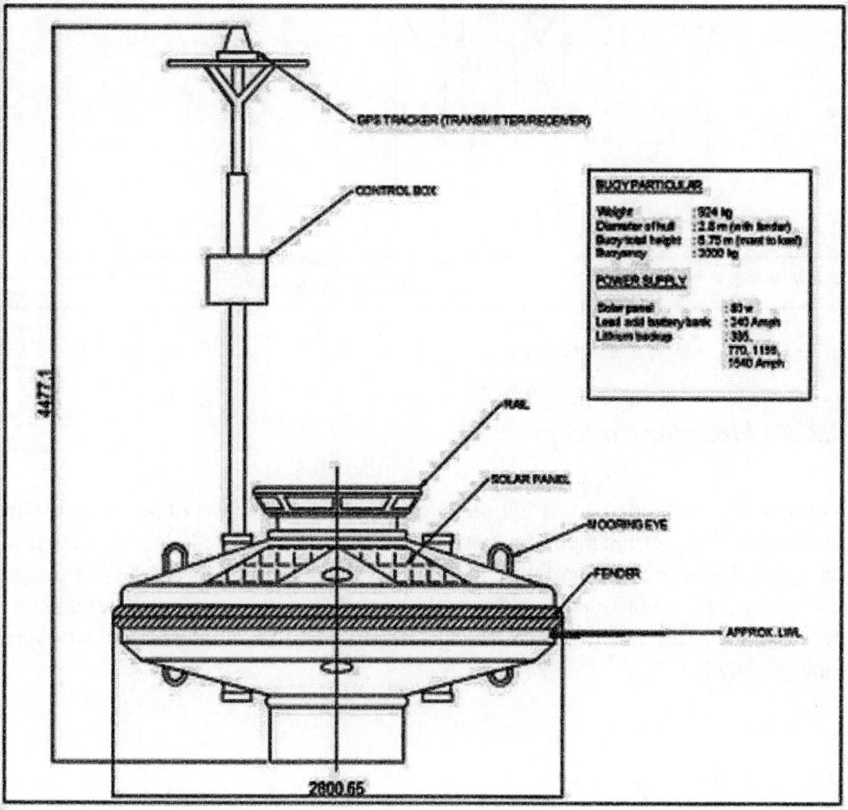

Fig. 33.1 Wavescan buoy

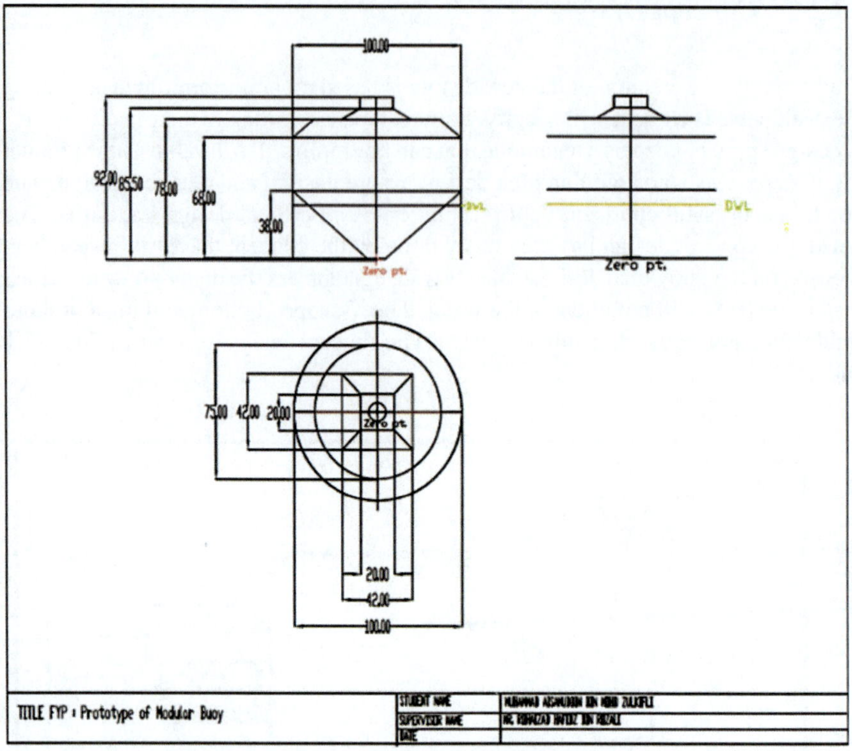

Fig. 33.2 Floating buoy concept design

33.2.3 Modular Concept

"Modular" is a method of construction involving the construction of portions distant from the building site and then transporting them to the designated place. Installation is completed on-site of the prefabricated units [7, 8]. The modules can be positioned side-by-side, end-to-end, or stacked to provide various configurations and styles. The inter-module linkages tie together the separate modules to build the overall structure of the product [9].

33.2.4 Device Systems

A device system consists of numerous sensors, storage data, and transmission data transmitted wirelessly to the ground workstation called a marine data buoy. The ground station is equipped with a forecasting system, remote monitoring system, communication module system, and communication network system. Basically, the requested data from the remote monitoring system is used to acquire, archive, access, and analyze the data. The process in order of data gathered will be supplied toward weather forecasting and analytical statistics [10–12]. The main components of a typical data buoy are sensory and buoy bodies, such as the base and tower. Then, the building is supported by the floor, and precisely the buoy placement layout parallel to a body of water and submerged devices [6, 13]. In summary, the construction process and proposed system will be discussed in this paper.

33.3 Methodology

The purpose of the present work is to develop the buoy using the modular method. Based on the concept design previously discussed, the step-by-step construction process will be shown for other researchers to apply for buoy development. In addition, the system will be attached to the buoy and tested in water.

33.3.1 Construction of the Buoy

The researcher changed the float body material from a conventional steel body to a fiberglass body in this study. The first step is to develop the mold for the bottom, body, and top module using timber. Then, using fiberglass on the top of the fiberglass. The result is shown in Figs. 33.3 and 33.4. The product will be pulled out from the mold. Figures 33.5 and 33.6 indicate that the hook eye and clips are installed as part of the fittings.

Fig. 33.3 Mold bottom and body module

Fig. 33.4 Mold for top module

Fig. 33.5 Hook eye attach body module

After finish, each module of the buoy will be installed together. Use a gel coat and resin mix method and hand lay-up process to combine two modules; module 1 is the bottom, and module 2 is the body of the modular buoy. Then, module 3 is for the tower to put all the electronic systems. The fabrication system is divided into electrical and electronic systems, which required simulation and programming processes to complete the buoy system.

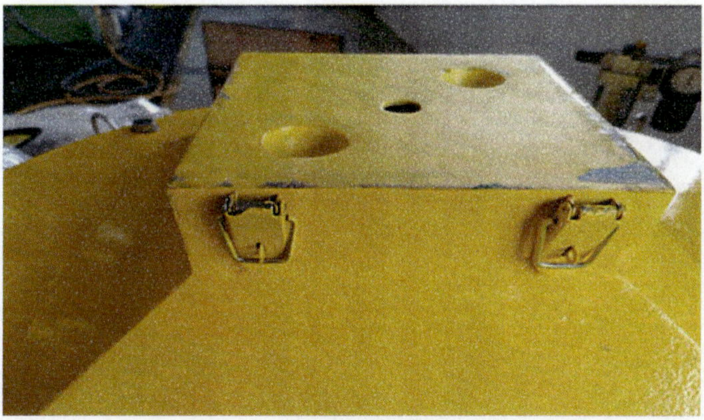

Fig. 33.6 Clips attach between top module and tower

33.4 Results and Discussion

The complete modular buoy, as shown in Fig. 33.7, is divided into three modules. The installation process of the modular buoy's bottom, body, and top module prototype

Fig. 33.7 Complete installation of modular buoy

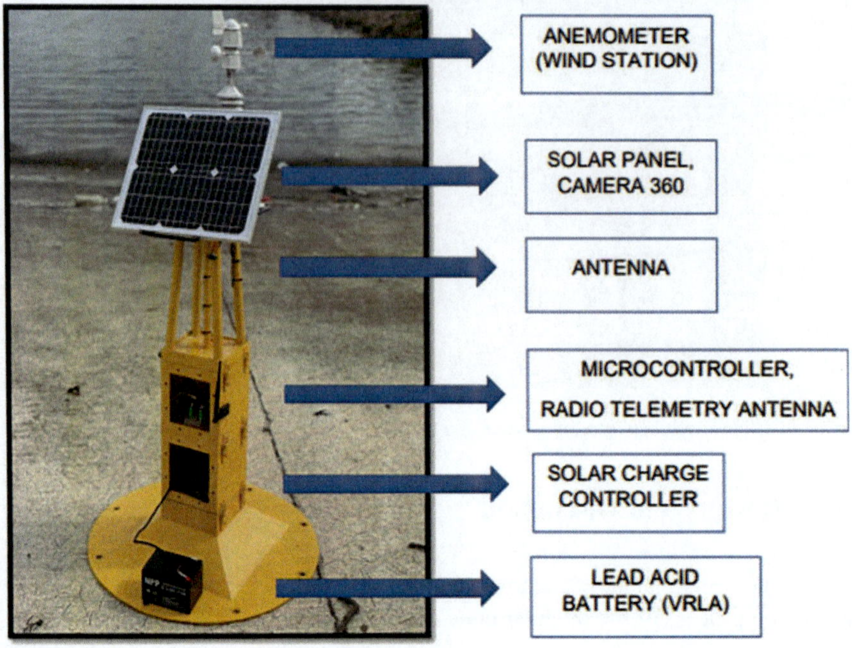

Fig. 33.8 System installation at modular buoy

have been combined. The body and top module are integrated using bolt and nut size 22 inch. The construction process took almost a month to be completed. The prototype weight of the modular buoy measures about 21.5 kg in lightweight condition. In addition, as previously mentioned, the system that can be installed, as shown in Fig. 33.8, consists of an anemometer, solar panel, camera, antenna, microcontroller, telemetry, solar charge controller, and battery. This modular buoy can suit to collect data at the ocean for research purposes. As the researcher figured out, the buoy needs to be added more weight to make it float more stable with a different system at module 3. The researchers add the sandbag inside module 1, about 35 kg, the result of the modular buoy with the system floating on the water is as shown in Fig. 33.9. The detailed technical specifications of the buoy are given in Table 33.1.

33.5 Conclusion

In conclusion, the project has covered developing the buoy and system for research purposes. However, since the buoy system usage is still in development for research purposes, this project needs an appropriate diagram and installation component to combine electrical and electronic systems inside the buoy. The appropriate diagram, such as block diagram and wiring diagram, is part of developing a monitoring system

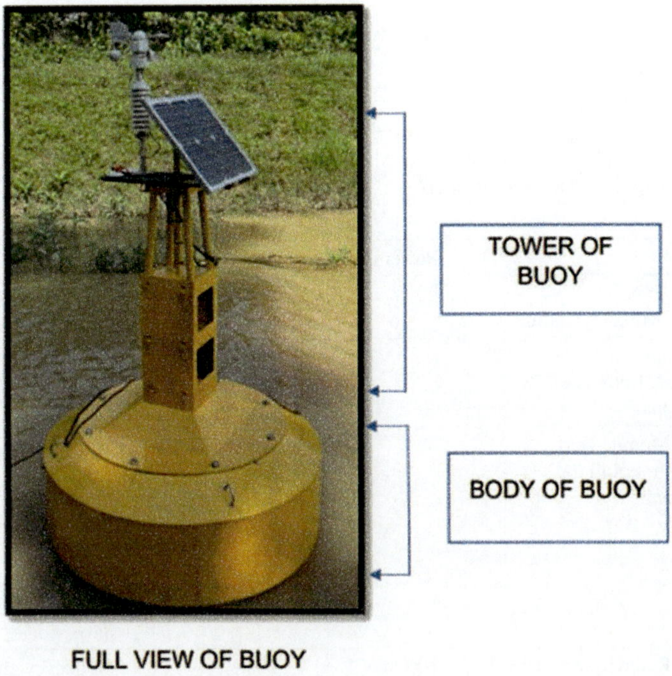

Fig. 33.9 Full modular buoy on the water

for the buoy. In addition, the diagram is essential for electrical and electronic to represent the control system sequentially with a connected system. This research already achieves its objectives: to develop the floating buoy and installing electrical, electronic, and wireless network communication systems. The system used a battery as the system's primary power and a solar charge controller as a secondary power. Finally, this system needs to be tested in actual sea conditions so that researchers can use the data from the buoy. The researcher can use this buoy for future research purposes by taking the data using a specific time.

Table 33.1 Floating Buoy Technical Specifications

Buoy overall dimensions					
Light Weight : 21.5 kg Dead Weight : 56.5 kg Diameter of the hull : 1.0 m (without fender) Buoy total height : 1.0 m (mast to keel)					
Components					
No	Item	Qty	Power (W)	Voltage (V)	Others
1	Solar Panel Type: Monocrystalline	1	20 W	18 V	Power Tolerance: 3 %
2	Antenna * Type: Immersion RC Antenna	2	–	12 V	Gain: 2.15 dBi Circular Polarization: RHCP
3	GPS Module Neo M8N + Folding Antenna	1	–	1.65–3.6 V	Max current: 5mA at 3.0 V Power Save Mode Range Accuracy: 0.6 m Running current: 23 mA at 3.0 V Power Save Mode
4	FPV Radio Telemetry Ground Module + Folding Antenna	1	100 MW	–	Frequency channel: 915 MHz Receiving sensitivity: -117 dBM
5	Solar Charge Controller	1	–	12 V / 24 V	Rated Current: 60 A
6	Lead Acid Battery *Type: VRLA	1	–	12 V	Capacity: 12 Ah 20 hours rate 3.4 kg
7	Lithium Polymer Battery	1	–	3 S (11.1 V)	Capacity: 2200 mAh Discharge Rate: 40 c/ 80 c
8	Battery Case *Type: Li-ion Cell Battery Case	1	–	7.4 V	–
9	Sensor - Wave - Pressure	1 1	–	–	Running current: 23 mA
10	Anemometer *Type: Digital weather station -Wind -Temperature - Humidity	1	–	–	Temperature accuracy: +- 1°C Humidity range: 20% to 90% Indoor temperature range: (0°C to 50°C)

(continued)

Table 33.1 (continued)

Buoy overall dimensions					
11	Anemometer Digital Receiver	1	–	–	Alkaline Battery 2 AAA 1.5 V
12	Microcontroller *Type: ArduPilot Mega 2560 V2.5 Flight Controller	1	–	–	Arduino Compatible Accelerometer, built-in barometer sensor, and GPS
13	LCD Monitor *Type: TFT color monitor	1	–	12 V– 24 V	Size: 7 inch
14	Waterproof run camera	1	DC 5-20 V (Non-direct power supply)	–	HD1080 P Working current: 270 mA

Acknowledgements The authors would like to thank Universiti Kuala Lumpur (UniKL) for supporting this research under the Short Term Research Grant (STR18014).

References

1. Park YW (2016) Design of Korean standard modular buoy body using polyethylene polymer material for ship safety. J Mater Sci Chem Eng 4(01):65
2. Milburn HB, McLain PD, Meinig C (1996) ATLAS buoy-reengineered for the next decade. Ocean Coast Manag 2:698–702
3. Aasen SE et al (2006) A deepwater tsunami surveillance system for Malaysia. OCEANS 2006-Asia Pacific, 2:1–7
4. Barker CH, Kourafalou VH et al (2020) Progress in operational modeling in support of oil spill response. J Mar Sci Eng 8(9):668
5. Żelazny K (2014) A method of calculation of ship resistance on calm water useful at preliminary stages of ship design. Sci J Marit Univ Szczec 38(110):125–130
6. Rozali RH et al (2019) Floating buoy technology for research purposes. Int J Innov Technol Explor Eng 8(12):5514–5520
7. Jolliff JV (1974) Modular ship design concepts. Nav Eng J 86(5):11–32
8. Fafandjel N, Rubeša R, Mrakovčić T (2008) Procedure for measuring shipbuilding process optimisation results after using modular outfitting concept. Strojarstvo 50(3):141–150
9. Lacey AW, Chen W et al (2019) Review of bolted inter-module connections in modular steel buildings. J. Build. Eng. 23:207–219
10. Nugroho WH, Kusnindar K et al (2017) An experimental investigation of a weather buoy-wireless data acquisition based on microcontroller. J Electr Electron Eng 1(2):54–60
11. Liu QK, Chen YH (2017) Research on sea surface drifting buoy based on Beidou communication System. FMSMT:1300–1304
12. Bahamon N, Aguzzi J et al (2011) The new pelagic operational observatory of the Catalan Sea (OOCS) for the multisensor coordinated measurement of atmospheric and oceanographic conditions. J Sens 11(12):11251–11272
13. García E, Quiles E et al (2018) Sensor buoy system for monitoring renewable marine energy resources. J Sens 18(4):945

Chapter 34
Numerical Simulation of Heat Generation During Plunging Stage in Orbital Friction Stir Welding on Pipe Aluminum Alloys AA6061-T6 Adapting a Pure Lagrangian Formulation

Kamal Ahmad, Mokhtar Awang, Srinivasa Rao Pedapati, Anuar Abu Bakar, and Zaimi Zainal Mukhtar

Abstract Friction stir welding consists of three operation stages operation beginning with plunging, pre-heating (dwelling) and welding along the joint. Each stage has its own functionality critically controlled by process parameters with a respective period for a successful welding process. This paper deals with numerical modeling of heat generation during the plunge stage using the fully coupled thermomechanical Abaqus/Explicit package. Rapidly rise temperature is observed once plunging tool shoulder established contact with pipe surface. Frictional dissipation energy is prominent over plastic deformation during plunge phase. Numerical thermal response is correlated with literature experimental data.

Keywords Finite element · Friction stir welding · Lagrangian · Eulerian

K. Ahmad (✉) · Z. Z. Mukhtar
Universiti Kuala Lumpur Malaysian Institute of Marine Engineering Technology, Jln Pantai Remis, 32200 Lumut, Perak, Malaysia
e-mail: kamalmat@unikl.edu.my

Z. Z. Mukhtar
e-mail: zaimi@unikl.edu.my

M. Awang · S. R. Pedapati
Universiti Teknologi PETRONAS, Seri Iskandar, Perak, Malaysia
e-mail: mokhtar_awang@utp.edu.my

S. R. Pedapati
e-mail: srinivasa.pedapati@utp.edu.my

A. Abu Bakar
Universiti Malaysia Terengganu, Kuala Terengganu, Terengganu, Malaysia
e-mail: anuarbakar@umt.edu.my

A. Ismail et al. (eds.), *Design in Maritime Engineering*, Advanced Structured Materials 167, https://doi.org/10.1007/978-3-030-89988-2_34

34.1 Introduction

Friction stir welding (FSW) is a relatively new welding process, and it was invented in 1991 at the Technical Welding Institute (TWI) of UK. It is a solid-state joining technique meaning that the formation of joints in the temperature is not exceeding the melting point, and the basic concept of FSW is remarkably simple. The result of the FSW joint is with high strength and high integrity particularly in aluminum alloys [1]. The very important parameters in FSW are tool rotation rate (v, rpm) in clockwise or anticlockwise direction and tool traverse speed (n, mm/min) along the line of joint [2]. It should be noted that frictional coupling of the tool surface with the workpiece is going to govern the primary heating, and the rest small percentage of heat generation is governed by plastic deformation [3]. Another important process parameter is the angle of spindle or tool tilt with respect to the workpiece flat surface, while tool offset setting is being used in geometrical pipes [4]. Further, the insertion depth of pin into the workpieces (also called target depth) is important for producing sound welds with smooth tool shoulders. The insertion depth of the pin is associated with the pin height.

Industrial plates, sections and pipes have a fundamental role in particular sectors such as chemical, petrochemical, nuclear, oil and gas industries and shipbuilding. Commonly used material is aluminum which is most performed by the conventional fusion welding. The main drawback of using this method is that the weld joint is produced with high heat input and high temperatures resulting in detrimental effect to the stress. This problem is exacerbated with multi-pass welding sequences [5, 6]. Therefore, the solid-state nature of FSW makes an attractive welding process to reduce high heat input generated.

Previous investigations have developed several experimental methods such as stress relaxation [7], X-ray diffraction [8–10], ultrasonic [11, 12] and cracking [13]. All these methods are investigated using conventional or fusion welding process with a variety of material types instead of using the FSW process. Recently [4] developed FSW expandable inner mandrel support is used onto a small diameter pipe to investigate its tensile strength and axial force with respect to tool rotation rate and pipe rotation speed. The tool offset setting is an equivalent function as the tool tilt angle in the plate; however, this parameter is not sufficiently reported and studied.

The measurement of thermal response is experimentally limited to the adjacent welding zone due to physical condition. Lammlein [4] applied a numerical CFD simulation completely using Eulerien motion to further investigate the thermal conditions and material flow at most contours level. However, adapting Eulerian approach is actually unable to track free surface deformation due to the fact that no mesh distortion is simulated. To gain physical insight of the FSW process and the evolution of the stress through its thickness in aluminum geometrical pipes and plates, numerical modeling and simulation adapting with the Lagrangian approach are believed the most effective method to handle problems with element distortion successfully.

The present work will use the finite element method adapted with the pure Lagrangian approach to model and simulate the thermomechanical analysis. Simulation process and parameters aspect are critically concentrated on mesh motion [14], material behavior, nonlinear dynamic system, coupled thermomechanical procedure [15] and contact friction interaction. Previous investigations focus on modeling procedures of geometrical plate surfaces [14–29] and will be adapted in this research with some adjustment in modeling concerning to its complexity needs.

34.2 Methodology—Finite Element Model

A. *Heat Transfer*

Derived from the energy balance of the system, the 3D heat transfer equation for FSW process can be written as:

$$\frac{\partial}{\partial x}\left(K_{xx}\frac{\partial T}{\partial x}\right)+\frac{\partial}{\partial y}\left(K_{yy}\frac{\partial T}{\partial y}\right)+\frac{\partial}{\partial z}\left(K_{zz}\frac{\partial T}{\partial z}\right)+Q=\rho c\frac{\partial T}{\partial t}$$

Frictional heat is using Coulomb's friction law.

Local friction force, $F_f = \mu F_n$.

Where μ is the friction coefficient between tool shoulder and workpiece and F_n is the normal force applied to workpiece.

Frictional heat, $q_f = F_f v = \mu F_n v$.

Where v is the rotational velocity of pin equal to 2πRN, R is the distance of the calculated point from the tool axis, and

N is the rotational speed of the tool.

Frictional heat, $q_f = F_f v = \mu F_n v = \mu F_n 2\pi \mathrm{RN}$.

Plastic deformation heat, q_p

$$q_p = n\sigma\dot{\varepsilon}^{pl}$$

where n is the fraction heat dissipates due to plastic straining (0.9), σ is the shear stress, and $\dot{\varepsilon}^{pl}$ is the plastic straining rate.

Heat generated, Q = frictional heat, q_f + plastic deformation heat, q_p

$$= q_f + q_p$$

B. *Model Description*

The numerical model for the deformable pipe used aluminum 6061-T6 and has a dimension of OD 90 × ID 80 × 5.0 mm wall thickness. The tool is modeled as rigid, and it has a plunging velocity of 3.5 mm/sec, while the rotation rate is 920 rpm. The tool insertion depth set into the pipe is 2.9 mm. The geometrical dimensions of the tool are shown in Fig. 34.1.

The deformable pipe was discretized using eight-nodes linear reduced integration (C3D8RT) elements, and the graded fine mesh is shown in Fig. 34.2 specifically at most severe distortion area to avoid volumetric locking. Total of 258,355 elements are applied on the pipe and were implemented with hourglass control. Mass scaling is implemented by artificially increasing the loading rates and density in order to

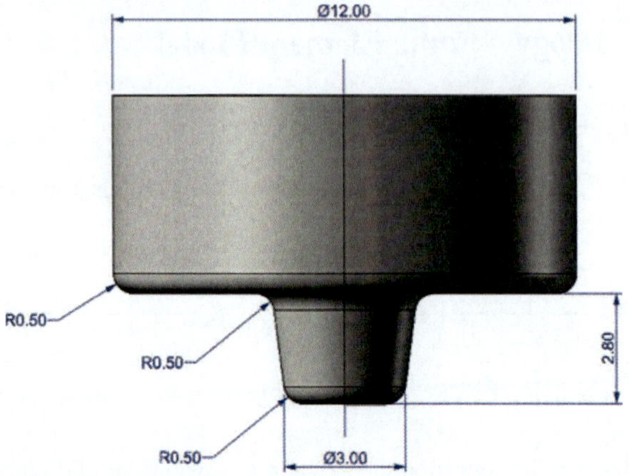

Fig. 34.1 Geometrical of the FSW tool

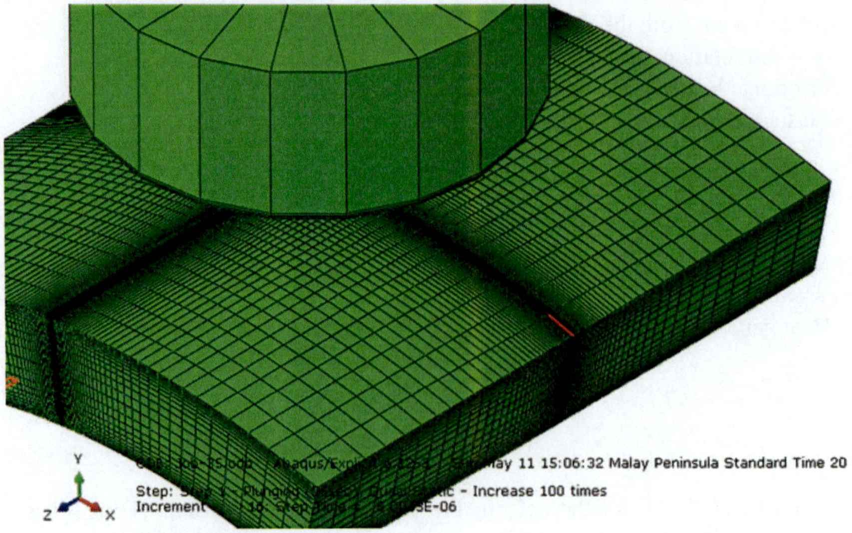

Fig. 34.2 Fine-graded mesh on pipe

handle the quasi-static problem of higher natural plunging FSW time. The tool is modeled as an isothermal body with ambient temperature of 25 °C.

Each side and bottom of the pipes is constrained in axis-symmetry except for the top surface allowing for material deformation from the tool penetration. Experimentally, the pipe is actually clamped for each side and inner in order to keep the contact of weld joint face and fix the position in place.

Heat convection coefficient of the pipe surfaces is assumed as 30 W/m^2 °C at an ambient temperature of 25^{0}C. Contact interaction between tool and plate surface is based on penalty method with constant friction coefficient of 0.3. Motion description of arbitrary Lagrangian-Eulerian (ALE) is implemented on pipe domain for handling severe mesh distortion.

C. *Material Model*

Thermal history is recorded at location of thermal point "A" as specified in Fig. 34.3. Thermal and mechanical properties of aluminum 6061-T6 [3] are modeled as a temperature dependent as shown in Fig. 34.4 and Table 34.1. With increasing temperature, the thermal parameters are increased, while mechanical parameters are decreased.

D. *Johnson–Cook Model*

The pipe deforms thermal plastically during plunging of FSW. The Johnson–Cook model is appropriately used for simulating the plastic stress. This model describes

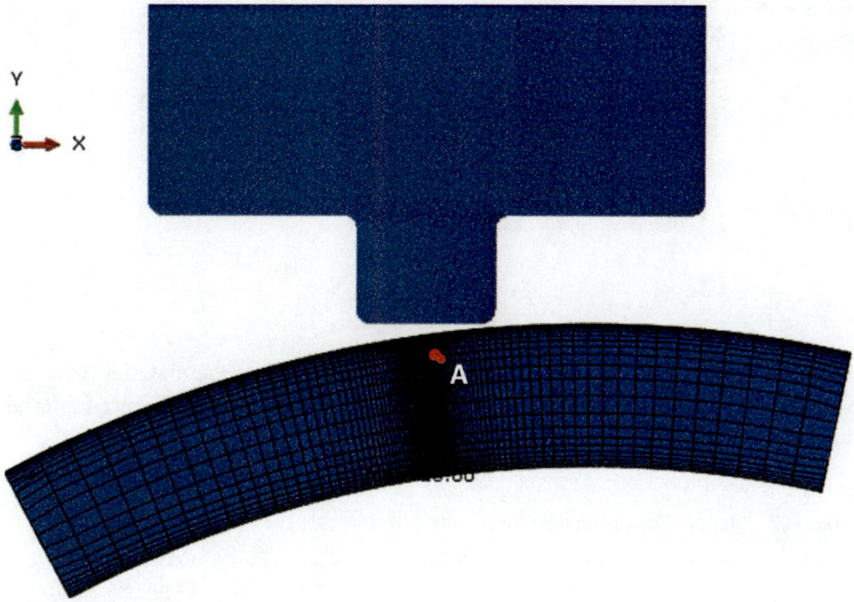

Fig. 34.3 Location of thermal point **A**

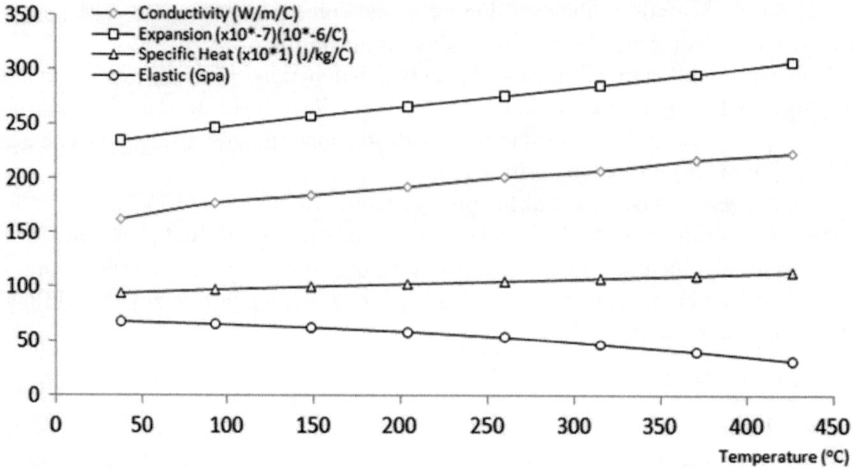

Fig. 34.4 Material properties of AA 6061-T6

Table 34.1 Density of AA6061-T6

Temp. (°C)	37.8	93.3	148.9	204.4	260	315.6	371.1	426.7
Density (kg/m^3)	2690	2690	2670	2660	2660	2630	2630	2600

that the stress (σ) is dependent on plastic strain (ε_p), plastic strain rate $\left(\dot{\varepsilon}_p\right)$ and temperature $\left(\dot{T}\right)$ as the following:

$$\sigma\left(\varepsilon_p, \dot{\varepsilon}_p, \dot{T}\right) = \left[A + B\left(\varepsilon_p\right)^n\right]\left[1 + Cln\left(\dot{\varepsilon}_p^*\right)\right]\left[1 - \left(\dot{T}^*\right)^m\right]$$

where $\dot{\varepsilon}_p = \frac{\dot{\varepsilon}_p}{\dot{\varepsilon}_\rho} = \frac{\text{plastic strain rate}}{\text{user defined plastic strain rate}}$

$$\dot{T}^* = \frac{T - T_{\text{room}}}{T_{\text{melt}} - T_{\text{room}}}$$

The Johnson–Cook parameter values [3] used to simulate the material behavior Al 6061-T6 are specified in Table 34.2. Parameter A, B, C, n and m are material constants which are determined through high strain rate deformation tests.

Table 34.2 Johnson–Cook parameter for AA 6061-T6

Initial yield stress, A (MPa)	Hardening modulus, B (MPa)	Work hardening exponent, n	Thermal softening coefficient, m	T_{melt} (°C)	T_{room} (°C)	Coefficient on strain rate, C (MPa)	$\dot{\varepsilon}_{po}$
293.4	121.26	0.23	1.34	582	25	0.002	1

34.3 Results and Discussion

A successful solid-state FSW process requires heat input with critical controlled temperature range just below the melting temperature in the region around of weld joint line. During numerical modeling, plunging tool velocity is set 3.5 mm/sec and the rotation rate is 920 rpm. A plunging period of 25 s is required to complete tool penetration to a depth of 2.9 mm of material thickness 3.2 mm.

Comparison of results for both numerical and experimental procedures is shown in Fig. 34.5. Both temperatures are measured at location point "A" as previously specified. The maximum value of FE temperature for complete plunging at 82.8 s is reached as 253.45 °C. From the FE curve, it is observed that with the begin of the initial ambient temperature of 25 °C, the temperature increased constantly until time 65 s above; there is a rapid rise of temperature. It means that the tool's shoulder surface has started to touch and plunge into the plate top surface. The material around the tool pin is already deformed by localized plastic and being pushed-up by tool down force. The experimentally temperature is recorded to steadily rise since the beginning until a rapid rise at around 60 s plunges. The difference temperature rise between experimental and FE result might be caused by time delay recorded by thermocouple response and heat transfer from heat source. The thermocouple K-type has a time constant of 0.25 s and delaying of 5 times to become 1.25 s time delay recorded. FE temperature curve at tool reference point has reached a maximum value of 427.1 °C, and its distribution shows uniformity and steadily rise.

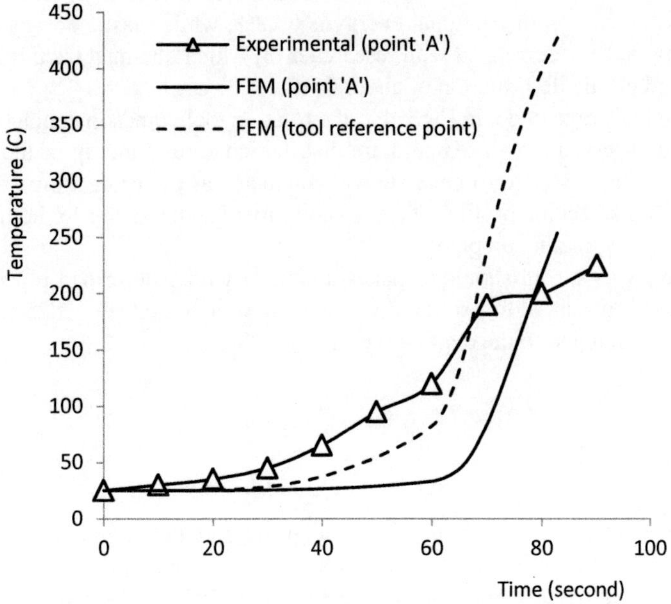

Fig. 34.5 Thermal history at point "A" and tool reference point

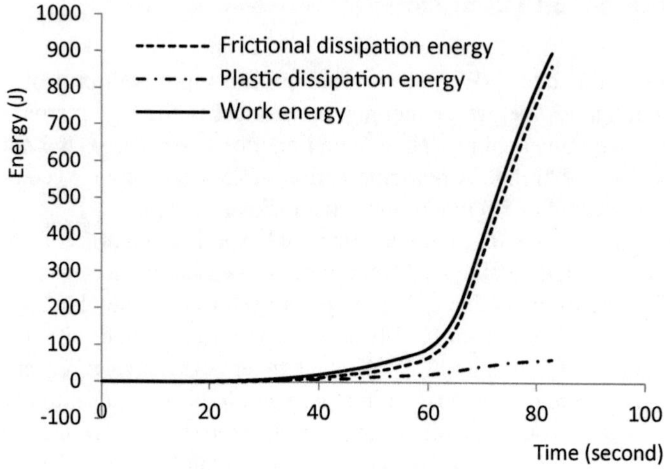

Fig. 34.6 Time-dependent energy dissipation for plunge phase FSW

The time-dependent energies involved in the process are plotted in Fig. 34.6. It shows the numerical result of the dissipation energy for both frictional and plastic deformation. Since the beginning of plunge phase, work energy is dissipated away as frictional heat, and plastic dissipation is steadily rise as the process advance. With plunge speed of 3.5 mm/sec and rotation rate 920 rpm, the generation of frictional heat energy is obtained 862.73 J, and plastic deformation energy is generated 62.85 J. Percentage generation of frictional energy is 93.2%, while plastic energy is 6.8%. This energy range is correlated with researcher [3] which obtained friction of about 90–95% and plastic deformation of about 5–10%.

At the initial stage of 5 s in Fig. 34.7, the rotating tool comes in contact with the plate and it observed a high-temperature distribution around the tip of the tool. At time 6.25 s, heat is started to generate and conducted away through pipe thickness in uniformly distribution profile. There is similarity of heat transfer for both sides of advancing and retreating on pipe.

At time 18.75 s, tool shoulder makes a contact with a deformed pipe surface. Temperature is observed to rise rapidly to a maximum 362.02 °C. At 25 s, the tool is completely plunged to the depth 4.0 mm.

34.4 Conclusion

Plunge stage of friction stir welding is a main phase and the initial condition for the next successful friction welding process. Numerical simulation was performed in order to investigate thermomechanical behavior specifically at the extreme region of the joint line. The main conclusions obtained in this investigation can be summarized as follows.

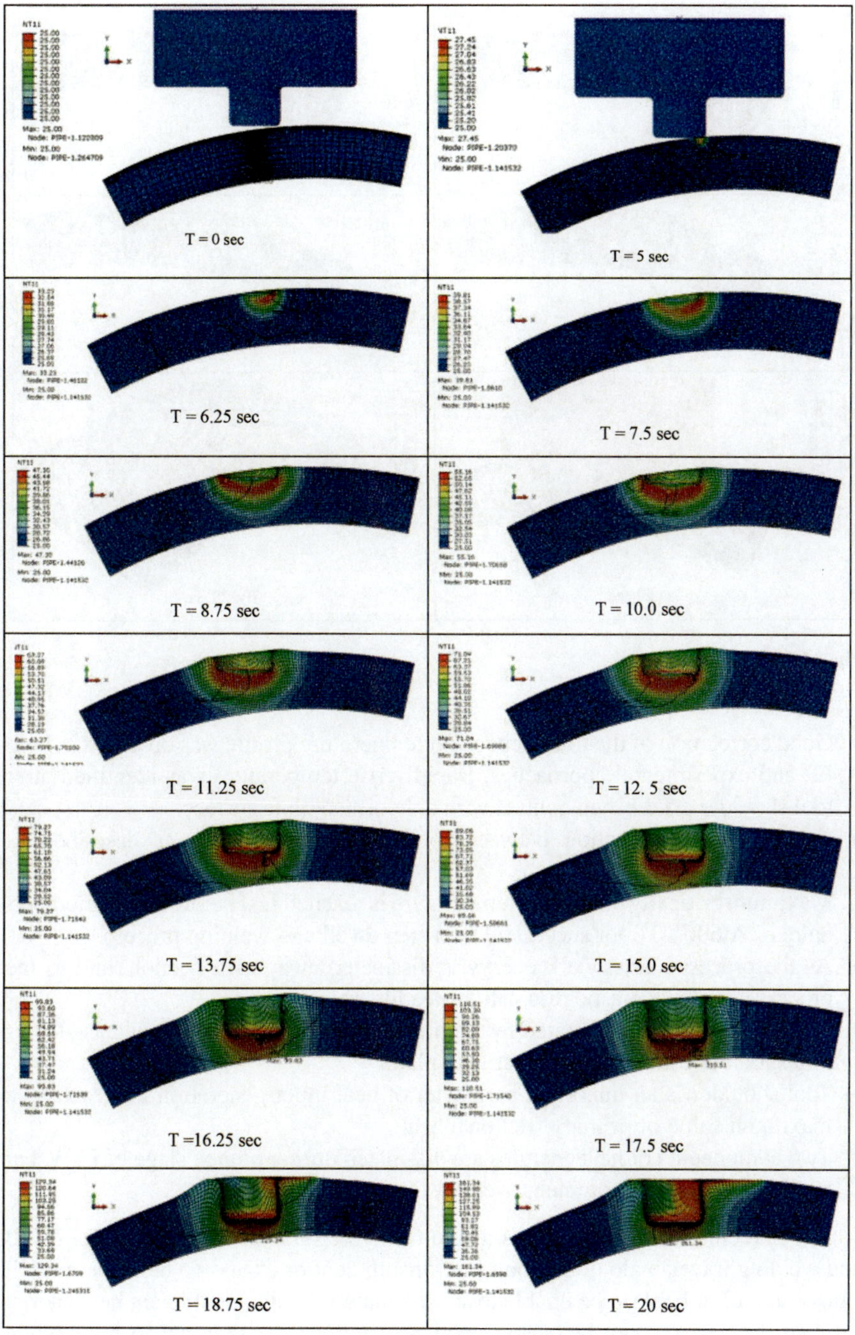

Fig. 34.7 Temperature distribution for plunge phase at different time periods

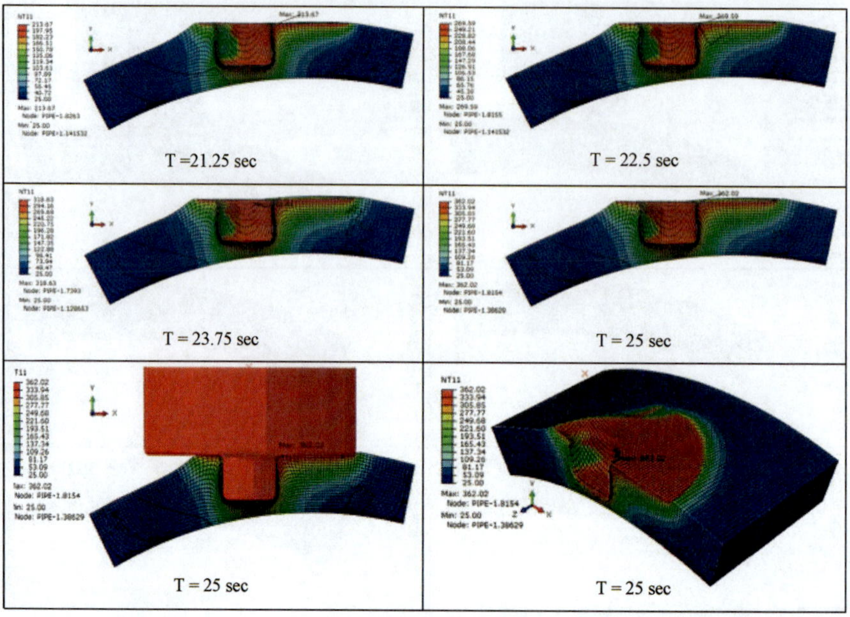

Fig. 34.7 (continued)

- Good correlation of the time-dependent temperature profile was observed in both FE and experimental approaches. Rapidly rise temperature was identified after tool shoulder established contact with a deformed plate surface.
- There is well-interaction between tool shoulder and plate as described as temperature rise rapidly.
- Maximum critically controlled temperature is reached just below melting temperature of Al6061-T6 for successful next step dwell and welding process.
- As the process starts, work energy is dissipated away as a friction, and as the process advances, plastic dissipation steadily rise.
- The amount of heat generated by friction and plastic is suddenly increased after tool shoulder established contact with plate.
- Tool shoulder is an important parameter of heat input generation for reaching a maximum value of primary frictional heat.
- Two components of heat energies are dissipated during plunge stage of FSW, but frictional energy is prominent over plastic deformation.

The outcome of this research is a significant step for the next process of dwell and welding traverse along the joint line. Insufficient or excessive heat energy will cause the tool of FSW to be unable to stir the material flow. Insufficient heat energy resulted the material not being hot enough to initiate the material to flow, while excessive energy will make the material too hot and exceeds the melting point, thus resulting the material starts to melt.

References

1. Kim D, Badarinarayan H, Kim JH et al (2010) Numerical simulation of friction stir butt welding process for AA5083-H18 sheets. Eur J Mech A Solids 29(2):204–215
2. Mishra RS, Ma ZY (2005) Friction stir welding and processing. Mater Sci Eng R 50:1–78
3. Awang M, Mucino V H, Feng Z, David S. A, (2005) Thermo-mechanical modeling of friction stir spot welding (FSSW) process: use of an explicit adaptive meshing scheme. SAE Tech Pap No. 2005–01–1251
4. Lammlein D H, Gibson B. T, et al (2012). The friction stir welding of small-diameter pipe: an experimental and numerical proof of concept for automation and manufacturing. Proceedings of the Institution of Mechanical Engineers, Part B: 2 Acad. J. Manuf. Eng. 26(3): 383–398
5. Sattari-Far I, Javadi Y (2008) Influence of welding sequence on welding distortions in pipes. Int J Press Vessel Pip 85:265–274
6. Deng D, Murakawa H (2006) Prediction of welding residual stress in multi-pass butt-welded modified 9Cr–1Mo steel pipe considering phase transformation effects. Comput Mater Sci 37:209–219
7. Pang HL, Pukas SR (1989) Residual stress measurements in a cruciform welded joint using hole drilling and strain gauges. Strain 25(1):7–14
8. Chen CM, Kovacevic R (2003) Finite element modeling of friction stir welding—thermal and thermo mechanical analysis. Int J Mach Tools Manuf 43:1319–1326
9. Cheng W, Finnie L (1985) A method for measurement of axisymmetric axial residual stresses in circumferentially welded thin-walled cylinders. J Eng Mater Technol 107:181–185
10. Chandra U (1985)"Determination of residual stress due to girth-butt welds in pipes", ASME J Pres Vessel Technol 107:178–184
11. Tanala E, Bourse G, Fremiot M, Belleval JFD (1994) Determination of near surface residual stresses on welded joints using ultrasonic methods. NDT E Int. 28(2):83–88
12. Chu S L, Peukrt H, Schnider E (1987) "Residual stress in a welded steel plate and their measurements using ultrasonic techniques." MRL Bull Res Dev 1(2):45–50
13. Masubuchi K, Martin D C (1961) "Investigation of residual stresses by use of hydrogen cracking". Welding J 40:553s–563s
14. Xu SW, Deng XM, Reynolds AP, Seidel TU (2001) Finite element simulation of material flow in friction stir welding. Sci Technol Weld Join 6:191–193
15. Buffa G, Hua J, Shivpuri R, Fratini L (2006) A continuum based FEM model for friction stir welding—model development. Mater Sci Eng A 419:389–396
16. Zhang HW, Zhang Z, Chen JT (2006) 3D modeling of material flow in friction stir welding under different process parameters. J Mater Process Technol 183:62–70
17. Vijay S, Srdja Z, Radovan K (2005) Thermo-mechanical model with adaptive boundary conditions for friction stir welding of Al 6061. Int J Mach Tools Manuf 45:1577–1587
18. Buffa G, Ducato A, Fratini L (2011) Numerical procedure for residual stresses prediction in friction stir welding. Finite Elem Anal Des 47:470–476
19. Donea J, Antonio H, Ponthot J P and Rodrıguez-Ferran A (2004) "Arbitrary lagrangian–eulerian methods". Encyclopedia Comput Mech
20. Zhang Z, Zhang HW (2009) Numerical studies on controlling of process parameters in friction stir welding. J Mater Process Technol 209:241–270
21. Rajamanickam N, Balusamy V, Reddy GM, Natarajan K (2009) Effect of process parameters on thermal history and mechanical properties of friction stir welds. Mater Des 30:2726–2731
22. Ravi SY (2010) "Thesis: experimental investigation of plastic deformation of Ti-6Al-4V under various loading conditions". The Ohio State University
23. Hirasawaa S, Badarinarayan H, Okamoto K, Kawanami TTT (2010) Analysis of effect of tool geometry on plastic flow during friction stir spot welding using particle method. J Mater Process Technol 210:1455–1463
24. Assidi M, Fourment L, Guerdoux S, Nelson T (2010) Friction model for friction stir welding process simulation: calibrations from welding experiments. Int J Mach Tools Manuf 50:143–155

25. Kim D, Badarinarayan H, Kim JH, Kim C, Okamoto K, Wagoner RH, Chung K (2010) Numerical simulation of friction stir butt welding process for AA5083-H18 sheets. Eur J Mech A Solids 29:204–215
26. Ulysse P (2002) Three-dimensional modeling of the friction stir-welding process. Int J Mach Tools Manuf 42:1549–1557
27. Mandal S, Rice J, Elmustafa AA (2008) Experimental and numerical investigation of the plunge stage in friction stir welding. J Mater Process Technol 203:411–419
28. Fratini L, Buffa G, Shivpuri R (2007) Improving friction stir welding of blanks of different thicknesses. Mater Sci Eng A 459:209–215
29. Bastier A, Maitournam MH, Roger F, Dang VK (2008) Modeling of the residual state of friction stir welded plates. J Mater Process Technol 200:25–37
30. Daryl LL (2007) A first course in the finite element method, 4th edn. University of Wisconsin-Platteville, Thompson